N-sided Polygons

Any polygon

$\text{Degrees}_{\text{total}} = 180°\,(N - 2)$

Regular polygon only

$\text{Degrees}_{\text{each angle}} = 180°\,(N - 2)/N$

Circles

$C = 2\pi r$

or

$C = \pi d$

$A = \pi r^2$

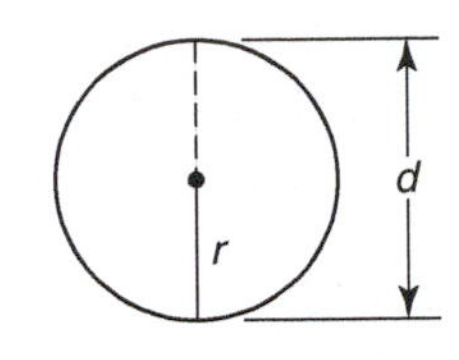

Volumes and Surface Areas

Prism

LSA = lateral surface area

TSA = total surface area

B = area of base

$V = B \times h$

LSA = area of all faces

TSA = LSA + 2B

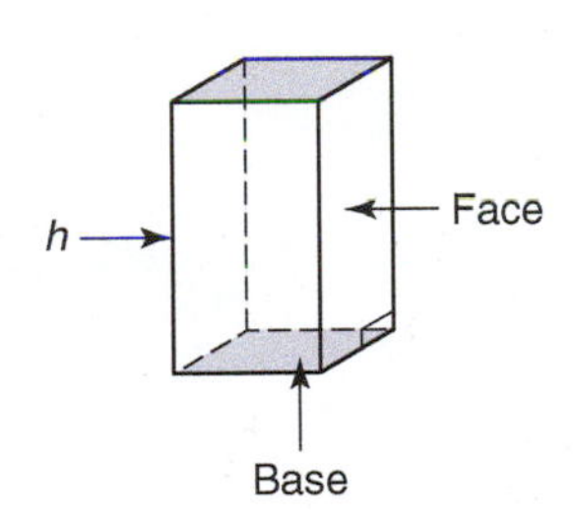

Cylinders

$V = \pi r^2 h$

LSA = $2\pi rh$

or

LSA = πdh

TSA = LSA + $2\pi r^2$

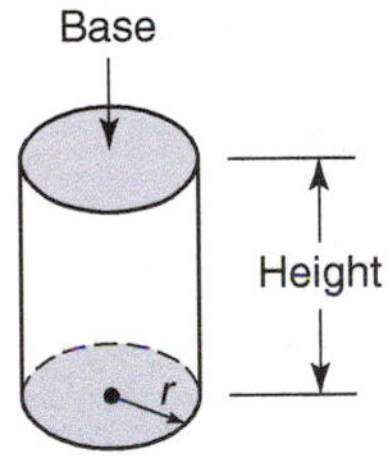

Cones

$V = \frac{1}{3}\pi r^2 h$

LSA = πrs

(s = slant height)

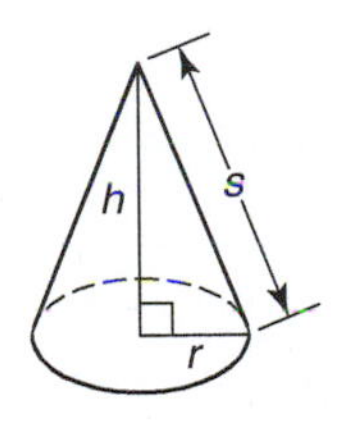

Temperature Conversions

C = 5/9(F − 32°) (Fahrenheit to Celsius)

F = 9/5C + 32° (Celsius to Fahrenheit)

Right Triangle Trigonometry Ratios

Sin θ = opp/hyp

Cos θ = adj/hyp

Tan θ = opp/adj

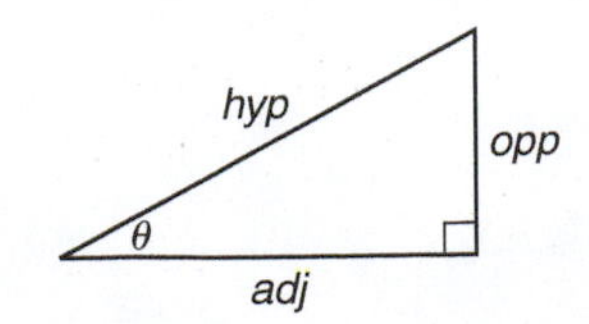

Continues inside back cover

Mathematics for Carpentry and the Construction Trades

THIRD EDITION

Alfred P. Webster
Kathryn E. Bright

Prentice Hall
Boston Columbus Indianapolis New York San Francisco Upper Saddle River
Amsterdam Cape Town Dubai London Madrid Milan Munich Paris Montréal Toronto
Delhi Mexico City São Paulo Sydney Hong Kong Seoul Singapore Taipei Tokyo

Editorial Director: Vernon R. Anthony
Acquisitions Editor: David Ploskonka
Editorial Assistant: Nancy Kesterson
Director of Marketing: David Gesell
Executive Marketing Manager: Derril Trakalo
Senior Marketing Coordinator: Alicia Wozniak
Marketing Assistant: Les Roberts
Senior Managing Editor: JoEllen Gohr
Associate Managing Editor: Alexandrina Benedicto Wolf
Operations Supervisor: Central Publishing
Operations Specialist: Laura Messerly
Art Director: Jayne Conte
Cover Designer: Suzanne Duda
Cover Image: Dreamstime
Full-Service Project Management: Revathi Viswanathan
Composition: PreMediaGlobal
Printer/Binder: LSC Communications
Cover Printer: LSC Communications
Text Font: 11/13 TimesNewRomanPS

Credits and acknowledgments borrowed from other sources and reproduced, with permission, in this textbook appear on the appropriate page within the text. Unless otherwise stated, all artwork has been provided by the author.

Library of Congress Cataloging-in-Publication Data
Webster, Alfred P.
Mathematics for carpentry and the construction trades / Alfred P. Webster, Kathryn B. Judy.—3rd ed.
p. cm.
Includes bibliographical references and index.
ISBN 978-0-13-511400-1
1. Carpentry—Mathematics. 2. Building—Mathematics. I. Judy, Kathryn Bright. II. Title.
TH5612.W43 2012
694.01'51—dc22

2010050932

Prentice Hall
is an imprint of

www.pearsonhighered.com

ISBN 10: 0-13-511400-4
ISBN 13: 978-0-13-511400-1

11 2020

Preface

This text is intended to meet the needs of a two-semester course for students of carpentry and building construction. Throughout, the emphasis has been on simplicity. Illustrated problems include step-by-step explanations of their solutions. Explanations have been presented in summary form with the necessary steps highlighted for easy identification. Drawings and photographs have been included to clarify the particular subject under discussion. In the interest of drawing the reader's attention to facts that affect our environment and remarks that provoke ways of making the greening of our world more of a conscious effort, at the beginning of each chapter is a "Did You Know" statement.

Chapters 1 through 10 cover the fundamental mathematics necessary to a broad range of skills. Although problems in these chapters apply to a variety of areas, the emphasis has been given to applications in the building construction field. We suggest that calculators not be used in the first sections in Chapters 1 through 3 to allow students to increase their mathematical skills in some basic areas. Thereafter, calculator usage is encouraged, with emphasis on efficiency and accuracy.

Chapters 11 through 27 cover matters of direct concern to the builder. The sequence of topics in these chapters follows the logical construction process insofar as is practical. Phases of construction normally relegated to subcontractors (including masonry, plumbing, heating, and electrical, among others) have not been covered. The occurrence of these phases of construction is of concern to the primary contractor, and their sequencing is alluded to in a summary chapter; however, we have made no attempt to include mathematics related to these areas.

Although this text is not intended to be a complete "how-to" manual with respect to building techniques, a certain amount of instruction has been included. In many areas of building, an understanding of the relevant mathematics is coupled with an understanding of how the construction is done. Furthermore, efficiency and accuracy (both highly desirable goals for the builder and estimator) are best achieved when an understanding of building methods has been reached.

Users of this text will find the topic sequence logical and explanations clear and concise. The problems are realistic and practical and typical of the types of calculations that builders can expect to encounter in practice. Answers to the odd-numbered exercises have been included in the back of the book. In addition, a solutions manual is available that contains many detailed solutions.

This text is the result of the depth of experience we bring to users of this material. Our extensive backgrounds as teachers of applied mathematics and practitioners in the building construction field should make this book valuable to its users.

We wish to thank the reviewers of this edition for their helpful comments: Stephen Csehoski, Pennsylvania Highlands Community College; Barbara Ries, Chippewa Valley Technical College; Joseph Ryan, Cape Cod Community College, Eastern Maine Community College.

Finally, we wish to express our appreciation to Sally Webster and Daisy Bright, both accomplished teachers of mathematics. Their identification of errors and suggestions for changes have been a valuable contribution to this effort.

A. P. Webster
K. E. Bright

Contents

CHAPTER 1

Whole Numbers

> **Did You Know**
>
> Energy-efficient windows (double, triple, and/or low E glass) can save 30% or more in residential heating and cooling costs compared to single-pane windows.

OBJECTIVES

Upon completing this chapter, the student will be able to:

1. Determine place value of numerals.
2. Round whole numbers to a given place value.
3. Distinguish between precision and accuracy (optional).
4. Set up problems involving more than one type of arithmetic operation.
5. Solve problems using the proper order of operations.
6. Add, subtract, multiply, and divide whole numbers and decimal numbers.

1.1 Place Value and Rounding

Place Value for Whole Numbers

To make sure that we understand the decimal system (the system we use), a brief review of place value is in order. A numeral such as 562 means 500 + 60 + 2. In other words, there are 5 hundreds, 6 tens, and 2 ones. Figure 1–1 indicates the place values for five hundred sixty-two (562), one thousand eight hundred fifty-three (1853), and five million, thirty thousand, four hundred six (5,030,406).

Figure 1–1

Ten-millions	Millions	Hundred-thousands	Ten-thousands	Thousands	Hundreds	Tens	Ones
					5	6	2
				1	8	5	3
	5	0	3	0	4	0	6

Exercise 1–1A

Determine the place value of the digits indicated, as in examples a and b.

	Number	Digit	Place Value
a.	8524	2	Tens place
b.	63,407	3	Thousands place
1.	42,527	7	
2.	165,329	3	
3.	4,061,755	0	
4.	8,262,419	2	
5.	5,317,204	5	

Exercise 1–1B

Determine the digit that is in each of the following places.

	Number	Digit	Place Value
a.	48,291	2	hundreds place
b.	53,649,720	5	ten-millions place
1.	4813		tens place
2.	598,372		thousands place
3.	15,144		ones place
4.	8,475,136		millions place
5.	36,721,540		ten-thousands place

Rounding Whole Numbers

Numbers can represent either exact quantities or approximate amounts. "There are five people in the room." Five represents an exact number in this situation. We know that there cannot be $4\frac{1}{2}$ or $5\frac{1}{4}$ people in the room. On the other hand, a "4-in. slab on grade" represents an approximate number. All measured quantities are approximate because it is impossible to measure exactly without error. A 4-in. slab is not exactly 4 in. thick—it may be $3\frac{7}{8}$ in. thick in one place and $4\frac{1}{16}$ in. thick in another. Frequently, it is not possible or even desirable to state a quantity exactly. Estimating, a very important part of building construction, involves rounding numbers to approximate quantities. The following examples illustrate the proper method of rounding whole numbers.

Examples

A. Round 48,623 to the hundreds place.

48,623	1. Locate the proper place. In this example 6 is in the hundreds place.
48,623 (↓ over the 2)	2. Examine the digit to the right of the digit to be rounded. Examine the 2 in this case.
48,623 → 48,600 (↓ over the 2)	3. If the digit to the right (the 2 in this example) is less than 5, that digit and all others to its right are replaced by zeros.

B. Round 48,623 to the thousands place.

48,623	1. Locate the proper place. In this example 8 is in the thousands place.
48,623 (↓ over the 6)	2. Examine the digit to the right of the digit to be rounded, a 6 in this example.
48,623 → 49,000 (+1 ↓ over the 8)	3. If the digit to the right is equal to or greater than 5, add 1 to the digit to be rounded (the 8 in this example), and replace all digits to the right with zeros.

C. The cost of constructing a small vacation cottage is $121,753. Round this to the nearest hundred dollars.

121,753	1. Identify the hundreds place. In this example 7 is in the hundreds place.
121,753 (↓ over the 5)	2. Examine the digit to the right of the digit to be rounded, 5 in this case.
121,753 → 121,800 (+1 ↓ over the 7)	3. Since 5 is 5 or greater, round up the 7 to the next higher number, and replace all digits to the right with zeros.
$121,800	4. This is the approximate cost of the newly constructed house.

Why replace the rounded numbers with zeros? Let's examine what would happen in example C above if the numbers are simply dropped without being replaced with zeros.

121,753 → 1218	1. This is the "rounded" number if the digits are dropped and not replaced by zeros.
$1218	2. A $121,753 house is *not* approximately $1218!

N O T E

Sometimes you do drop the digits without replacing them with zeros. This occurs only when the digits being dropped are positioned to the right of the decimal point (discussed in section 3.1).

Exercise 1–1C

Using the rules for rounding whole numbers, round the following numbers to the place indicated.

1. 862,451 to the nearest ten ________; hundred ________.
2. 5899 to the nearest ten ________; thousand ________.
3. 541,722 to the nearest thousand ________; ten-thousand ________.
4. 5428 to the nearest ten ________; thousand ________.
5. 25,478,491 to the nearest hundred ________; million ________.
6. A small guest house costs $127,459 to construct. Round this to the nearest thousand dollars.
7. An excavator rents for $755 per day. Round this to the nearest hundred dollars.
8. There are 16,093,440 centimeters in 100 miles. Round the number of centimeters to the nearest hundred-thousand.
9. A building lot contains 294,500 square feet (approximately 7 acres). Round to the nearest ten thousand square feet.
10. A four-story building contains 23,047 square feet of office space. Round this to the nearest ten square feet.

NOTE

Section 1.2 is optional and can be omitted with no disruption in the continuity of the book. It is designed for students requiring a rudimentary understanding of precision and accuracy.

If section 1.2 is omitted, also omit problems 25, 27.c, and 27.d in the review exercises.

1.2 Accuracy and Precision

The terms **accuracy** and **precision** are frequently used interchangeably, although they are actually different concepts. The distinctions and definitions of the two terms are very important to the scientist, but carpenters and builders usually do not need to be as concerned about the technical differences. These definitions are not complete: they are the "practical" or working definitions of **precision** and **accuracy.**

Precision: precision refers to the place value to which something is measured. It also means "getting very close to the same value every time the specific quantity is measured" (repeatability).

Accuracy: providing a measurement that is "correct". (Remember, no measurement can ever be *exactly* correct.)

We will revisit each of these terms in the chapters on fractions and decimals. Let's take some examples of each of these:

Precision: A room has been studded with nominal 2″ × 4″ studs (actual dimensions $1\frac{1}{2}$″ × $3\frac{1}{2}$″. A carpentry student is supposed to measure the inside dimensions of a room, and reads $142\frac{1}{16}''$ on the tape measure. The student carefully takes two more readings and each time reads $142\frac{1}{16}''$. The room has been measured **precise** to the nearest $\frac{1}{16}$ inch. However, if the student was erroneously measuring the *outside* rather than the *inside* dimensions of the room, the measurements would be too large by approximately 7″, the width of 2 studs. Thus, the measurements are not accurate, even though they were measured **precise** to the nearest $\frac{1}{16}$ inch. We could say the measurements are *precisely wrong*.

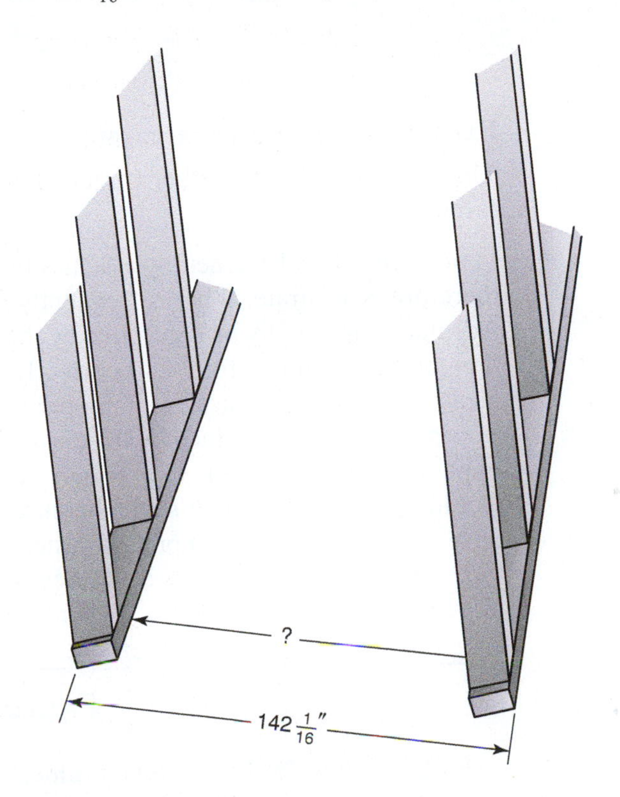

Accuracy: A foundation contractor has one of his workers measure the length of the foundation. The foundation measurement is found to be $\frac{7}{8}$ inch less than 40 feet. A second measurement shows it to be 1 inch over 40 feet, and the third measurement reads very close to 40 feet. If the actual dimension is 40 feet, then these measurements are **accurate,** because the average is close to 40 feet. However, the measurements are not very precise, because the readings varied by almost 2 inches.

A target is sometimes used to illustrate precision and accuracy.

Example 1. Precision without accuracy.

Remember, for a measurement to be precise, the results must vary by only a small amount, even if they are not correct.

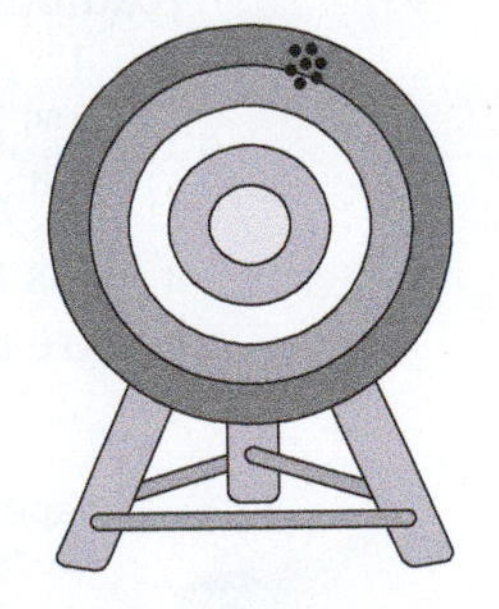

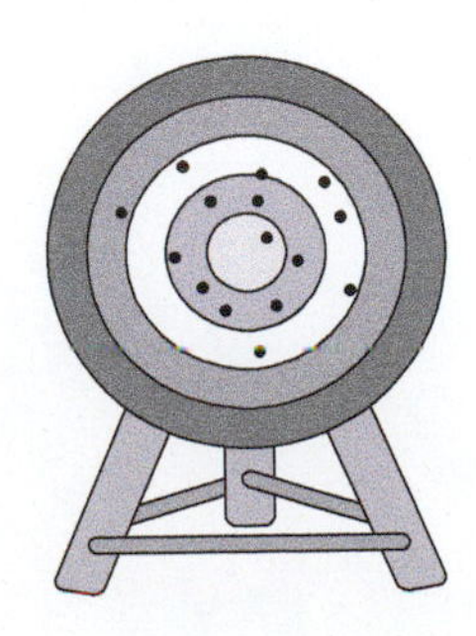

Example 2. Accuracy without precision.

The hits are widely spaced, but they all tend to center around the bull's-eye.

Example 3. Neither precise nor accurate.

The hits are not close together (hence not precise), and they are not close to the bull's-eye (hence not accurate).

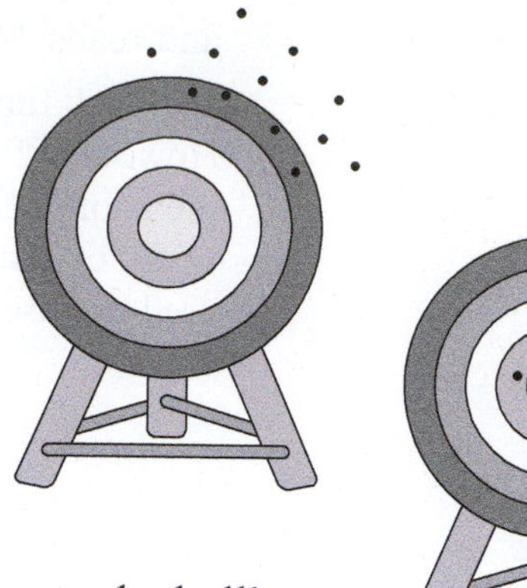

Example 4. Precise and accurate.

The shots are close together (precise) and close to the bull's-eye (accurate).

In practice, a builder generally assumes that a measurement is being taken correctly, and therefore is accurate. What is frequently referred to as ***accuracy*** is actually precision. Accuracy can usually be controlled by the person taking the measurement. If the measurement is taken carefully and correctly, then it will usually be ***accurate*** (unless the measuring instrument itself is improperly calibrated).

Precision is determined by the measuring instrument. A bathroom scale that weighs in pounds will not be able to precisely determine the weight of a handful of nails. That would best be done with a scale that measures in ounces and/or grams. The odometer on a car will not be able to precisely measure a distance of a hundred feet. A surveyor's tape would be an appropriate instrument for that measurement.

Precision will be revisited in more detail in Chapters 2 and 3.

Exercise 1–2

1. A building lot has accepted dimensions of 1255 inches wide by 2643 inches deep. Several student surveyors measure the lot, and each one reads the results to be 1255″ by 2643″ to the nearest inch.
   a. Are the students' measurements precise to the nearest inch?
   b. Are the students' measurements precise to the nearest 1/16th inch?
   c. Are the students' measurements accurate?
2. The length of a new driveway is 55 feet–8 inches. A builder's apprentice inadvertently uses a damaged 100-foot tape measure which has had the first 5 feet of the tape removed. He measures the driveway three times and gets the following readings:

   1st reading: 60 ft–8 in.
   2nd reading: 60 ft–8 in.
   3rd reading: 60 ft–8 in.

   a. Has the length been measured precise to the nearest inch?
   b. Are the measurements accurate?

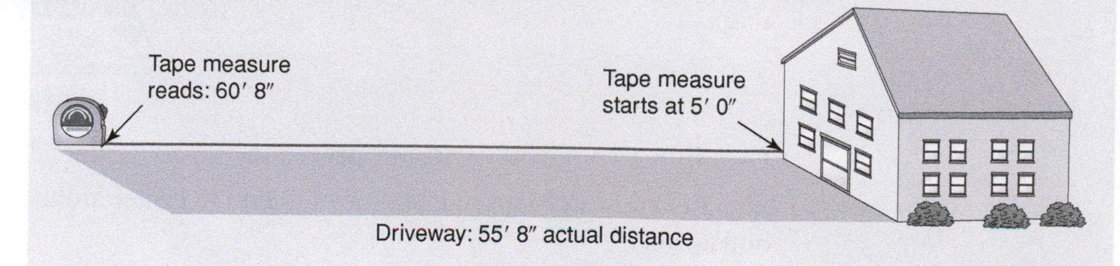

1.3 Order of Operations

In solving problems with more than one operation, it is important to follow a certain order, as shown below. Arithmetic operations fall into several categories:

- *CATEGORY A:* squares, square roots, and other exponents
- *CATEGORY B:* multiplication and division
- *CATEGORY C:* addition and subtraction

When solving a problem without parentheses, the following order should be followed:

Step 1. Perform operations in category A (squares, square roots, and other exponents).
Step 2. Perform operations in category B (multiplication and division).
Step 3. Perform operations in category C (addition and subtraction).

When parentheses are present, perform the operations in the parentheses first, then follow the order outlined above.

The default order of operations is left to right $(L \rightarrow R)$. If all operations are of the same category, always solve left to right.

Examples

A. This example illustrates what could happen if calculations are not done left to right:

$13 - 8 - 3.$

With no parentheses and both calculations being of equal priority, we start from the left.

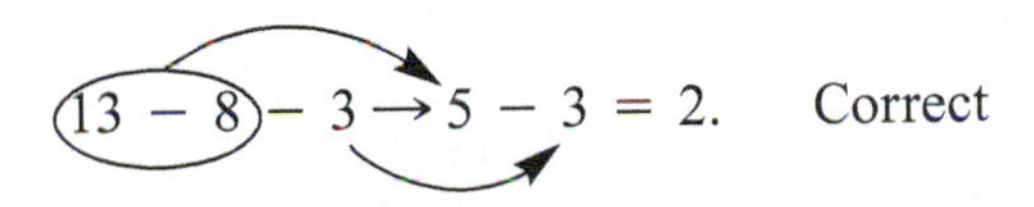

Correct

We first subtract 8 from 13, and then subtract 3 from the 5. This produces the correct answer of 2.

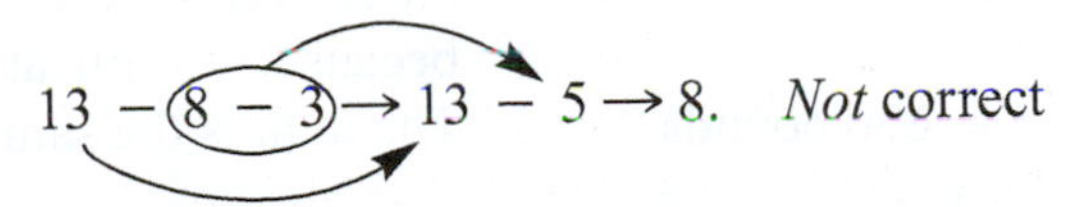

Not correct

(If we had done the subtractions from right to left, we would have done the ***incorrect*** calculation shown at left. As you can see, this incorrect method yields an incorrect answer.) In mathematical terms we are saying that subtraction is not *commutative*.

B. $8 - 6 \div 2 + 4 \times 2^2$

$8 - 6 \div 2 + 4 \times \underbrace{2^2}$ — 1. Clear all exponents.

$8 - 6 \div 2 + 4 \times 4$

$8 - \underbrace{6 \div 2} + \underbrace{4 \times 4}$ — 2. Perform all multiplication and division from left to right.

$8 - 3 + 16$ — 3. Perform all addition and subtraction from left to right.

$8 - 3 + 16 = 21$

C. $(8 - 6) \div 2 + (4 \times 2)^2$. Note that this problem is identical to example A except for the parentheses.

$\underbrace{(8 - 6)} \div 2 + \underbrace{(4 \times 2)}^2$ — 1. Perform operations inside parentheses.

$2 \div 2 + 8^2$ — 2. Evaluate exponents.

$2 \div 2 + 64$
$\underbrace{2 \div 2} + 64$
$1 + 64 = 65$

3. Perform all multiplication and division from left to right.
4. Add and subtract from left to right.

Note how different the answers are despite the fact that exactly the same numbers and operations occur in both problems. Order of operation is important!

D. A contractor orders the following 2 × 4s: twenty-five 8-ft studs and thirty 12-ft studs. How many lineal feet of 2 × 4s has he ordered?

$$25 \times 8 + 30 \times 12$$
$$= 200 + 360$$
$$= 560 \text{ lineal feet}$$

These are frequently written with parentheses for clarification. Do the parentheses, as they are shown, change the answer?

$$(25 \times 8) + (30 \times 12) =$$

E. Drill presses weighting 72 pounds each are to be shipped in crates weighing 13 pounds each. If 10 drill presses are to be shipped, what is the total shipping weight?

$10 \times (72 + 13)$

1. Find the total shipping weight of each drill press (the amount in parentheses).

$10 \times 85 = 850$ pounds

2. Multiply by the number being shipped.

Or:

$\underbrace{10 \times 72}_{720} + \underbrace{10 \times 13}_{130}$

1. Find the weight of all 10 drills and the weight of all 10 crates. Parentheses are not necessary because multiplication is done before addition.

$720 + 130 = 850$ pounds

2. The total is the same with either method.

F. A builder is constructing 25 identical apartments, each with 425 square feet of space downstairs and 380 square feet upstairs. How much is the total area of the 25 apartments?

$25 \times (425 + 380)$

1. Perform the operation in parentheses first; that is, find the area of each apartment.

$25 \times 805 = 20{,}125$
$20{,}125 \text{ ft}^2$

2. Multiply times the number of apartments.
3. Note that square feet is abbreviated as ft^2.

Alternate method:

$\underbrace{(25 \times 425)}_{10{,}625} + \underbrace{(25 \times 380)}_{9500}$

1. Parentheses are not really necessary in this case, but are used for clarification.

$10{,}625 + 9500 = 20{,}125 \text{ ft}^2$

2. Finding the total downstairs area for all 25 apartments, the total upstairs area for all apartments, and then adding them together gives the same result.

Exercise 1–3

Perform each operation in the correct order.

1. $5 + 8 \times 2$
2. $12 - 6 \times 2$
3. $8 + 6 \div 3 - 4$
4. $5 + 3 \times 2^2$
5. $8 \times (5 - 3) + 6$
6. $(3 + 2)^2 \div 5$
7. A trucker drives 30 mph for 1 hr and 55 mph for 2 hr to deliver prefabricated materials for a house. How far has he traveled?
8. Determine the total wattage in an electrical circuit with this load: five 175-watt (W) lamps, three 150-W lamps, two 75-W lamps, and four 60-W lamps.
9. To excavate a foundation, a backhoe is rented for 2 days at $125 per day, and a bulldozer is rented for 4 hours at $115 per hour. Find the total cost of renting the equipment.
10. Fifteen houses are being constructed in a development. Seven identical houses have 720 ft^2 of space on the first floor and 435 ft^2 on the second floor. Five identical one-story houses have 1250 ft^2 of area, and the remaining three houses have 825 ft^2 on the first floor, 255 ft^2 on the second floor, and 180 ft^2 of living space above the garage. Find the total square footage of the 15 houses being built.
11. Rechargeable hand-held drills weigh 3 pounds each without the battery pack. The battery packs weigh 2 pounds each. How much would 7 drills with their battery packs weigh? Work this problem two ways, using examples D and E above.

REVIEW EXERCISES

1. Find the total cost if the following expenses are incurred by a contractor: framing lumber $855, nails $38, shingles $152, siding $278.
2. A table saw weighs 278 lb. If it is to be shipped in a wooden crate weighing 26 lb, what is the total weight of the saw when prepared for shipping?
3. An electrician uses the following lengths of Number 12 wire: 82 in., 185 in., and 1461 in. How much wire, in inches, did the electrician use?
4. A water meter installed at a construction site read 8351 ft^3 on July 1 and 15,823 ft^3 on August 1. How many cubic feet of water were used in the month of July?
5. A total of 384 yd^3 of earth must be excavated for a basement. If 195 yd^3 have been removed, how many more cubic yards must be removed?
6. If holes spaced 17 in. center-to-center are to be drilled in a stud, how far apart are the two holes indicated on the diagram below?

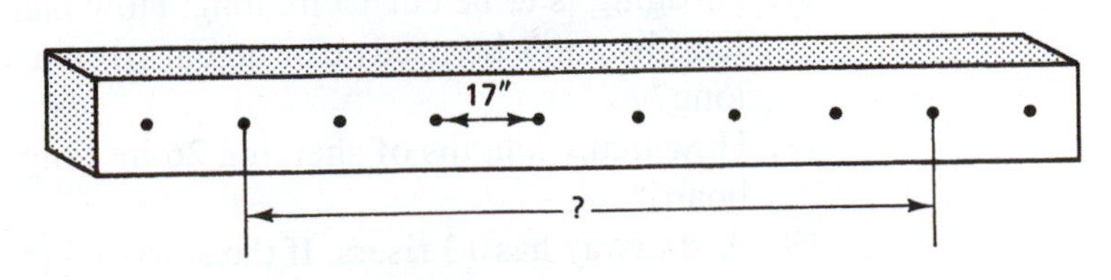

7. If lumber costs $439 per mbf (thousand board feet), what is the cost of 78 mbf?
8. A contractor's pickup truck averages 19 mpg. How far can he drive on a full tank of gas if the tank holds 27 gal?

9. A cabinetmaker can cut 325 wooden plugs from an 8-ft board. How many plugs can he cut from 71 boards of the same size?
10. Boards are cut 21 in. long to serve as fire stops between studs. If 386 fire stops are to be cut, how many inches of boards are needed? (Ignore waste due to cutting.)
11. A contractor purchases 42 mbf (thousand board feet) of lumber at a cost of $15,372. What is the price of 1 mbf?
12. How many joists spaced 16″ o.c. (on center) are required for a floor that is 32 ft long (384 in.)? (*Hint:* Note in the diagram that there is one more joist than the number of spaces. Because you are actually determining the number of spaces, you will need to add one joist.)

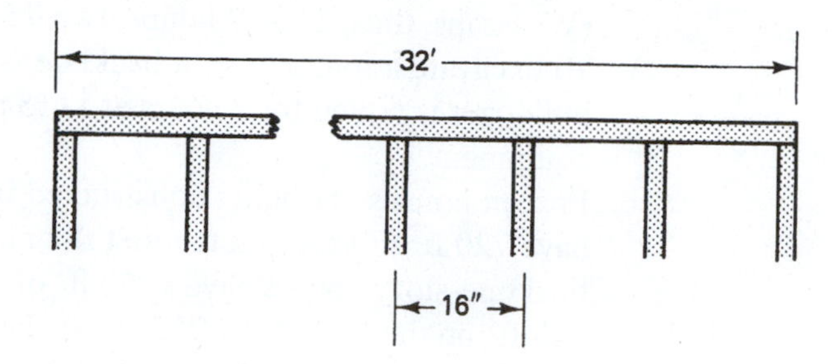

13. A set of basement stairs has an opening 81 in. long. How wide is each tread if there are 9 treads? (Assume there is no overhang on the treads.) Stairs will be covered in detail in Chapter 23.
14. A mason lays an average of 121 bricks per hour. How many hours will it take him to lay 4356 bricks?
15. A roll of copper tubing is 50 ft (600 in.) long. If gas range connectors must be 30 in. long, how many connectors can be cut from the roll of tubing? (Ignore cutting waste.)
16. The structural steel I-beam is to have 18 equally spaced holes drilled in it as shown. How far apart are the holes center-to-center? (*Hint:* The number of spaces is one less than the number of holes.)

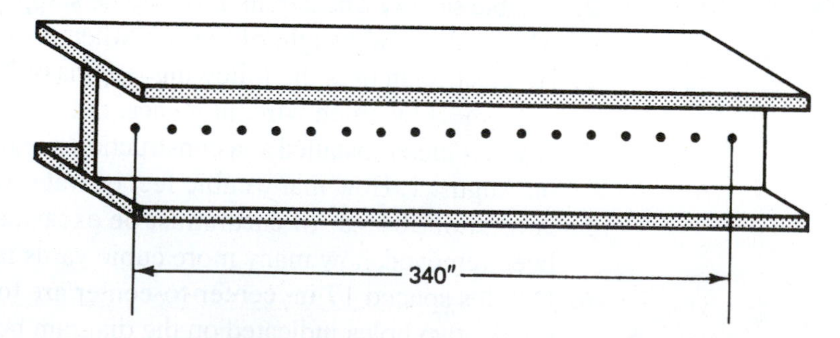

17. Bridging is to be cut 15 in. long. How many whole lengths of bridging (partial lengths can't be used) can be cut from a 1 × 3 piece of strapping that is 8 ft long?
18. How many lengths of shelving 26 in. long can be cut from an 8-ft 1 in. × 8-in. board?
19. A stairway has 13 risers. If the story height (distance from the top of the first to the top of the second floor) is 104 in., what is the height of each riser?
20. Bridging is to be 17 in. long. Ignoring cutting waste, how many pieces of bridging can be cut from a 12-ft length of 1 in. × 3 in. strapping? (Only whole pieces of bridging count!)

21. Shapers weighing 320 lb each are packed in shipping crates that each weigh 45 lb. What is the total shipping weight of 14 shapers?
22. Find the perimeter (distance around) of the floor plan shown.

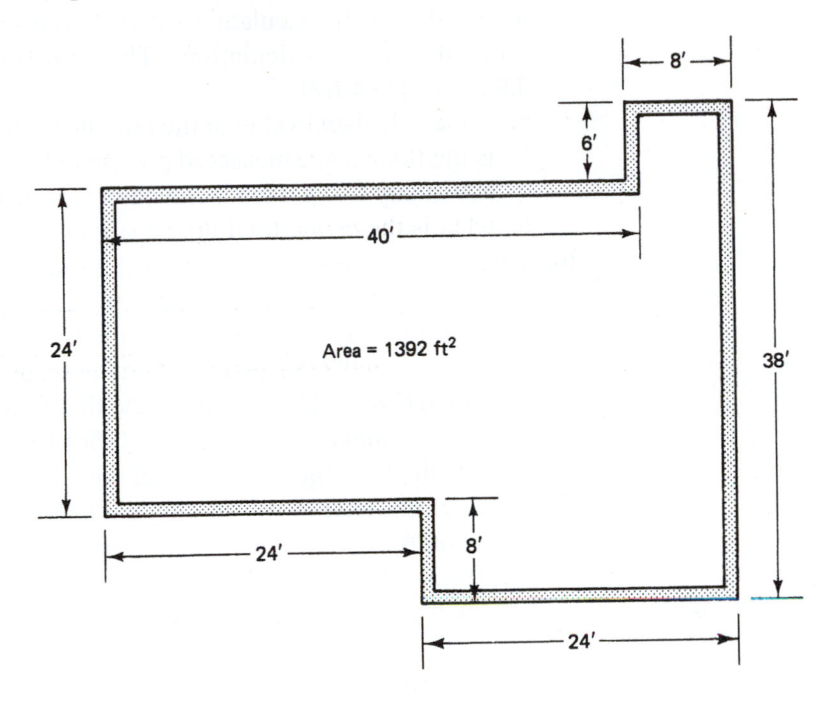

23. The floor plan shown has the same area (1392 ft) as the one in exercise 22. What is the perimeter of the floor plan? Which foundation and shell would be less expensive to erect? Why?

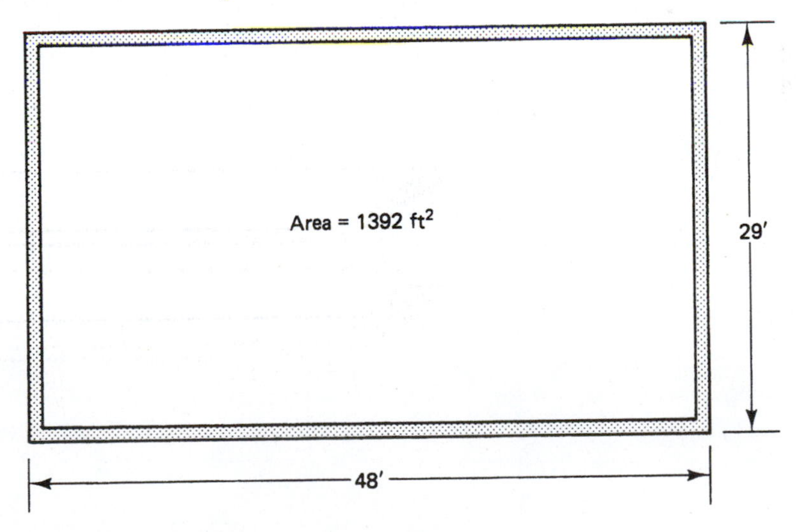

24. A builder has a subdivision with 4 different house plans. Each style has a different kitchen with different amounts of counter space.

    Style ***A*** kitchens have 16 feet of counter
    Style ***B*** kitchens have 22 feet of counter
    Style ***C*** kitchens have 31 feet of counter
    Style ***D*** kitchens have 19 feet of counter.

    All countertops will be surfaced with white laminate. Find the total number of lineal feet of laminate needed for all 84 homes in the subdivision. (Ignore any allowance for waste.) There are 17 houses with ***A*** kitchens, 22 houses with ***B*** kitchens, 37 houses with ***C*** kitchens, and 8 with ***D*** kitchens.

25. A large bank building is measured to have 47,507 square feet of area on each identical floor. In calculating the total floor space, the top floor was inadvertently omitted from the calculations. The total floor space was incorrectly stated as 237,535 square feet.
    a. is the calculated value of the total floor space accurate?
    b. is the floor space measured precise to the nearest square foot?
    c. how many floors were included in the calculation?
    d. what is the ***actual*** total floor space of the building?
26. Three house plans have the following square footage:
    Plan A is a one-floor home with 1232 square feet.
    Plan B is a two-story home with 955 square feet on the first floor and 735 square feet on the second floor.
    Plan C has 1225 sq. feet on the first floor, 622 sq. feet on the second floor, and a 430 square foot office space above the garage.
    What is the total square footage of 6 plan A, 5 plan B, and 8 plan C homes?
27. An electric meter installed on a construction site read 14,335 kwh (kilowatt-hours) on April 1. A month later, the meter read 15,199 kwh. On July 1, the meter read 16,375 kwh.
    a. how many kwh of electricity were used in the month of April?
    b. how many kwh of electricity were used during the 3 months from April 1 to July 1?
    c. are these readings precise to the nearest kwh?
    d. should we assume these readings are accurate?

CHAPTER 2

Fractions

Did You Know

Compact fluorescent light bulbs use about one-third the electricity of standard incandescent bulbs.

OBJECTIVES

Upon completing this chapter, the student will be able to:

1. Reduce fractions to lowest terms.
2. Find equivalent fractions with the same denominators.
3. Change improper fractions to mixed numbers, and vice versa.
4. Read a carpenter's rule, precise to the nearest $\frac{1}{16}$ inch.
5. Read a carpenter's rule, precise to the nearest $\frac{1}{32}$ inch.
6. Find the lowest common denominator of fractions.
7. Determine the equivalent fractions with the lowest common denominators.
8. Order fractions according to size.
9. Add fractions and reduce to lowest terms.
10. Add mixed numbers and reduce to lowest terms.
11. Solve problems involving addition of fractions and mixed numbers.
12. Subtract fractions and express answers in simplest terms.
13. Subtract mixed numbers and express answers in simplest terms.
14. Solve problems involving subtraction of fractions and mixed numbers.
15. Multiply fractions.
16. Multiply mixed numbers.
17. Solve problems involving multiplication of fractions and mixed numbers.
18. Divide fractions.
19. Divide any combination of mixed numbers and fractions.
20. Solve problems involving division of fractions and mixed numbers.
21. Add, subtract, multiply, and divide fractions and mixed numbers, and solve problems using a combination of operations.
22. Change improper fractions to mixed numbers and vice versa.

2.1 Equivalent Fractions

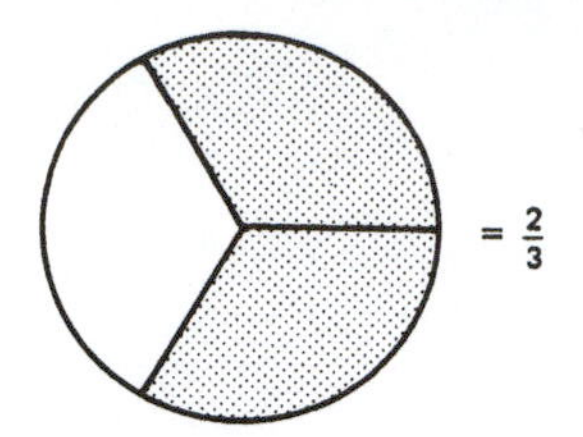

Figure 2–1

Fractions are used to express parts of a whole. The fraction $\frac{2}{3}$ indicates that 3 equal parts make up the whole and that 2 of these are being considered (Figure 2–1). The top number is the *numerator,* the *fraction bar* indicates division, and the bottom number is the *denominator* (remember D for "Down").

A proper fraction has a smaller numerator than denominator. Proper fractions represent less than a whole quantity. For example, in a quantity divided into 8 equal parts, if 3 are being used or considered, the fraction $\frac{3}{8}$ represents this concept; $\frac{8}{8}$ would represent a whole quantity in this case.

An improper fraction has a numerator larger than the denominator. Its value represents more than a whole quantity; $\frac{9}{5}$ is an example of an improper fraction (a misnomer, since there is really nothing improper about improper fractions). All improper fractions can be converted to mixed numbers.

If a quantity is divided into four equal parts and one of these parts is used, this is equivalent to dividing the quantity into 16 equal parts and using four of them (Figure 2–2). Notice that exactly the same portion of the square is darkened in both cases. $\frac{1}{4}$ and $\frac{4}{16}$ are equivalent fractions.

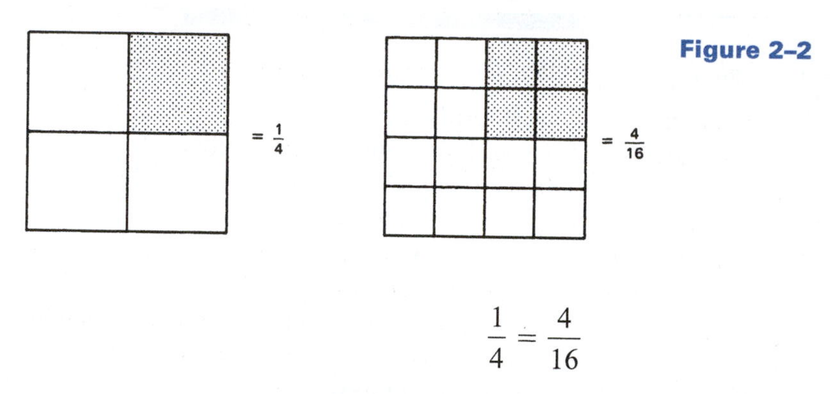

Figure 2–2

$$\frac{1}{4} = \frac{4}{16}$$

Another example of equivalent fractions is shown in Figure 2–3:

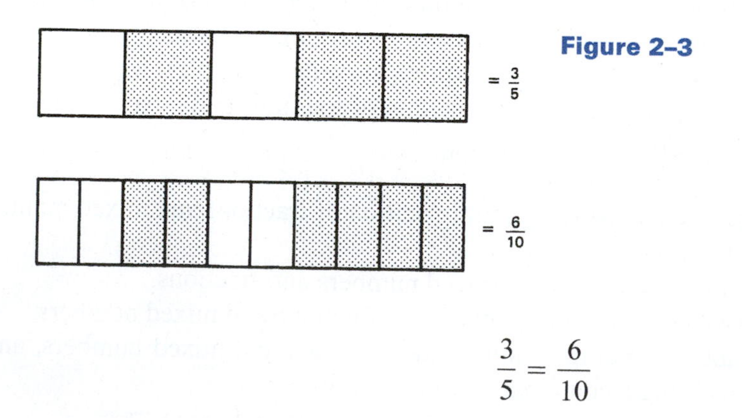

Figure 2–3

$$\frac{3}{5} = \frac{6}{10}$$

When an equivalent fraction is given with the smallest possible whole-number numerator and denominator, it is said to be *reduced to lowest terms.* For example,

$$\frac{25}{75} = \frac{1}{3}; \quad \frac{35}{49} = \frac{5}{7}$$

To reduce fractions to lowest terms, determine the largest number that can divide a whole number of times into both the numerator and the denominator, and reduce the fraction by that amount.

Examples

A. $\frac{25}{75} = \frac{25 \times 1}{25 \times 3} = \frac{\cancel{25} \times 1}{\cancel{25} \times 3} = \frac{1}{3}$

(divide numerator and denominator by 25)

B. $\frac{35}{49} = \frac{7 \times 5}{7 \times 7} = \frac{\cancel{7} \times 5}{\cancel{7} \times 7} = \frac{5}{7}$

(divide numerator and denominator by 7)

C. A 10-foot × 10-foot room is to be tiled with square tiles one foot on each side. Forty tiles have been laid. What fraction of the floor has been tiled?

$\frac{\text{40 tiles laid}}{\text{100 tiles total}} = \frac{40}{100} = \frac{4}{10} = \frac{2}{5}$

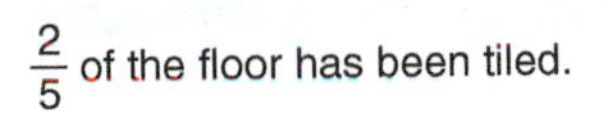

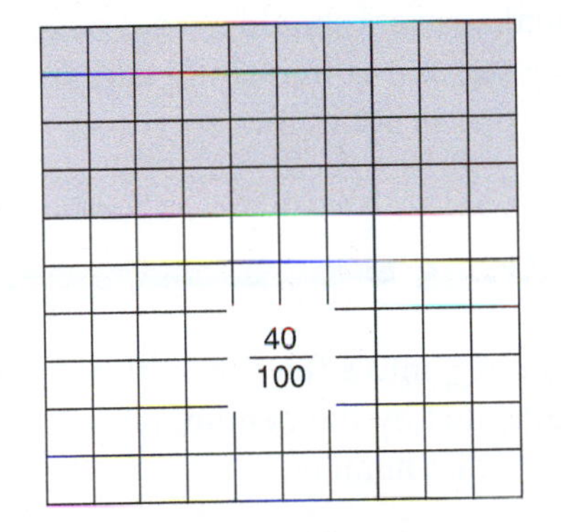

=

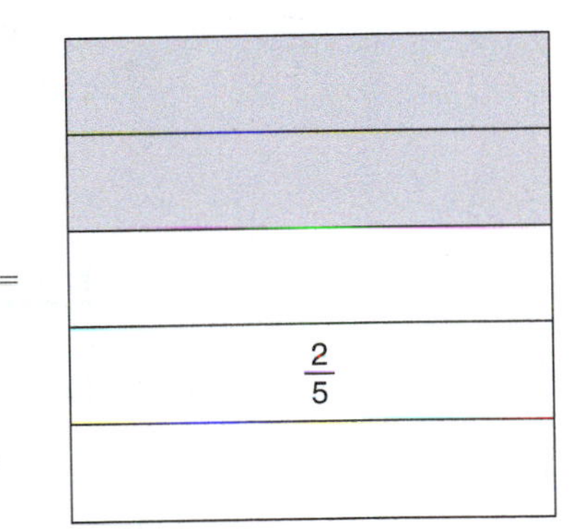

Exercise 2–1A

Reduce the following fractions to equivalent fractions in lowest terms.

1. $\frac{28}{36}$
2. $\frac{4}{30}$
3. $\frac{32}{64}$
4. $\frac{3}{39}$
5. $\frac{7}{21}$
6. $\frac{48}{64}$
7. $\frac{4}{17}$
8. $\frac{5}{45}$

9. $\frac{3}{24}$

10. $\frac{14}{26}$

11. $\frac{5}{32}$

12. $\frac{6}{16}$

13. $\frac{2}{8}$

14. $\frac{3}{16}$

15. $\frac{24}{32}$

16. A total of 80 rafters must be cut to size at the proper angle. Twenty-five of the rafters have already been cut. What fraction of the rafters has been cut?
17. A carpenter has used all but 5 drywall screws from a box that originally contained 100 screws. What fraction of the drywall screws remains?

Final answers should always be reduced to lowest terms. However, to perform intermediate steps, it is frequently necessary to change a reduced fraction into an equivalent fraction that is not in lowest terms.

Examples

D. Change $\frac{3}{4}$ into a fraction with a denominator of 36. Determine what number must be multiplied by the denominator 4, and then multiply both denominator and numerator by that number.

$$\frac{3}{4} = \frac{3 \times 9}{4 \times 9} = \frac{27}{36}$$

(numerator and denominator must be multiplied by 9)

E. What fraction with a denominator of 24 is equivalent to $\frac{5}{8}$?

$$\frac{5}{8} = \frac{5 \times 3}{8 \times 3} = \frac{15}{24}$$

(numerator and denominator must be multiplied by 3)

There is a "shortcut" way to do this:

$$\frac{5}{8} = \frac{?}{24} \rightarrow 24 \div 8 \times 5 = 15 \rightarrow \frac{5}{8} = \frac{15}{24}$$

There are many instances like this where you are really performing the same operation, but viewing it in a different way.

F. Use the shortcut method to find the following equivalent fraction: $\frac{3}{8} = \frac{?}{72}$

$$\frac{3}{8} = \frac{?}{72} \rightarrow 72 \div 8 \times 3 = 27 \rightarrow \frac{3}{8} = \frac{27}{72}$$

Exercise 2–1B

Find the equivalent fractions with the denominators indicated.

1. $\frac{5}{9} = \frac{?}{27}$
2. $\frac{5}{12} = \frac{?}{24}$
3. $\frac{3}{16} = \frac{?}{64}$
4. $\frac{5}{8} = \frac{?}{32}$
5. $\frac{1}{2} = \frac{?}{56}$
6. $\frac{5}{4} = \frac{?}{16}$
7. $\frac{9}{16} = \frac{?}{64}$
8. $\frac{35}{32} = \frac{?}{64}$
9. $\frac{5}{5} = \frac{?}{35}$
10. $\frac{2}{3} = \frac{?}{12}$
11. $\frac{5}{8} = \frac{?}{16}$
12. $\frac{1}{4} = \frac{?}{16}$
13. $\frac{3}{8} = \frac{?}{32}$
14. $\frac{1}{2} = \frac{?}{8}$
15. $\frac{7}{16} = \frac{?}{32}$

Changing Mixed Numbers to Improper Fractions

A mixed number is the sum of a whole number and a proper fraction. The mixed number $2\frac{1}{4}$ represents $2 + \frac{1}{4}$. Any improper fraction can be changed into a mixed number, and vice versa.

Example

G. Change the mixed number $2\frac{1}{4}$ into the equivalent improper fraction.

$$2\frac{1}{4} = 2 + \frac{1}{4} = \frac{2}{1} + \frac{1}{4}$$

1. Any whole number can be represented as a numerator over 1.

$$\frac{2 \cdot 4}{1 \cdot 4} + \frac{1}{4}$$

2. Change $\frac{2}{1}$ to its equivalent fraction with a denominator of 4.

$$\frac{8}{4} + \frac{1}{4} = \frac{9}{4}$$

3. Add the numerators to determine the improper fraction equivalent to the mixed number. *Note:* Do *not* add the denominators.

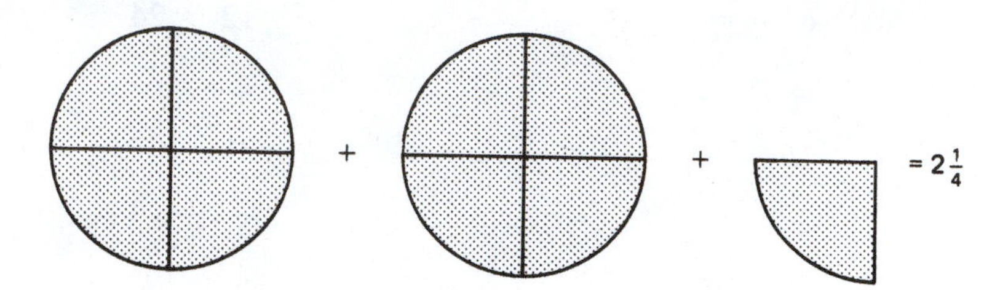

Examples H, I, and J demonstrate a shortcut method for changing any mixed number to an improper fraction:

Examples

H. Change the mixed number $5\frac{2}{3}$ into an improper fraction.

$$5\frac{2}{3} = 3 \cdot 5 + \cdots$$

1. Multiply the denominator of the fraction times the whole number. In this example the denominator 3 is multiplied by the whole number 5.

$$5\frac{2}{3} = (3 \cdot 5) + 2$$

2. Add the original numerator to the product. In this example the numerator 2 is added to the product 15.

$$\frac{15 + 2}{3} = \frac{17}{3}$$

3. The resulting amount is the numerator of the improper fraction. The denominator is not changed.

I. Convert $3\frac{2}{3}$ to an improper fraction.

$$3\frac{2}{3} = \frac{(3 \cdot 3) + 2}{3} = \frac{9 + 2}{3} = \frac{11}{3}$$

J. PT (pressure-treated) decking measures $1\frac{1}{4}''$. This is usually stated as an improper fraction. What is the measurement as an improper fraction?

$1\frac{1}{4} \rightarrow 4 \times 1 + \cdots$ 1. Multiply the denominator 4 times the whole number 1.

$1\frac{1}{4} \rightarrow (4 \times 1) + 1$ 2. Add the original numerator 1 to the product 4.

$\frac{(4 + 1)}{4} = \frac{5}{4}$ 3. The resulting amount is the numerator of the improper fraction. The denominator does not change.

$1\frac{1''}{4} = \frac{5''}{4}$ 4. $1\frac{1}{4}$ decking is commonly referred to as $\frac{5}{4}$ (five-quarter) decking.

Exercise 2–1C

Change the following mixed numbers to improper fractions.

1. $5\frac{1}{8}$
2. $3\frac{2}{5}$
3. $1\frac{7}{8}$
4. $7\frac{2}{3}$
5. $6\frac{3}{8}$
6. $4\frac{1}{5}$
7. $8\frac{3}{8}$
8. $7\frac{1}{9}$
9. $5\frac{2}{5}$
10. $3\frac{1}{4}$
11. Interlocking plywood underlayment measures $1\frac{1}{4}''$ thick. What is the thickness as an improper fraction?
12. A drill bit has a $1\frac{3}{8}''$ diameter. What is the equivalent improper fraction?
13. A countertop is $1\frac{1}{2}$ inch thick. State the thickness as an improper fraction to the nearer $\frac{1}{16}$ of an inch. (Note: this requires 2 operations: Changing to an improper fraction and then changing to the equivalent fraction.)
14. A $2'' \times 4''$ board is actually $1\frac{1}{2}'' \times 3\frac{1}{2}''$ in dimension. State the actual dimensions as improper fractions.
15. Give the actual dimensions of a $2'' \times 6''$ board as improper fractions. (See problem 14.)

Changing Improper Fractions to Mixed Numbers

An improper fraction can be changed to an equivalent mixed number by the following method.

Example

K. Change $\frac{14}{3}$ to a mixed number.

$\frac{14}{3} = 3\overline{)14}$

1. Remember: a fraction always indicates division. Divide the denominator into the numerator.

$$\begin{array}{r} 4 \text{ R } 2 \\ 3\overline{)14} \\ \underline{12} \\ 2 \end{array}$$

2. The whole number of times the denominator divides into the numerator is the whole-number part of the mixed number.

$$\begin{array}{r} 4 \text{ R } 2 \\ 3\overline{)14} \end{array} = 4\frac{2}{3}$$

3. The remainder becomes the numerator of the fraction part of the mixed number. Note that the denominator is the same in both the mixed number and the improper fraction.

Here is a shortcut method that works *only* in one special, but frequently occurring, case:

Example

L. Convert $\frac{29}{21}$ to a mixed number.

$\frac{29}{21} = 1\frac{?}{21}$

1. By inspection, observe that the denominator 21 will divide into the numerator 29 only one time.

$$\begin{array}{rr} 29 \rightarrow & 29 \\ \overline{21} \rightarrow & \underline{-\ 21} \\ & 8 \end{array} \qquad \frac{29}{21} = 1\frac{8}{21}$$

2. Whenever the denominator divides *only one time,* the new numerator can be found by subtracting the denominator from the numerator of the improper fraction. *Note:* This method cannot be used if the whole-number part is any number other than 1.

Exercise 2–1D

Change the following improper fractions to mixed numbers and reduce to lowest terms where necessary. Use the shortcut method whenever possible.

1. $\frac{8}{5}$
2. $\frac{9}{4}$
3. $\frac{17}{16}$
4. $\frac{38}{32}$
5. $\frac{25}{16}$
6. $\frac{54}{8}$
7. $\frac{62}{15}$
8. $\frac{35}{32}$
9. $\frac{56}{8}$
10. $\frac{40}{16}$

Reading a Carpenter's Rule or Measuring Tape

When reading a carpenter's rule or tape measure marked 16 divisions per inch, a measurement precise to the nearest $\frac{1}{16}$ inch can be read. Be careful to read the correct whole number of inches. When concentrating on the fractional measurement, it's surprisingly easy to misread the number of whole inches. $15\frac{11}{16}''$ is measured precise to the nearest $\frac{1}{16}$ inch, but it is not accurate if the real reading is $14\frac{11}{16}''$!

Learning to quickly and correctly read a carpenter's rule is a very important skill, and there is no substitute for practice. There will be several opportunities for practice in this chapter, and many more on-the-job opportunities. The carpenter's rule shown is divided into inches, half-inches, $\frac{1}{4}$ inches, $\frac{1}{8}$ inches, and $\frac{1}{16}$ of an inch, the smallest division on the ruler. The following measurements are located on the ruler shown below.

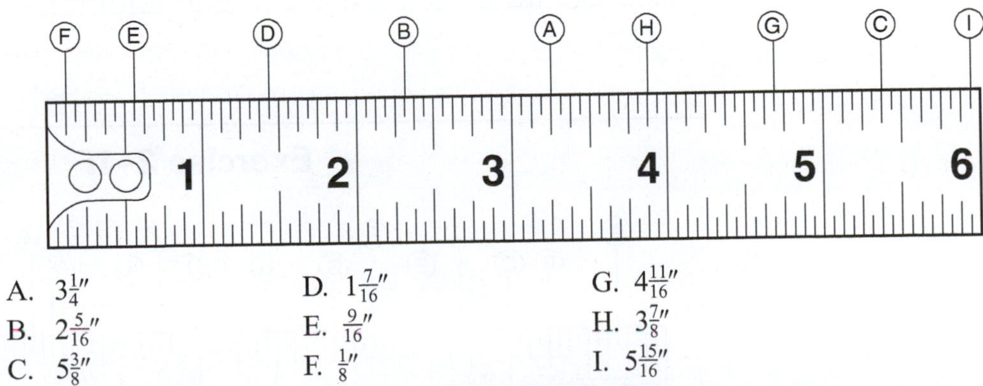

A. $3\frac{1}{4}''$
B. $2\frac{5}{16}''$
C. $5\frac{3}{8}''$
D. $1\frac{7}{16}''$
E. $\frac{9}{16}''$
F. $\frac{1}{8}''$
G. $4\frac{11}{16}''$
H. $3\frac{7}{8}''$
I. $5\frac{15}{16}''$

NOTE

Any time the measurement is an *even* number of marks (16^{ths} of an inch) to the right of the whole number, the fraction will be reducible to 8^{ths}, 4^{ths}, or $\frac{1}{2}$ inch.

A. This mark is between 3 and 4. Therefore we know the measurement is 3 inches plus some fraction of an inch. Each mark represents $\frac{1}{16}''$; therefore A is $3\frac{4}{16}''$. This reduces to $3\frac{1}{4}''$.

B. This mark is between 2 and 3. It is 5 marks beyond 2; therefore B represents the measurement $2\frac{5}{16}''$.

C. This mark is 5″ plus $\frac{6}{16}$ of an inch, or $5\frac{6}{16}''$. This reduces to $5\frac{3}{8}''$. Notice that only the shortest of the marks will not reduce. All others can be reduced to 8ths, 4ths, or $\frac{1}{2}$ inches.

D. This point is 7 small marks beyond the 1. Thus this measurement is $1\frac{7}{16}''$.

E. This measurement can be determined by counting 9 small marks, indicating $\frac{9}{16}''$. Notice that this is one mark beyond the large $\frac{1}{2}''$ mark. With practice, one can mentally change the $\frac{1}{2}''$ to $\frac{8}{16}''$, and counting one mark past that yields $\frac{9}{16}''$. This is usually faster than counting all the $\frac{1}{16}''$ marks. With a little practice, it becomes automatic.

F. This is the $\frac{1}{8}''$ mark. The smallest marks all represent $\frac{1}{16}''$. The next longer marks are $\frac{1}{8}$ inches, the next longer are $\frac{1}{4}$ inches, and the middle unit is $\frac{1}{2}''$.

G. We know this measurement is 4″ plus a fraction. G is 3 marks ($\frac{3}{16}''$) past the $\frac{1}{2}''$ mark. Change the $\frac{1}{2}''$ mentally to $\frac{8}{16}''$ and count 3 past that to yield $\frac{11}{16}''$ for a total value of $4\frac{11}{16}''$.

H. This measurement is 3″ plus $\frac{14}{16}''$ or $3\frac{7}{8}''$ in reduced form.

I. This measurement is one mark before the 6″ mark. Therefore it reads $5\frac{15}{16}''$.

Exercise 2–1E

What measurement does each of the locations A to P on the ruler below indicate? Give each answer as a mixed number reduced to lowest terms.

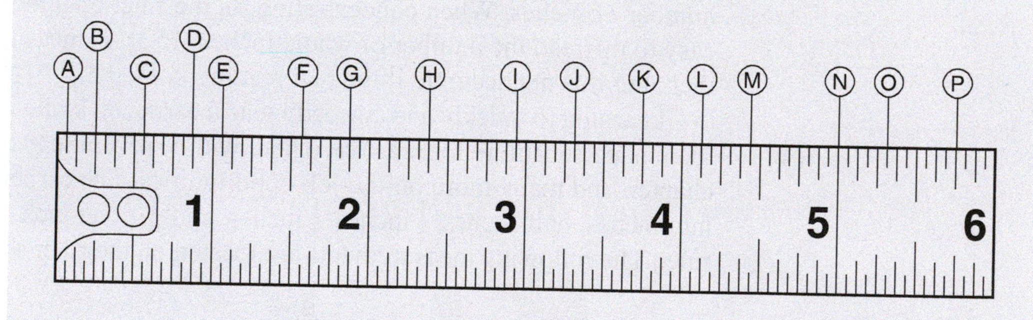

Exercise 2–1F

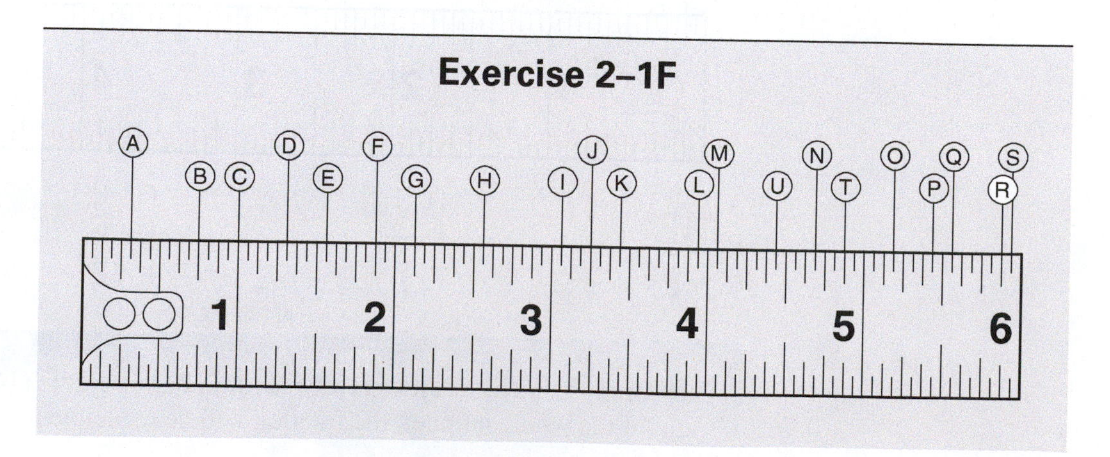

Find each measurement A–U. Give answer as a mixed number reduced to lowest terms.

A.	H.	O.
B.	I.	P.
C.	J.	Q.
D.	K.	R.
E.	L.	S.
F.	M.	T.
G.	N.	U.

Some tape measures have $\frac{1}{32}''$ division marks. The principle is the same, but there are 32 marks instead of 16 per inch. Study the tape measure below, and see why the units are read as shown. Try reading each point before looking at the answer below. Are you getting the correct answers?

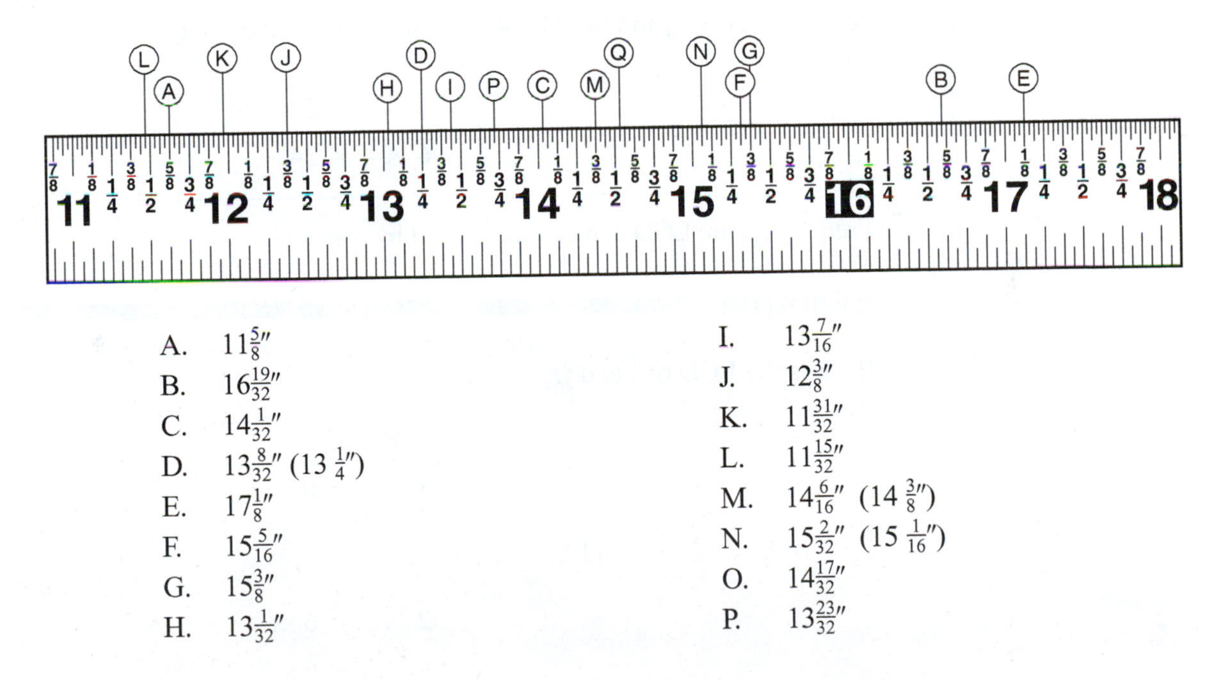

A. $11\frac{5}{8}''$
B. $16\frac{19}{32}''$
C. $14\frac{1}{32}''$
D. $13\frac{8}{32}''$ ($13\ \frac{1}{4}''$)
E. $17\frac{1}{8}''$
F. $15\frac{5}{16}''$
G. $15\frac{3}{8}''$
H. $13\frac{1}{32}''$

I. $13\frac{7}{16}''$
J. $12\frac{3}{8}''$
K. $11\frac{31}{32}''$
L. $11\frac{15}{32}''$
M. $14\frac{6}{16}''$ ($14\ \frac{3}{8}''$)
N. $15\frac{2}{32}''$ ($15\ \frac{1}{16}''$)
O. $14\frac{17}{32}''$
P. $13\frac{23}{32}''$

A tape measure marked with 16 divisions per inch is ***precise*** to the nearest $\frac{1}{16}$ inch. A proper reading will never be off by more than one-half that division. In other words, if a reading is precise to the nearest $\frac{1}{16}$ inch, it will be off by no more than $\frac{1}{32}$ inch. A tape measure marked with 32 divisions per inch is precise to the nearest $\frac{1}{32}$ inch. Any measurement falling between marks will be read as the value corresponding to the closer mark. If read correctly, the measurement will never be off by more than half the distance between the $\frac{1}{32}$ inch marks. Therefore a measurement made on a tape measure that is precise to the $\frac{1}{32}$ inch should be off by no more than $\frac{1}{64}$ inch.

NOTE

When a measurement of $11\frac{5}{8}''$ is read on a tape measure marked in $\frac{1}{16}$ of an inch, the measure is still precise to the nearest $\frac{1}{16}$ of an inch. If $11\frac{5}{8}''$ is read on a measure marked in $\frac{1}{32}$ of an inch, it is precise to the nearest $\frac{1}{32}$nd of an inch.

Non-scientists frequently refer to this as "accurate to the nearest $\frac{1}{16}$ (or $\frac{1}{32}$) of an inch," but technically it is "precise to the nearest $\frac{1}{16}$ (or $\frac{1}{32}$) of an inch."

2.2 Finding the Lowest Common Denominator and Equivalent Fractions

The lowest common denominator, usually referred to as the LCD, is the smallest number that is a multiple of each denominator.

Example

A. The LCD of $\frac{3}{8}$ and $\frac{5}{6}$ is 24. 24 is the smallest number that is a multiple of both 8 and 6.

$$8 \times 3 = 24$$
$$6 \times 4 = 24$$

Frequently, the LCD can be determined by inspection.

Examples

B. 8 is the LCD of $\frac{1}{4}$ and $\frac{3}{8}$.

$$4 \times 2 = 8$$
$$8 \times 1 = 8$$

C. 6 is the LCD of $\frac{1}{2}$ and $\frac{2}{3}$.

$$2 \times 3 = 6$$
$$3 \times 2 = 6$$

Here is one method that can be used when the LCD is not obvious by inspection. This method involves prime numbers. A prime number is any number that has factors of itself and 1 only. Here are the first 10 prime numbers:

2, 3, 5, 7, 11, 13, 17, 19, 23, 29

Examples

D. Find the LCD of the fractions $\frac{1}{5}$, $\frac{5}{9}$, and $\frac{4}{15}$.

$$\begin{aligned} 5 &= 5 \\ 9 &= 3 \cdot 3 \\ 15 &= 3 \cdot 5 \end{aligned}$$

1. Line up the denominators vertically and write them as prime factors.

$$\begin{aligned} \rightarrow 5 &= 5 \\ 9 &= 3 \cdot 3 \\ 15 &= 3 \cdot 5 \\ \text{LCD} &= 5 \cdot \underline{\qquad\qquad} \end{aligned}$$

2. Starting at the top, include in the LCD every factor in the first denominator.

$$\begin{aligned} 5 &= 5 \\ \rightarrow 9 &= 3 \cdot 3 \\ 15 &= 3 \cdot 5 \\ \text{LCD} &= 5 \cdot 3 \cdot 3 \end{aligned}$$

3. Move to the next denominator (9 in this example). *All* factors *not already included* must be included in the LCD.

$$\begin{aligned} 5 &= 5 \\ 9 &= 3 \cdot 3 \\ \rightarrow 15 &= 3 \cdot 5 \\ \text{LCD} &= 5 \cdot 3 \cdot 3 \end{aligned}$$

4. Neither the 5 nor the 3 in the denominator 15 should be included since both those factors are already in the LCD.

$$\text{LCD} = 5 \cdot 3 \cdot 3 = 45$$

5. Multiply the factors to determine the LCD.

E. Find the LCD and equivalent fractions of $\frac{5}{9}$, $\frac{11}{12}$, and $\frac{4}{15}$.

$$\begin{aligned} 9 &= 3 \cdot 3 \\ 12 &= 3 \cdot 2 \cdot 2 \\ 15 &= 3 \cdot 5 \end{aligned}$$

1. List the denominators and factor into prime factors.

$$\begin{aligned} \rightarrow 9 &= 3 \cdot 3 \\ 12 &= 3 \cdot 2 \cdot 2 \\ 15 &= 3 \cdot 5 \\ \text{LCD} &= 3 \cdot 3 \cdot \underline{\qquad\qquad} \end{aligned}$$

2. Starting at the top, include in the LCD all factors in the first number (9).

$$\begin{aligned} 9 &= 3 \cdot 3 \\ \rightarrow 12 &= 3 \cdot 2 \cdot 2 \\ 15 &= 3 \cdot 5 \\ \text{LCD} &= 3 \cdot 3 \cdot 2 \cdot 2 \cdot \underline{\qquad\qquad} \end{aligned}$$

3. Move to the next number (12). Include *only* those numbers that are not already included in the LCD. (Note that the 3 from the 12 is not included.)

$$\begin{aligned} 9 &= 3 \cdot 3 \\ 12 &= 3 \cdot 2 \cdot 2 \\ \rightarrow 15 &= 3 \cdot 5 \\ \text{LCD} &= 3 \cdot 3 \cdot 2 \cdot 2 \cdot 5 \end{aligned}$$

4. Move to the last number (15). Include only the 5; the 3 is already included in the LCD.

$$\text{LCD} = 3 \cdot 3 \cdot 2 \cdot 2 \cdot 5 = 180$$

5. Multiply all the factors to determine the LCD.

$$\frac{5}{9} = \frac{100}{180}$$

$$\frac{11}{12} = \frac{165}{180}$$

$$\frac{4}{15} = \frac{48}{180}$$

6. Find the equivalent fractions using the LCD as the denominator.

Exercise 2–2A

Either by inspection or by the process of finding LCDs, determine the equivalent fractions with the lowest common denominator for the following pairs of fractions.

1. $\frac{3}{8}, \frac{5}{6}$
2. $\frac{3}{8}, \frac{5}{32}$
3. $\frac{2}{3}, \frac{3}{4}$
4. $\frac{3}{16}, \frac{5}{8}$
5. $\frac{5}{8}, \frac{2}{3}$
6. $\frac{1}{7}, \frac{2}{3}$
7. $\frac{9}{64}, \frac{3}{32}$
8. $\frac{5}{16}, \frac{1}{4}$
9. $\frac{5}{18}, \frac{2}{3}$
10. $\frac{7}{12}, \frac{11}{18}$
11. $\frac{3}{4}, \frac{1}{2}$
12. $\frac{5}{16}, \frac{5}{12}$
13. $\frac{1}{9}, \frac{2}{3}$
14. $\frac{3}{5}, \frac{2}{3}$
15. $\frac{11}{12}, \frac{5}{9}$
16. $\frac{3}{8}, \frac{5}{64}$
17. $\frac{5}{6}, \frac{1}{3}$
18. $\frac{2}{6}, \frac{1}{4}$
19. $\frac{3}{8}, \frac{5}{16}$
20. $\frac{5}{24}, \frac{3}{16}$
21. $\frac{3}{4}, \frac{9}{16}$
22. $\frac{11}{16}, \frac{5}{8}$
23. $\frac{3}{8}, \frac{1}{2}$
24. $\frac{3}{4}, \frac{15}{16}$

Most fractions that carpenters work with have denominators that are multiples of 2. These denominators can be found by inspection.

Examples

F. A $\frac{3}{8}$-in. drill bit is too small and a $\frac{1}{2}$-in. bit is too large for a particular job. Which bit should be tried next?

$\frac{3}{8} \quad \frac{1}{2}$

1. The next bit to be tried should be halfway between the $\frac{3}{8}$- and the $\frac{1}{2}$-in. bits.

$\frac{3}{8} = \frac{6}{16}$

$\frac{1}{2} = \frac{8}{16}$

2. Change both fractions to equivalent fractions with a denominator of 16. The new denominator should be twice the size of the larger denominator. In this example 8 is the larger denominator; therefore, the new denominator is $8 \times 2 = 16$.

$\frac{6}{16}, \frac{8}{16} \rightarrow \frac{7}{16}$

3. The numerator is halfway between the two numerators.

$\frac{7}{16}$ in.

4. The appropriate bit to try is $\frac{7}{16}$ in.

G. Given a 64-piece set of drill bits with every size from $\frac{1}{64}$ to $\frac{64}{64}$ in., what drill bit is one size larger than $\frac{5}{32}$ in.?

$\frac{5}{32} = \frac{10}{64}$

1. Change to the equivalent fraction with a denominator of 64.

$\frac{10}{64} + \frac{1}{64}$

2. Since the next larger size will be $\frac{1}{64}$ in. larger, add $\frac{1}{64}$ to the bit size. (Add the numerators only.)

$\frac{11}{64}$ in.

3. An $\frac{11}{64}$-in. bit is one size larger than a $\frac{5}{32}$-in. bit.

H. Find the bit size that is halfway between a $\frac{3}{32}$-in. bit and a $\frac{1}{8}$-in. bit.

$\frac{3}{32} = \frac{6}{64}$

1. Find equivalent fractions with denominators twice the larger denominator ($32 \times 2 = 64$).

$\frac{1}{8} = \frac{8}{64}$

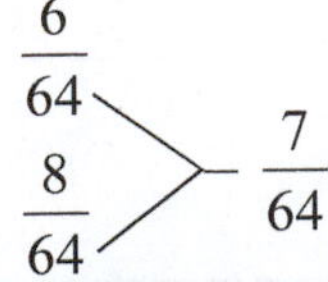

2. Find the number between the two numerators.

$\frac{7}{64}$ in.

3. This is the drill bit size halfway between a $\frac{3}{32}$-in. bit and a $\frac{1}{8}$-in. bit.

Exercise 2–2B

Assuming that all bits between $\frac{1}{64}$ and $\frac{64}{64}$ in. are available, determine which drill bit size is halfway between the following sizes.

1. $\frac{7}{32}$ and $\frac{1}{4}$ in.
2. $\frac{9}{32}$ and $\frac{5}{16}$ in.
3. $\frac{5}{8}$ and $\frac{3}{4}$ in.
4. $\frac{3}{8}$ and $\frac{1}{2}$ in.
5. $\frac{1}{2}$ and $\frac{3}{4}$ in.
6. $\frac{1}{16}$ and $\frac{1}{8}$ in.
7. $\frac{11}{16}$ and $\frac{3}{4}$ in.
8. $\frac{3}{32}$ and $\frac{1}{16}$ in.
9. $\frac{25}{32}$ and $\frac{13}{16}$ in.
10. $\frac{15}{32}$ and $\frac{1}{2}$ in.

Assuming that all bits between $\frac{1}{64}$ and $\frac{64}{64}$ in. are available, find the next size larger and the next size smaller than each given size.

Size	Next size larger	Next size smaller
11. $\frac{5}{32}$ (in.)		
12. $\frac{3}{16}$ (in.)		
13. $\frac{31}{32}$ (in.)		
14. $\frac{9}{32}$ (in.)		
15. $\frac{5}{8}$ (in.)		
16. $\frac{1}{8}$ (in.)		
17. $\frac{3}{4}$ (in.)		

18. $\frac{1}{16}$ (in.)		
19. $\frac{3}{8}$ (in.)		
20. $\frac{9}{32}$ (in.)		

2.3 Addition of Fractions and Mixed Numbers

In order to be added, fractions must have the same denominators. The numerators are added, and the fraction is simplified if necessary.

Examples

A. $\frac{5}{8}+\frac{2}{8}=\frac{7}{8}$

1. Denominators are the same.
2. Add the numerators.

B. $\frac{1}{9}+\frac{2}{9}=\frac{3}{9}=\frac{1}{3}$

1. Denominators are the same.
2. Add the numerators and reduce the fraction to lowest terms.

C. $\frac{3}{4}+\frac{1}{8}$

1. Denominators are not the same.

$$\frac{3}{4}=\frac{6}{8}$$

2. Determine the LCD and the equivalent fractions.

$$+\frac{1}{8}=\frac{1}{8}$$

3. Add the numerators.

$$\frac{7}{8}$$

D. $\frac{2}{3}+\frac{5}{8}$

1. Denominators are not the same.

$$\frac{2}{3}=\frac{16}{24}$$

$$+\frac{5}{8}=\frac{15}{24}$$

$$\frac{31}{24}=1\frac{7}{24}$$

2. Determine the LCD and find the equivalent fractions.
3. Add the numerators.
4. Convert to a mixed number.

E.
$$\begin{array}{r} 5\frac{5}{8} = 5 + \frac{5}{8} \\ + 3\frac{1}{8} = 3 + \frac{1}{8} \\ \hline 8 + \frac{6}{8} = 8\frac{3}{4} \end{array}$$

1. Add the whole numbers.
2. Add the fractions.
3. Reduce the fraction.

F.
$$\begin{array}{r} 3\frac{3}{4} \\ + 1\frac{1}{4} \\ \hline 4\frac{4}{4} = 4 + 1 = 5 \end{array}$$

1. Add the whole numbers.
2. Add the fractions.
3. Simplify ($\frac{4}{4} = 1$).

G.
$$\begin{array}{r} 2\frac{5}{9} = 2\frac{5}{9} \\ + 5\frac{2}{3} = 5\frac{6}{9} \\ \hline 7\frac{11}{9} \end{array}$$

1. Determine the LCD and change to equivalent fractions.
2. Add the whole numbers.
3. Add the fractions.

$$7\frac{11}{9} = 7 + 1\frac{2}{9}$$

4. Change the improper fraction to a mixed number.

$$7 + 1\frac{2}{9} = 8\frac{2}{9}$$

5. Add the whole number and the mixed number.

H. In section 2.1, you learned to read a carpenter's rule by counting over the number of 16ths past the whole number and reducing where appropriate. An easier way to do this is by addition, a process which becomes intuitive with a little practice.

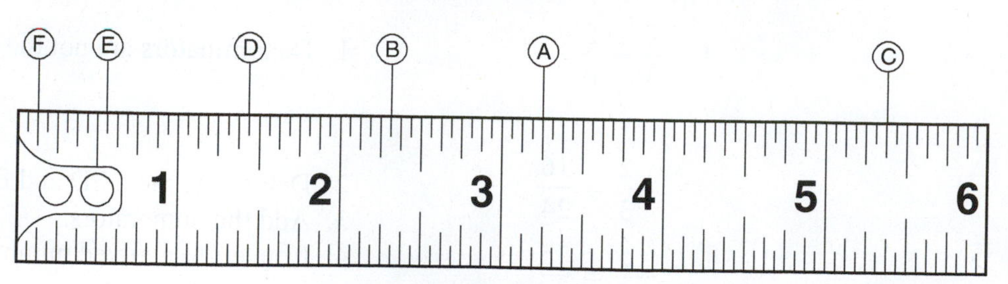

Using the addition method, find measurement B.

$2\frac{1}{4}'' + \frac{1}{16}'' = \ldots$ 1. B is $\frac{1}{16}''$ beyond (greater than) $2\frac{1}{4}''$.

$2\frac{4}{16}'' + \frac{1}{16}'' = \ldots$ 2. Find equivalent fractions.

$2\frac{4}{16}'' + \frac{1}{16}'' = 2\frac{5}{16}''$ 3. Measurement B is $2\frac{5}{16}''$.

I. Using addition, find measurement D.

$1\frac{3}{8}'' + \frac{1}{16}'' = \ldots$ 1. D is $\frac{1}{16}''$ beyond $1\frac{3}{8}''$.

$1\frac{6}{16}'' + \frac{1}{16}'' = 1\frac{7}{16}''$ 2. Find equivalent fractions and add.

J. Here is another way to find measurement D.

$1\frac{1}{4}'' + \frac{1}{8}'' + \frac{1}{16}''$ 1. D is $\frac{1}{8}''$ plus $\frac{1}{16}''$ beyond $1\frac{1}{4}''$.

$1\frac{4}{16}'' + \frac{2}{16}'' + \frac{1}{16}'' = 1\frac{7}{16}''$ 2. Find equivalent fractions and add.

K. Shelving is made from $\frac{3}{4}''$ plywood laminated on both sides with $\frac{1}{16}''$ plastic laminate. What is the total thickness of the shelf?

$\frac{3}{4}'' + \frac{1}{16}'' + \frac{1}{16}''$ 1. Remember to add the $\frac{1}{16}''$ twice.

Or

$\frac{3}{4}'' + 2\left(\frac{1}{16}''\right)$ 2. Or double the laminate first.

$\frac{3}{4}'' + \frac{1}{8}''$ 3. Note that doubling the numerator is equivalent to taking $\frac{1}{2}$ of the denominator.

$\frac{6}{8}'' + \frac{1}{8}'' = \frac{7}{8}''$ 4. Find equivalent fractions and add.

L. A house framed with 2 × 6 studs (which are actually $1\frac{1}{2}'' \times 5\frac{1}{2}''$) has the following layers, starting with the inside layer: $\frac{3}{16}''$ wood paneling, $\frac{3}{8}''$ sheetrock, 1-inch rigid foam panels, 2 × 6 studs ($5\frac{1}{2}''$ deep), $\frac{1}{2}''$ exterior plywood sheathing, and shingles with a thickness of $\frac{9}{16}''$ when overlapped. Find the total thickness of the wall.

$$\frac{3}{16}'',\ \frac{3}{8}'',\ 1 \text{ inch}\ \ 5\frac{1}{2}'',\ \frac{1}{2}'',\ \frac{9}{16}''$$

$$\downarrow\ \ \downarrow\ \ \downarrow\ \ \downarrow\ \ \downarrow\ \ \downarrow$$

$$\frac{3}{16}\ \ \frac{6}{16}\ \ 1 \text{ inch}\ \ 5\frac{8}{16}\ \ \frac{8}{16}\ \ \frac{9}{16}$$

1. Find the LCD (16 in this example).

2. Find the equivalent fractions.

$$\frac{3}{16}'' + \frac{6}{16}'' + 1 \text{ inch} + 5\frac{8}{16}'' + \frac{8}{16}'' + \frac{9}{16}'' = 6\frac{34}{16}''$$

3. Add the fractions and whole numbers.

$$6\frac{34}{16} \rightarrow 6 + 2\frac{1}{8} \rightarrow 8\frac{1}{8}''$$

4. Change the improper fraction to a mixed number and add to the whole number 6.

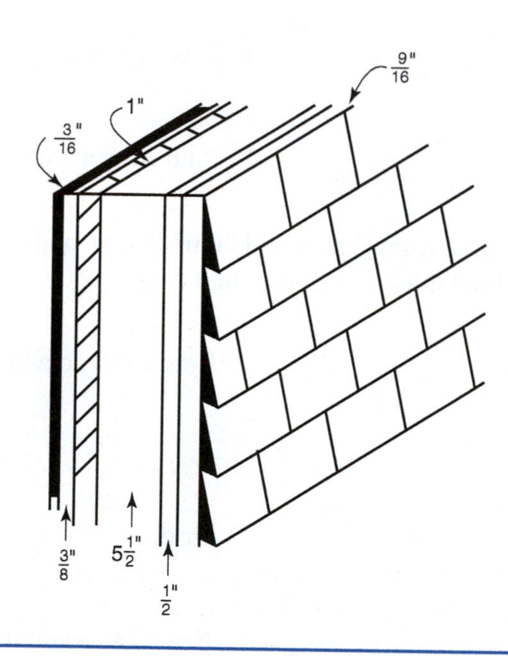

Exercise 2–3

1. $\begin{array}{r} \frac{2}{3} \\ +\frac{1}{4} \\ \hline \end{array}$

2. $\begin{array}{r} \frac{5}{8} \\ +\frac{1}{8} \\ \hline \end{array}$

3. $\frac{9}{10} + \frac{4}{15}$

4. $\frac{3}{4} + \frac{5}{16}$

5. $\begin{array}{r} 2\frac{3}{16} \\ +\ 5\frac{5}{16} \\ \hline \end{array}$

6. $\begin{array}{r} 6\frac{1}{5} \\ +\ 3 \\ \hline \end{array}$

7. $8\frac{3}{4} + 5\frac{24}{32}$

8. $3\frac{3}{4} + 7\frac{3}{16} + 5\frac{3}{8}$

9. $\begin{array}{r} 2\frac{5}{6} \\ +\ 7\frac{1}{9} \\ \hline \end{array}$

10. $\begin{array}{r} 3\frac{7}{8} \\ +\ 4\frac{3}{4} \\ \hline \end{array}$

11. A contractor acquires three adjoining lots with frontages of $131\frac{1}{3}$ ft, $85\frac{3}{4}$ ft, and $165\frac{5}{6}$ ft. What is the total frontage of the three lots?
12. A desk is made of $\frac{3}{4}$-in. particleboard and is covered on both sides with plastic laminate $\frac{1}{32}$ in. thick. What is the thickness of the desktop?
13. A 2 × 4 stud is actually $1\frac{1}{2} \times 3\frac{1}{2}$ in. How thick is a wall constructed of 2 × 4s that is covered on both sides by $\frac{3}{8}$-in. sheetrock? (The thickness is determined by the $3\frac{1}{2}$-in. dimension.)
14. Square ceramic floor tiles measure $3\frac{3}{8}$ in. on a side. If grout is to be $\frac{3}{16}$ in. wide, determine the total length needed to install four tiles as shown.

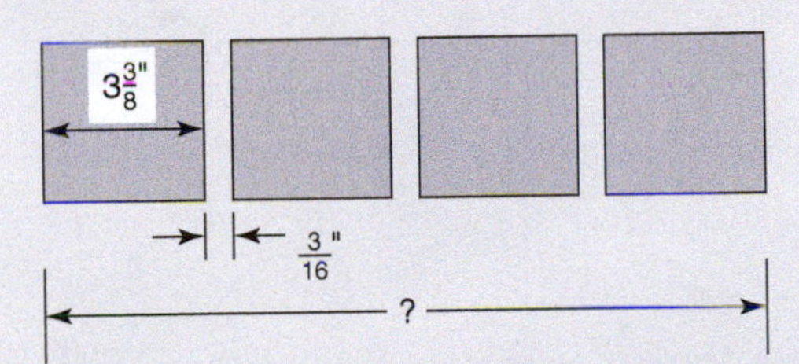

15. An outside wall is made up of the following: $\frac{3}{8}$-in. sheetrock, $\frac{15}{16}$-in. Styrofoam insulation board, $5\frac{1}{2}$-in. studs, and $\frac{3}{4}$-in. exterior sheathing. Determine the total thickness of the wall.
16. The top of a dresser is made of $\frac{3}{4}$-in. boards covered with a $\frac{1}{32}$-in. veneer on one side. What is the thickness of the top?
17. A partition between the living room and dining area is made of 2 × 4s covered with sheetrock on one side and sheetrock and paneling on the other. What is the total thickness of the wall if the 2 × 4s are actually $3\frac{1}{2}$ in., the sheetrock is $\frac{3}{8}$ in. thick, and the paneling is $\frac{1}{4}$ in. thick?
18. Using the addition method, read the following measurements from the ruler shown below:

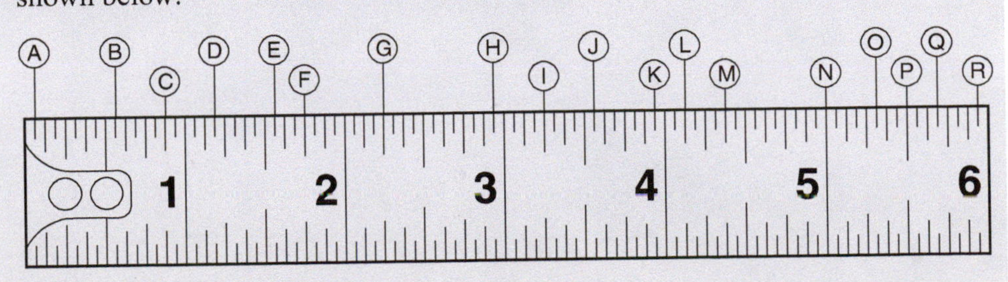

A.

B.

C.

D.

E.

F.

G.

H.

I.

J.

K.

L.

M.

N.

O.

P.

Q.

R.

19. Read the following measurements, reducing where possible.

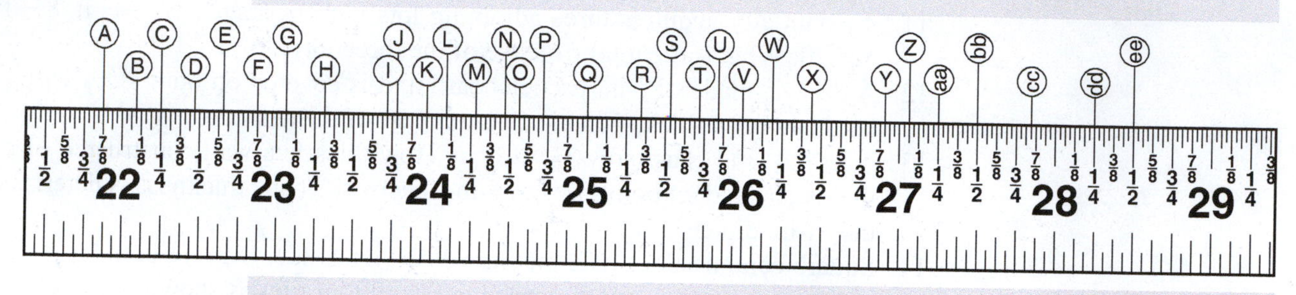

A.

B.

C.

D.

E.

F.

G.

H.

I.

J.

K.

L.

M.

N.

O.

P.

Q.

R.

S.

T.

U.

V.

W.

X.

Y.

Z.

aa.

bb.

cc.

dd.

ee.

2.4 Subtraction of Fractions and Mixed Numbers

Fractions and mixed numbers must have the same denominators in order to be subtracted. The numerators are subtracted, and fractions are simplified if necessary.

Examples

A.
$$\begin{array}{r} \frac{5}{9} \\ -\frac{1}{9} \\ \hline \frac{4}{9} \end{array}$$

Numerators are subtracted and the difference is written over the denominator.

B.
$$\begin{array}{rcr} \frac{3}{4} & = & \frac{6}{8} \\ -\frac{1}{8} & = & \frac{1}{8} \\ \hline & & \frac{5}{8} \end{array}$$

1. Determine the LCD and find the equivalent fractions.
2. Subtract the numerators.

C.
$$\begin{array}{rcr} 3\frac{7}{16} & = & 3\frac{7}{16} \\ -1\frac{1}{8} & = & 1\frac{2}{16} \\ \hline & & 2\frac{5}{16} \end{array}$$

1. Determine the LCD and the equivalent mixed numbers.
2. Subtract the fractions and subtract the whole numbers.

D.
$$\begin{array}{r} 3\frac{1}{4} \\ -1\frac{3}{4} \\ \hline \end{array}$$

In this example the fraction $\frac{3}{4}$ is too large to be subtracted from the fraction $\frac{1}{4}$.

$$\begin{array}{rcl} 3\frac{1}{4} & = & 2 + 1 + \frac{1}{4} = 2\frac{5}{4} \\ -1\frac{3}{4} & = & 1 + \quad \frac{3}{4} = 1\frac{3}{4} \end{array}$$

$$\begin{array}{r} 2\frac{5}{4} \\ -\ 1\frac{3}{4} \\ \hline 1\frac{2}{4} = 1\frac{1}{2} \end{array}$$

1. Rename $3\frac{1}{4}$ as $2 + 1 + \frac{1}{4}$.
2. Rename 1 as $\frac{4}{4}$ (any nonzero number over itself equals 1).
3. $2 + \frac{4}{4} + \frac{1}{4} = 2 + \frac{5}{4} = 2\frac{5}{4}$.
4. Subtract.

E. $$\begin{array}{r} 5\frac{1}{8} \\ -\,3\frac{5}{8} \\ \hline \end{array}$$

In this example the fraction $\frac{5}{8}$ is too large to be subtracted from the fraction $\frac{1}{8}$. Here is a simplified method for performing this subtraction:

$$\begin{array}{rcl} \overset{-1}{\not{5}}\frac{1}{8} & = & 4\frac{9}{8} \quad (1+8) \\ -\,3\frac{5}{8} & = & 3\frac{5}{8} \\ \hline & & 1\frac{4}{8} = 1\frac{1}{2} \end{array}$$

1. Reduce the whole number by 1.
2. *Add* the *denominator* and *numerator* of the fraction $\frac{1}{8}$ to obtain the *new numerator* 9.
3. Subtract and simplify the answer.

F. $$\begin{array}{rcccl} 14\frac{3}{16} & = & \overset{-1}{\not{14}}\frac{3}{16} & = & 13\frac{19}{16} \quad (3+16) \\ -9\frac{5}{8} & & -\,9\frac{10}{16} & = & 9\frac{10}{16} \\ \hline & & & & 4\frac{9}{16} \end{array}$$

1. Find the LCD and determine the equivalent fractions.
2. Subtract 1 from the 14.
3. Add the denominator and numerator to determine the new numerator: $16 + 3 = 19$.
4. Perform the subtraction.

G. $$\begin{array}{rcr} 8 & = & 7\frac{16}{16} \\ -\,3\frac{5}{16} & = & -\,3\frac{5}{16} \\ \hline & & 4\frac{11}{16} \end{array}$$

1. Change 8 into $7\frac{16}{16}$.
2. Subtract.

H. A $1\frac{1}{2}$ in.-thick board is run through a planer, which removes $\frac{1}{16}''$. What is the thickness of the board after being planed?

$$\begin{array}{rcr} 1\frac{1}{2} & \rightarrow & 1\frac{8}{16} \\ -\,\frac{1}{16} & \rightarrow & -\,\frac{1}{16} \\ \hline & & 1\frac{7}{16}'' \end{array}$$

1. Find the LCD.
2. Determine equivalent fractions.
3. Perform the subtraction.

I. A $1\frac{1}{2}''$ rough board has a total of $\frac{9}{16}''$ planed off its surfaces. A $\frac{1}{32}''$-thick laminate is then applied to one surface. What is the final thickness of the board?

$$1\frac{1}{2} = 1\frac{8}{16} = \frac{24}{16}$$
$$-\frac{9}{16} = \frac{9}{16} = \frac{9}{16}$$
$$\frac{15''}{16}$$

1. Find the LCD and determine equivalent fractions.
2. Perform the subtraction.

$$\frac{15''}{16} = \frac{30''}{32}$$
$$+\frac{1''}{32} = \frac{1''}{32}$$
$$\frac{31''}{32}$$

3. Find equivalent fractions and perform the addition.

J. A finish carpenter is building kitchen cabinets with panel doors. To calculate the width of the panel in the doors, the width of the rails must be deducted (subtracted) from the total width of the finished door, and the depth of the routed grooves must be added to the panel width (see illustration). Let's work through this example: A raised panel door is to be $14\frac{1}{4}''$ wide. The door stiles are $3\frac{3}{8}''$ wide, and the routed grooves on the stiles are $\frac{1}{4}''$ deep. Find the width of the raised panel.

$$3\frac{3''}{8}$$
$$+3\frac{3''}{8}$$
$$6\frac{6''}{8} = 6\frac{3''}{4}$$

1. Find the total width of the stiles and reduce to lowest terms.

$$14\frac{1''}{4}$$
$$-6\frac{3''}{4}$$

2. The width of the stiles is to be subtracted from the total width of the door.

$$14\frac{1''}{4} \rightarrow 13\frac{5''}{4}$$
$$-6\frac{3''}{4}$$

3. Change the $14\frac{1}{4}''$ to its equivalent fraction of $13\frac{5}{4}''$.

$$\begin{array}{r} 13\frac{5''}{4} \\ -\ 6\frac{3''}{4} \\ \hline 7\frac{2''}{4} = 7\frac{1''}{2} \end{array}$$

4. Subtract and reduce to lowest terms.

$$\begin{array}{r} \frac{1''}{4} \\ +\ \frac{1''}{4} \\ \hline \frac{1''}{2} \end{array}$$

5. Double the depth of the grooves on the stiles. These must be added back to the width of the raised panel.

$$\begin{array}{r} 7\frac{1''}{2} \\ +\ \frac{1''}{2} \\ \hline 8'' \end{array}$$

6. For this $14\frac{1}{4}''$-wide door, the width of the raised panel must be 8″.

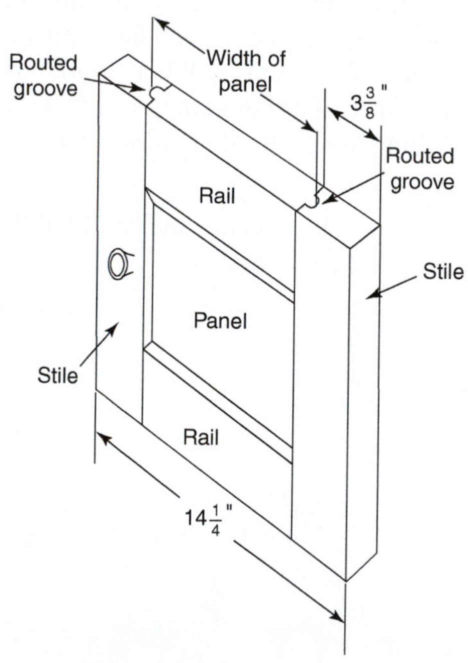

To find the height of the panel, the same method is used, starting with the finished height of the door.

Exercise 2–4

Use the simplified method for subtracting mixed numbers where appropriate.

1. $\begin{array}{r} \frac{3}{4} \\ -\ \frac{5}{8} \\ \hline \end{array}$

2. $\begin{array}{r} \frac{5}{9} \\ -\ \frac{1}{6} \\ \hline \end{array}$

3. $\begin{array}{r} 3\frac{1}{8} \\ -\ 2\frac{15}{16} \\ \hline \end{array}$

4. $\begin{array}{r} 4\frac{1}{2} \\ -\ 2\frac{3}{8} \\ \hline \end{array}$

5. $\begin{array}{r} 6\frac{5}{9} \\ -\ 2\frac{2}{3} \\ \hline \end{array}$

6. $\begin{array}{r} 4\frac{5}{16} \\ -\ 2 \\ \hline \end{array}$

7. $\begin{array}{r} 3\frac{1}{4} \\ -\ \frac{5}{8} \\ \hline \end{array}$

8. $\begin{array}{r} 9\frac{3}{16} \\ -\ 3\frac{1}{8} \\ \hline \end{array}$

9. $\begin{array}{r} 24\frac{5}{8} \\ -\ 23\frac{25}{32} \\ \hline \end{array}$

10. $\begin{array}{r} 14 \\ -\ 5\frac{3}{8} \\ \hline \end{array}$

11. A steel lintel above a fireplace is $46\frac{5}{16}$ in. long. From the diagram shown, determine the width of the fireplace opening.

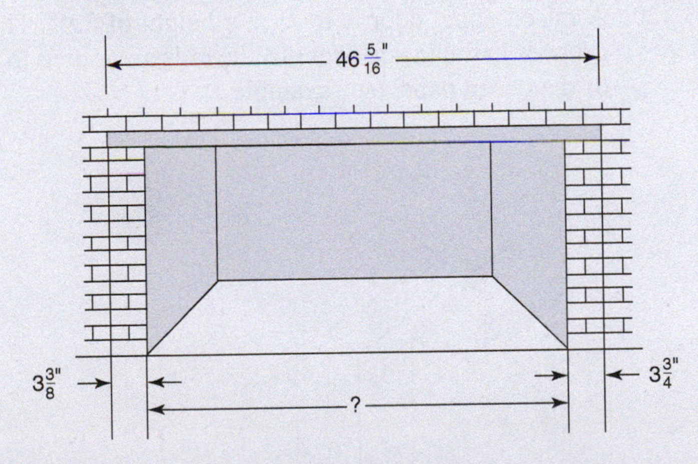

12. A planer takes a $\frac{3}{64}$-in. cut off a board that was $1\frac{7}{8}$ in. thick. What is the final thickness of the board?
13. A board 85 in. long is cut in two lengths. The saw kerf wastes $\frac{3}{16}$ in. If one length is $37\frac{3}{8}$ in. long, what is the other length?

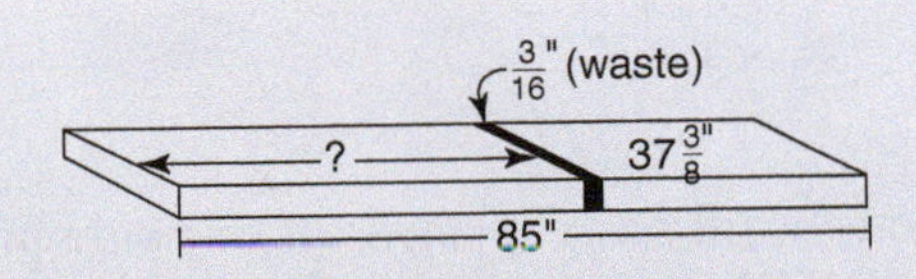

14. A jointer is set to remove $\frac{5}{64}$ in. from the width of an oak board. If the board was $3\frac{7}{8}$ in. wide, what is its width after jointing once?

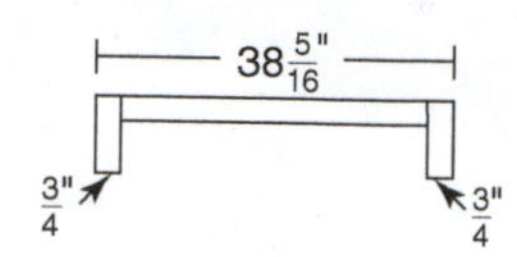

15. The outside width of a bookcase is $38\frac{5}{16}$ in. If the sides are made of $\frac{3}{4}$-in. stock, what is the inside width?
16. What is the final thickness of a $3\frac{1}{2}$-in. piece of stock if $\frac{3}{8}$ in. is planed off one side and $\frac{5}{16}$ in. is planed off the opposite side?
17. A board must be $46\frac{5}{16}$ in. long. How much should be cut from a 4-ft board?
18. The outside diameter (O.D.) of a pipe is $3\frac{1}{8}$ in. What is the inside diameter (I.D.) of the pipe if its walls are $\frac{3}{16}$ in. thick. (*Hint:* The wall thickness must be considered on both sides of the pipe.)

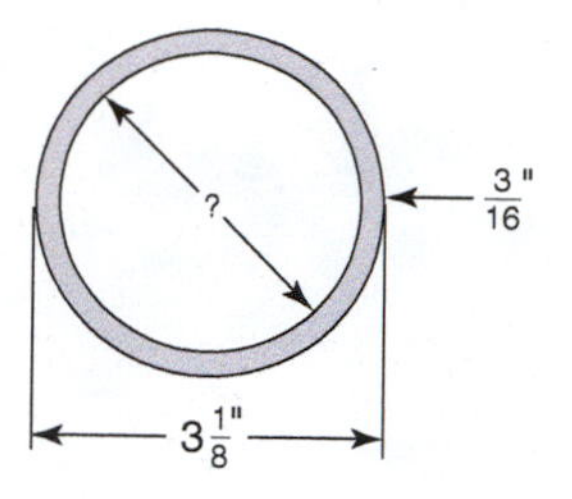

19. A panel door on a cabinet is to have a finished width of $22\frac{1}{2}''$. The stiles on the door are $3\frac{3}{8}''$ wide, and the grooves in the inside edges of the stiles are routed $\frac{5}{16}''$ deep. Find the width of the panel (see example J).

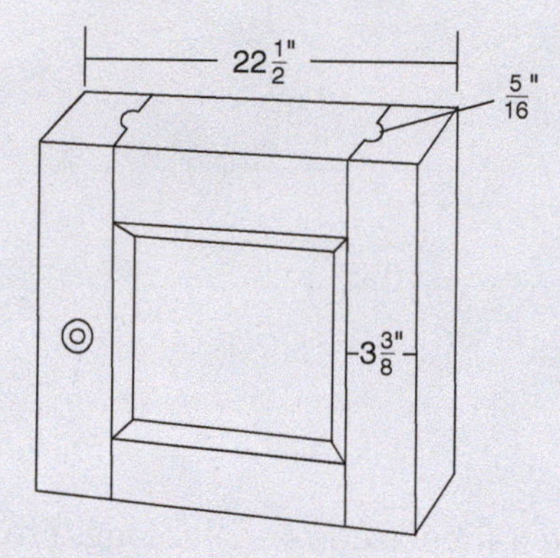

20. A raised panel door is to have a height of $39\frac{1}{4}''$. The rails are $3\frac{7}{8}''$ wide, and the grooves in the inside edges of the rails are routed to a depth of $\frac{1}{4}''$. Find the height of the raised panel (see example J).

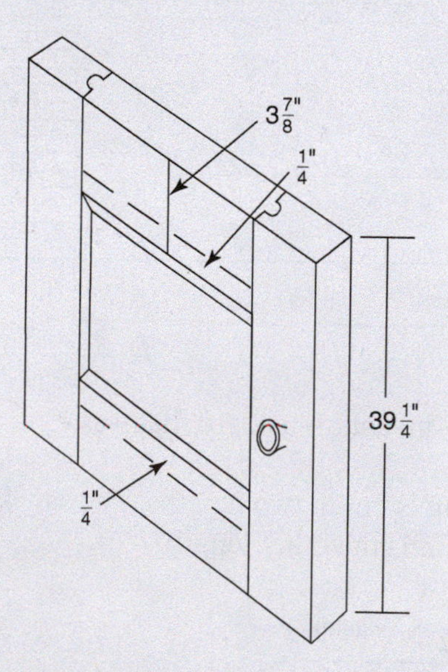

21. A $1\frac{1}{4}''$ rough-cut board has $\frac{3}{8}''$ planed off one side and $\frac{5}{16}''$ from the other side. A high-quality $\frac{3}{64}''$-thick veneer is then applied to one side. What is the final thickness of the board?

2.5 Multiplication of Fractions and Mixed Numbers

Multiplication of fractions is generally simpler than addition and subtraction of fractions. There is no need to find the LCD when multiplying. In multiplication, the numerators are multiplied together, and the denominators are multiplied together.

Examples

A. $\frac{3}{4} \times \frac{5}{8} = \frac{15}{32}$

1. Multiply the numerators together.
2. Multiply the denominators together.

B. $\frac{3}{8} \times \frac{16}{21} = \frac{3 \cdot 16}{8 \cdot 21} = \frac{48}{168}$

$\frac{48}{168} = \frac{2}{7}$

1. Multiply the numerators.
2. Multiply the denominators.
3. Reduce to lowest terms by dividing numerator and denominator by 24.

In example B it would have been easier to reduce the fractions before multiplying the numerators and denominators.

$\frac{\cancel{3}^{1}}{8} \times \frac{16}{\cancel{21}_{7}}$

1. 3 divides into both the 3 in the numerator and the 21 in the denominator.

$\frac{1}{\cancel{8}_{1}} \times \frac{\cancel{16}^{2}}{7}$

2. 8 divides into both the 16 in the numerator and the 8 in the denominator.

$\frac{1}{1} \times \frac{2}{7} = \frac{2}{7}$

3. If all reducing is done before the multiplication, the answer will not need to be reduced further.

Examples

C. $5\frac{1}{4} \times \frac{3}{7}$

1. Mixed numbers cannot be multiplied directly.

$\frac{21}{4} \times \frac{3}{7}$

2. Change the mixed number to an improper fraction.

$\frac{\cancel{21}^{3}}{4} \times \frac{3}{\cancel{7}_{1}}$

3. Reduce by dividing the 21 in the numerator and the 7 in the denominator by 7.

$\frac{3}{4} \times \frac{3}{1} = \frac{9}{4}$

4. If no units of measurement are involved, it is best to leave the answer as an improper fraction. When units are involved (inches, feet, pounds, etc.), the answer should be given as a mixed number.

$\frac{9}{4} = 2\frac{1}{4}$

5. Because most of the problems occurring in building construction involve units, in this book the answers will be converted to a mixed number.

D. $5\frac{3}{16} \times 4$

$\frac{83}{16} \times \frac{4}{1}$

1. Convert to an improper fraction.
2. A whole number should be put over a denominator of 1 when being multiplied by a fraction.

$\frac{83}{\not{16}_4} \times \frac{\not{4}^1}{1}$

3. Reduce.

$\frac{83}{4} \times \frac{1}{1} = \frac{83}{4} = 20\frac{3}{4}$

4. Multiply and convert to a mixed number.

E. A set of stairs has 13 risers each with a rise of $8\frac{7}{16}''$. Find the total rise (the distance from the first floor to the second). (Stairs will be discussed in detail in Chapter 23.)

$8\frac{7}{16}'' \times 13 = \ldots$

1. Multiply the rise (height) of each step by the number of steps.

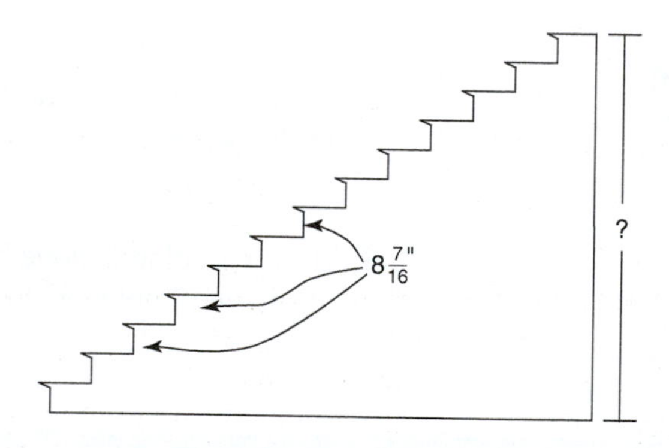

$\frac{135}{16} \times 13 = \frac{1755}{16}$

2. Change the $8\frac{7}{16}''$ to an improper fraction and perform the multiplication.

$\frac{1755}{16} = 109\frac{11}{16}''$

3. Change to a mixed number. The total rise of the stairs is $109\frac{11}{16}''$ (9 ft $1\frac{11}{16}''$).

F. One-quarter ($\frac{1}{4}$) of an order of lumber is used to build steps. Of the remaining $\frac{3}{4}$ of the order, $\frac{2}{3}$ is used to build a small deck. What fraction of the original order is still left?

$\frac{3}{4}$ left

1. $\frac{3}{4}$ of original stack of boards are left after building steps.

$\frac{1}{4}$ of lumber used for steps

$\frac{3}{4}$ of lumber left after building steps

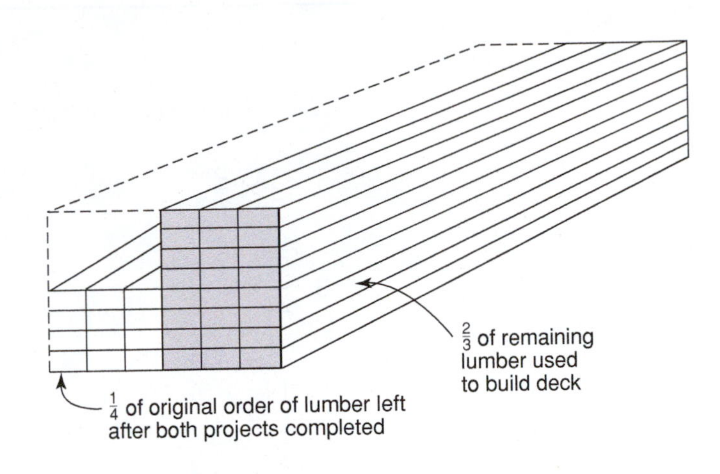

$\frac{1}{3}$ left	2. $\frac{1}{3}$ of the *remaining* boards are left after building the deck.
$\frac{1}{3} \times \frac{3}{4} = \ldots.$	3. $\frac{1}{3}$ of the remaining $\frac{3}{4}$ is left. *Of* means to multiply.
$\frac{1}{3} \times \frac{3}{4} = \frac{1}{4}$	4. $\frac{1}{4}$ of the *original* order of lumber is left.

G. In the preceding example, if there were 480 8-ft 2 × 4s in the original order, how many are left?

$\frac{1}{4} \times 480 = \ldots$	1. $\frac{1}{4}$ *of* the 480 means multiply.
$\frac{1}{4} \times 480 = 120$	2. 120 8-ft 2 × 4s are left.

H. The lumberyard sells 6″-wide finish oak boards in the following lengths: 4-ft (48″), 6-ft (72″), 8-ft (96″), 10-ft (120″), 12-ft (144″), 14-ft (168″), 16-ft (192″). Find the shortest length of stock that will yield 9 sections of board each $17\frac{1}{2}$″ long if the saw blade has $\frac{1}{4}$″ of kerf. (The kerf is the thickness of the cut, and this amount of wood is wasted with each cut.)

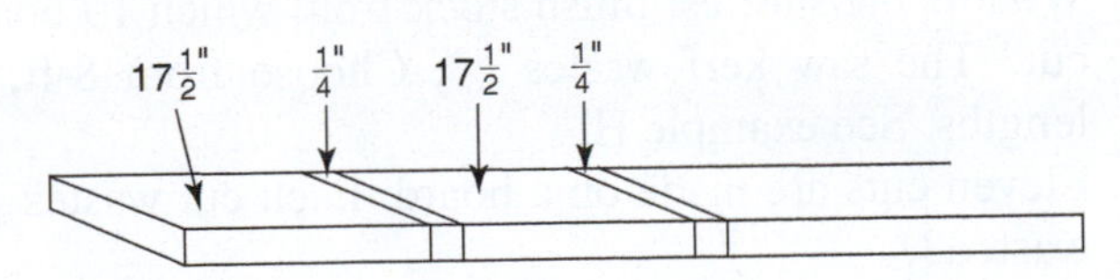

$17\frac{1}{2}'' + \frac{1}{4}'' = 17\frac{3}{4}''$	1. Find the length removed by each cut.
$17\frac{3}{4}'' \times 9 = 159\frac{3}{4}''$	2. Change to an improper fraction and multiply.
$159\frac{3}{4}'' \rightarrow 168''$	3. Round up to the next length. The 14-ft board is needed.

Exercise 2–5

Give all answers in reduced form. Change improper fractions to mixed numbers.

1. $\frac{4}{5} \times \frac{5}{6}$
2. $\frac{17}{18} \times \frac{12}{17}$
3. $\frac{3}{4} \times \frac{5}{7}$
4. $5\frac{1}{2} \times 3\frac{1}{4}$
5. $14\frac{7}{8} \times 4$
6. $3\frac{2}{3} \times 4\frac{1}{2}$
7. $2\frac{3}{16} \times \frac{4}{5}$
8. $\frac{7}{8} \times 5\frac{5}{7}$
9. $8 \times \frac{3}{5}$
10. $7\frac{2}{9} \times \frac{3}{10}$

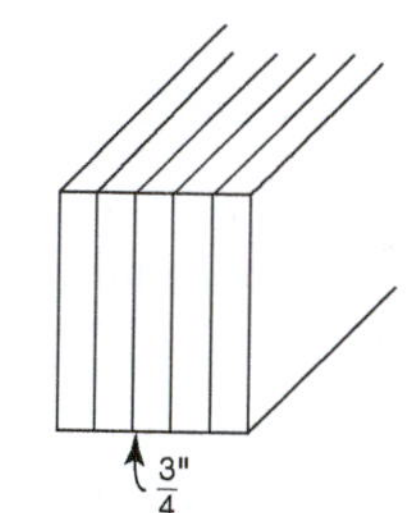

11. A beam is made of built-up 1 in. × 8 in. boards. If 5 boards are laminated together, what is the thickness of the beam? (1 × 8s are actually $\frac{3}{4}$ in. thick; see diagram).
12. Five boards each $7\frac{5}{8}$ in. long are needed. If they are to be cut from one board, how long must the board be? Allow 1 in. for waste.
13. A flight of stairs has 13 risers. If each riser is $8\frac{3}{16}$ in. high, what is the total height of the stairs? Give the answer in inches.
14. A stack of 4 ft × 8 ft sheets of $\frac{3}{8}$-in.-thick wallboard contains 55 sheets. What is the height of the stack?
15. A board is $86\frac{1}{4}$ in. long. What is $\frac{3}{4}$ of its length? (See diagram.)
16. It takes a quarter of an hour ($\frac{1}{4}$ hr) to replace 1 ft of a special ceiling trim in a remodeling job. How many hours will it take to replace $37\frac{1}{2}$ ft of trim? Give the answer as a mixed number.
17. Cedar shingles are laid to expose $\frac{5}{12}$ ft per course. How high is a wall that requires 22 courses of shingles? Give the answer as a mixed number.
18. Seven lengths of board are to be removed from an 8-ft 2 × 4. Each length is to be $5\frac{3}{8}$ in., and each cut wastes $\frac{1}{16}$ in. How many inches are removed from the board? (*Hint:* Because the entire board is not used, seven lengths will require seven cuts.)
19. What is the shortest finish stock from which 15 board sections each $11\frac{3}{4}''$ can be cut? The saw kerf wastes $\frac{3}{16}''$. Choose from 8-ft, 10-ft, 12-ft, 14-ft, or 16-ft lengths. See example H.
20. Eleven cuts are made on a board. Each cut wastes $\frac{3}{16}''$. What is the total amount wasted?
21. A building's staircase has 22 risers, with a unit rise of $7\frac{3}{8}''$ per riser. Find the total rise (in inches) for the staircase. (Stairs will be covered in detail in Chapter 23.)

2.6 Division of Fractions and Mixed Numbers

Division problems involving fractions are converted to multiplication problems. To convert the division problem to a multiplication problem, invert the divisor and change the division sign to a multiplication sign.

Examples

A. $\frac{3}{4} \div \frac{5}{8}$

1. The divisor is always the number after the division sign. In this example the divisor is $\frac{5}{8}$.

$\frac{3}{4} \times \frac{8}{5}$

2. Invert (turn upside down) the divisor and change the division sign to a multiplication sign.

$\frac{3}{\cancel{4}_1} \times \frac{\cancel{8}^2}{5}$

3. Reduce the numerator and denominator. *Caution!* Do *not* attempt to reduce until the divisor has been inverted.

$\frac{3}{1} \times \frac{2}{5} = \frac{6}{5} = 1\frac{1}{5}$

4. Multiply and change to a mixed number.

B. Dividing a fraction by a whole number:

$\frac{3}{8} \div 5$

$\frac{3}{8} \div \frac{5}{1}$

1. Write the whole number as a fraction over 1.

$\frac{3}{8} \times \frac{1}{5}$

2. Invert the divisor.

$\frac{3}{8} \times \frac{1}{5} = \frac{3}{40}$

3. Multiply.

C. Dividing mixed numbers:

$4\frac{1}{8} \div 1\frac{5}{16}$

$\frac{33}{8} \div \frac{21}{16}$

1. Change the mixed numbers to improper fractions.

$\frac{33}{8} \times \frac{16}{21}$

2. Invert the divisor.

$\frac{{}^{11}\cancel{33}}{\cancel{8}_1} \times \frac{\cancel{16}^2}{\cancel{21}_7}$

3. Reduce the numerators and denominators.

$\frac{11}{1} \times \frac{2}{7} = \frac{22}{7} = 3\frac{1}{7}$

4. Multiply and convert to a mixed number.

D. $\frac{5}{9} \div \frac{3}{5}$

1. *Caution!* The numerators and denominators cannot be reduced in a division problem.

$\frac{5}{9} \times \frac{5}{3} = \frac{25}{27}$ 2. Now no reducing can occur.

E. You are using a saw blade with a $\frac{7}{64}''$ kerf (cutting waste). How many $9\frac{3}{4}''$ board sections can be cut from a board that is 120″ long?

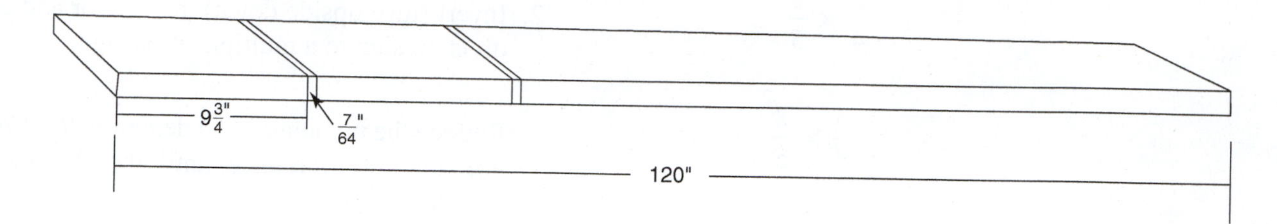

$9\frac{3}{4}'' + \frac{7}{64}'' = 9\frac{55}{64}''$ 1. Add the saw kerf to the length of each section.

$120'' \div 9\frac{55}{64}'' =$ 2. Divide the board length by the length of each section and kerf.

$12\frac{108}{631}$ 3. This is the number of board sections that can be cut.

12 sections 4. Only whole sections are usable, so round down.

F. A carpenter has completed $\frac{3}{4}$ of a job and has used 240 bf (board feet) of lumber. How much lumber is needed for the entire job?

$240 \text{ bf} \div \frac{3}{4} =$ 1. Divide the number of board feet used by the fraction of the job completed.

$240 \div \frac{3}{4} = 320 \text{ bf}$ 2. When dividing by a fraction less than 1, the result will be greater than the 240 bf already used.

G. A set of porch stairs has a total rise of $34\frac{1}{4}''$. Each step has a unit rise of $8\frac{9}{16}''$. How many steps are in the set?

$34\frac{1}{4}'' \div 8\frac{9}{16}'' =$ 1. Divide the total rise by the unit rise to find the number of steps.

$\frac{137}{4} \div \frac{137}{16} =$ 2. Convert to improper fractions.

$\frac{\overset{1}{\cancel{137}}}{\underset{1}{\cancel{4}}} \times \frac{\overset{4}{\cancel{16}}}{\underset{1}{\cancel{137}}} = 4$ 3. Invert and multiply.

4 = 4 steps 4. The set of porch stairs has 4 steps.

Exercise 2–6

1. $\frac{5}{8} \div 2\frac{3}{4}$
2. $\frac{4}{9} \div \frac{2}{3}$
3. $\frac{3}{5} \div \frac{5}{6}$
4. $1\frac{3}{4} \div 2\frac{5}{8}$
5. $6\frac{3}{4} \div 2\frac{5}{8}$
6. $9 \div \frac{3}{4}$
7. $\frac{5}{16} \div 15$
8. $11\frac{5}{16} \div 11\frac{5}{16}$
9. $1\frac{1}{2} \div 3\frac{3}{8}$
10. $4\frac{1}{16} \div 6\frac{1}{4}$
11. A wall is 8 ft high and $142\frac{3}{8}$ in. wide. It is to be covered by vertical boards 8 ft long and $3\frac{3}{4}$ in. wide. How many 8-ft boards are needed to cover the wall? (*Hint:* When determining the amount of material needed, always round *up*.)
12. How many sheets of $\frac{1}{32}$-in.-thick plastic laminate are in a pile $6\frac{3}{8}$ in. high?
13. Banister posts are cut from an 8-ft (96-in.) piece of cylindrical stock. How many posts $25\frac{3}{8}$ in. long can be cut from the stock? (*Hint:* When calculating the number of pieces that can be cut from a given length, always round *down.*)
14. The living room ceiling in a new house is to be taped and mudded. A $94\frac{1}{2}$-ft^2 area has already been done, and this represents $\frac{3}{8}$ of the job. How large is the ceiling?
15. An 8-ft board is to be sanded. If $2\frac{1}{2}$ ft has been done, what fraction of the board must still be sanded?
16. How many sheets of $\frac{3}{8}$-in. paneling are in a stack of sheets 3 ft high?
17. How many shelves for a bookcase can be cut from a board $15\frac{3}{4}$ ft long if each shelf is to be $3\frac{1}{4}$ ft?
18. How long does it take to cut through metal stock $4\frac{1}{2}$ in. thick if the cutting tool cuts at the rate of $\frac{3}{16}$ in./min?
19. A cabinetmaker has completed $\frac{2}{3}$ of a job. He has used 250 board feet in the project. How many board feet are required for the entire job?
20. A staircase has a total rise of $95\frac{1}{16}''$. Each step has a unit rise of $7\frac{5}{16}''$. Find the number of treads in the staircase. (*Hint:* The number of treads is always one less than the number of risers; see the diagram at the top of page 48.)
21. A finish carpenter is fitting the woodwork around a number of identical windows in an office building. It is estimated that $\frac{3}{8}$ of the job is completed, and 480 lineal feet of trim boards has been used. How many lineal feet of trim boards are required for the entire job?

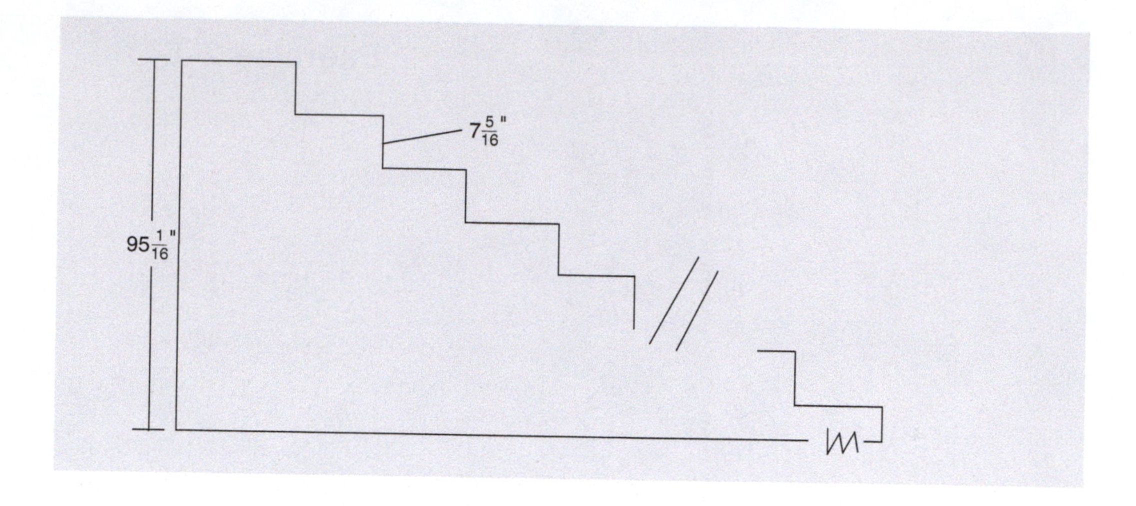

REVIEW EXERCISES

1. A bookcase $56\frac{3}{8}$ in. high is made of $\frac{3}{4}$-in. stock with 5 equally spaced shelves. How much space is between each shelf?

2. A $2\frac{1}{4}$-in.-diameter hole in a 2 in. × 6 in. stud must be enlarged to allow a pipe to fit through it. If the pipe has a $3\frac{3}{16}$-in. (O.D.) diameter, by how much must the hole be enlarged? Make the hole $\frac{1}{4}$" larger than the pipe.

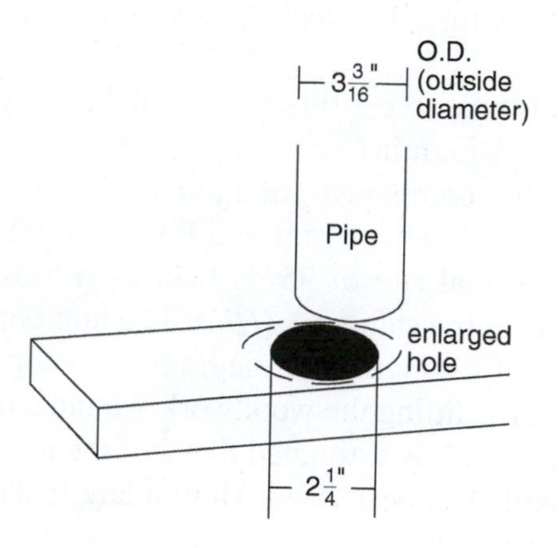

3. A roof is covered with shingles weighing 255 lb per square. The nails used per square weigh $4\frac{1}{4}$ lb. What will be the weight of a roof requiring $3\frac{3}{4}$ squares?

4. The following colors are being mixed to produce a certain shade of paint: $4\frac{1}{4}$ gal of white, $1\frac{1}{8}$ gal of gray, $3\frac{5}{8}$ gal of blue, and $\frac{1}{4}$ gal of green. If each gallon of paint covers 320 ft^2, how many square feet will the mixture cover?
5. A cabinetmaker is cutting a large number of shelves $17\frac{1}{2}$ in. long. He has stock available in 2-, 4-, and 6-ft lengths. Which size should he choose in order to minimize waste? (Ignore saw kerf.)

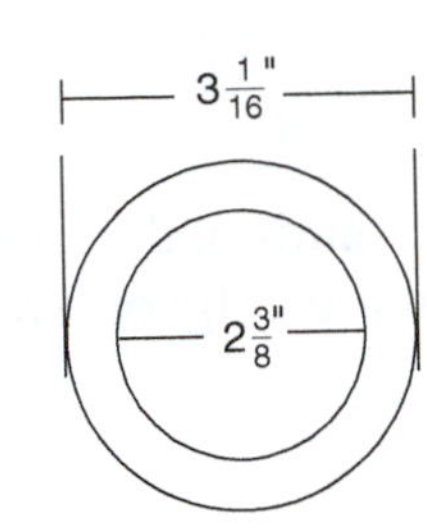

6. What is the thickness of a pipe that is $3\frac{1}{16}$ in. O.D. and $2\frac{3}{8}$ in. I.D.? (See diagram.)
7. A dimension stick is $21\frac{1}{4}$ in. long. If it is to be divided into five equal spaces, what will be the length of each space? (See diagram.)
8. How many additional nails will be needed if $33\frac{1}{2}$ lb is sufficient to complete $\frac{3}{4}$ of a job?
9. An exterior wall has the following layers: $\frac{1}{4}$-in. paneling, $\frac{5}{8}$-in. firecode sheetrock, $5\frac{1}{2}$-in. studs, $\frac{1}{2}$-in. CDX plywood sheathing, $1\frac{1}{4}$-in. rigid insulation, and $\frac{5}{8}$-in. cedar shingles. What is the total thickness of the wall?

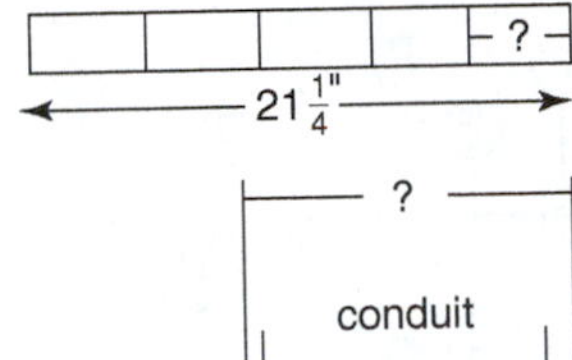

10. A construction corporation sells shares of stock for $12\frac{3}{8}$ dollars per share. If 11 investors jointly invest in 15 shares, what is the value of each investor's stock?

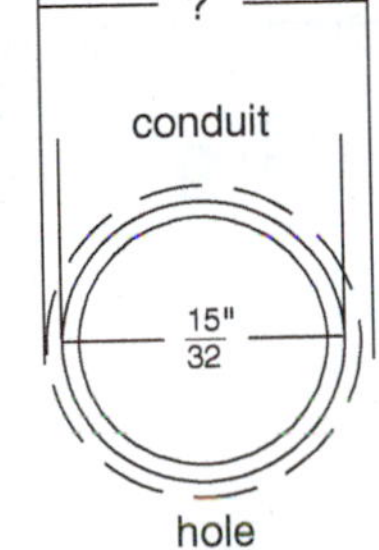

11. Order the following drill bits in size from smallest to largest: $\frac{7}{8}$, $\frac{11}{16}$, $\frac{43}{64}$, and $\frac{25}{32}$ in.
12. A hole is to be drilled $\frac{5}{64}$ in. oversized to accommodate a conduit with a $\frac{15}{32}$-in. O.D. What size drill bit should be used? (See diagram.)
13. A small pickup truck has $\frac{3}{8}$ of a tank of gas. If there are $4\frac{11}{16}$ gallons of gas left, how many gallons will the tank hold?
14. A 48-in. board is to be divided into four equal sections. If each cut wastes $\frac{1}{8}$ in., determine the length of each section. (*Hint:* Subtract the waste from the total length first. Remember, four sections require three cuts.) Give the answer to the nearest $\frac{1}{32}$ of an inch.

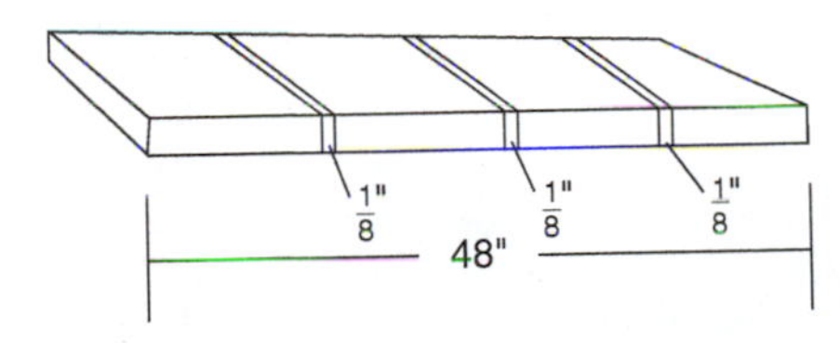

15. Four shims are used to raise a floor $1\frac{29}{32}$ in. in one corner. If two shims are each $\frac{13}{16}$ in. thick and one shim is $\frac{5}{32}$ in., how thick must the fourth shim be?
16. A pipe has a wall thickness of $\frac{3}{16}''$. The O.D. of the pipe is $1\frac{1}{2}''$. Find the I.D. of the pipe.

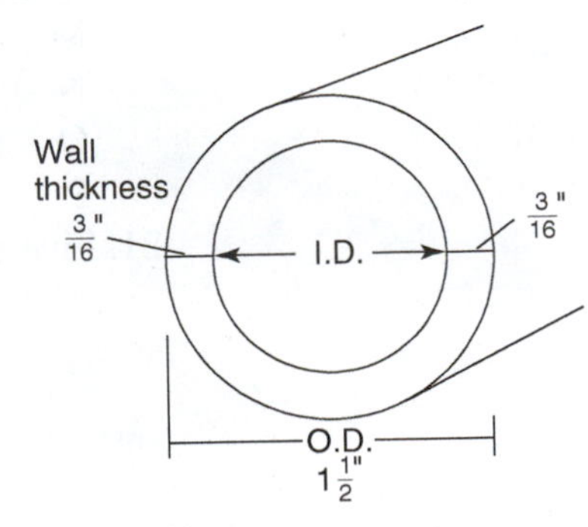

17. (Challenge problem) A hole with an $18\frac{3}{16}''$ diameter must be enlarged to allow a pipe to fit through it. The hole should be $\frac{5}{8}''$ larger than the diameter of the pipe. The pipe has wall thickness of $\frac{5}{16}''$ and an I.D. of $21\frac{1}{2}''$. By how much must the hole be enlarged?

18. A 4-foot (48″) board has $\frac{3}{4}$″ trimmed from each end. The board is then carefully sawn into 3 ***equal*** sections. Each saw cut wastes $\frac{3}{16}$″. How long is each board section?

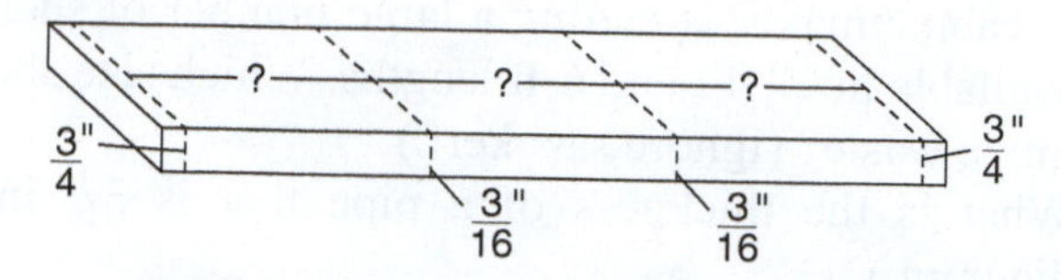

19. Baseboards are being installed in 24 identical cubicles. Two-thirds of the cubicles have been completed, and 764 feet of trim boards have been used. (Ignore waste.)
    a. How many feet are needed for the entire job?
    b. How many feet are needed to complete the job?
    c. How many feet are needed for each cubicle?
20. Read the following measurements, reducing where possible:

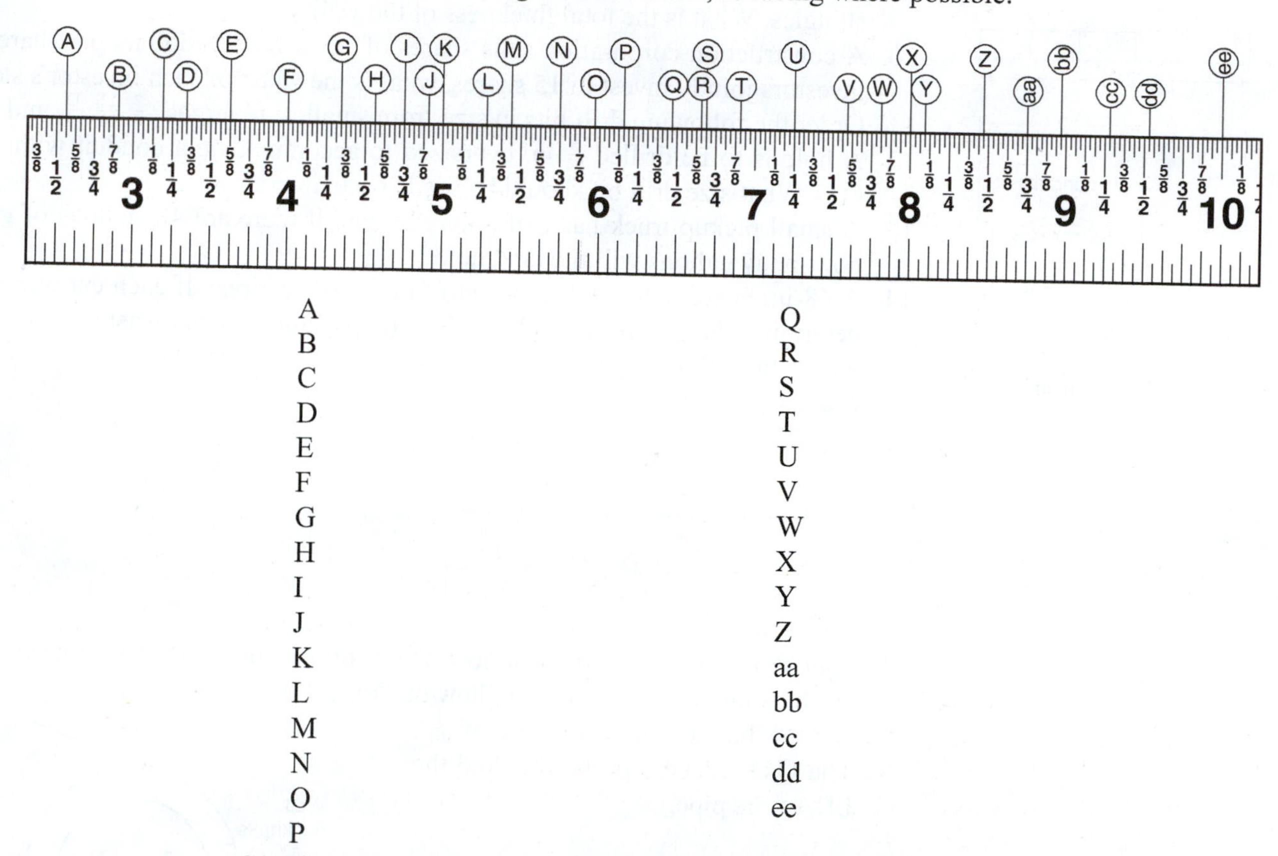

A
B
C
D
E
F
G
H
I
J
K
L
M
N
O
P

Q
R
S
T
U
V
W
X
Y
Z
aa
bb
cc
dd
ee

21. Five solar panels are equally spaced along a roof area 141 $\frac{1}{2}$″ long. There is a gap of 5 $\frac{3}{8}$″ between each panel. Determine the width of each panel.

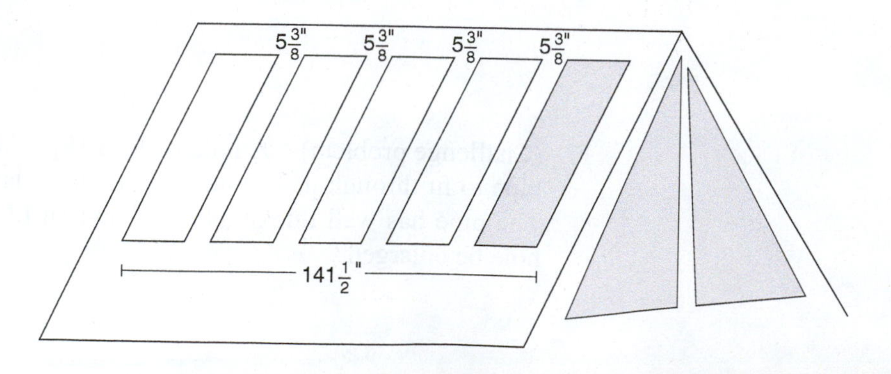

CHAPTER 3

Decimal Fractions

Did You Know

A front-loading clothes washer uses less electricity, less water, and less soap than a top-loading washer.

OBJECTIVES

Upon completing this chapter, the student will be able to:

1. Identify decimal fractions and their fractional equivalents.
2. Identify the place value of a decimal fraction.
3. Round decimal fractions to a specified place.
4. Determine precision of a decimal value.
5. Add decimal fractions.
6. Solve problems involving the addition of decimals.
7. Subtract decimal fractions.
8. Solve problems involving subtraction of decimal fractions.
9. Multiply decimal fractions.
10. Solve problems involving multiplication of decimals.
11. Divide decimal fractions.
12. Solve problems involving division of decimal fractions.
13. Mentally multiply numbers by powers of 10.
14. Mentally divide numbers by powers of 10.
15. Convert common fractions to decimal fractions.
16. Convert mixed number fractions to mixed decimals.
17. Convert decimal fractions to common fractions.
18. Convert mixed decimals to mixed fractions.
19. Convert repeating decimals to fractions.
20. Convert decimal inches to the nearest $\frac{1}{2}$, $\frac{1}{4}$, $\frac{1}{8}$, $\frac{1}{16}$, $\frac{1}{32}$, and $\frac{1}{64}$ in.
21. Convert fractions to decimals and vice versa using a table.

3.1 Place Value and Rounding

Any common fraction with a denominator of 10, 100, 1000, and so on (known as powers of 10) can be written as a decimal fraction. Usually, a *decimal fraction* is called simply a *decimal.* Below are some common fractions and their decimal equivalents:

Common Fraction	Decimal Equivalent	Read as:
$\frac{9}{10}$	0.9	Nine-tenths
$\frac{83}{100}$	0.83	Eighty-three hundredths
$\frac{521}{1000}$	0.521	Five hundred twenty-one thousandths

Any common fraction with a denominator of 10 (or a power of 10) can be converted to a decimal by moving the decimal point in the numerator and denominator an equal number of spaces in the same direction until the denominator = 1. Moving the decimal point is equivalent to dividing or multiplying by 10 or a power of 10. Examples A and B illustrate this.

Examples

A. $\frac{9}{10} = \frac{9.}{10.} = \frac{.9.}{1.0.}$ 1. Move the decimal point in the denominator to the left until the denominator equals 1.

$\frac{.9}{1.0} = \frac{.9}{1}$ 2. Move the decimal point in the numerator to the left the same number of places. (When not shown, the decimal point is always understood to be to the right of the rightmost digit.)

$\frac{.9}{1} = .9$ or 0.9 3. Drop the denominator. (A denominator that equals 1 can always be omitted.) A leading zero is placed to the left of the decimal point if there is no whole number.

B. $\frac{521}{1000} = \frac{.521.}{1.000.}$ 1. Move the decimal point in the denominator and numerator three places to the left.

$\frac{.521}{1} = 0.521$ 2. Since the denominator is now equal to 1, it is dropped. *Note:* For each place the decimal point is moved to the left, the number is actually being divided by 10.

A terminating decimal can be written as a common fraction by reversing the process.

Examples

C. Convert 0.7 to a common fraction.

$0.7 = \frac{.7}{1}$ 1. Any number can be written as a fraction with denominator of 1.

$\frac{.7}{1.0} = \frac{7.}{1.0.}$ 2. Move the decimal point in the numerator to the right as many places as necessary to make the numerator a whole number (one place to the right in this example).

$\frac{7.}{10.} = \frac{7}{10}$ 3. Move the decimal point in the denominator the same number of places to the right, placing a zero for each move to the right.

D. Convert 0.41 to a common fraction.

$0.41 = \frac{.41.}{1.00.} = \frac{41}{100}$ Move to the decimal point two places to the right in the numerator and the denominator. Place zeros in the denominator for each move to the right. (For each place the decimal point is moved to the right, the number is actually being multiplied by 10.)

Figure 3–1 indicates the place value for decimal fractions. A number such as 0.531 means 0.5 ($\frac{5}{10}$ or five-tenths) + 0.03 ($\frac{3}{100}$ or three hundredths) + 0.001 ($\frac{1}{1000}$ or one thousandth). Shown in the figure are the numbers fifty-three thousandths (0.053 or $\frac{53}{1000}$), four hundred eighty-one thousandths (0.481 or $\frac{481}{1000}$), and two hundred fifteen and thirty-five hundredths (215.35 or $215\frac{35}{100}$). When reading a decimal number, the decimal point is always read as "and." The following examples will illustrate the method for rounding decimal fractions.

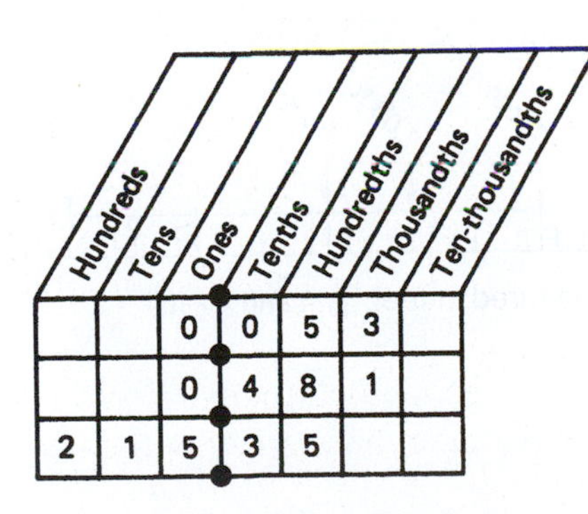

Figure 3–1

Examples

E. Round 0.2839 to two decimal places (to the nearest hundredths place).

0.2839 1. Locate the digit to be rounded, 8 in this example.

0.2839 2. Examine the digit just to the right of the digit to be rounded, 3 in this example.

0.28~~39~~ 3. If the digit to the right is less than 5, drop that digit and all other digits to the right of it.

0.28 4. Since the answer is rounded to two decimal places, only two digits are shown to the right of the decimal point.

F. Round 2.57381 to three decimal places (to the nearest thousandth).

2.57381	1. Locate the digit in the thousandths place, in this example the 3.
↓ 2.57381	2. Examine the digit just to the right of the digit to be rounded, in this case the 8.
+1 2.57381	3. If the digit to the right is 5 or greater, increase the digit to be rounded by 1, and drop all digits to its right.
2.574	4. *Note:* Do *not* replace the dropped digits with zeros. Replacement applies only to whole-number rounding.

G. There are three common types of surveyor's tapes: (a) metric, (b) feet, inches, and fractional inches, and (c) feet and decimal feet (tenths and hundredths of a foot). Carpenter's tapes are measured in feet, inches, and fractional inches. In the illustration, note the differences between the surveyor's tape in decimal feet and the carpenter's tape in feet, inches, and fractional inches.

Carpenter's rule or tape

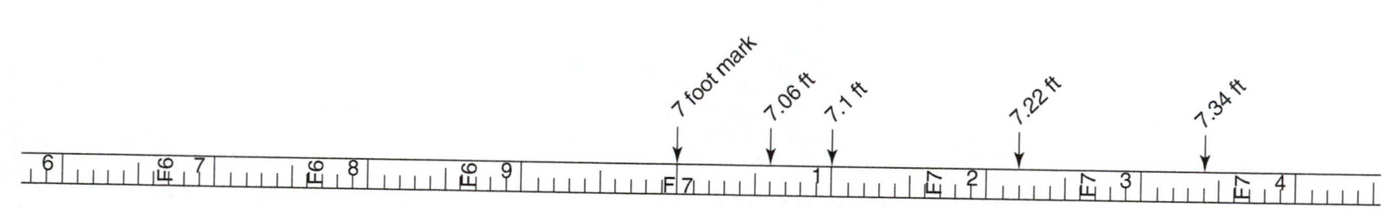

Surveyor's tape measured in feet, $\frac{1}{10}$ ft and $\frac{1}{100}$ ft.

A lot measures 523.76 feet deep. Round this dimension to the nearest tenth of a foot.

523.76 feet	1. Locate the digit to be rounded, in this example, the 7.
↓ 523.76 feet	2. Examine the digit just to the right of the digit to be rounded, 6 in this example.
+1 523.76 feet	3. If the digit to the right is 5 or greater, increase the digit to be rounded by 1 and drop all digits to its right.
523.8 feet	4. Do *not* replace the dropped digits with zeros. Replacement applies only to whole-number rounding.

Exercise 3–1

Round to the place value indicated.

1. 83.4067 to the nearest hundredths ________, thousandths ________
2. 0.0053 to the nearest hundredths ________, thousandths ________
3. 42.613 to the nearest ones ________, tenths ________
4. 24,992 to the nearest hundredths ________, ones ________
5. 0.5038 to the nearest tenths ________, hundredths ________
6. 0.598 to the nearest ones ________, hundredths ________
7. 4.600 to the nearest tenths ________, hundredths ________
8. 0.8066 to the nearest ones ________, hundredths ________
9. 0.4251 to the nearest ones ________, tenths ________
10. 3.5012 to the nearest tenths ________, thousandths ________
11. A rectangular building lot measures 228.47 feet deep by 115.50 feet wide. Round these dimensions to the nearest whole number of feet.
12. A surveyor's tape is used to measure the length of a foundation. The measurement reads 121.83 feet. Round this to the nearest tenth of a foot.

3.2 Determining Precision of Decimal Fractions

When a measurement is given to a certain place value, the precision of that measurement is the same as the place value. In the number 18.43, the rightmost digit, 3, is in the hundredths place. Therefore, the measurement is precise to the nearest 0.01 (hundredths). The measurement is off by no more than half the most precise place value. Thus, in 18.43, the measurement is off by no more than 0.005 (half of 0.01). The number must be equal to or larger than 18.425, and the number must be less than 18.435 in order to round to 18.43. This can be written as: $18.425 \leq 18.23 < 18.435$. The symbol $\leq$ means "less than or equal to." The symbol $<$ means "less than." For practical applications, this is more often written as 18.43 ± 0.005. This can be read as "18.43 plus or minus 0.005" or "18.43 give or take 0.005."

Here are examples:

2.714 is precise to the nearest 0.001 (thousandths place) and therefore will have a measuring error of no more than 0.0005. This can be written as 2.714 ± 0.0005.

0.0058 is precise to the nearest 0.0001 (ten-thousandths place) and therefore will have an error of no more than 0.00005. Therefore the value can be written as 0.0058 ± 0.00005.

25.3 is precise to the nearest 0.1 (tenths) and therefore will have an error of no more than 0.05. This can be written as 25.3 ± 0.05.

14 is precise to the nearest unit, therefore this measurement will have an error of no more than 0.5, and can be written as 14 ± 0.5.

14.0 is precise to the nearest 0.1 (tenths place) and therefore this measurement will have an error of no more than 0.05. This is written as 14.0 ± 0.05.

14.00 is precise to the nearest 0.01 (hundredths place) and therefore has an error of no more than 0.005. This can be written as 14.00 ± 0.005.

If a measurement is given as 18.3 ± 0.05, the measurement must be at least 18.25, and less than 18.35.

A measurement shown as 4.29 ± 0.005 is at least 4.285 and less than 4.295.

Exercise 3–2

Determine the precision of the following problems, and use the ± notation to show the maximum variation of the measurements. The solution to the first problem is shown.

1. 15.9	Precise to the nearest	0.1	(tenths)	15.9 ± 0.05
2. 84	Precise to the nearest	______	( )	______
3. 14.82	Precise to the nearest	______	( )	______
4. 88.527	Precise to the nearest	______	( )	______
5. 105.2	Precise to the nearest	______	( )	______

3.3 Addition of Decimals

Line up the decimal points. This is the key rule in addition (and subtraction) of decimals.

Examples

A. Add 5.82 + 30.6 + 0.005.

$$\begin{array}{r} 5.82 \\ 30.6 \\ +\ 0.005 \end{array}$$

1. Position the decimal points in a vertical line. The decimal point in the answer will be in the same line.

$$\begin{array}{r} 5.82 \\ 30.6 \\ +\ 0.005 \\ \hline 36.425 \end{array}$$

2. Add the numbers, positioning the digits properly on either side of the decimal point.

B. \$315 + \$0.82

$$\begin{array}{r} \$315.00 \\ +\quad 0.82 \\ \hline \$315.82 \end{array}$$

Places without digits can be considered to be zeros.

C. A custom-designed girder is built up of plywood layers and 2 × 8s as shown. Determine the total width of the girder.

$$
\begin{array}{rl}
0.25 + 0.25 = 0.5 & \rightarrow 0.5 \\
1.5 + 1.5 = 3 & \rightarrow 3 \\
0.5 & \rightarrow 0.5 \\
0.375 + 0.375 = 0.75 & \rightarrow \underline{0.75} \\
 & 4.75''
\end{array}
$$

1. Group like values that can be combined mentally.
2. Add.
3. The total thickness of the girder is 4.75″.

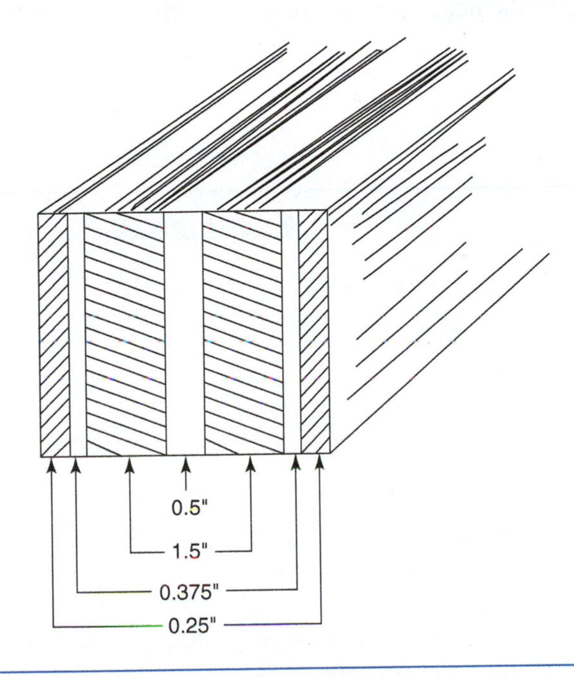

The rules for determining precision and accuracy of calculated numbers can become quite involved. We will use a simplified and not always technically correct rule:

When adding or subtracting numbers, the solution can be no more precise than the least precise number being added or subtracted.

Example: 5.4 + 2.33 = 7.73 → 7.7. The solution should be rounded to 7.7 (precise to the nearest tenth), because 5.4 is precise only to the nearest tenth. Just as a chain is no stronger than its weakest link, a solution can be no more precise than the least precise number used in the calculation.

Another example: a wall that is 6″ thick is painted with 1 coat of paint, which adds 0.01″ to the thickness of the wall. However, the wall is *still* 6″ thick, because the added thickness of the paint is too small to be relevant.

Some numbers are *exact.* Saying "5 people are in a room," 5 is an exact number. There are not 4.73325 people or 5.3 people in the room. There are exactly 5 people. Exact numbers do not change the precision of a problem.

Calculations involving multiplication and division use a different set of rules for rounding. The number of significant digits, rather than precision, determines the final rounded answer in such problems. These rules can become quite complicated and are not particularly useful in most building applications. Generally, a builder should be able to use a calculator, round "reasonably," and get a sufficiently close answer.

NOTE

All non-zero numbers are *significant*, that is, the rightmost value determines precision. However, either zeros can be *significant*, or they can simply serve as place holders, in which case they do not determine precision. For example, \$100 does not necessarily mean "exactly \$100" because the zeros are simply place holders. However, \$1.00 means exactly one dollar to the penny, because the zeros are not placeholders. In general, if removing the zeros will alter the essential value, the zeros are place holders (\$100 ≠ \$1). If removing the zeros does *not* alter the value (such as \$1.00 = \$1), then the zeros denote the precision of the measurement, in this case to the nearest penny. Refer to section 1.1 (rounding).

Exercise 3–3

1. 8.25 + 23.4 + 0.15
2. 0.03 + 85 + 4.2
3. 4.45 + 0.02 + 12.00
4. 0.068 + 0.2 + 3.41
5. 5.612 + 62.62 + 3.84
6. 0.422 + 4.81 + 53
7. 6.72 + 67.2 + 0.672
8. 0.5 + 0.03 + 0.008
9. 2.03 + 0.003
10. 7.49 + 6.21 + 0.315
11. Find the thickness of a sheet of 7-ply plywood. The plies have thicknesses of 0.03125″, 0.0625″, 0.1250″, 0.2500″, 0.1250″, 0.0625″, and 0.03125″.
12. A cabinetmaker purchases the following: plywood \$148.73, molding \$58.62, glue \$8.59, hinges \$25.77, drawer pulls \$25.80, and finish nails \$5.16. What is the total cost of the purchases?
13. A shelf is covered with plastic laminate on both sides. The laminate is 0.03125 in. thick and the plywood shelf is 0.3125 in. thick. What is the total thickness of the shelf?
14. A survey of an odd-shaped lot indicates that the five sides have the following dimensions: 225.62 ft, 78.25 ft, 145.62 ft, 138.45 ft, and 82.24 ft. Determine the distance around the lot.
15. A flyer for a lumberyard advertises the following: circular saws at \$89.99, utility cement mixers at \$239.99, cement mixing tubs at \$15.89, and exterior stain at \$29.95 per gallon. Find the total cost of purchasing a saw, a cement mixer, a tub, and 3 gallons of exterior stain.
16. During construction of a house, the rainfall was 7.28 in. during April, 8.61 in. in May, and 7.42 in. in June. What was the total rainfall during the three months of construction?
17. A shim 0.0067 in. thick is placed under a plywood subfloor 0.75 in. thick. What is the thickness of the floor and shim?
18. Find the total thickness of a built-up girder that has plywood and 2 × 8 layers (1.5″ thick) as given: 0.5″, 1.5″, 0.75″, 1.5″, 0.5″. (See diagram.)

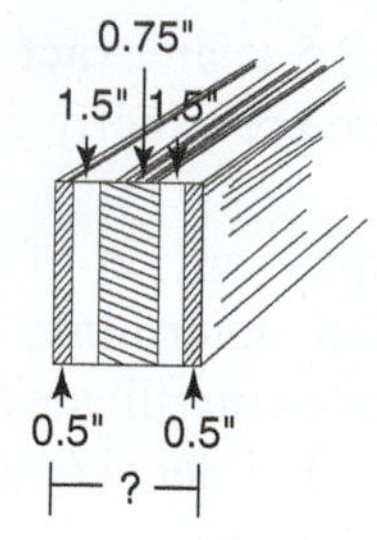

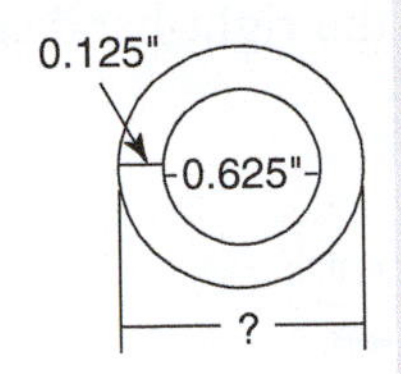

19. The inside diameter (I.D.) of a pipe measures 0.625". The wall thickness is 0.125". Find the outside diameter (O.D.) of the pipe (see diagram). (Remember, the wall thickness must be added twice.)
20. A rectangular lot is to be fenced with 182.44′ of fencing on each of the side boundaries, 114.31′ of fencing across the back of the lot, and 98.48′ across the front. (Less fencing is used in the front of the lot due to the driveway opening.) Determine the amount of fencing required, and then round this number to the nearest whole foot. Redo the addition, rounding each number before adding. Are these amounts equal?

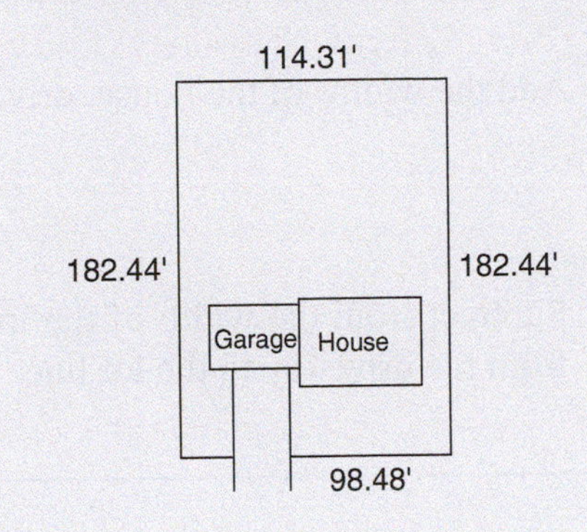

3.4 Subtraction of Decimals

Subtraction of decimals, like addition, requires that decimal points be lined up vertically.

Examples

A. 8.68 − 0.45

$$\begin{array}{r} 8.68 \\ -0.45 \\ \hline 8.23 \end{array}$$

Line up decimal points and subtract.

B. 9.2 − 3.825

$$\begin{array}{r} 9.200 \\ -3.825 \\ \hline 5.375 \end{array}$$

When necessary, fill decimal places with zeros before subtracting. The same method is used for decimals and whole numbers when borrowing or renaming.

C. A carpenter purchases a new tape measure for \$16.99 plus \$.85 tax. He pays for his purchase with a \$20 bill. How much change does he receive?

$$\begin{array}{r} \$16.99 \\ +\ 0.85 \\ \hline \$17.84 \end{array}$$

1. Line up the zeros and add the sales tax to the purchase price.

$$\begin{array}{r} \$20.00 \\ -\ 17.84 \\ \hline \$\ 2.16 \end{array}$$

2. Line up the decimal points, fill places with zeros, and subtract.
3. \$2.16 in change.

D. Determine the distance the edge of the driveway is from the right-hand side of the lot line, as shown.

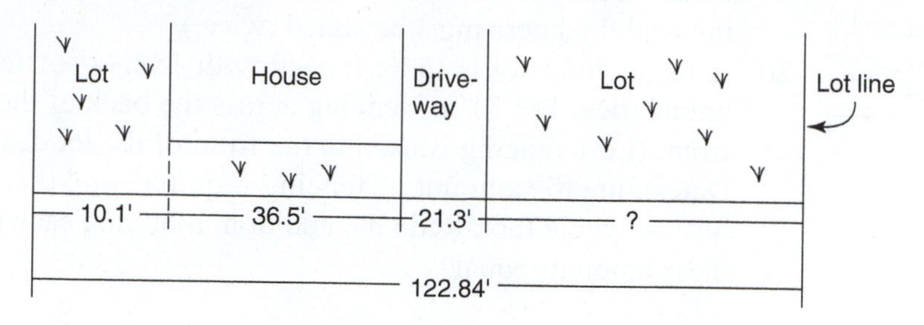

36.5′ 21.3′ + 10.1′ 67.9′	1. Add the widths of the house, driveway, and left side of the lot.
122.84′ −67.9′ 54.94′	2. Subtract from the width of the lot to determine the distance from the driveway to the lot line.

Exercise 3–4

1. 4.215 − 3.6
2. 93 − 14.21
3. 0.812 − 0.0053
4. 58.3 − 58.03
5. 0.297 − 0.0925
6. 4.62 − 3.1
7. 10.37 − 4.005
8. 8 − 7.99
9. 0.02 − 0.009
10. 13.51 − 4.11
11. A board 0.5625 in. thick has 0.0123 in. planed off it. What is the final thickness?
12. A lally column is positioned on a concrete footing as shown in the diagram. Determine the length of the lally column

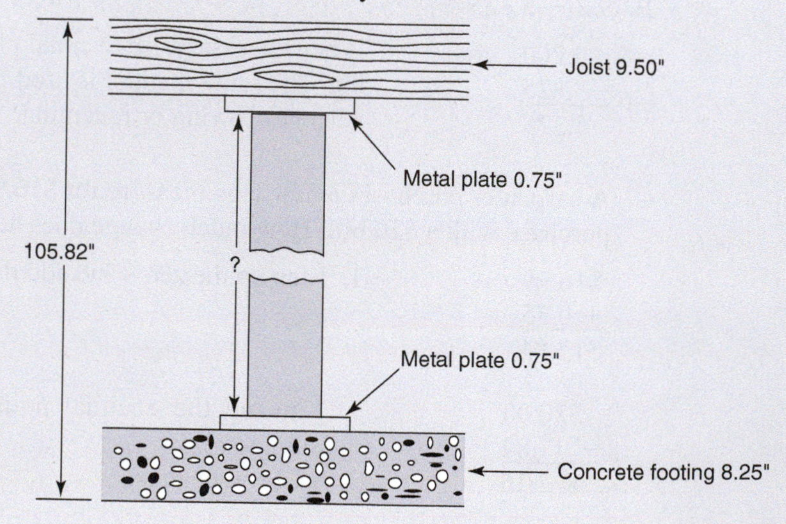

13. A pipe measures 1.0825 in. O.D. If the wall thickness is 0.052 in., what is the I.D. of the pipe?

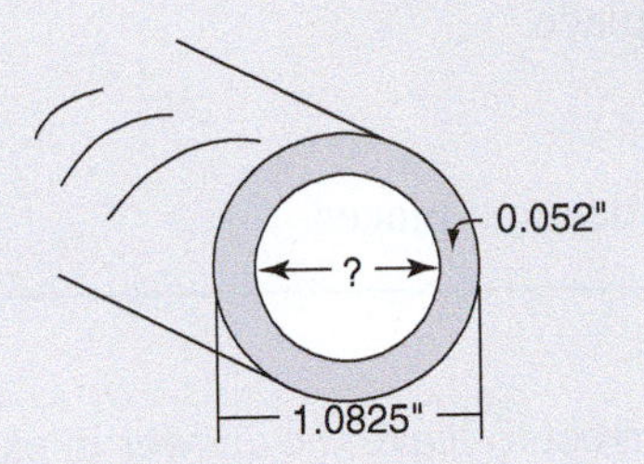

14. A template with a total length of 37.765 in. is divided into seven various lengths. Six of the lengths are 4.285, 3.715, 4.155, 5.025, 5.135, and 7.325 in. What is the missing length?
15. A jointer takes off 0.128 in. from the side of an oak board. If the board was 5.500 in. wide, what is the width after jointing?
16. A carpenter charges a customer \$256.80 for a small job. How much of that price is for labor if the materials cost \$135.66?
17. Two drill bits have diameters of 0.1875 and 0.625 in. What is the difference in the diameters of the two bits?
18. A carpenter charges \$8247.50 to build a deck. If the cost of materials is \$4122.41, and labor and overhead are \$3576.37, what profit does the carpenter make on the job?
19. A finish carpenter runs the edge of a board 5.5″ wide through a jointer twice. Each time, 0.05″ is removed from the edge. What is the width of the board after jointing?
20. Rough lumber 2.01″ thick is planed down to a thickness of 1.78″. How much did the planer remove?

3.5 Multiplication of Decimals

When multiplying decimals, the decimal points are initially ignored and the numbers are multiplied as if they were whole numbers. The placement of the decimal point is then determined, and rounding is done where necessary.

Examples

A. 45.3 × 2.06

$$\begin{array}{r} 45.3 \\ \times\ \ 2.06 \\ \hline 2718 \\ +\ 9060 \\ \hline \end{array}$$

1. Vertically line up the rightmost digits of each number.
2. Perform the indicated multiplication and addition, disregarding the decimal points.

$$\begin{array}{r} 45.3 \\ \times\ \ 2.06 \\ \hline 2718 \\ +\ 9060 \\ \hline 93.318 \end{array}$$

3. The placement of the decimal point is determined by the total number of decimal places in the two multipliers. 45.3 has one decimal place and 2.06 has two decimal places. Therefore, the decimal point in the answer must be positioned three places to the left.

B. 1.32×4.3

$$\begin{array}{r l} 1.32 & \leftarrow 2 \text{ decimal places} \\ \times\ 4.3 & \leftarrow 1 \text{ decimal place} \\ \hline 396 & \\ +\ 528 & \\ \hline 5.676 & \leftarrow 2 + 1 = 3 \text{ decimal places} \end{array}$$

Technically, it is not correct to leave the answer in example B as 5.676, because that answer is more accurate than either of the multipliers. Because the answer cannot be more accurate than the least accurate number that produced it, the answer should be rounded to one decimal place. 5.676 rounds to 5.7 (the same accuracy as 4.3).

It is not the intent of this book to teach the sometimes confusing rules of precision and accuracy. Instead, "reasonable rounding" will be used. Especially with calculator usage, it will almost never be necessary for a carpenter to employ the accuracy that can be obtained on a calculator. Rounding to two, three, or sometimes four decimal places is generally sufficient.

Example

C. Variable-speed $\frac{1}{4}$-in. drills are on sale for \$29.59 each. What is the price of three drills? (Disregard sales tax.)

$$\begin{array}{r l} \$29.59 & \leftarrow 2 \text{ decimal places} \\ \times\ \ \ 3 & \\ \hline \$88.77 & \leftarrow 2 \text{ decimal places} \end{array}$$

Here 3 is an exact number, so the accuracy is determined by the \$29.59.

Exercise 3–5

1. 4.28×5.3
2. 1.85×6
3. 0.43×2.1
4. 5.7×3.29
5. $\$4.98 \times 6$
6. 3.81×4.2
7. 0.05×6.1
8. 3.3×4.8
9. 6.2×23
10. 31×6.2
11. A contractor purchases 14.853 gal of gas to be burned in a generator used for power construction equipment. Super-unleaded gas sells for \$3.219 per gallon. Determine the cost of the gas for the generator. Round to two decimal places (the nearest penny).
12. A contractor hires five carpenters at \$18.23 per hour. They are to be paid time and a half for all hours worked over 40. How much must the contractor pay in wages if the five carpenters each work 49.5 hr? (*Hint*: Find the amount earned for 40 hr and then $1\frac{1}{2}$ times the \$18.23 rate for the remaining hours.)
13. A sheet of plywood has five plies that are each 0.125 in. thick. What is the thickness of the plywood?
14. A furnace burns 0.83 gal of oil per hour at a certain setting. If No. 2 heating oil sells for \$3.189 per gallon, what is the cost of running the furnace for 6.3 hr?

15. Find the total height of a seven-story building if each story measures 11.85 feet.
16. Scrap metal from a construction site sells for $0.051 per pound. What is a load of scrap metal worth that weighs 2380 lb?
17. A submersible pump in a well pumps 32.8 gal of water per minute. If water weighs 8.3 lb/gal, how many pounds of water does the pump deliver per minute?
18. A carpenter can lay a square of roofing shingles in 2.4 hr. How long would it take to lay 4.7 squares of shingles at that rate?

3.6 Division of Decimals

After manipulating the decimal points, division of decimals is similar to division of whole numbers.

Examples

A. Dividing a decimal by a whole number: $8\overline{)1.6}$

$$\begin{array}{r} \text{quotient} \\ \text{divisor}\overline{)\text{dividend}} \end{array}$$

$$8\overline{)\dot{1.6}}$$

1. Position the decimal point in the quotient just above its location in the dividend.

$$\begin{array}{r} 0.2 \\ 8\overline{)1.6} \end{array}$$

2. Divide as you would for whole numbers.

B. $14\overline{)0.395}$

$$\begin{array}{r} 0.0282 \\ 14\overline{)0.3950} \\ \underline{0\,0} \\ 39 \\ \underline{28} \\ 115 \\ \underline{112} \\ 30 \\ \underline{28} \end{array}$$

1. Position the decimal point in the quotient.
2. Place extra zeros after the decimal point in the dividend as necessary for placeholders.
3. Divide to one place beyond the desired accuracy.

$$\begin{array}{r} 0.028\not{2} \\ 14\overline{)0.3950} \end{array} \rightarrow 0.028$$

4. Round appropriately.

C. Dividing a decimal by a decimal: $4.23\overline{)30.4137}$

$$\begin{array}{r} 7.19 \\ 4.23.\overline{)30.41.37} \\ \underline{29\,61} \\ 80\,3 \\ \underline{42\,3} \\ 38\,07 \\ \underline{38\,07} \end{array}$$

1. Move the decimal point in the divisor as many places as necessary to make it a whole number (two places in this example).
2. Move the decimal point in the dividend the same number of places.
3. Locate the decimal point in the quotient, and divide.

D. Dividing a whole number by a decimal to one decimal place accuracy: $8.3\overline{)57}$

$8.3.\overline{)57.0.}$

1. Move the decimal point in the divisor one place to make the divisor a whole number.

$$\begin{array}{r} 6.86 \\ 8.3.\overline{)57.0.00} \\ \underline{49\ 8} \\ 7\ 2\ 0 \\ \underline{6\ 6\ 4} \\ 5\ 60 \\ \underline{4\ 98} \\ 62 \end{array}$$

2. Move the decimal point in the dividend the same number of places, using zeros to hold the place.
3. Place the decimal point in the quotient above the adjusted decimal point in the dividend.
4. For one-decimal-place accuracy, there should be two decimal places in the dividend.

6.86 rounds to 6.9

5. Round the answer to one decimal place.

E. A carpenter purchases 125 bf (board feet) of planed oak boards for $510.67. Find the price per board foot. (Board feet will be discussed in detail in Chapter 11.)

$$\begin{array}{r} 4.085\ .\ .\ . \\ 125\overline{)\$510.670\ .\ .\ .} \\ \underline{500} \\ 10\ 6 \\ \underline{00\ 0} \\ 10\ 67 \\ \underline{10\ 00} \\ 670 \\ \underline{625} \\ 45 \end{array}$$

1. Position the decimal point in the quotient just above its location in the dividend, and divide.

$4.085 → $4.09

2. Round to two places after the decimal, since this is in dollars. (There are exceptions to this rule—gasoline, for instance, may cost $1.599/gal.)

Exercise 3–6

1. $42\overline{)4.158}$
2. $3.84 \div 0.6$
3. $7.975 \div 2.75$
4. $0.09\overline{)1.62}$
5. $5.3\overline{)48.336}$
6. $4.82 \div 0.01$
7. $0.7 \div 35$
8. $0.14\overline{)0.0154}$
9. $0.062\overline{)62}$
10. $46.618 \div 3.7$
11. A pound of 2-in. sheetrock screws costs $5.28. If that amounts to $0.022 per sheetrock screw, how many are in a pound?
12. A carpenter paid $123.20 for lumber priced at $0.56 per board foot. How many board feet of lumber did he purchase?
13. Each tread in a stairway measures 9.3125 in. If the total run is 111.75 in., how many treads are there?
14. A carpenter uses 3.25 gal of paint to paint 1153.75 ft^2. How many square feet will one gallon of paint cover?
15. A project requires 36 lineal feet of lumber. What is the price per lineal foot if the lumber costs $15.12?

16. A fill-up for a contractor's truck cost $87.17. If the tank took 28.13 gal of gas, what was the price per gallon? Give answer to the nearest tenth of a penny (3 places after the decimal).
17. Cut boards each measure 8.625 in. If the boards are set end-to-end, their total length is 163.875 in. How many cut boards are there?
18. How many pieces of oak flooring 2.25 in. wide are needed to cover a hall floor 38.25 inches wide?
19. A stack of 0.625-inch-thick plywood is piled 51.875 in. high. How many sheets of plywood are in the stack?
20. A builder purchases 1500 bdf (board feet) of lumber for $3795.00 What is the price per board foot?

3.7 Multiplying and Dividing by Powers of 10

Being able to mentally multiply and divide by powers of 10 has always been a useful skill. As the United States increasingly adopts the metric system, it will become an *essential* skill. Using powers of 10 involving the metric system is covered in Chapter 10.

Some common powers of 10 are as follows:

Greater Than 1	Less Than 1
10	0.1 or ($\frac{1}{10}$)
100	0.01 or ($\frac{1}{100}$)
1000	0.001 or ($\frac{1}{1000}$)
10,000	0.0001 or ($\frac{1}{10,000}$)
100,000	0.00001 or ($\frac{1}{100,000}$)
1,000,000	0.000001 or ($\frac{1}{1,000,000}$)

Note that each number differs from the number 1 only by the location of the decimal point. This fact can be used to easily multiply and divide numbers by a power of 10 by shifting the decimal point. The numerals in the original number will not be changed. Study the following examples to determine one method of mentally multiplying and dividing by powers of 10.

Examples

A. 5.83×100

1. First determine whether the answer is larger or smaller than 5.83. Whenever the answer is *larger,* the decimal point is moved to the *right.*

1.00.

2. Observe that the number 100 differs from the number 1 by *two decimal places.*

$5.83 \times 100 = 583$

3. Move the decimal point in the number 5.83 to the *right two places.*

B. 47.3×0.001

1. Determine whether the answer is larger or smaller than 47.3. Whenever the answer is *smaller*, the decimal point is moved to the *left*.

0.001.

2. Determine the number of decimal places 0.001 differs from the number 1, namely *three places*.

$47.3 \times 0.001 = 0.047$

3. Move the decimal point in the number 47.3 to the *left three places*.

C. $85 \div 1000$

1. Determine whether the answer is larger or smaller than 85. Remember for a smaller answer the decimal point is moved to the left.

1.000.

2. Note that the number 1000 is 3 decimal places from the number 1. (The decimal point is always to the right of the last digit if it is not shown.)

$85. \div 1000 = 0.085$

3. Move the decimal point in the 85 to the *left three decimal places*.

D. $47 \div 0.01$

1. Determine whether the answer will be larger or smaller than 47. *Careful!* Dividing by a number between 0 and 1 makes the answer *larger*.

0.01.

2. Determine the number of decimal places 0.01 is from 1, namely 2.

$47. \div 0.01 = 4700$

3. Move the decimal point in the 47 to the *right two decimal places*.

E. A cabinetmaker purchases 1000 bf (board feet) of oak boards at $3.58 per board foot. Find the total price of the order.

$3.58 × 1000

1. Determine whether the answer is larger or smaller than $3.58, and move the decimal point accordingly.

$3.580. → $3580.

2. The decimal point is moved to the right 3 places.

$3580.00

3. When dealing with money, it is customary to include 2 places after the decimal (cents) even with a whole number of dollars.

F. A builder purchases 10,000 bf of framing lumber at a price of $2550. What is the price per board foot?

$2550 ÷ 10,000

1. Determine whether the answer is larger or smaller than $2550, and move the decimal point accordingly.

$2550 → $.2550

2. The decimal is moved to the left 4 places.

$0.255 or 25.5¢ per bf

3. Large quantities are sometimes priced in fractions (or decimal parts) of a cent.

Exercise 3–7

1. $83{,}000 \div 0.1$
2. 4.87×1000
3. $0.05 \div 0.001$
4. 0.05×0.01
5. $4800 \div 100$
6. $853 \times .01$
7. $38 \div 1000$
8. 0.005×100
9. $6200 \div 100$
10. 4.72×10
11. A builder purchases 100 pounds of roofing nails at \$73.00. What is the price per pound of roofing nails?
12. A framer buys 1000 10-foot 2″ × 6″ studs. Find the purchase price if each stud is priced at \$7.19.
13. In problem 12, each of the 10-foot 2″ × 6″ studs contains 10 board feet of lumber. Find the price per board foot for the studs.
14. Siding costs \$2.85/ft^2. Find the cost of 1000 square feet of siding.
15. What is the price of a thousand gallons of oil at \$3.19/gal?

3.8 Decimal and Fraction Conversions

Converting Common Fractions to Decimal Fractions

To convert a common fraction to its equivalent decimal, divide the numerator of the fraction by its denominator.

Examples

A. $\frac{3}{4}$ becomes $3 \div 4 = 0.75$.

$$\begin{array}{r} .75 \\ 4\overline{)3.00} \\ \underline{2\,8} \\ 20 \end{array}$$

B. $\frac{5}{8}$ becomes $5 \div 8 = 0.625$.

C. $\frac{1}{3}$ becomes $1 \div 3 = 0.3333. \ . \ .$ This is called a *repeating decimal.* No matter how far this division is taken, the same pattern will repeat. Repeating decimals are usually shown as $0.333 \ . \ . \ .$ or by putting a bar over the repeating digit(s), such as $0.\overline{3}$.

$$\begin{array}{r} .333. \ . \ . \\ 3\overline{)1.000. \ . \ .} \\ \underline{9} \\ 10 \\ \underline{9} \\ 10 \end{array}$$

$$0.333. \ . \ . = 0.\overline{3}$$

Converting Mixed Fractions to Decimals

When converting a mixed fraction to a decimal, keep in mind that the whole-number part of the mixed fraction will be exactly the same in the mixed decimal. Therefore, the whole number does not need to be converted.

Examples

D. $3\frac{7}{16}$ becomes $3 + (7 \div 16)$. Convert only the $\frac{7}{16}$ because the whole number 3 will be the same in both forms.

$$3 + \frac{7}{16} = 3 + (0.4375) = 3.4375$$

E. $4\frac{5}{12}$ becomes $4 + 5 \div 12$.

$$4 + (5 \div 12) = 4 + 0.416666\ldots$$

$$4.41666\ldots \quad or \quad 4.41\overline{6}$$

Note that the bar is over the 6 only, because that is the only digit that repeats.

Converting Decimals to Fractions

To convert a nonrepeating decimal to a fraction, put the decimal value over a denominator of 1, adjust the decimal point, and reduce, if necessary.

Examples

F. $0.375 = \frac{0.375}{1}$ 1. Convert to a fraction with the denominator of 1.

$\frac{.375}{1.000} = \frac{375.}{1000.}$ 2. Adjust the decimal point in the numerator to make it a whole number. Adjust the decimal point in the denominator the same number of places.

$\frac{375}{1000} = \frac{3}{8}$ 3. Reduce where possible.

Converting Mixed Decimals to Mixed Fractions

Example

G. $121.35 = 121 + \frac{.35}{1}$

1. Separate out the whole number since only the fraction needs to be converted.

$\frac{.35}{1.00} = \frac{35}{100} = \frac{7}{20}$

2. Convert the decimal to a fraction and reduce.

$121 + \frac{7}{20} = 121\frac{7}{20}$

3. Don't forget to add on the whole number!

Converting Repeating Decimals to Fractions

Any repeating decimal can be converted to a fraction with a denominator of 9, 99, 999, and so on, depending on how many digits are repeating.

Examples

H. $0.\overline{5} = \frac{5}{9}$

1. Since 5 is the only repeating digit, the denominator is only one digit, a 9.

$\left(0.5 = \frac{5}{10}\right)$

2. *Careful!* 0.5 and $0.\overline{5}$ are not equal. The terminating decimal 0.5 equals $\frac{5}{10}$.

I. $0.83838383... = 0.\overline{83}$

1. This decimal has two repeating digits.

$0.\overline{83} = \frac{83}{99}$

2. The denominator is 99 because there are two repeating digits in the numerator.

J. $2.\overline{3} = 2 + .\overline{3}$

$.\overline{3} = \frac{3}{9}$

1. Separate the whole number and convert only the repeating decimal.

$\frac{3}{9} = \frac{1}{3}$

2. Reduce.

$2.\overline{3} = 2\frac{1}{3}$

3. Don't forget to include the whole number in your answer.

NOTE

The method described in examples H to J works *only* if the repeating digits start just to the right of the decimal point (in the tenths place). For example, this method would *not* work for 0.8333. . . because the nonrepeating digit 8 is in the tenths place.

Converting Decimal Inches to Special Fractions

The most useful fractions for carpenters are those having denominators of 2, 4, 8, 16, 32, and 64, because these are the same divisions that are on a carpenter's rule and/or the diameters of non-metric drill bits. Especially with the common usage of calculators on the job, calculations are frequently done in decimal form and must be converted to the nearest useful equivalent fraction. Carefully study the examples below—this is a very important tool for the carpenter. You are expected to use a calculator for these problems.

Examples

K. Convert 0.831 in. to the nearest $\frac{1}{16}$ in.

$$\frac{.831}{1} \times \frac{16}{16}$$

1. Multiply 0.831 times the fraction $\frac{16}{16}$.

$$\frac{.831}{1} \times \frac{16}{16} = \frac{13.296}{16}$$

2. Perform the multiplication in the numerator, leaving the denominator as 16.

$$\frac{13.296 \text{ in.}}{16} \doteq \frac{13}{16} \text{ in.}$$

3. Round the numerator to the nearest whole number.

L. Convert 6.894 in. to the nearest $\frac{1}{32}$ in.

$$\frac{.894}{1} \times \frac{32}{32}$$

1. Temporarily ignoring the whole number, multiply 0.894 times the fraction $\frac{32}{32}$.

$$\frac{.894}{1} \times \frac{32}{32} = \frac{28.608}{32}$$

2. Find the product of the numerators, leaving the denominator as 32.

$$\frac{28.608}{32} \doteq \frac{29}{32}$$

3. Round the numerator to the nearest whole number.

$$6.894 \text{ in.} \doteq 6\frac{29}{32} \text{ in.}$$

4. Add the whole number of inches to the fraction. (The symbol $\doteq$ means "approximately equal to.")

M. Convert 14.629 in. to the nearest $\frac{1}{64}$ in.

$$\frac{.629}{1} \times \frac{64}{64}$$

1. Multiply by $\frac{64}{64}$.

$$\frac{.629}{1} \times \frac{64}{64} = \frac{40.256}{64}$$

$$\frac{40.256}{64} \doteq \frac{40}{64}$$

2. Round numerator to the nearest whole number.

$$\frac{40}{64} = \frac{5}{8}$$

3. When possible, reduce the fraction to lowest terms.

$$14.629 \text{ in.} \doteq 14\frac{5}{8} \text{ in.}$$

4. Add the whole number to the reduced fraction.

NOTE

It is sometimes more useful to leave the fraction in the unreduced form. For instance, $\frac{40}{64}$ in. indicates to the carpenter that the measurement is precise to the nearest $\frac{1}{64}$ in. Reducing to $\frac{5}{8}$ in. suggests that the measurement is precise only to the nearest $\frac{1}{8}$ in. If knowing the degree of precision is important, the fraction should not be reduced. It should be reduced, however, before an actual measurement is made: It is much simpler to measure $\frac{5}{8}$ in. on a carpenter's rule than to measure $\frac{40}{64}$ in.

Exercise 3–8A

Convert to equivalent decimals.

1. $\frac{5}{8}$
2. $\frac{3}{25}$
3. $\frac{2}{3}$
4. $\frac{8}{15}$
5. $1\frac{7}{8}$
6. $22\frac{25}{32}$
7. $8\frac{2}{3}$
8. $3\frac{5}{64}$
9. $19\frac{9}{16}$
10. $8\frac{3}{8}$

Convert to equivalent fractions, reduced to lowest terms.

11. 0.825
12. 0.135
13. 0.41
14. 4.825
15. 15.425
16. 7.82
17. 4.235
18. 9.17
19. 26.24
20. 11.006

Convert the repeating decimals to simplified fractions.

21. 0.666. . .
22. $0.\overline{936}$
23. $2.\overline{45}$
24. 6.18181818. . .
25. 25.4444. . .
26. $3.\overline{39}$
27. 3.66666. . .
28. $5.\overline{6}$
29. $35.\overline{2}$
30. 14.212121. . .

Convert these decimals to the nearest fraction equivalent, as specified, as shown in the example. Reduce to lowest terms whenever possible.

	Decimal	Half	Fourth	8th	16th	32nd	64th
a.	2.39 in.	$2\frac{1}{2}$ in.	$2\frac{1}{2}$ in.	$2\frac{3}{8}$ in.	$2\frac{3}{8}$ in.	$2\frac{3}{8}$ in.	$2\frac{25}{64}$ in.
31.	15.891 in.						
32.	35.319 in.						
33.	7.299 in.						
34.	121.814 in.						
35.	7.629 in.						
36.	38.79 in.						
37.	48.584 in.						
38.	134.327 in.						
39.	13.622 in.						
40.	58.05 in.						

41. Measurements of 28.365, 26.891, and 18.372 in. are to be totaled and converted to an equivalent fraction. What is the total of the three lengths to the nearest $\frac{1}{16}$ in.? (*Hint:* Add the decimal numbers and then convert the final answer.)
42. A blueprint specification calls for a steel rod to have a 0.429-in. diameter. If this rod is to fit snugly through a hole to be drilled in a stud, what size drill bit, to the nearest $\frac{1}{16}$ in., should be used?
43. What is the smallest drill bit, to the nearest $\frac{1}{32}$ in., that can be used to drill a hole that must be at least 0.139 in. in diameter?
44. A length is measured on a decimal ruler (a ruler with inches divided into 10 equal units). If the decimal ruler reads 8.9 in., what is the closest equivalent value in $\frac{1}{8}$ in.?
45. The footing for a lally column is 8.235 in. thick. What does that equal to the nearest $\frac{1}{8}$ in.?
46. A desktop is made of $\frac{5}{8}$-in. plywood covered on one side with a laminate 0.03 in. thick. What is the thickness of the desktop to the nearest $\frac{1}{32}$ in.?
47. A hole must be drilled to allow a pipe with 0.30″ O.D. to fit snugly through it. To the nearest $\frac{1}{32}$ inch, what size drill bit should be used? Remember, the hole needs to be as large as or larger than the O.D.
48. A laminated beam has layers of 0.250″, 1.750″, 0.500″, and 0.225″. Find the thickness of the beam to the nearest $\frac{1}{32}$ inch.
49. A measurement of 2.94″ is taken on a decimal ruler. What would the measurement read on a ruler marked in feet and inches to the nearest $\frac{1}{16}$ inch?
50. Add the following and give answer to the nearest $\frac{1}{16}$ of an inch:
    0.144 in. + $\frac{17}{32}$ in. + 1.18 in. + $\frac{5}{16}$ in.+ 0.19 in. + 0.815 in. = ________

Many carpenters use tables to convert from fraction to decimal and vice versa. A laminated copy of the conversion table is frequently taped to planers, sanders, and other pieces of equipment in the shop. Although tables are not always as easy to use as a calculator, they are frequently readily available when a calculator is not.

METRIC AND CUSTOMARY DECIMAL EQUIVALENTS
FOR FRACTIONAL PARTS OF AN INCH

Fraction	Decimal Equivalent		Fraction	Decimal Equivalent	
	Customary (in.)	Metric (mm)		Customary (in.)	Metric (mm)
1/64	0.015625	0.3969	33/64	0.515625	13.0969
1/32	0.03125	0.7938	17/32	0.53125	13.4938
3/64	0.046875	1.1906	35/64	0.546875	13.8906
1/16	0.0625	1.5875	9/16	0.5625	14.2875
5/64	0.078125	1.9844	37/64	0.578125	14.6844
3/32	0.09375	2.3813	19/32	0.59375	15.0813
7/64	0.109375	2.7781	39/64	0.609375	15.4781
1/8	0.1250	3.1750	5/8	0.6250	15.8750
9/64	0.140625	3.5719	41/64	0.640625	16.2719
5/32	0.15625	3.9688	21/32	0.65625	16.6688
11/64	0.171875	4.3656	43/64	0.671875	17.0656
3/16	0.1875	4.7625	11/16	0.6875	17.4625
13/64	0.203125	5.1594	45/64	0.703125	17.8594
7/32	0.21875	5.5563	23/32	0.71875	18.2563
15/64	0.234375	5.9531	47/64	0.734375	18.6531
1/4	0.250	6.3500	3/4	0.750	19.0500
17/64	0.265625	6.7469	49/64	0.765625	19.4469
9/32	0.28125	7.1438	25/32	0.78125	19.8438
19/64	0.296875	7.5406	51/64	0.796875	20.2406
5/16	0.3125	7.9375	13/16	0.8125	20.6375
21/64	0.328125	8.3384	53/64	0.828125	21.0344
11/32	0.34375	8.7313	27/32	0.84375	21.4313
23/64	0.359375	9.1281	55/64	0.859375	21.8281
3/8	0.3750	9.5250	7/8	0.8750	22.2250
25/64	0.390625	9.9219	57/64	0.890625	22.6219
13/32	0.40625	10.3188	29/32	0.90625	23.0188
27/64	0.421875	10.7156	59/64	0.921875	23.4156
7/16	0.4375	11.1125	15/16	0.9375	23.8125
29/64	0.453125	11.5094	61/64	0.953125	24.2094
15/32	0.46875	11.9063	31/32	0.96875	24.6063
31/64	0.484375	12.3031	63/64	0.984375	25.0031
1/2	0.500	12.7000	1	1.00	25.4000

Examples

N. Using the table, convert $\frac{25}{32}''$ to the equivalent decimal.

$\frac{25}{32} \rightarrow 0.78125$

1. Find the fraction, and read the appropriate decimal equivalent.

O. Using the table, convert $7\frac{5}{8}''$ to the equivalent decimal.

$\frac{5}{8} \rightarrow 0.6250$

1. Convert only the fraction.

$7 + 0.6250 \rightarrow 7.6250''$

2. Add the whole number to the fraction for the full value.

P. Using the table, convert 2.8750 to the equivalent fraction.

$0.8750 \rightarrow \frac{7}{8}$	1. Look up only the decimal value.
$2.8750 \rightarrow 2\frac{7}{8}$	2. Add the whole number to the fraction for the full value.

Q. Using the table, convert 0.1094″ to the equivalent fraction.

	1. This decimal value is not listed in the table.
0.109375—0.1094—0.1250	2. 0.1094 is between 0.109375 and 0.1250.
$0.109375 < 0.1094 < 0.1250$	3. These symbols are the "greater than/less than" symbols. This would be read "0.1094 is greater than 0.109375 and less than 0.1250."
$0.1094 - 0.109375 = 0.000025$	4. The desired value is 0.000025 greater than the smaller decimal number on the table.
$0.1250 - 0.1094 = 0.0156$	5. The desired value is 0.0156 less than the larger decimal number on the table.
$\mathbf{0.000025} < 0.0156$	6. Choose the table equivalent with the smaller variation from the desired number.
$0.1094 \doteq 0.109375 \rightarrow \frac{7}{64}$	7. 0.1094 is approximately $\frac{7}{64}$.

Exercise 3–8B

Convert the following fractions to decimals using the conversion table.

1. $\frac{5}{16}''$
2. $5\frac{7}{8}''$
3. $\frac{11}{16}''$
4. $\frac{3}{8}''$
5. $19\frac{3}{4}''$
6. $\frac{3}{32}''$
7. $\frac{7}{64}''$
8. $\frac{57}{64}''$
9. $28\frac{1}{2}''$
10. $42\frac{1}{4}''$

Convert the following decimals to fractions using the conversion table.

11. 0.359375″
12. 0.046875″
13. 14.4375″
14. 1.015625″
15. 30.625″
16. 8.2500″
17. 0.671875″
18. 5.21875″
19. 4.500″
20. 2.1875″

Convert the following decimals to the closest equivalent fractions using the conversion table.

21. 4.871″
22. 0.04257″
23. 1.877″
24. 5.259″
25. 0.78121″

Exercise 3–8C

Convert the following fractions to decimals.

1. $5\frac{3''}{8}$
2. $\frac{2''}{3}$
3. $7\frac{1''}{4}$
4. $2\frac{5''}{16}$
5. $\frac{5''}{64}$
6. $3\frac{5''}{8}$
7. $\frac{9''}{16}$
8. $\frac{3}{25}$
9. $\frac{43}{80}$
10. $\frac{9}{28}$

Convert the following decimals to exact fractions, and also to approximate equivalents to the precision indicated. The solution to 11. is shown.

11. 0.825 $(\frac{1}{16})$
12. 0.7″ $(\frac{1}{8}'')$
13. 0.03125″ $(\frac{1}{32}'')$
14. 5.275″ $(\frac{1}{16}'')$
15. 2.08 $(\frac{1}{16})$

11. $0.825 = \frac{33}{40}$, $0.825 \doteq \frac{13}{16}$
16. 0.15625″ $(\frac{1}{32}'')$
17. 3.5625″ $(\frac{1}{16}'')$
18. 4.3125″ $(\frac{1}{16}'')$
19. 0.28 $(\frac{1}{16}'')$
20. 0.1375 $(\frac{1}{32}'')$

21. A decimal ruler reads 8.900 in. What is that equal to as an exact fraction? What is it approximately equal to precise to the nearest $\frac{1}{32}''$?
22. A blueprint specification calls for a steel rod to have a 0.3125-in. diameter. If this rod is to fit snugly through a hole to be drilled in a stud, what size drill bit should be used? (Assume all drill bits between $\frac{1}{64}''$ and $\frac{64}{64}''$ are available.)
23. What is the smallest drill bit that can be used to drill a hole that must be at least 0.6875 in. in diameter?
24. The footing for a lally column is 8.235 in. thick. What does that equal in fractional inches? Give answer to the nearest $\frac{1}{32}''$, reduced if possible.
25. A bookcase shelf is made of $\frac{5}{8}''$ plywood covered on both sides with a laminate 0.03″ thick. What is the thickness of the shelf to the nearest $\frac{1}{16}''$?

CHAPTER 4

Weights, Measures, and Conversions

Did You Know

Using local materials in house building can conserve by minimizing transportation costs and possibly some processing expenses.

OBJECTIVES

Upon completing this chapter, the student will be able to:

1. Solve problems involving units of length.
2. Convert from one unit of length to an equivalent unit.
3. Solve addition and subtraction problems involving mixed units.
4. Solve multiplication and division problems involving mixed units.
5. Perform one-step conversions using the unit conversion method.
6. Perform two- and three-step conversions using the unit conversion method.
7. Convert from one unit of area to another.
8. Convert from one unit of volume to another.
9. Solve problems involving unit conversions.
10. Convert rates from one unit to another.
11. Solve problems involving rate conversions.

4.1 Linear Measure

The common units of length used in the United States are the inch, foot, yard, and mile. Rods, chains, fathoms, and hands are examples of other units of length that are used for specific purposes. Rods and chains are used in surveying, fathoms are used to measure nautical depths, and hands measure the height of horses. Following is a partial listing of linear equivalents. Chapter 10 covers the metric system. You will discover that the metric system eliminates virtually all of the cumbersome conversions that are necessary in the "American" system of weights and measures (i.e., feet, inches, miles, pounds, quarts, etc.).

1 ft = 12 in.	1 yd = 3 ft	1 mile = 320 rods
1 mi = 5280 ft	1 rod = $16\frac{1}{2}$ ft	1 mile = 80 chains
1 hand = 4 in.	1 chain = 66 ft	1 chain = 4 rods
1 fathom = 6 ft	1 yd = 36 in.	

Every carpenter must be able to readily convert units from feet to inches, and vice versa.

To convert feet to inches: multiply the number of feet by 12.

Examples

A. Convert 18 ft to inches.

$$18 \text{ ft} \times 12 \text{ in./ft} = 216 \text{ in.}$$

B. Convert 3.75 ft to inches.

$$3.75 \text{ ft} \times 12 = 45 \text{ in.}$$

To convert inches to feet: divide the number of inches by 12.

Multiply to convert to a smaller unit **(*example:* converting feet to inches).**

Divide to convert to a larger unit **(*example:* converting inches to feet).**

Examples

C. Convert 36 in. to feet.

$$36 \text{ in.} \div 12 \text{ in./ft} = 3 \text{ ft}$$

D. Change 18 in. to feet.

$$18 \text{ in.} \div 12 = 1.5 \text{ ft}$$

In example D the answer is in decimal feet. In surveying, decimal feet are routinely used, but the carpenter generally works in feet and inches. Study the following examples for changing inches into feet and inches, and decimal feet into feet and inches.

Examples

E. Change 437 in. into feet and inches. It is recommended that these problems be worked on a calculator.

$$437 \text{ in.} \div 12 = 36.41\overline{6} \text{ ft}$$

1. Dividing inches by 12 yields feet.

$$36.41\overline{6} \text{ ft} - 36 \text{ ft} = 0.41\overline{6} \text{ ft}$$

2. Write down the 36 whole feet and then subtract longhand or with the calculator. The decimal part that is left is also in feet.

$0.41\overline{6} \times 12 = 5$ in.

3. Multiply the remaining decimal by 12. This converts the decimal feet back to inches.

437 in. = $36.41\overline{6}$ ft = 36 ft 5 in.

4. It is a good idea to write down the whole number of feet as they are found.

$$\begin{array}{r} \text{feet} \quad \text{inches} \\ 36 \;\; \text{R} \; \textcircled{5} \\ 12\overline{)437} \\ \underline{36} \\ 77 \\ \underline{72} \\ \textcircled{5} \end{array}$$

36 R5 = 36 ft 5 in.

5. If this problem is worked without a calculator, the 36 in the quotient equals the number of feet and the remainder 5 is inches.

F. Change 19.58333. . . ft into feet and inches.

$19.58\overline{3}$ ft − 19 ft = $0.58\overline{3}$ ft

1. The 19 is the whole number of feet. Subtract the 19 and write it down as feet. *Note:* Key 19.583333 into the calculator. Rounding the number when entering it in the calculator can cause a significant error in accuracy in some cases.

$0.58\overline{3}$ ft × 12 = 7 in.

2. Multiply the decimal feet by 12 to convert to inches.

$19.58\overline{3}$ ft = 19 ft 7 in.

3. Write down the inches.

Exercise 4–1A

Convert the following into feet and inches. Whenever a repeating decimal is to be converted, enter the repeating digits as many times as possible into the calculator. Rounding off can significantly alter the accuracy of the answer.

1. 2.75 ft
2. $14.58\overline{3}$ ft
3. 9.5 ft
4. $8.1\overline{6}$ ft
5. $7.08\overline{3}$ ft
6. 131 in.
7. 85 in.
8. 29 in.
9. 74 in.
10. 17 in.

Frequently, measurements do not come out to a whole number of inches. In such cases, the carpenter will generally change decimal inches to the nearest $\frac{1}{8}$, $\frac{1}{16}$, etc., in., depending on the accuracy required. Study the following examples.

Examples

G. Find 15.671 in. to the nearest $\frac{1}{16}$ in.

$15.671 \text{ in.} - 15 \text{ in.} = 0.671 \text{ in.}$

1. Record and subtract 15, the whole number of inches.

$\frac{0.671 \text{ in.}}{1} \times \frac{16}{16} = \frac{10.736 \text{ in.}}{16}$

2. Multiply the decimal inches by $\frac{16}{16}$ since $\frac{1}{16}$ in. is the desired precision.

$\frac{10.736 \text{ in.}}{16} \doteq \frac{11}{16} \text{ in.}$

3. Round the numerator to the nearest whole number.

$15.671 \text{ in.} \doteq 15\frac{11}{16} \text{ in.}$

4. This is the fractional equivalent to the nearest $\frac{1}{16}$ in.

H. Surveyors' tapes are measured in feet and decimal feet to the nearest hundredth of a foot. If a surveyors' tape is used to measure a length of 7.43 ft, what is that length in feet and inches to the nearest $\frac{1}{8}$ in.?

$7.43 \text{ ft} - 7 \text{ ft} = 0.43 \text{ ft}$

1. Record and subtract the whole number of feet.

$0.43 \text{ ft} \times 12 = 5.16 \text{ in.}$

2. Change to inches.

$5.16 \text{ in.} - 5 \text{ in.} = 0.16 \text{ in.}$

3. Record and subtract the whole number of inches.

$\frac{0.16 \text{ in.}}{1} \times \frac{8}{8} = \frac{1.28 \text{ in.}}{8}$

4. Multiply by $\frac{8}{8}$ since that is the desired precision.

$\frac{1.28 \text{ in.}}{8} \doteq \frac{1}{8} \text{ in.}$

5. Round the numerator to the nearest whole number.

$7.43 \text{ ft} \doteq 7 \text{ ft } 5\frac{1}{8} \text{ in.}$

6. This is the fractional equivalent to the nearest $\frac{1}{8}$ in.

I. Machinists use decimal inches in many of their measurements. A precision-milled part used in making furniture is 49.1782 in. long. Determine the length in feet and inches to the nearest $\frac{1}{32}$ in. Reduce the fraction if possible.

$49.1782 \text{ in.} \div 12 = 4.0981833 \text{ ft}$

1. Convert to feet.

$4.0981833 \text{ ft} - 4 \text{ ft} = 0.0981833 \text{ ft}$

2. Record and subtract the whole number of feet.

$0.0981833 \text{ ft} \times 12 = 1.1782 \text{ in.}$

3. Convert the decimal feet back to inches.

$1.1782 \text{ in.} - 1 \text{ in.} = 0.1782 \text{ in.}$

4. Record and subtract the whole inches.

$$0.1782 \text{ in.} \times \frac{32}{32} = \frac{5.7024}{32} \text{ in.}$$

5. Multiply by $\frac{32}{32}$ since that is the desired precision.

$$\frac{5.7024}{32} \doteq \frac{6}{32} = \frac{3}{16} \text{ in.}$$

6. Round the numerator to the nearest whole number and reduce to lowest terms.

$$49.1782 \text{ in.} = 4 \text{ ft } 1\frac{3}{16} \text{ in.}$$

7. These are equivalent to the nearest $\frac{1}{32}$ in.

Exercise 4–1B

Change to feet and inches to the precision indicated.

1. 4.1785 ft to the nearest $\frac{1}{16}$ in.
2. 121.8316 in. to the nearest $\frac{1}{32}$ in.
3. 7.19932 ft to the nearest $\frac{1}{8}$ in.
4. 41.3167 in. to the nearest $\frac{1}{4}$ in.
5. 19.3751 in. to the nearest $\frac{1}{64}$ in.
6. 4.1923 ft to the nearest $\frac{1}{16}$ in.
7. 85.2914 in. to the nearest $\frac{1}{16}$ in.
8. 4.6281 ft to the nearest $\frac{1}{32}$ in.
9. 4.7522 ft to the nearest $\frac{1}{16}$ in.
10. 16.3343 ft to the nearest $\frac{1}{32}$ in.
11. $44.\overline{83}$ in. to the nearest $\frac{1}{32}$ in.

12. $5.58\overline{3}$ ft to the nearest $\frac{1}{4}$ in.

13. 88.417 in. to the nearest $\frac{1}{8}$ in.

14. 93.7157 in. to the nearest $\frac{1}{32}$ in.

15. 6.572 ft to the nearest $\frac{1}{8}$ in.

16. 1.899 ft to the nearest $\frac{1}{8}$ in.

17. A carpenter uses a surveyor's tape to measure a length of 52.81 feet. Convert this to feet and inches to the nearest $\frac{1}{4}$ in.

18. A window trims out with a toal width of 2 ft. $5\frac{1}{4}''$. Find the width in decimal feet.

19. A surveyor's tape measures the width of a custom-made door at $4.08\overline{3}$ ft. Change to feet and inches precise to the nearest $\frac{1}{16}''$.

20. What is $8.458\overline{3}$ ft in feet and inches? Round to the nearest $\frac{1}{4}''$.

4.2 Operations with Mixed Units

Addition and subtraction of measurements can be accomplished in either decimal feet or feet and inches. Which method is simpler depends on the form of the measurements. Study the following examples.

Examples

A. Add 3 ft 8 in. + 5 ft 9 in. + 6 ft 3 in.

3 ft 8 in. 5 ft 9 in. + 6 ft 3 in. 14 ft 20 in.	1. Add the feet and inches separately.
14 ft + 1 ft 8 in.	2. If the total number of inches equals or exceeds 12, convert to feet and inches.
15 ft 8 in.	3. Combine the feet.

B. Subtract 5 ft 7 in. from 8 ft 3 in.

8 ft 3 in. − 5 ft 7 in.	1. The 7 in. cannot be subtracted directly from 3 in.

$$\begin{array}{r} \overset{\text{12 in.}}{\cancel{8}\,7\text{ ft} \quad 15\text{ in.}} \\ -\ 5\text{ ft} \quad 7\text{ in.} \\ \hline \end{array}$$

2. Borrow 1 ft from the 8 ft, convert it to 12 in., and add to the existing 3 in.

$$\begin{array}{r} 7\text{ ft} \quad 15\text{ in.} \\ -\ 5\text{ ft} \quad 7\text{ in.} \\ \hline 2\text{ ft} \quad 8\text{ in.} \end{array}$$

3. Perform the subtraction.
4. 2 ft 8 in. is the difference.

C. Add 3.25 ft + 4.41$\overline{6}$ ft and give the answer in feet and inches.

$$\begin{array}{r} 3.25\text{ ft} \\ +4.41\overline{6}\text{ ft} \\ \hline 7.\overline{6}\text{ ft} \end{array}$$

1. Perform the addition. When using a calculator, do not round 4.41$\overline{6}$ to 4.417. The answer will be more accurate if entered into the calculator as 4.41666666.

$7.\overline{6}\text{ ft} - 7\text{ ft} = 0.\overline{6}\text{ ft}$

2. Record and subtract the whole number of feet.

$0.\overline{6}\text{ ft} \times 12 = 8\text{ in.}$

3. Multiply the decimal feet by 12 to convert to inches.

$7.\overline{6}\text{ ft} = 7\text{ ft } 8\text{ in.}$

4. Give the answer in feet and inches.

D. Using the diagram of the partial blueprint, find the missing dimension.

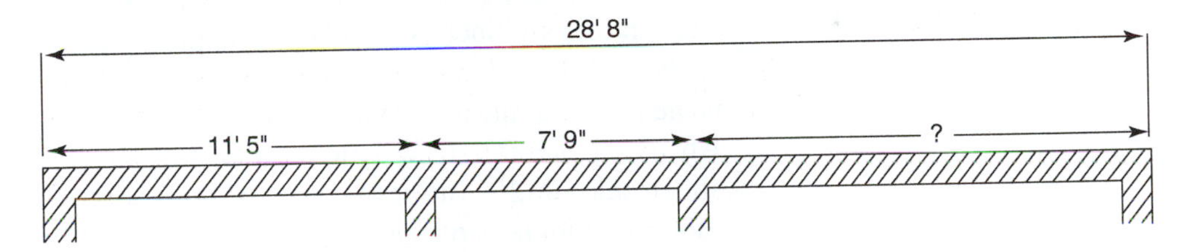

$$\begin{array}{r} 11'5'' \\ 7'9'' \\ \hline 18'14'' \end{array}$$

1. Add the feet and inches separately.

$$\begin{array}{rcr} 28' \ \ 8'' & \rightarrow & 28' \ 8'' \\ -18' \ 14'' & \rightarrow & -19' \ 2'' \\ \hline (+1')\ (-12'') & & 9' \ 6'' \end{array}$$

2. Set up to subtract from the total length.
3. Deduct 12 in. from the 14 in., and add the resulting 1 ft to the existing 18 ft.
4. Subtract to determine the missing dimension.

Alternate method:

11 ft 5 in. → 132″ + 5″ → 137″

1. Change all units to inches.

7 ft 9 in. → 84″ + 9″ → 93″ (sum 230″)

2. Add intermediate dimensions.

28 ft 8 in. → 336″ + 8″ → 344″ − 230″ = 114″

3. Subtract from total length.

114″ → 9.5 ft → 9 ft 6 in.

4. Convert back to feet and inches. *Careful!* 9.5 ft = 9 ft 6 in., not 9 ft 5 in.

Exercise 4–2A

Perform the indicated addition and subtraction. Give answers in feet and inches.

1. 8 ft 9 in. + 9 ft 2 in.
2. 3 ft 7 in. − 2 ft 10 in.
3. $4.\overline{6}$ ft − 2 ft 7 in.
4. 3.285 ft + 6.715 ft
5. 4 ft 11 in. + 5 ft 9 in. + 8 ft 11 in.
6. 16 ft 5 in. − 4 ft 10 in.
7. $3.41\overline{6}$ ft − 2 ft 10 in.
8. 7 ft 11 in. + 0 ft 9 in. + 3 ft 7 in.
9. 4.583 ft − 2.916 ft
10. 3 ft 5 in. + 1 ft 8 in.
11. A porch extends 18 ft 10 in. across the front of a house. An additional 5 ft 10 in. is not covered by the porch. What is the total length of the front of the house?
12. A hole 6 feet deep must be dug for concrete posts. How much deeper do you need to dig if your hole is 4-ft 9 in. deep?
13. A 12-foot board will have 10 sections cut from it, each section being $9\frac{3}{4}''$ long. Each cut wastes $\frac{1}{8}''$ (saw kerf). How much is left on the original board? Give answer in feet and inches to the nearest quarter inch.
14. A shed 12.0 feet long is covered by a roof that has $8\frac{3}{8}''$ overhang on each end. What is the length of the roof? Give answer in feet, inches, and fraction of an inch.
15. An experimental straw bale wall is constructed in the following fashion: $\frac{1}{2}''$ sheetrock, .0625″ vapor barrier, 16″ thick straw bale, and $\frac{3}{4}''$ stucco compound. Find the thickness of the wall in feet and inches to the nearest quarter inch.

Most addition, subtraction, and measuring is done with feet and inches. Multiplication and division, however, is best done with decimal feet or whole inches. The following examples demonstrate how to convert from feet and inches to decimal feet, and from feet and inches to whole inches.

Examples

E. Convert 5 ft 9 in. to decimal feet.

9 in. ÷ 12 = 0.75 ft — 1. Divide the inches by 12 to convert to feet.

0.75 ft + 5 ft = 5.75 ft — 2. Add the decimal feet to the whole number of feet.

5 ft 9 in. = 5.75 ft

F. Change 18 ft 5 in. to decimal feet.

5 in. ÷ 12 = $0.41\overline{6}$ ft — 1. Divide the inches by 12 to convert to feet. Do *not* round off.

$0.41\overline{6}$ ft + 18 ft = $18.41\overline{6}$ ft — 2. Add the decimal feet to the whole number of feet.

18 ft 5 in. = $18.41\overline{6}$ ft

G. Convert 9 ft 8 in. to inches.

$9 \text{ ft} \times 12 = 108 \text{ in.}$ — 1. Multiply the feet by 12 to convert to inches.

$108 \text{ in.} + 8 \text{ in.} = 116 \text{ in.}$ — 2. Add to the existing 8 in.

$9 \text{ ft } 8 \text{ in.} = 116 \text{ in.}$

H. A wall requires 19 studs each 7 ft 8 in. long. What would be the total length of the studs placed end-to-end? Give the answer in feet and inches.

$7 \text{ ft } 8 \text{ in.} \times 19$ — 1. This is a multiplication problem that will be solved by first converting the 7 ft 8 in. to decimal feet.

$8 \text{ in.} \div 12 = 0.\overline{6} \text{ ft}$ — 2. Convert feet and inches to decimal feet.

$0.\overline{6} \text{ ft} + 7 \text{ ft} = 7.\overline{6} \text{ ft}$ — 3. Add decimal feet to the whole number of feet.

$7.\overline{6} \text{ ft} \times 19 = 145.\overline{6} \text{ ft}$ — 4. Multiply to obtain the total decimal feet. *Important:* When done on a calculator, do *not* remove these figures from the calculator or round. Accuracy will be lost.

$145.\overline{6} \text{ ft} - 145 = 0.\overline{6} \text{ ft}$ — 5. Record and subtract the whole feet.

$0.\overline{6} \text{ ft} \times 12 = 8 \text{ in.}$ — 6. Convert the decimal feet to inches. *Note:* Some calculators will show the result as 7.999999. This should be interpreted as 8.

$145.\overline{6} \text{ ft} = 145 \text{ ft } 8 \text{ in.}$ — 7. This problem can also be solved by first changing feet to inches and adding to the existing inches.

I. Find $\frac{7}{8}$ of the length 5 ft 4 in.

$5 \text{ ft } 4 \text{ in.} = 5.\overline{3} \text{ ft}$ — 1. Convert 5 ft 4 in. to decimal feet.

$5.\overline{3} \text{ ft} \times \frac{7}{8} = 4.\overline{6} \text{ ft}$ — 2. Without removing 5.33333333 ft from the calculator, multiply by $\frac{7}{8}$ (multiply by 7 and then divide by 8 or use the fraction key).

$4.\overline{6} \text{ ft} = 4 \text{ ft } 8 \text{ in.}$ — 3. Convert 4.6 ft to feet and inches.

Alternate method: Change all units to inches before multiplying.

Find $\frac{7}{8}$ of the length 5 ft 4 in.

$5 \times 12 = 60 \rightarrow 60 + 4 = 64''$ — 1. Convert the measurement to inches.

$64'' \times \frac{7}{8} = 56''$ — 2. Find $\frac{7}{8}$ of 64″ by multiplication.

$56'' \div 12 = 4.\overline{6}$ ft	3. Change to feet by dividing.
$4.\overline{6}$ ft $-$ 4 ft $= 0.\overline{6}$ ft	4. Record the 4 ft and subtract it from the value.
$0.\overline{6}$ ft $\times$ 12 $= 8''$	5. Convert the $0.\overline{6}$ ft to inches.
$4.\overline{6}$ ft = 4 ft 8 in.	6. Remember to add back the 4 feet.

J. A board 10 ft 6 in. is to be divided into 9 equal segments. Find the length of each segment.

10 ft 6 in. = ? in.	1. Convert to inches before dividing.
10 ft $\times$ 12 = 120 in.	2. Convert 10 ft to inches.
120 in. + 6 in. = 126 in.	3. Add to the existing 6 in.
126 in. $\div$ 9 = 14 in.	4. Divide.
14 in. = 1 ft 2 in.	5. Convert the answer to feet and inches.

Exercise 4–2B

Perform the indicated operations, giving answers in feet and inches.

1. 8 ft 9 in. $\times$ 5
2. 4 ft 9 in. $\times \frac{2}{3}$
3. 8 ft 2 in. $\div$ 7
4. 4 ft 1 in. $\div$ 7
5. 3 ft 8 in. $\div$ 11
6. 2 ft 6 in. $\div$ 5
7. 4 ft 6 in. $\div$ 9
8. 5 ft 3 in. $\times$ 4
9. 16 ft 8 in. $\times \frac{3}{5}$
10. 5 ft 1 in. $\times$ 6
11. Three boards are each 8 ft 2 in. long. Find the total length of the boards.
12. A board 6 ft 3 in. long is to be divided equally into three board sections. Find the length of each section. (Ignore cutting waste.)
13. A trimmed board is to be cut into 3 equal pieces. Each cut wastes $\frac{3}{16}''$. The board is 6 ft $4\frac{7}{8}$ in. long. What is the length of each cut? (*Hint*: 3 pieces require 2 cuts.) Give answer in feet and inches and fractional inches.
14. Seventeen shelves have each been cut to be 1 ft $7\frac{1}{2}$ inches long. What is the combined length of the shelves? Give answer in feet and inches to the nearest $\frac{1}{2}$ inch.
15. Five board sections are each $8\frac{3}{4}''$ long and 3 sections are $9\frac{3}{8}''$ each. Find the combined length of all sections in feet and inches to the nearest $\frac{1}{8}$ inch.

4.3 Unit Conversions

Because carpenters so frequently need to convert from feet to inches and vice versa, it is important to know the fast way to accomplish these conversions (covered in section 4.1): feet to inches—multiply by 12; inches to feet—divide by 12. Many other conversions are also used in building construction, and sometimes it is not obvious whether the conversion factor should be multiplied or divided. The unit conversion method solves this problem. In the unit conversion method, the quantity to be converted is always multiplied by a fraction known as a *unity ratio.* In a unity ratio, the numerator and denominator are equivalent but are different units of measurement. Here are some examples of unity ratios:

$$\frac{1 \text{ pound}}{16 \text{ ounces}} \quad \frac{16 \text{ ounces}}{1 \text{ pound}} \quad \frac{1 \text{ foot}}{12 \text{ inches}}$$

$$\frac{12 \text{ inches}}{1 \text{ foot}} \quad \frac{1 \text{ yard}}{3 \text{ feet}} \quad \frac{36 \text{ inches}}{1 \text{ yard}}$$

Multiplying by a unity ratio *does not change the original amount. It does change the type of units.* Study the following examples to determine how to convert from one unit to another using unity ratios.

The following table gives a number of useful conversions.

1 ft = 12 in.	1 yd = 3 ft	1 mile = 320 rods
1 mi = 5280 ft	1 rod = $16\frac{1}{2}$ ft	1 mile = 80 chains
1 hand = 4 in.	1 chain = 66 ft	1 chain = 4 rods
1 fathom = 6 ft	1 yd = 36 in.	

Examples

A. Convert 9336 ft to miles.

$$1 \text{ mi} = 5280 \text{ ft}$$

1. Using the conversion factor 1 mi = 5280 ft, set up the two unity ratios.

$$\frac{1 \text{ mi}}{5280 \text{ ft}} \text{ or } \frac{5280 \text{ ft}}{1 \text{ mi}}$$

$$\frac{9336 \text{ ft}}{1} \times \frac{5280 \text{ ft}}{1 \text{ mi}}$$

2. This unity ratio will not work, because the final units would be (feet × feet)/mile, or square feet/miles.

$$\frac{9336 \not{\text{ft}}}{1} \times \frac{1 \text{ mi}}{5280 \not{\text{ft}}}$$

3. This is the proper unity ratio to use. Units of feet cancel, leaving only the desired unit of miles.

$$\frac{9336}{1} \times \frac{1 \text{ mi}}{5280} \doteq 1.8 \text{ mi}$$

4. All units in the numerator are multiplied together. Divide by all units in the denominator. Multiplication and division by 1 can be ignored.

B. Convert 18.5 yards to rods. For this conversion, the following equivalents are used: 1 yd = 3 ft, and 16.5 ft = 1 rod.

$$18.5 \text{ yd} \times \frac{?}{?} = \frac{? \text{ rods}}{1}$$

1. It is easiest to set up a problem like this if the original units and final units are written down first.

$$\frac{18.5 \not{\text{yd}}}{1} \times \frac{3 \text{ ft}}{1 \not{\text{yd}}} = \frac{? \text{ rods}}{1}$$

2. Since the direct conversion factor between yards and rods is not given, yards must first be converted to feet.

$$\frac{18.5}{1} \times \frac{3 \not{\text{ft}}}{1} \times \frac{1 \text{ rod}}{16.5 \not{\text{ft}}} = \frac{? \text{ rods}}{1}$$

3. The resulting feet are converted to rods by multiplying by the unity ratio 1 rod/16.5 ft.

$$\frac{18.5}{1} \times \frac{3}{1} \times \frac{1 \text{ rod}}{16.5} \doteq 3.36 \text{ rods}$$

4. When the only unit left is the unit desired, perform the multiplication and division.

$$18.5 \text{ yd} \doteq 3.36 \text{ rods}$$

C. A surveyor measures a tract of land to be 38.25 chains long. What is the length of the land in rods?

$$\frac{38.25 \not{\text{chains}}}{1} \times \frac{66 \not{\text{ft}}}{\not{\text{chain}}} \times \frac{1 \text{ rod}}{16.5 \not{\text{ft}}} = 153 \text{ rods}$$

$$38.25 \text{ chains} = 153 \text{ rods}$$

1. Set up the unit conversion using two conversions.
2. Cancel units where appropriate.
3. Perform the appropriate multiplication and division.

D. A road going into a new subdivision is to be 0.5 miles long. How long is the road in yards?

$$\frac{0.5 \not{\text{mi}}}{1} \times \frac{5280 \not{\text{ft}}}{1 \not{\text{mi}}} \times \frac{1 \text{ yd}}{3 \not{\text{ft}}} = 880 \text{ yds}$$

$$0.5 \text{ miles} = 880 \text{ yards}$$

1. Set up the unit conversion and cancel units where appropriate. Do multiplication or division on each fraction in the same order as it occurs in the problem.

Exercise 4–3

Using the unit conversion method, convert to the units desired.

1. 8141 ft to miles
2. 121 chains to miles
3. 420 ft to fathoms
4. 8.3 ft to hands
5. 2.5 rods to yards
6. 82 ft to rods
7. 4.5 chains to feet
8. A town street is 17.5 chains long. What is its length in feet?
9. A house lot measures 17.28 rods by 21.55 rods. What are the dimensions in decimal feet?
10. Sewer pipelines to a new development measure 180 rods. How many yards of pipe are needed?

Problems 11–15 use the following imaginary conversion factors:

1 chunk = 22 scoops
3 scoops = 5.5 blobs
5 blobs = 18 clumps

Here is a sample problem:

Convert 14.7 chunks to clumps. Round clumps to the nearest tenths.

$$\frac{14.7 \text{ chunks}}{1} \times \frac{22 \text{ scoops}}{1 \text{ chunk}} \times \frac{5.5 \text{ blobs}}{3 \text{ scoops}} \times \frac{18 \text{ clumps}}{5 \text{ blobs}} = \frac{? \text{ clumps}}{1}$$

$$\frac{14.7 \text{ chunks}}{1} \times \frac{22 \text{ scoops}}{1 \text{ chunk}} \times \frac{5.5 \text{ blobs}}{3 \text{ scoops}} \times \frac{18 \text{ clumps}}{5 \text{ blobs}} = \frac{2134.44 \text{ clumps}}{1}$$

Rounded to the tenths (0.1), 14.7 chunks = 2134.4 clumps.

Which is the larger unit, chunks or clumps? (answer: chunks are the larger unit, because it takes fewer of them to measure the same quantity. In fact, it takes 145.2 clumps to equal one chunk.)

Note how the chunks, scoops, and blobs units cancel out, leaving only the equivalent number of clumps.

11. 1 scoop = ? chunks
12. 128 blobs = ? chunks
13. 158 clumps = ? scoops
14. 35 scoops = ? blobs
15. Which is the greater quantity: 363 clumps or $2\frac{1}{2}$ chunks?

Do you find that problems using imaginary equivalents (like chunks and scoops) are more difficult to solve? Once you rely on the method, rather than previously known conversion equivalents, imaginary and real conversions should become equally easy.

4.4 Area and Volume Conversions

The following are some common conversions for area and volume.

1 square foot = 144 square inches
1 square yard = 9 square feet
1 acre = 43,560 square feet
1 square mile = 640 acres
1 cubic foot = 1728 cubic inches
1 cubic yard = 27 cubic feet
1 gallon = 231 cubic inches
1 cubic foot = 7.5 gallons

Square units are units of area. The simplest types of areas to find are rectangular shapes, but the areas of triangles, circles, and irregular figures can also be measured and computed. For a rectangle, the area in square units is found by multiplying *length* × *width*. The length and width are in linear units, but the resulting area is in square units. This is because *inches* × *inches* = *square inches*, *feet* × *feet* = *square feet*, and so on.

Examples

A. How many square feet are there in a rectangular ranch house that measures 24 ft by 36 ft?

36 ft
Area = 864 ft²   24 ft

$$\text{area} = \text{length} \times \text{width}$$
$$= 36 \text{ ft} \times 24 \text{ ft} = 864 \text{ ft}^2$$

B. A certain square is 1 yd long by 1 yd wide. This means that it has an area of 1 yd^2. How many square feet of area does the square have? Study the figure to understand why 1 yd^2 = 9 ft^2. (Carpeting, vinyl, and other types of flooring are usually priced and sold by the square yard.)

$$1 \text{ yd}^2 = 1 \text{ yd} \times 1 \text{ yd}$$
$$= 3 \text{ ft} \times 3 \text{ ft}$$
$$= 9 \text{ ft}^2$$

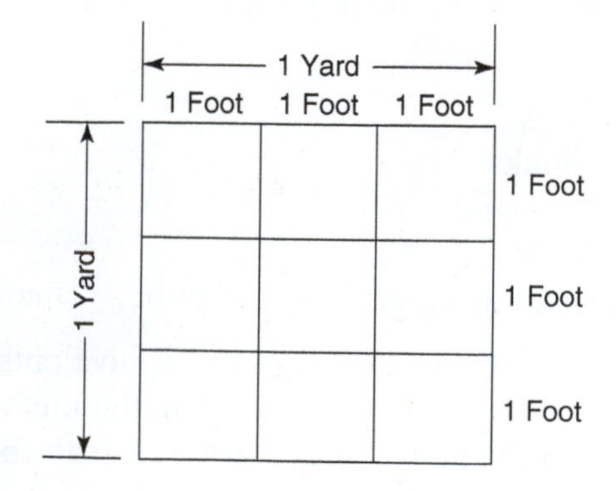

It is useful to know some of the more common square and cubic conversions. The conversion 1 yd^3 = 27 ft^3 is used frequently in building construction. Other commonly used conversions are 1 yd^2 = 9 ft^2 and 144 in.^{2} = 1 ft^2. It is unnecessary to memorize all the square and cubic conversions. If the linear conversions are known, the square and cubic conversions can be found. Study the following examples.

Examples

C. Determine the number of square inches in 1 yd^2.

1 yd = 36 in.	1. Write down the linear conversion.
$(1\text{ yd})^2 = (36\text{ in.})^2$	2. Square each side of the conversion.
$1\text{ yd}^2 = 1296\text{ in}^2$	3. 1 yd × 1 yd = 1 yd^2, and 36 in. × 36 in. = 1296 in^2.

D. How many cubic feet are there in 1 yd^3?

1 yd = 3 ft	1. Write down the linear conversion.
$(1\text{ yd})^3 = (3\text{ ft})^3$	2. Cube both sides of the conversion since the cubic conversion is desired.
$1\text{ yd}^3 = 27\text{ ft}^3$	3. 1 yd × 1 yd × 1 yd = 1 yd^3, and 3 ft × 3 ft × 3 ft = 27 ft^3.

CAUTION

Be careful to recognize the difference between a certain number of square feet and that same number of feet square. For example, *4 square feet* and *4 feet square* are not the same thing! (Figure 4–1). An area of 4 square feet can have any shape as long as it has a total area of 4 ft^2. An area 4 ft square has the specific shape of a square and actually has an area of 4 ft × 4 ft or 16 ft^2.

Figure 4–1

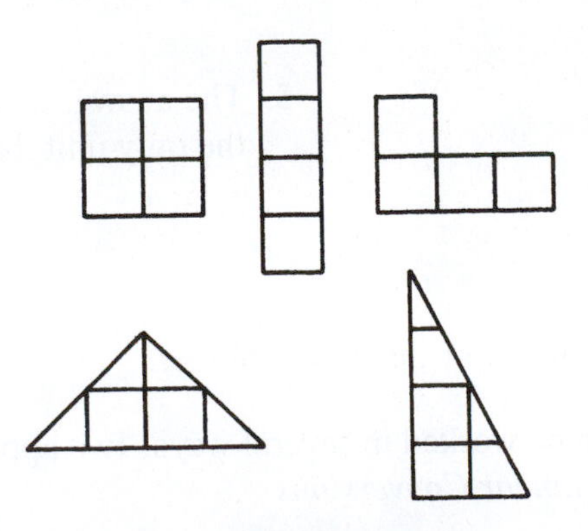

All of these figures have an area of ***4 square feet***

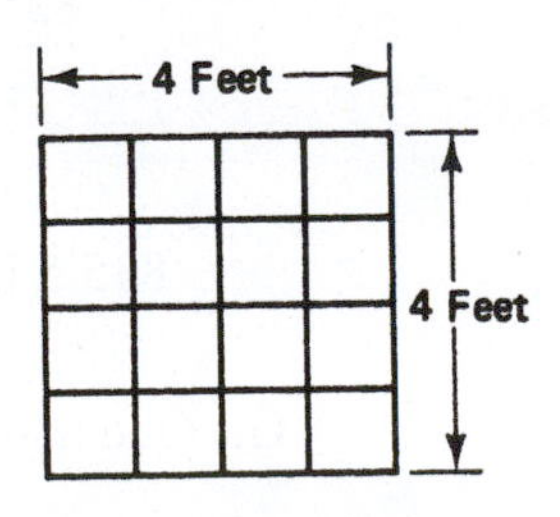

A figure ***4 feet square*** has an area of ***16 square feet***

Examples

E. A rectangular cellar excavation is to be 40 ft long, 28 ft wide, and 8 ft deep. Determine the number of cubic yards of dirt to be excavated.

First the volume in cubic feet must be determined. For a rectangular solid, *volume* = *length* × *width* × *height* (or *depth*).

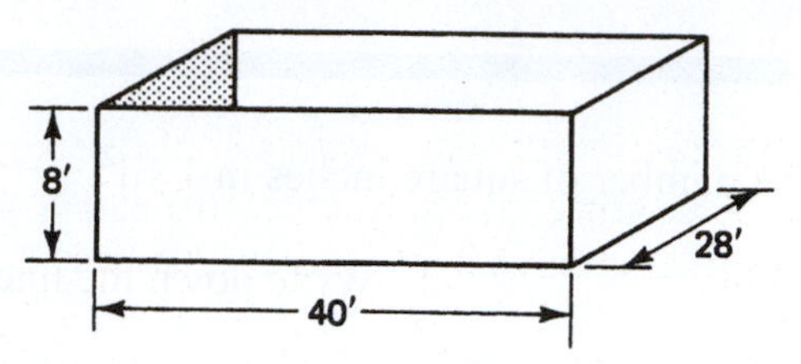

$40 \text{ ft} \times 28 \text{ ft} \times 8 \text{ ft} = 8960 \text{ ft}^3$

1. Determine the volume in cubic feet.

$\dfrac{8960 \cancel{\text{ft}^3}}{1} \times \dfrac{1 \text{ yd}^3}{27 \cancel{\text{ft}^3}}$

2. Using the unit conversion method, multiply by the conversion factor of $\frac{1}{27}$ to determine the equivalent number of cubic yards.

$\dfrac{8960}{1} \times \dfrac{1 \text{ yd}^3}{27} = 331.85 \text{ yd}^3$

3. Observe that the unwanted units cancel.

F. A stair tread has an area of 835.5 in.^{2}. Convert this to square feet.

$1 \text{ ft} = 12 \text{ in.}$

1. Assuming that the conversion between square feet and square inches is not known, write down the linear conversion.

$(1 \text{ ft})^2 = (12 \text{ in.})^2$

2. Square both sides.

$1 \text{ ft}^2 = 144 \text{ in.}^2$

3. This is the conversion in square units.

$\dfrac{835.5 \cancel{\text{in.}^2}}{1} \times \dfrac{1 \text{ ft}^2}{144 \cancel{\text{in.}^2}}$

4. Multiply by the unit conversion 1 ft^2/144 in.^{2}, canceling the units of square inches.

$\dfrac{835.5}{1} \times \dfrac{1 \text{ ft}^2}{144}$

5. The answer is in square feet, since that is the only unit that did not cancel.

$835.5 \text{ in.}^2 = 5.8 \text{ ft}^2$

G. One square mile contains how many square feet?

This problem can be worked in several ways. Two approaches are shown here.

Method 1: Squaring Linear Conversions

$$1 \text{ mi} = 5280 \text{ ft}$$
$$(1 \text{ mi})^2 = (5280 \text{ ft})^2$$
$$1 \text{ mi}^2 = 27{,}878{,}400 \text{ ft}^2$$

Method 2: Unit Conversion Method. For this method the following conversions must be known:

$$1\text{ mi}^2 = 640\text{ acres}$$
$$1\text{ acre} = 43{,}560\text{ ft}^2$$

$$\frac{1\text{ }\cancel{\text{mi}^2}}{1} \times \frac{640\text{ }\cancel{\text{acres}}}{1\text{ }\cancel{\text{mi}^2}} \times \frac{43{,}560\text{ ft}^2}{1\text{ }\cancel{\text{acre}}} = 27{,}878{,}400\text{ ft}^2$$

Observe that the unit fractions are set up in such a way as to cancel out all unwanted units. The problem started with square miles and ended up with square feet. Therefore, square miles and acres had to cancel out, leaving only the square feet.

H. A cellar slab 24′ by 44′ is 6″ thick. Find the number of cubic yards of concrete required for the slab.

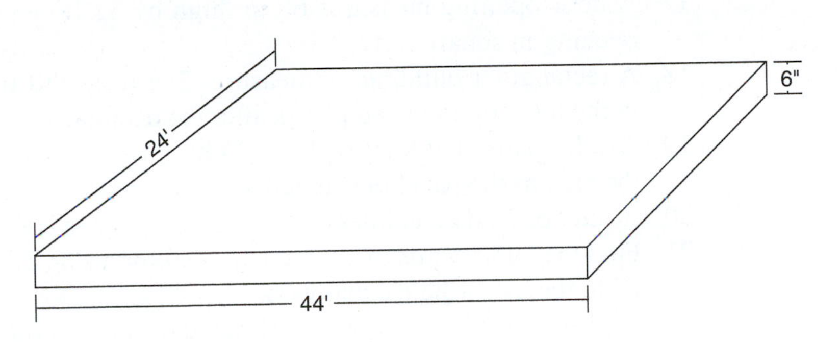

1. $6'' \div 12''/\text{ft} = 0.5\text{ ft}$ — 1. Convert the thickness in inches to feet.

2. $V = 24' \times 44' \times 0.5' = 528\text{ ft}^3$ — 2. Find the volume in cubic feet.

3. $\frac{528\text{ }\cancel{\text{ft}^3}}{1} \cdot \frac{1\text{ yd}}{27\text{ }\cancel{\text{ft}^3}} = 19.56\text{ yd}^3$ — 3. Convert the cubic feet to cubic yards.

4. $19.56\text{ yd}^3 \doteq 20\text{ yd}^3$ — 4. 20 yd^3 are needed for the slab.

Exercise 4–4

Convert to the units indicated.

1. 1 yd^3 to cubic inches
2. 8.56 yd^2 to square feet
3. 1426 acres to square miles
4. 6.8 ft^2 to square inches
5. 58,935 in.^3 to cubic yards
6. 141 ft^2 to square yards
7. 574 ft^2 to square rods (1 rod = 16.5 ft)
8. 3.4 mi^2 to acres
9. 84,000 yd^2 to acres
10. 165 ft^3 to cubic yards

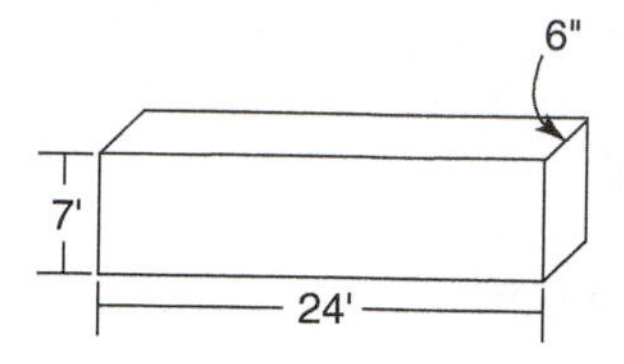

11. A driveway requires 1500 ft^3 of gravel. How many cubic yards is that? Round to the nearest cubic yard.
12. A foundation wall is to be 7 ft high, 24 ft long, and 6 in. thick. How many cubic yards of concrete are needed for the wall? (*Hint:* First change the 6 in. to feet and find the number of cubic feet in the wall. Then convert to cubic yards.)
13. A building lot contains 35,200 ft^2. What part of an acre is this? Round to two decimal places.
14. When pressurizing a building to test for air leaks, the volume of air in the building must be determined. A large meeting hall is 44 feet wide, 60 feet long, and has a 12-foot ceiling. How many cublic yards of air are in the building? (The dimensions given are inside dimensions.)
15. A kitchen is 10 ft 5 in. long by 7 ft 9 in. wide. What is the area of the kitchen in square feet? (*Hint:* Change the units to decimal feet before multiplying.)
16. A square tile is 9 in. × 9 in. How many of these tiles are required to cover 1 yd^2?
17. A door opening measures 68 in. high by 32 in. wide. What is the area of the opening in square feet?
18. A rectangular building lot measures 295 ft by 438 ft. Find the number of acres in the lot. Round to two places after the decimal.
19. A solar panel 1 ft 8 in. wide by 33 in. long is to be installed on a roof. What is the area of the panel in square feet?
20. A tract of land contains 2112 acres. What size is the tract in square miles?
21. Find the cubic yards of concrete needed for a slab 20′ by 30′ by 9″ thick. Round up to the next whole cubic yard.
22. High-density cluster housing is being developed on lots that are 55 ft by 50 ft. There are 240 lots in the development. Additionally, roads require 331,260 square feet, and green space accounts for 0.074 square miles. What is the total acreage of the development? Give answer to the nearest tenth of an acre.
23. How many cubic yards of cement are needed for a foundation wall 6 ft 4 in. high by 42 ft 8 in. wide by $9\frac{3}{4}$ in. thick? (*Hint:* change all linear units to decimal feet before multiplying.)
24. A room is 22 ft 5 in. by 14 ft 7 in. with a 6 ft × 6 ft. alcove. Ignoring cutting and fitting waste, determine the square yards of carpet needed to cover the area wall-to-wall. Ignore cutting and fitting waste; round up to the next whole number of square yards.

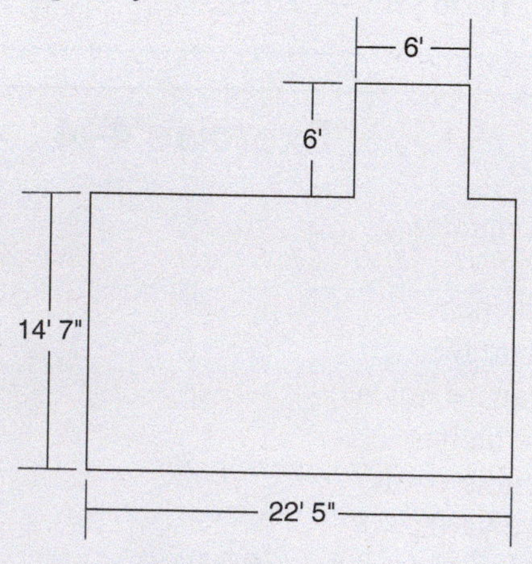

25. A 2.217 square mile tract of land contains 150 5-acre ranchettes, 521,000 square feet of roads, and a 42.1-acre pond. The remainder of the land is to be left as natural pasture. How many acres are set aside for pasture? Round to the nearest acre.

4.5 Rate Conversion Problems

Rates can frequently be thought of as "per" ratios: miles *per* hour, pounds *per* square inch, miles *per* gallon, dollars *per* pound, and so on. "Per" always can be written as a fraction bar with the unit before the "per" in the numerator and the unit after "per" in the denominator. Miles per hour can be written miles/hour. Pounds per square foot can be written pounds/square foot. Rates are unity ratios that are equal for a specific situation only. The unity ratio 12 in./1 ft is always true, because 12 in. always equals 1 ft. The rate 45 miles per hour is a unity ratio *only* for the specific instance when that is the speed. If it takes more or less time to travel 45 mi, then 45 miles does not equal 1 hour. Following are some unit conversions involving rates:

Examples

A. A car travels for 3.5 hr at 52 mph (miles per hour). How far has it gone?

$$\frac{3.5\text{ hr}}{1} \times \frac{?}{?} = \frac{?\text{ mi}}{1}$$

1. Write down the given units and those desired.

$$\frac{3.5\text{ \cancel{hr}}}{1} \times \frac{52\text{ mi}}{1\text{ \cancel{hr}}} = \frac{?\text{ mi}}{1}$$

2. The rate 52 mph means 52 miles/hours. "Per" always means a fraction bar.

$$\frac{3.5}{1} \times \frac{52\text{ mi}}{1} = 182\text{ mi}$$

3. When only the unit desired is left, perform the multiplication.

B. 3520 feet per minute = ? miles per hour.

$$\frac{3520\text{ ft}}{1\text{ min}} \times \frac{?}{?} = \frac{?\text{ mi}}{\text{hr}}$$

1. Write down the units given and those desired.

$$\frac{3520\text{ \cancel{ft}}}{1\text{ min}} \times \frac{1\text{ mi}}{5280\text{ \cancel{ft}}} \times \frac{?}{?} = \frac{?\text{ mi}}{\text{hr}}$$

2. Change feet to miles with the unit ratio 1 mi/5280 ft.

$$\frac{3520}{1\text{ \cancel{min}}} \times \frac{1\text{ mi}}{5280} \times \frac{60\text{ \cancel{min}}}{1\text{ hr}} = \frac{?\text{ mi}}{\text{hr}}$$

3. Convert minutes to hours using the unity ratio 60 min/1 hr.

$$\frac{3520}{1} \times \frac{1\text{ mi}}{5280} \times \frac{60}{1\text{ hr}} = \frac{40\text{ mi}}{\text{hr}}$$

4. When only the desired units are left in the numerator and denominator, perform the multiplication and division.

3520 feet per minute = 40 mph

C. On a certain trip, a small plane used 8.9 gal of gas per hour while averaging 120 mph. What was the gas consumption in miles per gallon?

$$\frac{?}{?} \times \frac{?}{?} = \frac{\text{mi}}{\text{gal}}$$

1. Write down the ratio desired. Note that miles must end up in the numerator and gallons in the denominator.

$$\frac{120\text{ mi}}{\cancel{\text{hr}}} \times \frac{1\ \cancel{\text{hr}}}{8.9\text{ gal}} = \frac{?\text{ mi}}{\text{gal}}$$

2. Notice that the 8.9 gal per hour ratio had to be inverted in order to achieve the desired units in the answer.

$$\frac{120\text{ mi}}{1} \times \frac{1}{8.9\text{ gal}} = \frac{13.5\text{ mi}}{\text{gal}}$$

3. When only the units desired are left, perform the required division.

13.5 miles per gallon

This could have been set up and worked as a division problem without using unity ratios, but most students find this method easier to understand.

D. Dry mahogany weighs 53 pounds per cubic foot (53 lb/ft^3). What is its weight in ounces per cubic inch (oz/in.^3)?

$$\frac{53\text{ lb}}{1\text{ ft}^3} \times \frac{?}{?} = \frac{?\text{ oz}}{\text{in}^3}$$

1. Set up the units given and those desired.

$$\frac{53\ \cancel{\text{lb}}}{1\text{ ft}^3} \times \frac{16\text{ oz}}{1\ \cancel{\text{lb}}} = \frac{?\text{ oz}}{\text{in}^3}$$

2. Multiply by the unity ratio 16 oz/lb. The units left at this point are in oz/ft^3.

$$\frac{53}{1\ \cancel{\text{ft}^3}} \times \frac{16\text{ oz}}{1} \times \frac{1\ \cancel{\text{ft}^3}}{1728\text{ in}^3} = \frac{?\text{ oz}}{\text{in}^3}$$

3. Multiply by the unity ratio 1 ft^3/1728 in^3.

$$\frac{53}{1} \times \frac{16\text{ oz}}{1} \times \frac{1}{1728\text{ in}^3} = \frac{0.49\text{ oz}}{\text{in}^3}$$

4. When only the correct units are left, perform the multiplication and division.

$$53\text{ lb/ft}^3 = 0.49\text{ oz/in}^3$$

Steps 2 and 3 can be reversed. The results would be the same.

E. A certain insulation has an *R* factor of 3.2 per inch of thickness. What is the *R* factor of batts $1\frac{1}{2}$ feet thick?

$$\frac{1.5\ \cancel{\text{ft}}}{1} \times \frac{12\ \cancel{\text{in.}}}{1\ \cancel{\text{ft}}} \times \frac{3.2}{1\ \cancel{\text{in.}}} = \frac{57.6}{1}$$

1. Set up the conversion, and cancel units.

$$R\text{ factor} = 57.6$$

2. There are no units attached to the *R* factor.

Exercise 4–5

Convert to the units indicated.

1. 25 mph (miles per hour) to feet per hour
2. \$3.52 per pound to dollars per ounce
3. 25.8 gpm (gallons per minute) of water to pounds per minute (1 gal of water $\doteq$ 8.3 lb)
4. 80 mph to yards per minute
5. 2.8 lb of nails per 400 board feet (bf) to ounces of nails per board foot
6. 55 mph to feet per second
7. The live-load rating in a building is 220 psf (pounds per square foot). What is this in psi (pounds per square inch)? (*Remember:* 1 ft^2 = 144 in^2.)
8. A carpenter's truck averages 14 mpg. In most of the countries in the world, this would be measured in kilometers per liter. Determine the truck's gas consumption in kilometers per liter. Use the conversions 1 gal = 3.785 liters and 1 mi = 1.61 km.
9. A water pump can pump 28 gpm. At that rate, how many pounds per minute can it pump? (1 gal of water $\doteq$ 8.35 lb.)
10. On a certain trip, a car averages 58 mph and gets 14.3 mpg. How much fuel does the car burn in gallons per hour?
11. A cast-iron pipe weighs 4.2 lb per foot of length. What does it weigh in ounces per inch of length?
12. A stack of dry maple boards weighs 225 lb. If dry maple weighs 49 lb/ft^3, how many cubic feet of maple are in the stack?
13. A certain insulation has an *R* factor of 4 per inch. What is the *R* factor per foot for this insulation?
14. A dry spruce 2 × 4 weighs approximately 1.5 oz per lineal inch. What is this equal to in pounds per lineal foot?
15. Insulation has an *R* factor of 2.1 per inch. Find the *R* factor in a wall 1 ft 2 in. thick.
16. Change 66 fps (feet per second) to mph (miles per hour).
17. A well yields 21 gpm (gallons per minute) of water. What is this equivalent to in cubic feet per hour?
18. On a particular flight, a small plane burns 11.3 gallons per hour and averages 94 mph (miles per hour). What is its fuel consumption in miles per gallon? Round to a whole number of miles.
19. A fuel-efficient car averages 18 kilometers per liter. Find the equivalent mpg (miles per gallon). See problem 8.
20. A pulley rotates at 20 rpm (revolutions per minute). With each revolution, it moves an assembly line belt forward by 4.4 feet. How fast is the belt moving forward in miles per hour?

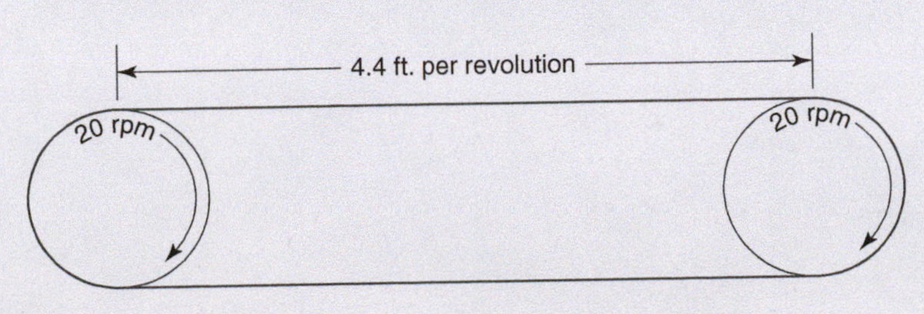

REVIEW EXERCISES

Convert to feet and inches, precise to the nearest $\frac{1}{16}''$ inch, reducing where possible:

1. 4.875ft.
2. 0.889 ft
3. $49\frac{3}{16}$ in.
4. $55.\overline{8}$ in.
5. $27.\overline{3}$ in.
6. A surveyor's tape reads 85.37 ft. Convert to feet and inches to the nearest $\frac{1}{4}$ inch.
7. A bookcase has five shelves, each 2 ft $3\frac{1}{2}$ in. long. What is the total length of all shelves in the bookcase? Give answer in feet and inches to the nearest $\frac{1}{2}$ inch.
8. Window trim requires two trim boards 5 ft $7\frac{1}{2}$ in. long and two boards 4 ft $5\frac{1}{4}$ in. long. Give the total length of the window trim, precise to the nearest $\frac{1}{4}$ inch.
9. The second floor of a home is 7 ft $11\frac{7}{8}$ in. above the finished first floor. If steps going to the second floor have risers $7\frac{3}{8}$ in. tall, how many steps are there?
10. Stairs going into an unfinished cellar have a $6\frac{3}{8}''$ rise. If there are 11 stairs, what is the height of the basement?
11. One rod (used primarily in surveying) equals $16\frac{1}{2}$ feet. One acre = 43,560 square feet. How many square rods are in one acre?

Use the following conversions for problems 12–15:
1 tub = $3\frac{1}{2}$ containers
5 containers = 2.2 boxes
4 boxes = 1 bag

12. 4 tubs = ? boxes
13. $7\frac{1}{2}$ containers = ? bags
14. 3 bags = ? tubs
15. $2\frac{1}{2}$ boxes = ? containers
16. A ship is traveling at 5 knots (nautical miles per hour). What is this in miles per hour? Use the conversion factor 55 knots = 63.4 mph.
17. A furlong is $\frac{1}{8}$ mile. How many yards are in one furlong?
18. A boat takes 15 min to go from one shore to another, a distance of 1850 feet. How fast is the boat traveling in knots? (see problem 16 above)
19. A 10″ pulley will lift a bucket approximately 31.4″ for each revolution it makes (ignore slippage). How high can the bucket be lifted with $7\frac{1}{2}$ revolutions of the pulley? Give answer in feet and inches to the nearest $\frac{1}{2}$ in.

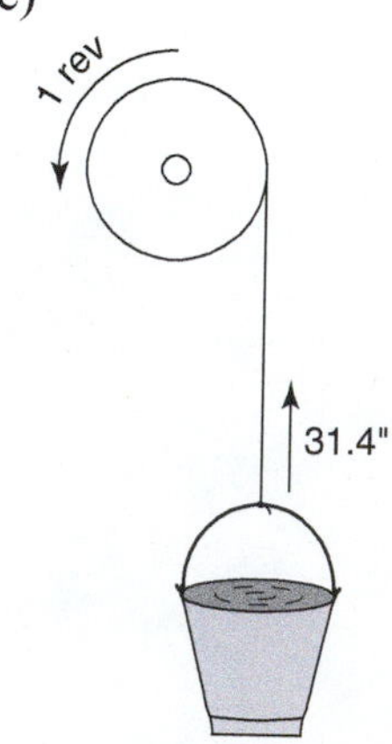

20. An unloaded diesel truck averages 17.5 miles per gallon (mpg), but mileage is reduced to 12.3 mpg when pulling a trailer. The truck makes a round trip of 100 miles to pick up the trailer (i.e., it is pulling the trailer one way). How many gallons of fuel are used in the round trip?

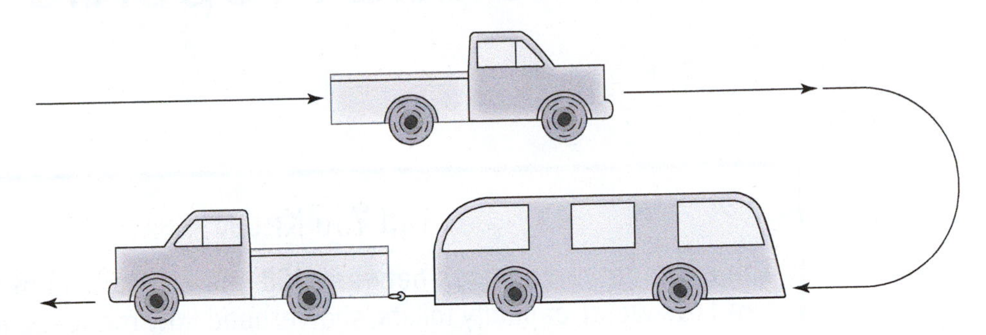

CHAPTER 5

Ratio and Proportion

Did You Know

The average person uses between 100 and 250 gallons of water per day. Low-water-capacity toilets, showerhead flow reducers, and faucet aerators can save money. Where codes allow, gray water diverted from bathing and washing machines can be used to water lawns.

OBJECTIVES

Upon completing this chapter, the student will be able to:

1. Identify and express ratios in several forms.
2. Express ratios in simplest terms.
3. Solve problems involving ratios.
4. Solve for the unknown in a proportion.
5. Identify and set up direct and inverse proportions.
6. Solve problems involving proportions.

5.1 Ratio

A ratio is a comparison of (usually) two quantities. A gas–oil mix of 16 to 1 is a ratio of the amount of gas compared to oil to be mixed for a certain engine. The Greek lowercase letter pi (π) represents the ratio of the circumference of a circle to its diameter. A blueprint may have a ratio (scale) of $\frac{1}{4}$ in. to 1 ft. The pitch of a roof is the ratio of the rise to the span. (This is the classic definition. In common usage, many builders now refer to pitch as the ratio of rise over run.) Here are several common ways to express ratios.

Examples

A. $\frac{2 \text{ in.}}{5 \text{ in.}}$

1. Any ratio can be expressed as a fraction.

$\frac{2 \text{ in.}}{5 \text{ in.}} = \frac{2}{5}$

2. Whenever possible, reduce to a unitless ratio by canceling identical units in the numerator and denominator.

B. 2:5

1. Any ratio expressed as a fraction can also be expressed in this form, read “the ratio of two to five.”

$\frac{2}{5} = 2:5$

2. The fraction bar is replaced by a colon; the numerator is written before the colon; the denominator is written after the colon.

C. A roof has a rise of 7 ft and a run of 14 ft. Express the rise over run in simplest terms.

7:14

1. Any ratio written in this form can be reduced like a fraction.

$\frac{7}{7}:\frac{14}{7}$

2. 7 divides into both the first and second numbers (numerator and denominator in fraction form).

1:2

3. The simplified ratio is therefore 1:2.

D. Express 16 qt to 5 gal as a ratio in simplest terms.

16 qt:5 gal

1. Write as a ratio.

4 gal:5 gal

2. Change to the same units. (The 5 gallons could have been changed to quarts instead for the same final result.)

$4\ \cancel{\text{gal}}:5\ \cancel{\text{gal}}$

3. Cancel identical units.

4:5

4. This is a unitless ratio reduced to lowest terms. Removing the units makes the ratio more useful. Now it could be used for measuring 4 qt to 5 qt or 4 cups to 5 cups, as well as 4 gal to 5 gal.

E. Simplify the ratio 2 ft 8 in. to 6 ft 8 in.

2 ft 8 in.:6 ft 8 in.

1. Simplify to a reduced, unitless ratio.

32 in.:80 in.

2. Change to inches.

32:80

3. Cancel the unit inches.

2:5

4. Reduce to lowest terms.

2 ft 8 in.:6 ft 8 in. = 2:5

5. The ratio in simplest terms is 2:5.

F. A line 15 in. on a blueprint represents a length of 40 ft on a house. To what scale was the blueprint drawn?

15 in. = 40 ft

1. For scales such as maps and blueprints, the = sign is often used instead of a colon.

$\frac{15\text{ in.}}{40} = \frac{40\text{ ft}}{40}$

2. Blueprints are generally shown as a scale of X in. = 1 ft. Therefore, divide both sides by 40 to obtain 1 ft on the right-hand side.

$\frac{3}{8}$ in. = 1 ft

3. Simplify, leaving the blueprint in fractions of an inch compared to 1 ft of house length.

G. Simplify the ratio 8.25:3.16.

8.25:3.16 — 1. When a ratio cannot easily be expressed as a ratio of two whole numbers, it is frequently expressed as a ratio of some quantity:1.

$\frac{8.25}{3.16}:\frac{3.16}{3.16}$ — 2. Divide both sides by the second number to obtain the number 1 on the right side.

2.61:1 — 3. Divide and round where necessary.

H. An engine has a compression ratio of 4.6. What does this mean?

Compression ratio = 4.6 — 1. Whenever a ratio is written as one number, the comparison is considered to be to 1.

$4.6 = 4.6:1 = \frac{4.6}{1}$ — 2. In this instance, there is 4.6 times as much volume in the cylinder at its maximum as there is at its minimum.

I. A concrete mix uses a mixture of cement, sand, and crushed stone in the ratio of 1:2:5 by weight. Thus, for every pound of cement used, 2 lb of sand and 5 lb of crushed stone are mixed with it. How much of each component is necessary if 4000 lb of concrete mix is needed for a job?

$1 + 2 + 5 = 8$ — 1. Add together to find the denominator. This means there are a total of 8 parts.

$\frac{1}{8}$ — 2. Since there is 1 part cement in the mixture, the fraction $\frac{1}{8}$ represents the ratio of cement in the mixture.

$\frac{2}{8} = \frac{1}{4}$ — 3. There are 2 parts sand; therefore, $\frac{2}{8}$ or $\frac{1}{4}$ represents the ratio of sand in the mixture.

$\frac{5}{8}$ — 4. This represents the ratio of crushed stone in the concrete mixture.

$\frac{1}{8} \times 4000 = 500$ lb cement

$\frac{1}{4} \times 4000 = 1000$ lb sand

$\frac{5}{8} \times 4000 = 2500$ lb crushed stone

5. Multiply the fraction that represents each component by the total amount of the mixture.

The mixture needs:

$$\begin{array}{r} 500 \text{ lb cement} \\ 1000 \text{ lb sand} \\ +\ 2500 \text{ lb crushed stone} \\ \hline 4000 \text{ lb total mixture} \end{array}$$

6. The sum of the individual components must equal the weight of the mixture.

J. Find the slope of a roof if the rise is 5 ft and the run is 15 ft.

$\text{Slope} = \dfrac{\text{rise}}{\text{run}}$ 1. This is the definition of slope.

$\dfrac{5 \text{ ft}}{15 \text{ ft}} = \dfrac{1}{3}$ 2. Reduce to a unitless ratio.

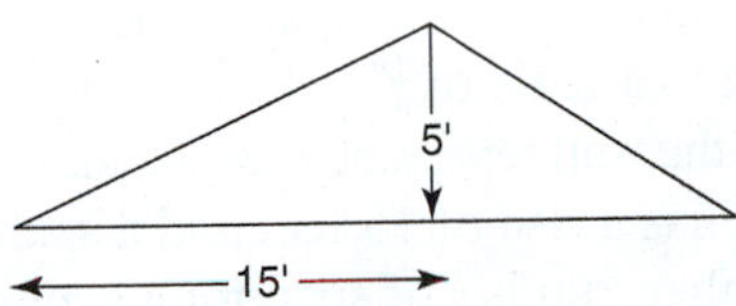

K. Find the unit rise or rise per foot run of a symmetrical roof that has a span of 24 ft and a rise of 4 ft.

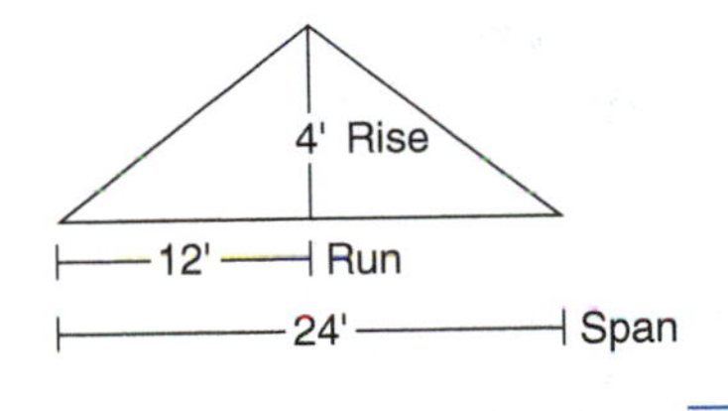

Rise = 4 ft, run = 12 ft 1. For a symmetrical roof, the run is $\frac{1}{2}$ the span.

$\dfrac{4 \text{ ft}}{12 \text{ ft}} = \dfrac{4}{12}$ 2. When using the *unit rise* terminology, the denominator is always 12 (meaning rise of 4 in. over run of 12 in. or 1 ft).

Exercise 5–1

Express as simplified, unitless ratios.

1. 3 ft : 6 in.
2. $\dfrac{25}{80}$
3. 2 ft 5 in. : 29 in.
4. 3 ft 8 in. : 9 ft 2 in.
5. 22 in. : 5 ft 6 in.
6. 25 lb cement : 50 lb sand : 75 lb crushed rock
7. 3 rejects to 18 good plumbing joints
8. Problem 7 as a ratio of rejects to total attempts
9. 40 oz gas to 2.5 oz oil
10. 15 qt to 3 gal
11. The blueprint of a house is drawn to a scale of $\frac{1}{4}$ in. = 1 ft. If the outside dimensions of the house measure 6 in. × 10 in. on the blueprint, what are the dimensions of the house?
12. A board measures $3\frac{3}{8}$ in. wide and 2 ft 3 in. long. Express the ratio of length to width in simplest terms.

13. Two gears have 32 teeth and 20 teeth. What is the ratio of the larger gear to the smaller?
14. Paint is mixed using 1 gal of base white, 1 qt of light blue, and 1 cup of gray. Determine the simplified, unitless ratio of the paint mixture, in the order given (1 gal = 4 qt; 1 qt = 4 cups).
15. Two partners in a construction company divide profits in the ratio 3:5. How much does each receive for a week when profits total $1152?
16. A concrete mix has a composition of cement to sand to crushed rock in the ratio 1:3:5. What fraction of the mixture is sand?
17. A roof has a rise of 3 ft and a run of 12 ft. Find the ratio of rise over run reduced to lowest terms. (Rise over run has commonly come to be known as pitch.)
18. A roof has a rise of 8 ft and a span of 24 ft. Find the ratio of rise over span reduced to lowest terms. (Rise over span is the classic definition of pitch.)
19. A blueprint is drawn to a scale of $\frac{1}{4}''$ = 1 foot. One line on a blueprint is $9\frac{3}{8}''$ long. What is the length of the wall represented by the line? Give answer in feet and inches.
20. A symmetrical roof has a rise of 15 feet and a span of 36 feet. Find the *unit rise* of the roof. (Remember, run is $\frac{1}{2}$ of span for a symmetrical roof.)

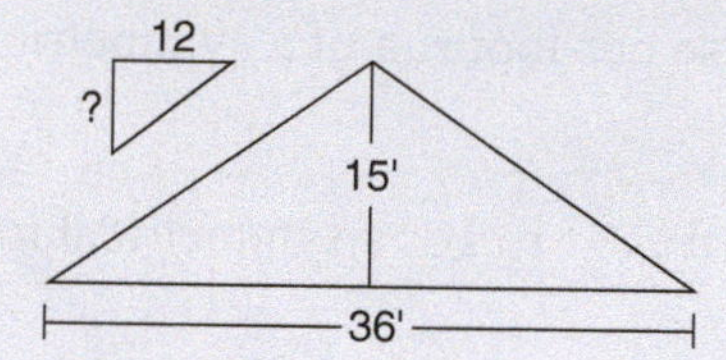

5.2 Proportion

Whenever two ratios can be set equal to each other, the equation formed is called a *proportion.* Many problems that can be solved by other means can be solved more easily using proportions.

Examples

A. Set up a proportion using the equal ratios (fractions) $\frac{2}{4}$ and $\frac{3}{6}$.

$$\frac{2}{4} = \frac{3}{6}$$

1. Since both $\frac{2}{4}$ and $\frac{3}{6}$ reduce to $\frac{1}{2}$, this is a true statement.

$$\frac{2}{4} \times \frac{3}{6}$$

2. In a proportion, the products of the diagonal numbers are always equal. This is called *cross-multiplication.*

$$2 \times 6 = 4 \times 3$$

3. Cross-multiply.

$$12 = 12$$

B. $$\frac{X}{24} = \frac{7}{3}$$

1. Two ratios set equal to each other form a proportion. If three of the four quantities are known, the fourth can always be found.

$$\frac{X}{24} \times \frac{7}{3}$$

2. Cross-multiply by finding the products of the diagonal numbers.

$3X = (7)(24)$	3. These products are always equal to each other.
$3X = 168$	4. Perform the multiplication.
$X = 168 \div 3$	5. Divide by 3 to solve for X.
$= 56$	

C. Solve for K in the following proportion.

$\frac{14.87}{3.91} = \frac{5.12}{K}$	1. The unknown can be in either the numerator or the denominator on either side of the equal sign.
$\frac{14.87}{3.91} \times \frac{5.12}{K}$	2. Cross-multiply.
$14.87K = (3.91)(5.12)$	
$14.87K = 20.0192$	3. Perform the multiplication.
$K = 20.0192 \div 14.87$	4. Divide by 14.87 to solve for K.
$= 1.35$	5. Divide and round.

D. When using a calculator, there is a shortcut method that works well for solving proportions.

$\frac{2}{15} = \frac{8}{?}$ (⊗ 8)	1. Cross-multiply the diagonal with both numbers given, and divide by the third number given. This yields the unknown.
$\frac{(2)}{15} = \frac{8}{?}$ (⊗ 8 = 120); $120 \div (2)$	2. $15 \times 8 \div 2 = (60)$. This method works regardless of the location of the unknown.

E. Solve for J in the following proportion.

$\frac{82.3}{5.2J} = \frac{19.5}{13.9}$	1. Cross-multiply.
$(5.2)(19.5)(J) = (82.3)(13.9)$	2. Perform the multiplication.
$101.4J = 1143.97$	
$J = 1143.97 \div 101.4$	3. Divide to solve for J.
$= 11.3$	4. Perform the division and round.

There are two types of proportions: direct proportions and inverse proportions. In a direct proportion, as one quantity increases, the corresponding quantity also increases. Or, if one quantity decreases, the other also decreases.

Examples

F. Three pounds of common bright 8d nails cost $1.19. What is the price of 7 lb of these nails?

As the pounds of nails increase, the cost of the nails will also increase. Therefore, this is a direct proportion.

G. A car averaging 55 mph can travel 225 mi in a certain length of time. How far can a car travel in the same length of time if it averages 62 mph?

This problem is also a direct proportion. As one quantity, miles per hour, increases, the other quantity, distance covered, will also increase.

If one quantity increases as the other quantity decreases, the proportion is an inverse proportion. Here are several examples of inverse proportions.

Examples

H. Two delivery trucks are traveling from a lumberyard to the same job site. Truck A reaches the site in 42 min, traveling an average speed of 38 mph. Truck B can average only 29 mph. How long does it take truck B to reach the site?

Since both trucks are traveling the same distance, the one that travels faster is going to take less time. Hence, as the speed increases, the time decreases. This is an inverse proportion since increasing one quantity causes the other quantity to decrease.

I. A 12-in. pulley is belted to an 8-in. pulley. If the larger pulley is rotating at 500 rpm, how fast is the smaller pulley rotating?

As the size of the pulley decreases, the speed at which it rotates increases. Since a decrease in one quantity causes an increase in the other, this is an inverse proportion.

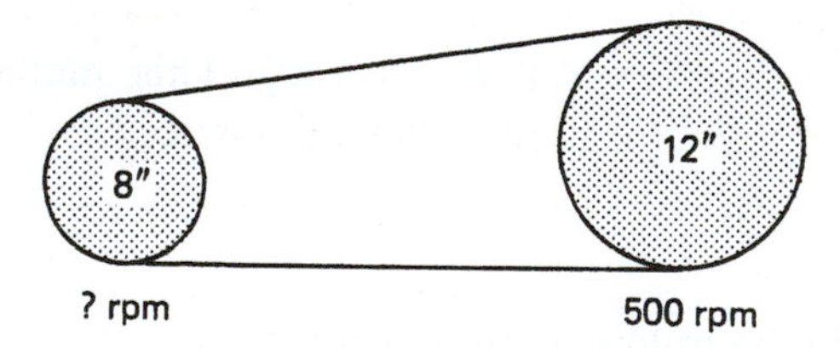

RULE

Proportions

All proportions, whether direct or inverse, can be set up in a similar fashion as long as two rules are observed:

1. *Always set up the fraction so that the same units are over each other.* For example: 5 in./8 in.; 6 ft/11 ft; or 15 workmen/22 workmen. If the same units are always over each other in a fraction, the units will cancel.
2. *Always set the smaller unit over the larger unit:*

$$\frac{\text{small}}{\text{large}} = \frac{\text{small}}{\text{large}}$$

Examples

J. A blueprint has a scale of $\frac{1}{4}$ in. = 1 ft. If the length of a house on a blueprint is $7\frac{1}{2}$ in., what is the actual length of the house?

$\frac{1}{4}$ in. = 1ft

$7\frac{1}{2}$ in. = ? ft

1. First determine what type of proportion is involved. This problem is a direct proportion since a longer line on the blueprint represents a longer wall on the house. Therefore, X, the unknown number of feet, is larger than 1 ft and must go in the denominator.

$$\frac{\frac{1}{4}\text{ in.}}{7\frac{1}{2}\text{ in.}} = \frac{1\text{ ft}}{X\text{ ft}}$$

2. Set up the ratios with inches/inches and feet/feet. The fractions are set up with the smaller number of inches and the smaller number of feet in the numerators.

$$\frac{\frac{1}{4}\text{ in.}}{7\frac{1}{2}\text{ in.}} \times \frac{1\text{ ft}}{X\text{ ft}}$$

3. Cross-multiply.

$$\tfrac{1}{4}X = (7\tfrac{1}{2})(1)$$

4. Perform the multiplication and divide by $\frac{1}{4}$ to solve for X.

$$X = (7\tfrac{1}{2})(1) \div (\tfrac{1}{4})$$

$$= 30\text{ ft}$$

5. $7\frac{1}{2}$ in. on the blueprint represents 30 ft on the house.

K. ABC Construction Company can build four identical ranch houses in 9 weeks. How many weeks would it take to build 18 houses?

4 houses = 9 weeks

18 houses = ? weeks

1. This is a direct proportion since the number of weeks required to build the houses increases as the number of houses increases. Therefore, X, the unknown number of weeks, is greater than 9 weeks and must go in the denominator.

$$\frac{4\text{ houses}}{18\text{ houses}} = \frac{9\text{ weeks}}{X\text{ weeks}}$$

2. Houses are over houses and weeks are over weeks. The smaller number of houses and the smaller number of weeks are both in the numerator.

$$\frac{4\text{ houses}}{18\text{ houses}} \times \frac{9\text{ weeks}}{X\text{ weeks}}$$

3. Cross-multiply.

$$4X = (9)(18)$$

$$X = (9)(18) \div (4)$$

4. Perform the multiplication and division to solve for X.

$$= 40.5\text{ weeks}$$

5. It would take $40\frac{1}{2}$ weeks to build 18 ranch houses at the same rate of construction.

L. A construction crew of 6 carpenters (all working at the same rate) can build a house in 5 weeks. How long would it take a crew of 10 carpenters to build the same house?

6 carpenters = 5 weeks
10 carpenters = ? weeks

1. This is an inverse proportion, since a larger crew should be able to do the job in less time. Increasing carpenters causes a decrease in weeks. Therefore, X, the unknown number of weeks, is less than 5, and must go in the numerator.

$$\frac{6 \text{ carpenters}}{10 \text{ carpenters}} = \frac{X \text{ weeks}}{5 \text{ weeks}}$$

2. Like units are over each other: carpenters over carpenters and weeks over weeks. The 6 carpenters and the X weeks are in the numerators since they are the smaller units.

$$\frac{6 \text{ carpenters}}{10 \text{ carpenters}} \times \frac{X \text{ weeks}}{5 \text{ weeks}}$$

3. Cross-multiply.

$10X = 30$

4. Divide by 10.

$X = 3$ weeks

5. It would take 3 weeks for 10 carpenters to do the same work that it would take 5 weeks for 6 carpenters to do.

M. Two pulleys are belted together. The larger pulley has a diameter of 24 in. and the smaller pulley has a 9-in. diameter. If the smaller pulley is rotating at 3600 rpm, how fast is the larger pulley rotating?

9 in. = 3600 rpm
24 in. = ? rpm

1. The larger pulley rotates slower than the smaller pulley; therefore, this is an inverse proportion. (As the diameter of a pulley increases, its rotational speed decreases.) Thus the unknown rpm is less than 3600.

$$\frac{9 \text{ in.}}{24 \text{ in.}} = \frac{X \text{ rpm}}{3600 \text{ rpm}}$$

2. The units inches are over inches and rpm are over rpm. The smaller number of inches and rpm are in the numerator.

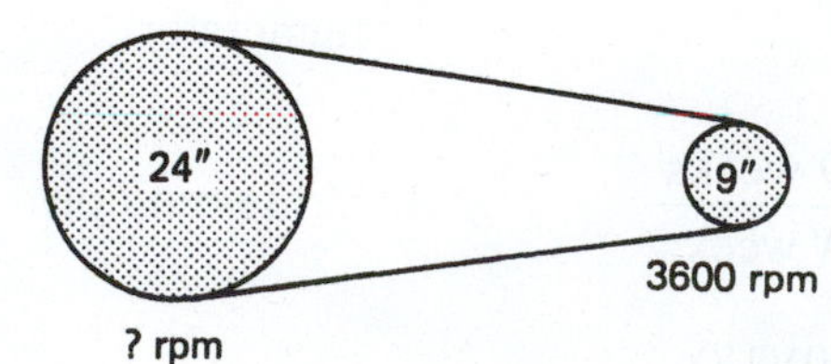

$$\frac{9 \text{ in.}}{24 \text{ in.}} \times \frac{X \text{ rpm}}{3600 \text{ rpm}}$$

3. Cross-multiply and solve for the unknown number of rpm.

$X = (3600)(9) \div (24)$

4. Divide by 24 to solve for X.

$= 1350$ rpm

5. The larger pulley rotates at 1350 rpm.

N. Find the unit rise (rise per foot run) of the symmetrical roof shown.

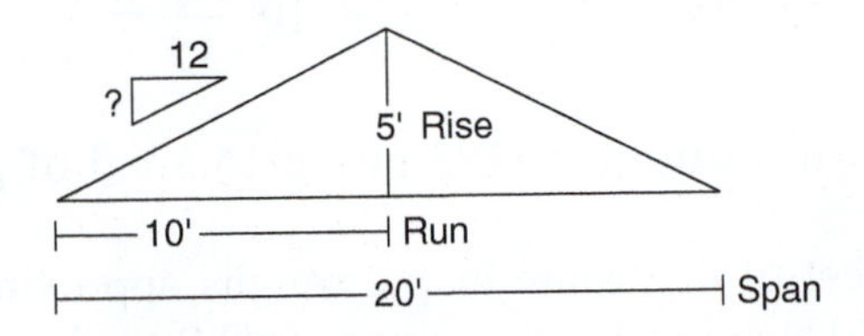

Span = 20 ft, run = 10 ft — 1. Use the run, not the span.

$\frac{5 \text{ ft}}{10 \text{ ft}} = \frac{? \text{ in.}}{12 \text{ in.}}$ — 2. The unit rise is the rise per foot (12 in.); therefore, the unit rise is *always* stated as rise/12.

$\frac{6}{12}$ = unit rise — 3. The unit rise (rise per foot run) is stated as $\frac{6}{12}$ (do not reduce to $\frac{1}{2}$).

O. Find the height of a symmetrical roof that has a unit rise of $\frac{4}{12}$ and a run of 20 ft.

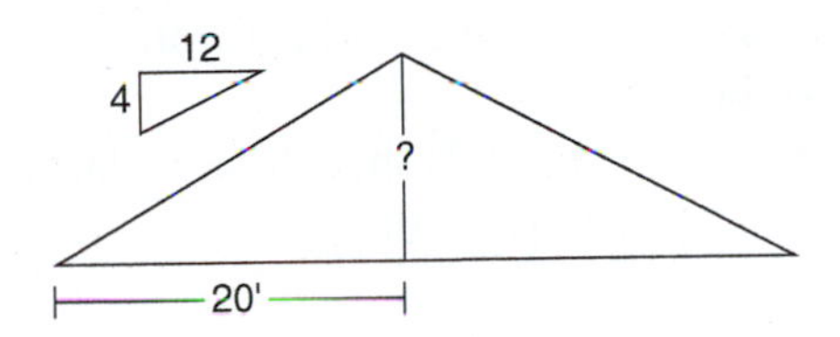

$\frac{4}{12} = \frac{X \text{ ft}}{20 \text{ ft}}$ — 1. The ratio of the unit rise equals the ratio of the rise to run.

$\frac{4}{12} = \frac{X}{20} \rightarrow X = 6.\overline{6}$ ft — 2. Cross-multiply and divide to find the rise.

$6.\overline{6}$ ft $\rightarrow$ 6 ft 8 in. — 3. Convert decimal feet to feet and inches.

Exercise 5–2

1. $\frac{3}{7} = \frac{X}{28}$
2. $\frac{2Y}{5} = \frac{8}{15}$
3. $\frac{3.08}{J} = \frac{1.45}{7.29}$
4. $\frac{3.6}{4.7} = \frac{6.3}{5.1X}$
5. $\frac{6.1}{2X} = \frac{8.3}{5.7}$
6. $\frac{4\frac{1}{4}}{X} = \frac{8\frac{1}{2}}{5\frac{1}{2}}$
7. $\frac{3\frac{3}{5}}{8\frac{2}{5}} = \frac{2X}{4\frac{1}{5}}$
8. $\frac{5.27}{3.1X} = \frac{8.64}{17.9}$

9. $\dfrac{2X}{5} = \dfrac{8}{27}$

10. $\dfrac{2\frac{1}{2}}{15} = \dfrac{X}{4}$

11. A lumber truck can travel 183 mi on 15.5 gal of gas. How far can it go on 12.2 gal?
12. $3\frac{1}{2}$ board feet of dry white maple weighs approximately 4.1 lb. What is the weight of 19 board feet of the same wood? Round to one decimal place.
13. The scale on a map is 1 in. = 25 mi. What is the distance from town A to town B if they are $2\frac{1}{8}$ in. apart on the map?
14. A room is 12 ft 6 in. wide by 18 ft 3 in. long. What would be the length and width of the blueprint of this room if a scale of $\frac{3}{8}$ in. = 1 ft is used? Give the answer to the nearest $\frac{1}{32}$ in.
15. A carpenter works 37 hr one week and makes \$711.51. At that same rate, how much would he earn in a workweek of 29 hr?
16. A 42-gal hot-water tank holds 351 lb of water. How many pounds of water can a 55-gal tank hold? (Round to the nearest whole pound.)
17. There are 27 lb of sand in 108 lb of concrete mix. How many pounds of sand are needed for 800 lb of the concrete mix?
18. A 1 in. × 4 in. dry white pine board 5-ft long weighs 3.5 pounds. What is the weight of an 8-ft board?
19. A rectangular pattern $3\frac{1}{2}$ in. × $5\frac{1}{2}$ in. is to be enlarged such that the width is $8\frac{3}{4}$ ft. What will the corresponding length be?

20. Seven masons can complete a job in $8\frac{1}{2}$ days. How many masons, working at the same rate, would be needed to do the job in $3\frac{1}{2}$ days?
21. A concrete wall 28 ft long weighs 23,000 lb. What is the weight of a concrete wall 41 ft long? (Assume that other dimensions are equal.)
22. A painter uses 3 qt of paint to cover 285 ft^2 of drywall. How many square feet of drywall can $5\frac{1}{2}$ gal cover?

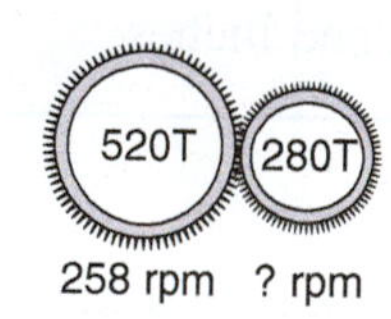

23. A large gear with 520 teeth meshes with a smaller gear with 280 teeth. What is the speed of the smaller gear if the larger gear is rotating at 258 rpm? (See diagram.)
24. Lumber costs \$13.02 for 14 board feet. At that rate, what is the price of 6 board feet?
25. Four carpenters can build storage units for a school in 78 hr. How long would it take three carpenters, working at the same rate, to build the units?
26. A lumber truck averaging 38 mph delivers building materials to a job site in 1 hr 15 min. What speed must the truck average to get the materials to the site in 55 min? (Round to the nearest whole number. *Hint:* change the 1 hr 15 min to min.)
27. Interior trim boards cost \$28.80 for 80 ft. What would 228 ft of trim boards cost?

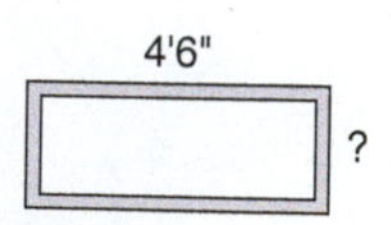

28. A picture frame has its width and length in the ratio 89 : 144. (The ratio 89 : 144 is considered to be the most pleasing rectangular proportion to the human eye.) What is the width of the frame if the length is 4 ft 6 in.? Give the answer in feet and inches to the nearest $\frac{1}{8}$ in. (*Hint:* change to inches to calculate). See diagram.

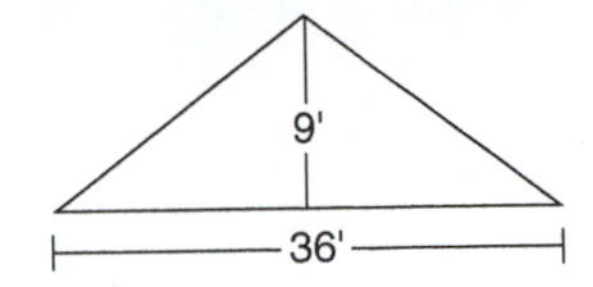

29. A symmetrical roof has a span of 36 ft and a rise (height) of 9 ft. Find the unit rise. (Remember, unit rise should always be stated as rise/12.) See diagram.
30. A roof has a unit rise of $\frac{5}{12}$ and a run of 16 ft. What is the height (rise) of the roof?

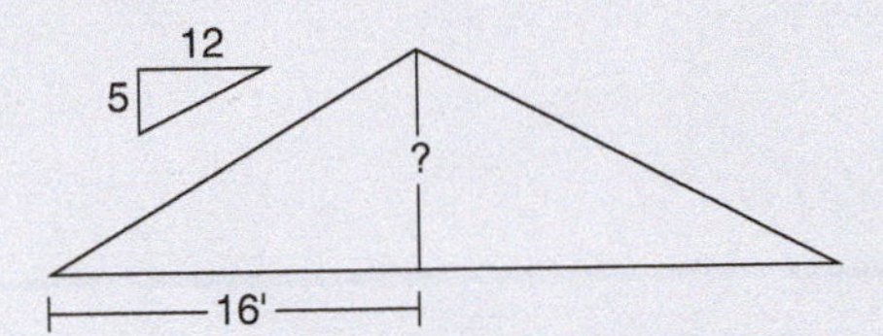

31. A delivery truck can travel 122 miles on 8.3 gallons. At the same rate, how far can it travel on a full tank? The fuel tank holds 33 gallons.
32. A delivery truck can travel from warehouse to job site in 47 minutes at an average of 42 mph. How long will the trip take if the truck averages 39 mph?
33. A blueprint uses the scale $\frac{3}{8}$″ = 1 foot. What are the dimensions of a room that is $4\frac{1}{4}$″ × $12\frac{3}{8}$″ on the blueprint? Give answer in feet and inches to the nearest $\frac{1}{4}$″.
34. Five workers can build 8 sheds in 6 days. How long would it take 3 workers to build 9 sheds? (*Hint:* work this in two steps. First determine the number of days it would take 3 workers to build 8 sheds, and then use that calculation to determine time needed to build 9 sheds.)
35. A 12″ pulley rotating at 85 rpm is belted to a $5\frac{1}{2}$″ pulley. How fast is the smaller pulley rotating? Round answer to 1 decimal place.
36. A pickup truck can get from point A to point B in 1 hr 22 min traveling at 62 mph. When pulling a trailer, the same trip takes 1 hr 55 min. At what average speed does the pickup truck travel when pulling a trailer? (*hint*: change times to minutes before calculating.)
37. A solar panel has a surface area of 7 square feet. If the panel produces 87.5 watts of electricity, how much power, in watts, can a bank of panels totaling 500 ft^2 produce?
38. An energy-efficient house remodel takes 3 carpenters 3 days. How long would it take 4 carpenters to do the remodel?
39. Energy-efficient windows are being installed in an older building. If 3 carpenters can install the windows in the building in $2\frac{1}{2}$ days, how long would it take 8 carpenters to complete the project? Assume all carpenters work at the same rate. Give answer in hours; days are 8-hr work days.
40. Man-hour are defined as the **number of workers × number of hours worked** per worker. Five carpenters each working 8 hr = 40 Man-hour. Ten carpenters each working 4 hr = 40 Man-hour also. Or 10 carpenters each working 2 hr + 5 carpenters working 4 hr = 40 Man-hour. If it takes 50 Man-hour to build a solar addition to a house, how long would it take 15 carpenters to build 60 identical solar additions? Give answer in 8-hr work days.

CHAPTER 6

Percents

Did You Know

Many appliances, computors, modems, and television sets use electricity just by being plugged in?

OBJECTIVES

Upon completing this chapter, the student will be able to:

1. Demonstrate an understanding of the meaning of percent.
2. Convert decimals to percents.
3. Convert fractions to percents.
4. Convert percents to decimals.
5. Convert percents to fractions.
6. Identify the base, percent (%), and amount in percent problems.
7. Find the amount when given the base and percent.
8. Find the percent when given the base and amount.
9. Find the base when given the percent and amount.
10. Solve "more than" and "less than" problems.
11. Identify the three parts of a word problem involving percent.
12. Set up and solve for the unknown.

6.1 Conversion of Fractions and Decimals

Percent or *per cent* can literally be translated as "*out of 100.*" When quantities are written as percents, they are being compared to a base of 100. The symbol for percent is %. Note the similarity between the symbols % and /100. Indeed, % is a shortcut way of writing /100. Thus, 15% means 15/100 or 15 out of 100.

Percents are used daily, and examples can be found everywhere. A power tool is on sale for 15% off. A sales tax of 5% is added to purchases. A waste allowance of 20% is figured for flooring. It will cost 35% more to build a particular house in New York than in Maine. The interest rate for home mortgages is $9\frac{1}{2}$%. These are a few of the types of percent problems carpenters and contractors deal with regularly.

Any decimal or fraction can be changed to a percent, and vice versa. Study the following examples carefully to determine the methods used for the various conversions.

Changing a Decimal to a Percent

Examples

A. Change 0.82 to a percent.

$0.82 = \frac{82}{100}$

1. Because % means "over 100," the % sign can be substituted for two decimal places or for the denominator /100.

$0.82. = 82\%$

2. Move the decimal point two places to the *right* and place the % sign after the number.

$\frac{82}{100} = 82\%$

3. Alternatively, replace the fraction bar and the denominator 100 with the % sign after the number.

B. Change 1.4 to a percent.

$1.4 = 1.40$

1. Position zeros after the decimal point as needed.

$1.40. = 140\%$

2. Move the decimal point two places to the *right* and include the percent sign.

C. Change 0.025 to a percent.

$0.02.5 = 2.5\%$ or $2\frac{1}{2}\%$

Move the decimal point two places to the *right* and include the percent sign. The percent can be written as a decimal or a fraction.

D. $1 = 100\%$

1. The number 1 is equal to 100%.

$0.8 = 80\%$

2. Any number smaller than 1 equals a percent less than 100%.

$2.3 = 230\%$

3. Any number greater than 1 equals a percent greater than 100%.

Changing a Fraction to a Percent

Examples

E. Change $\frac{5}{8}$ to a percent.

$\frac{5}{8} = 0.625$

1. A fraction should first be changed to a decimal.

$0.625 = 62.5\%$

2. Convert the decimal to a percent.

F. Change $1\frac{3}{4}$ to a percent.

$1\frac{3}{4} = 1.75$ — 1. Convert to a decimal.

$1.75 = 175\%$ — 2. Convert the decimal to a percent.

G. Change $\frac{1}{3}$ to a percent.

$\frac{1}{3} = .333...$ — 1. Convert to a repeating decimal.

$0.333... = 33.33...\%$ — 2. Convert to a percent.

$33.333...\% = 33\frac{1}{3}\%$ — 3. A common repeating decimal such as 0.333. . . ($\frac{1}{3}$) or 0.666. . . ($\frac{2}{3}$) is usually converted to the equivalent fraction.

H. Change $2\frac{5}{8}$ to a percent using the calculator.

$2\frac{5}{8} \rightarrow 2.625$ — 1. Use the F↔D key to convert from a fraction to a decimal.

$2.625 \rightarrow 262.5\%$ or $262\frac{1}{2}\%$ — 2. Manually convert decimal to percent.

Changing Percents to Decimals

Examples

I. Change 85% to a decimal.

$85\% = 85.\%$ — To convert a percent to a decimal, move the decimal point two places to the *left* and drop the % sign.

$0.85. = 0.85$

J. Change $22\frac{1}{4}\%$ to a decimal.

$22\frac{1}{4}\% = 22.25\%$ — 1. Change the fraction portion to a decimal.

$.22.25 = 0.2225$ — 2. Move the decimal point two places to the *left* and drop the % sign.

K. Change 0.05% to a decimal.

$0.05\% = 00.05\%$ — 1. Position zeros to the left of the decimal point, as needed. *Careful!* There is a tendency to move the decimal point incorrectly to the right when the percent itself is a decimal.

$.00.05 = 0.0005$

2. Always move the decimal point to the *left* to convert a percent to a decimal.

L. Convert 135% to a decimal using the calculator.

$135\% \rightarrow 1.35$

1. Using the percent key on the calculator (frequently a second key), input 135 and then hit the % key. The percent key automatically converts the percent to the equivalent decimal. *Caution!* The percent (%) key can be used *only* to go from percent to decimal, not from decimal to percent. In other words, the % key can be used whenever the percent is given.

Converting a Percent to a Fraction

Examples

M. Convert 42% to the equivalent common fraction.

$$42\% = \frac{42}{100}$$

1. Drop the percent sign and show as a fraction over 100. (*Remember:* % means /100.)

$$\frac{42}{100} = \frac{21}{50}$$

2. Reduce where possible.

N. Change 150% to a fraction.

$$150\% = \frac{150}{100}$$

1. Show as a fraction over 100.

$$\frac{150}{100} = \frac{3}{2} = 1\frac{1}{2}$$

2. Reduce and convert to a mixed number.

O. Change $166\frac{2}{3}\%$ to a fraction.

$$166\frac{2}{3}\% = \frac{500}{3}\%$$

1. Convert the mixed number to an improper fraction. Note that the improper fraction is still a percent.

$$\frac{500\%}{3} \times \frac{1}{100} =$$

2. Multiply by $\frac{1}{100}$ and drop the % sign. (This is equivalent to setting $\frac{500}{3}$ over a denominator of 100.)

$$\frac{\cancel{500}}{3} \times \frac{1}{\cancel{100}} = \frac{5}{3}$$

3. Reduce and multiply.

$$166\frac{2}{3}\% = \frac{5}{3} \quad \text{or} \quad 1\frac{2}{3}$$

4. Leave as an improper fraction or change to a mixed number.

P. Change $\frac{1}{4}\%$ to a fraction.

$\frac{1}{4}\% = \frac{1}{4} \times \frac{1}{100}$ 1. Replace the % sign by multiplying by $\frac{1}{100}$.

$\frac{1}{4} \times \frac{1}{100} = \frac{1}{400}$ 2. Multiply.

$\frac{1}{4}\% = \frac{1}{400}$ 3. $\frac{1}{4}\%$ is equivalent to $\frac{1}{400}$.

Q. Convert 38% to the equivalent fraction using the calculator conversion keys.

$38\% \rightarrow 0.38$ 1. Key in 38 and then the % key. This automatically converts 38% to the equivalent decimal value.

$0.38 \rightarrow \frac{19}{50}$ 2. Use the F↔D key to convert the decimal to a fraction.

R. A stamp of MC15 on lumber means the lumber has a maximum moisture content of 15%. State the moisture content as a fraction.

$15\% \rightarrow 0.15$ 1. Input 15 and hit the % key to convert to a decimal.

$0.15 \rightarrow \frac{3}{20}$ 2. Use the F↔D key to convert to a fraction.

Exercise 6–1

Complete the following table, converting all improper fractions to mixed numbers.

	Fraction	Decimal	Percent
1.	$1\frac{4}{5}$		
2.		0.3	
3.			182%
4.	$\frac{3}{8}$		
5.		0.25	
6.			0.04%
7.	$\frac{1}{1000}$		
8.		1.25	
9.			$\frac{1}{2}\%$
10.	$2\frac{2}{3}$		
11.		0.025	

12.			24%
13.	$\frac{2}{3}$		
14.		1.00	
15.			2.5%
16.	$\frac{3}{5}$		
17.		0.62	
18.			143%
19.	$\frac{3}{5000}$		
20.			$266\frac{2}{3}\%$

6.2 Percent Problems

Every percent problem has three parts: the base, the percent, and the amount. (Traditionally, the three parts have been called *base, rate* (percent), and *percentage* (amount). (This terminology tends to be confusing, and thus the more straightforward *bases, percent,* and *amount* will be used in this book.) In order to set up and solve percent problems, we must first be able to identify the components. Study the following examples.

Examples

A. What is *18%* of 52?

1. The 18% is the percent; therefore, it can be identified by the % sign.

What is 18% *of* 52?

2. The *base* is the quantity after the word *of.*

What is 18% of 52?

3. Since both the % and base have been identified, *what* refers to the *amount.* Hence we are looking for the amount in this example.

In this problem, the *amount* is the unknown.

B. 25 is *what percent* of 38?

1. *What percent* indicates that we are looking for the % in this example.

25 is what percent *of 38*?

2. The *base* is 38, since it is the quantity that comes after the word *of.*

25 is what percent of 38?

3. Since the % and base have both been identified, 25 represents the *amount.*

In this problem, the *percent* is the unknown.

C. 22 is *62%* of what number?

1. 62% is the percent.

22 is 62% *of what number*?

2. *Of* identifies *what number* as the *base.* Therefore, we are looking for the base.

22 is 62% of what number?

3. Since the % and base have been identified, 22 is the *amount.*

In this problem, the *base* is the unknown.

D. 15 is *what percent* of 32? — The % is the unknown.
35 is 82% of *what number*? — The *base* is the unknown.
What is 21% of 85? — The *amount* is the unknown.

Exercise 6–2A

Identify which of the three parts (base, percent, or amount) is the unknown.

1. 22% of 58 equals what number?
2. 48 is what percent of 19?
3. 45 is 14% of what number?
4. 38 out of 200 represents what percent?
5. 125 is 20% of what number?
6. 25 is what percent of 125?
7. 42 is 18% of what number?
8. What is 155% of 24?
9. 150 is 25% of what number?
10. What is $\frac{1}{2}$% of 1000?
11. What is $14\frac{1}{2}$% of 85?
12. 35 is 84% of what number?
13. 55 is what percent of 88?
14. 175 is what percent of 150?
15. 54 is 39% of what number?
16. What is 50% of 50?
17. What is 75% of 75?
18. 42.7 is 100% of what number?

There are several ways to approach percent problems. The *percent proportion* method is discussed here because it works equally well for all types of percent problems.

Percent Proportion Method

The problem is set up as a proportion using the following form:

$$\frac{\text{percent (\%)}}{100\%} = \frac{\text{amount}}{\text{base}}$$

This method will work regardless of which part of the problem is unknown. The following examples refer back to examples A, B, and C.

Examples

E. Set up and solve example A: What is 18% of 52?

$$\frac{18\%}{100\%} = \frac{\text{amount}}{52}$$

percent ↘ (18%); This is always 100% ↗ (100%); unknown amount ↙ (amount); base ↖ (52)

1. 18% is the percent, 52 is the base, and the amount is unknown.

$$\frac{18}{100} \times \frac{X}{52}$$

2. Cross-multiply.

$$100X = (18)(52)$$

3. Solve for the unknown amount X.

$$X = \frac{936}{100} = 9.36$$

9.36 is 18% of 52

4. 9.36 is the amount.

F. Set up and solve example B: 25 is what percent of 38?

$$\frac{X\%}{100\%} = \frac{25}{38}$$

(unknown percent → $X\%$; amount → 25; base → 38)

1. Here the percent is the unknown.

$$\frac{X}{100} \times \frac{25}{38}$$

2. Cross-multiply.

$$38X = (25)(100)$$

3. Solve for X and round.

$$X = \frac{2500}{38} \doteq 65.8$$

4. 65.8% is the percent.

25 is 65.8% of 38

G. Set up and solve example C: 22 is 62% of what number?

$$\frac{62\%}{100\%} = \frac{22}{X}$$

(percent rate → 62%; amount → 22; unknown base → X)

1. Here the base is the unknown.

$$\frac{62}{100} \times \frac{22}{X}$$

2. Cross-multiply.

$$62X = (22)(100)$$

$$X = \frac{2200}{62} \doteq 35.5$$

3. Solve for X and round.

22 is 62% of 35.5

4. 35.5 is the base.

Exercise 6–2B

Problems 1–18 are identical to Exercise 6–2A. Set up and solve for the unknown.

1. 22% of 58 equals what number?
2. 48 is what percent of 19?
3. 45 is 14% of what number?
4. 38 out of 200 represents what percent?

5. 125 is 20% of what number?
6. 25 is what percent of 125?
7. 42 is 18% of what number?
8. What is 155% of 24?
9. 150 is 25% of what number?
10. What is $\frac{1}{2}$% of 1000?
11. What is $14\frac{1}{2}$% of 85?
12. 35 is 84% of what number?
13. 55 is what percent of 88?
14. 175 is what percent of 150?
15. 54 is 39% of what number?
16. What is 50% of 50?
17. What is 75% of 75?
18. 42.7 is 100% of what number?
19. $79 is $\frac{1}{4}$% of a certain number. Without calculating, answer the following: is the number more likely the price of a. a stick of gum, b. a mansion on the coast, c. a new car, d. a kayak, e. a pound of steak?
20. Calculate the unknown value in problem 19.

"More Than" and "Less Than" Problems

Frequently, percent problems involve "more than" and "less than." Here are two examples of more than/less than problems:

1. 25 is 14% *more than* what number?
2. What is 18% *less than* 52?

More than/less than problems are usually changed to "of" problems and then worked as percent proportions.

Examples

H. What is 19% more than 83?

What is 19% *more than* 83?

1. Convert this "more than" into an "of" problem.

$$\text{What is } \begin{array}{r} 100\% \\ +\ \ 19\% \\ \hline 119\% \end{array} \text{ of } 83?$$

2. For a "more than" problem, *add* the percent given to 100%, and change *more than* to *of.* "More than" implies that the entire 100% is to be used plus 19% more. In this example, the entire 83 *plus* 19% of 83 is desired.

What is 119% of 83?

3. This is now similar to other percent problems in which the amount is unknown.

$$\frac{119\%}{100\%} = \frac{X}{83}$$

4. Solve by percent proportion.

$$100X = (119)(83)$$

5. In a "more than" problem, the amount is greater than the base.

$$X = \frac{9877}{100}$$

98.77 is 19% more than 83.

I. 225 is 28% less than what number?

225 is 28% *less than* what number?

1. Convert this "less than" problem to an "of" problem.

225 is 100% of what number?

$$\begin{array}{r} -\ 28\% \\ \hline 72\% \end{array}$$

2. For a "less than" problem, *subtract* the percent given from 100% and change the *less than* to *of.*

225 is 72% of what number?

3. Set up and solve by percent proportion.

$$\frac{72\%}{100\%} = \frac{225}{X}$$

4. Cross-multiply.

$$72X = (225)(100)$$

$$X = \frac{22{,}500}{72} = 312.5$$

5. In a "less than" problem, the amount is less than the base.

225 is 28% less than 312.5

J. Here are several examples of "more than" and "less than" problems rewritten as "of" problems.

1. *Original:* What is 12% more than 55?
   Rewritten: What is 112% of 55?
   (Add to 100% for a "more than" problem.)
   Setup: $\frac{112\%}{100\%} = \frac{X}{55}$

2. *Original:* What is 22% less than 39?
   Rewritten: What is 78% of 39?
   (Subtract from 100% for a "less than" problem.)
   Setup: $\frac{78\%}{100\%} = \frac{X}{39}$

3. *Original:* 15 is 32% more than what number?
   Rewritten: 15 is 132% of what number?
   ("More than": add to 100%)
   Setup: $\frac{132\%}{100\%} = \frac{15}{X}$

4. *Original:* 88 is 10% less than what number?
Rewritten: 88 is 90% of what number?
("Less than": subtract from 100%)
Setup: $\frac{90\%}{100\%} = \frac{88}{X}$

K. 139 is what percent more than 121?

$\frac{X\%}{100\%} = \frac{139}{121}$

1. If the percent is the unknown, the adjustment is made *after* the percent is determined.

$121X = 13{,}900$

2. 139 is 114.9% of 121. Since 121 = 100%, 139 is 14.9% *more than* 121.

$X\% = 114.9\%$

139 is 14.9% more than 121.

Exercise 6–2C

Solve the following more than/less than problems.

1. What is 32% more than 130?
2. 128 is what percent less than 155?
3. 392 is 18% less than what number?
4. What is 4% less than 226?
5. 553 is what percent more than 421?
6. 15 is 25% more than what number?
7. What is 47% more than 56?
8. What is 20% less than $89.95?
9. $455.92 is 14% less than what number?
10. What is $21\frac{1}{2}\%$ more than 162?
11. What is $\frac{1}{4}\%$ less than $1000?
12. The cost of a certain tool increased $18, which is a 25% increase. Find the original price and the new price.
13. 94 is what percent less than 118?
14. 94 is what percent of 118?
15. What number is $23\frac{1}{2}\%$ more than 100?
16. What number is $23\frac{1}{2}\%$ less than 100?
17. 100 is $23\frac{1}{2}\%$ less than what number?
18. $241 is 15% less than what number?
19. Increase $241 by 15%. Do you get the same answer as in 18? Why or why not?
20. What is 25% more than 100? Do this problem without a calculator!

6.3 Construction Applications Involving Percent

Applications (word problems) involving percent frequently do not have the words *of, more than,* or *less than* to identify the base. Therefore, other ways of determining the base must be considered. Study the following examples.

Examples

A. A large construction firm has 150 workers. If 20% of the workers have at least 15 years of carpentry experience, how many workers have at least 15 years of experience?

20% is the percent.

1. Percent is determined by the % sign.

150 is the base.

2. If there is a number that represents the *total* (in this case total workers), that is the base.

The unknown is the amount.

3. The amount is the number (unknown in this example) that represents *part* of the total. Part of the workers have 15 or more years of experience.

$$\frac{20\%}{100\%} = \frac{X}{150}$$

4. Solve as a percent proportion.

$X = 30$ workers

5. 30 out of 150 workers is equivalent to 20% of 150 workers.

B. A carpenter is making \$10.78 per hour after an increase of 10%. What was he making per hour before the raise?

$10\% + 100\% = 110\%$
The percent is 110%.

1. A 10% raise represents a "more than" problem. (Any increase can be thought of as "more than.") Therefore, the percent is 110%.

The base is the unknown.

2. Whenever there is a time difference, the *original quantity,* before any increase or decrease, is the base. The carpenter's wage before the raise is the original wage; therefore, it is the base.

\$10.78 is the amount.

3. The present quantity is the amount.

$$\frac{100\%}{100\%} = \frac{\$10.78}{X}$$

4. Solve as a percent proportion.

$X = \$9.80$

5. The carpenter was making \$9.80 before the raise.

C. A builder charges \$3850 to build a deck. If the lumber for the deck cost \$1225, what percent of the price was for lumber?

\$3850 is the base.

1. The base represents the *total* price.

\$1225 is the amount.

2. The amount represents a specific *part* of the total.

The unknown is the percent.

3. We are asked to find the percent.

$$\frac{X\%}{100\%} = \frac{\$1225}{\$3850}$$

4. Set up and solve as a percent proportion.

$$X = 31.81\% \doteq 31.8\% \quad \text{or} \quad 32\%$$

5. Approximately 32% of the total price charged for the deck is to cover the lumber.

D. A home that cost \$85,000 in 1992 now costs \$122,000 to build. What is the percent increase in cost?

\$85,000 is the base.

1. If two quantities are being compared, the quantity that occurred first is the base. This is true in the majority of situations. (Exception: If the problem had been stated, *What percent of today's price would you have paid in 1992?*, then the current price would be the base.) The quantity being used as the basis of comparison is the base.

$$\frac{X\%}{100\%} = \frac{\$122{,}000}{\$85{,}000}$$

2. Set up percent proportion with amount over base.

amount

$$\frac{144\%}{100\%} = \frac{122{,}000}{85{,}000}$$

base

3. 144% is approximately the percent today's house costs compared to the cost in 1992.

$$144\% - 100\% = 44\%$$

4. This is the increase in cost since 1992.

E. A builder receives a 15% discount on large quantities of lumber and materials at a building supply warehouse. If the builder's purchases price out at \$28,570.00 (retail price), how much must he pay?

$$100\% - 15\% = 85\%$$

1. This is a discount (less than) problem, so the percent the builder pays is 85% of the retail price.

amount

$$\frac{85\%}{100\%} = \frac{X}{\$28{,}570.00}$$

base

2. The retail price of \$28,570 is the base, and the amount is the unknown.

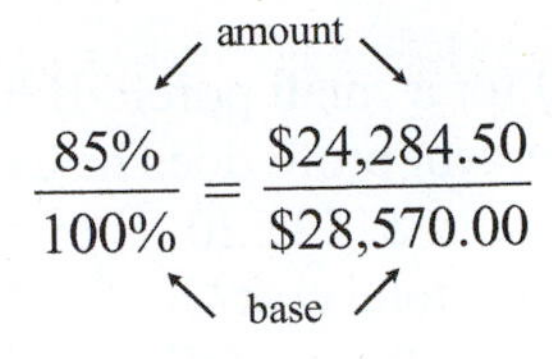

3. The amount must always represent the same quantity as the percent. *The builder paid 85%; the builder paid $24,284.50.* These statements say the same thing in two different ways. The base is always 100%.

SUMMARY

Guidelines for identifying parts of an application problem involving percents.

1. The percent is identified by the % sign.
2. The base is identified in a variety of ways:
   a. the total quantity
   b. the earlier quantity if one quantity came before the other one time wise
   c. the quantity after the word *of*
   d. in *more than/less than* problems, the quantity after the *more than/less than*
   e. the quantity being used as the basis of comparison
   f. the original quantity before it is divided up, increased, or decreased
   g. the original price before discounts are taken or taxes are added
3. The amount is (usually) the remaining quantity after the percent and base have been identified. The amount must always represent the same quantity as the percent.

Exercise 6–3

Set up and solve the following word problems.

1. On a bathroom remodeling job, 53% of the expenses incurred by a contractor are for lumber and materials. If the expenses total $8250 for the job, what is the cost of the lumber and materials?
2. $2\frac{1}{4}$-in. red oak flooring boards are to be used in a living room. 1350 lineal feet are required to cover the floor. How many lineal feet should be ordered if a 15% allowance is made for waste? (*Hint:* The order should include 15% more than the amount needed to cover the floor.)
3. A circular saw is on sale for $147.96. If this is 20% less than the regular price, what is the regular price of the saw?
4. The price of $\frac{3}{4}$-in. CDX plywood increased from $33.95 to $35.98 per sheet at a certain lumberyard. What percent increase is this? Round the percent to one decimal place.
5. An architect charges 8% of the estimated cost of the house for her services. If the architect's fees are $20,000, what is the estimated cost of the house?

6. A contractor charges $12,520 for a small porch. If he estimates that his total expenses will be $9850, what percent profit does he expect to earn on the job?
7. Labor costs on a house amount to $142,520. If this represents 43% of the total cost to the contractor, what is his total cost?
8. A contractor gets a $5\frac{1}{2}$% discount from retail price at a building supplies company. If the retail price of his purchase totals $8520.50, what does he pay for the materials? Round to the nearest cent.
9. A keg of nails costs $155. If there is a $4\frac{1}{2}$% sales tax, what must the purchaser pay for the nails? Round to the nearest cent.
10. A developer made a profit of $32,500 on the sale of a new house. If the house sold for $331,450, what percent of the sale price was profit?
11. A 7.25-ft^2 door is cut from a 4 ft $\times$ 8 ft sheet of cabinet-grade plywood. Ignoring waste, what percent of the plywood is used in the door?
12. 65% of all belt sanders manufactured by a certain company are still working 8 years later. What fraction of these sanders is still working after 8 years?
13. It is estimated that a 25% savings on energy costs can be had with an energy-efficient retrofit for a certain building. The retrofit is estimated to cost $28,500, and the annual energy costs are currently $37,580. Determine the number of years it will take to recover the initial retrofit expense.
14. A carpenter's income for July is $4840. If he deposits $6\frac{1}{2}$% of that in a savings account, how much does he save during July?
15. A builder receives a $7\frac{1}{2}$% discount on plumbing and electrical supplies. If she purchases supplies that would retail for $5846, how much money does the discount save her?
16. Allowing 15% for waste, how many board feet of lumber must be ordered for a project requiring 450 board feet?
17. A building lot is exactly 1 acre (43,560 ft^2). A one-story house has 1750 ft^2, and a detached garage has 672 ft^2. What percent of the lot is taken up by buildings?
18. A bank requires a 12% down payment on a new house costing $243,550. How much is the down payment?
19. A cottage costs $82,530 to construct. If the foundation costs $12,225, what percent of the total cost was the foundation?
20. A certain lot of lumber is 65% *clear* (that is, free of defects and knots). If there are 85,000 lineal feet of lumber in the lot, how much clear lumber is in the lot?
21. A carpenter purchases a power tool that regularly sells for $239.95. The tool is on sale for 15% off, and there is a sales tax of $4\frac{1}{2}$%. How much does the carpenter pay for the tool?
22. A framer grabs a pizza for lunch at Joe's Pizzeria. Joe advertises that his pizzas cost 10% less than his competitor across the street. If the pizzas across the street cost $9.50, what's the price of Joe's pizzas?
23. A builder estimates that 1850 board feet of lumber are required for a heat-collecting sun room. Determine the board feet to be ordered if 10% is added for waste.
24. In a labor-intensive project, skilled labor was 82% of the expense. If the *other* expenses came to $4525.19, what was the cost of skilled labor? Round to the nearest cent. (*Hint:* first figure the percent of *other* expense, and calculate from there.)
25. The width of a door is 45% of its height. If the width is 36 inches, find the height of the door. Give answer in feet and inches.
26. A two-story house has 3815 square feet. If 64% of the space is on the first floor, how many square feet are on each floor?

27. Monthly utilities in an energy-efficient building cost 38% less than those in a traditionally built building. If energy costs were $185.50 less in the energy-efficient building, what were the monthly costs in the traditionally built building?
28. In problem 27 above, what were the monthly utility costs in the energy-efficient building?
29. The energy-efficient "green" construction option costs 12.5% more than construction costs for a traditionally built home. If the "green" option costs an additional $27,580, what is the cost of the "non-green," traditionally built house?
30. In problem 29 above, what is the cost of the "green" house?

CHAPTER 7

Angles and Triangles

> **Did You Know**
>
> The payback period for installing a faucet aerator or low-flow showerhead is two weeks. For a 1.6 gallon flush toilet the payback time is $2\frac{1}{2}$ years, and for an Energy Star washing machine it is 4 years (for the average family of four).

OBJECTIVES

Upon completing this chapter, the student will be able to:

1. Identify different types of angles.
2. Solve problems involving different types of angles.
3. Identify the different types of triangles.
4. Identify congruent triangles.
5. Identify similar triangles.
6. Solve problems involving congruent triangles.
7. Solve problems involving similar triangles.
8. Solve right triangles using the Pythagorean theorem.
9. Solve problems involving right triangles.
10. Solve 45°–45°–90° triangles (isosceles right triangles).
11. Solve 30°–60°–90° triangles.
12. Solve 3–4–5 and 5–12–13 right triangles.
13. Solve problems involving special right triangles.
14. Determine the perimeter of triangles.
15. Solve for the area of right triangles.
16. Solve for the area of oblique triangles.
17. Solve problems involving perimeter and area of triangles.

7.1 Angle Measure

Angles are formed when two lines meet or intersect. The point at which they meet is called the *vertex* of an angle. The angle shown in Figure 7–1 can be written either as $\angle ABC$, $\angle CBA$, or simply as $\angle B$. If there is no possibility of confusing it with other angles, an angle is usually indicated by just the letter at its vertex. (B is at the vertex in this example.)

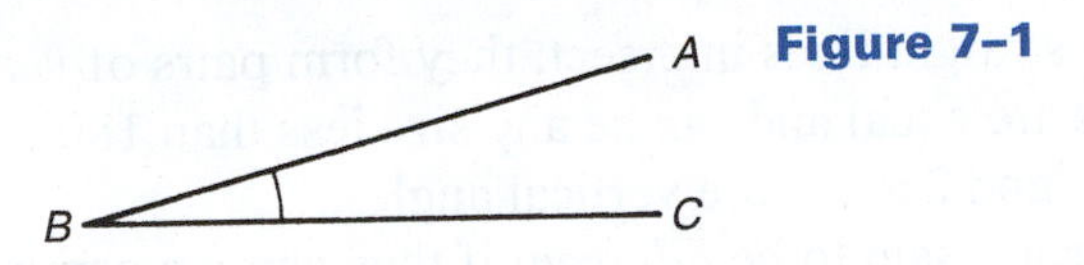

Figure 7–1

The designation $\angle J$ is not sufficient in Figure 7–2. The three-letter label $\angle MJN$ or $\angle NJM$ must be given to distinguish this angle from $\angle NJK$, $\angle KJL$, and so on.

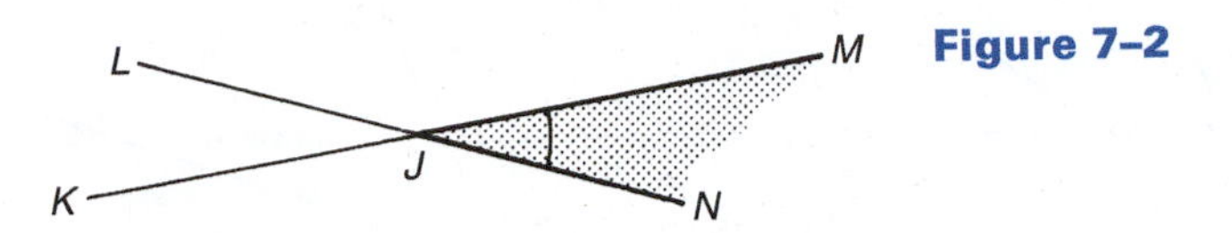

Figure 7–2

The size of an angle is measured in degrees. A degree is a unit of rotation. There are 360° in one full rotation. A small raised circle, °, is the symbol for degree. A full circular rotation is 360° (see Figure 7–3a).

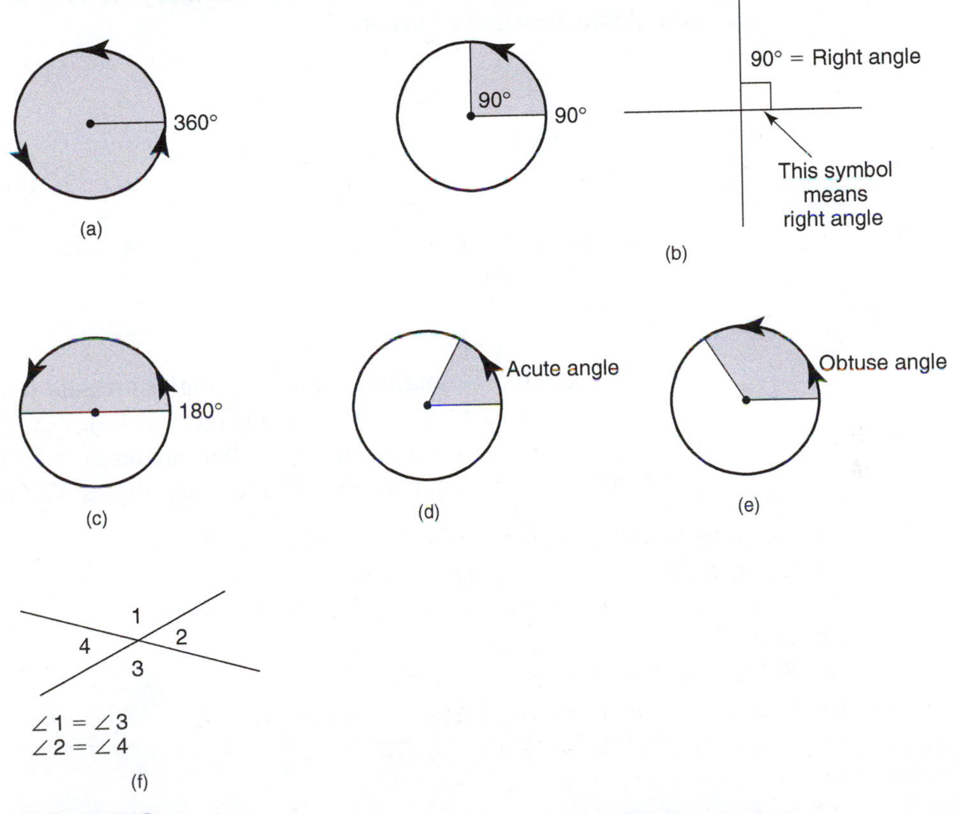

Figure 7–3

A quarter circular rotation is 90° (Figure 7–3b). The angle formed is called a *right angle.* Whenever two lines meet in such a way that they form a right angle, the lines are said to be *perpendicular.*

A half circular rotation is 180° ($\frac{1}{2}$ of 360°). A 180° angle, called a *straight angle,* does not look like an angle at all, but like a straight line. This is illustrated in Figure 7–3c.

Any angle less than 90° is called an *acute angle* (Figure 7–3d). Angles of 10°, 45°, 50°, and 85° are examples of acute angles.

Any angle greater than 90° but less than 180° is called an *obtuse angle* (Figure 7–3e). Angles of 120°, 150°, 93°, and 175° are examples of obtuse angles.

When two straight lines intersect, they form pairs of *vertical angles* (Figure 7–3f). Vertical angles are equal and can be any size less than 180°. In the diagram, 1 and 3 are vertical angles, and 2 and 4 are vertical angles.

Two angles are said to be *adjacent* if they share a common side between them and their vertices are located at the same point (Figure 7–4).

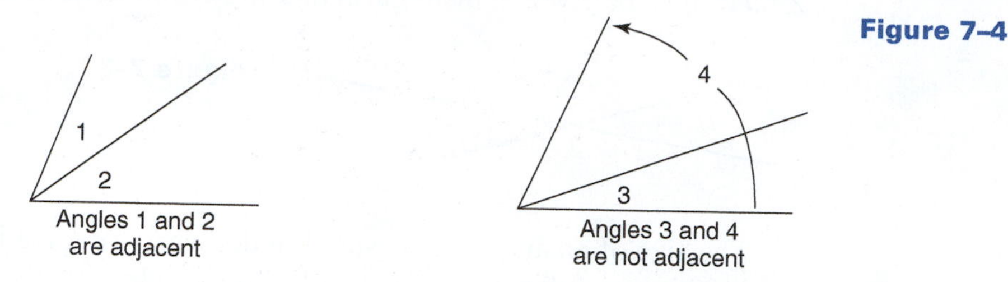

Figure 7–4

If the sum of two angles is 90°, the angles are said to be *complementary.* ∠*B* and ∠*C* shown in Figure 7–5a are complementary; ∠*XYZ* and ∠*ZYW* (Figure 7–5b) are complementary and adjacent.

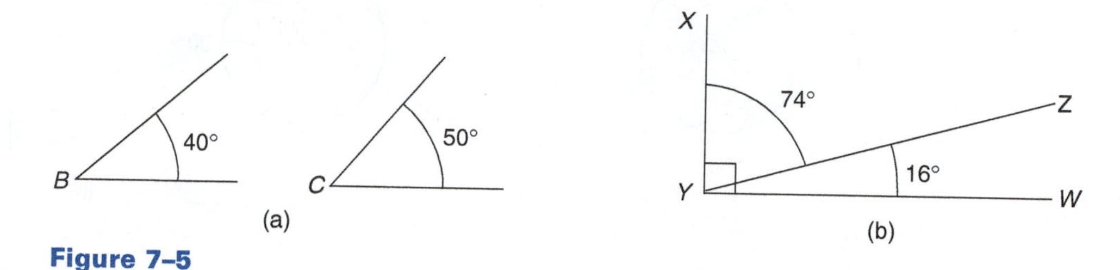

Figure 7–5

If the sum of two angles is 180°, the angles are said to be *supplementary* (see Figure 7–6). ∠*K* is supplementary to ∠*J* (Figure 7–6a). ∠*XYZ* and ∠*ZYW* (Figure 7–6b) are supplementary and adjacent. Note that one letter is sufficient to identify angles *K* and *J,* but three letters must be used to identify angles *XYZ* and *ZYW.* Why?

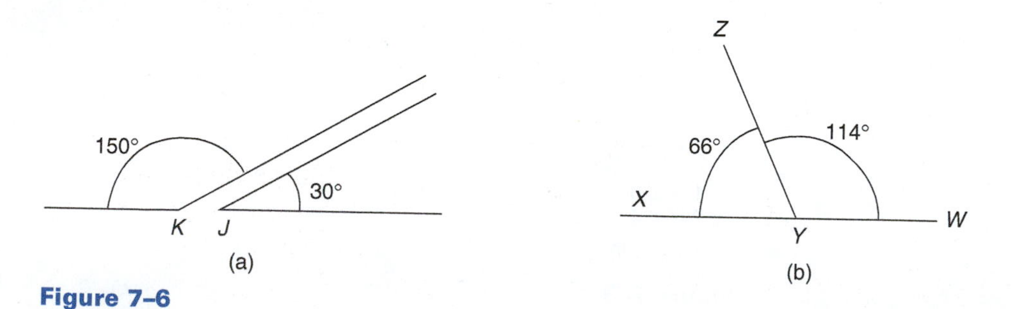

Figure 7–6

Examples

A. Two angles are complementary. If one measures 82°, what size is the other angle? (See Figure 7–7)
   Complementary angles add to 90°. Therefore, 90° − 82° = 8°. The other angle is 8°.

Figure 7–7

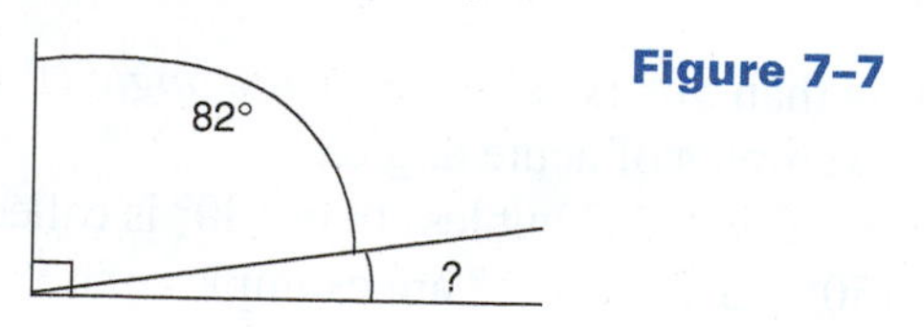

B. Two angles are supplementary. The smaller angle is acute. Classify the other angle. (See Figure 7–8.)
The other angle must be obtuse and supplementary. If one angle is less than 90° (acute), the other angle must be greater than 90° (obtuse) in order for them to add to 180° (which makes them supplementary angles).

Figure 7–8

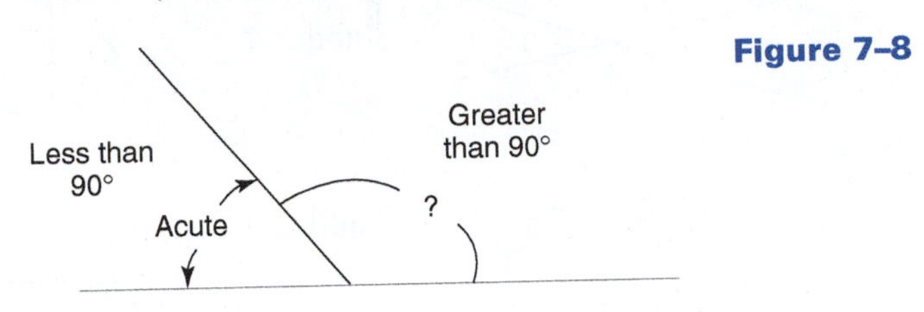

The size of an angle is not determined by the length of its sides. The two angles shown in Figure 7–9a are both 30°. Because the degree of rotation is the same for both of these angles, $\angle B = \angle E$. The sides of $\angle B$ could be extended without changing its size (Figure 7–9b).

Figure 7–9

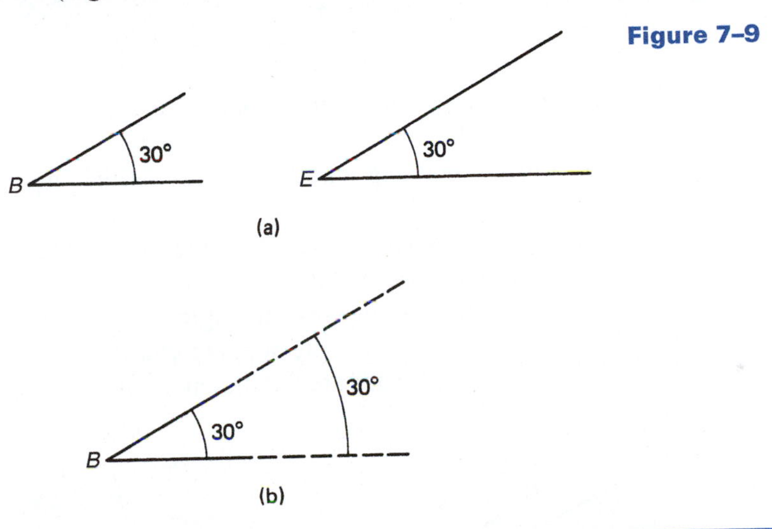

Exercise 7–1

Identify the angles shown below. Be sure to include all correct classifications. See problems a and b as examples.

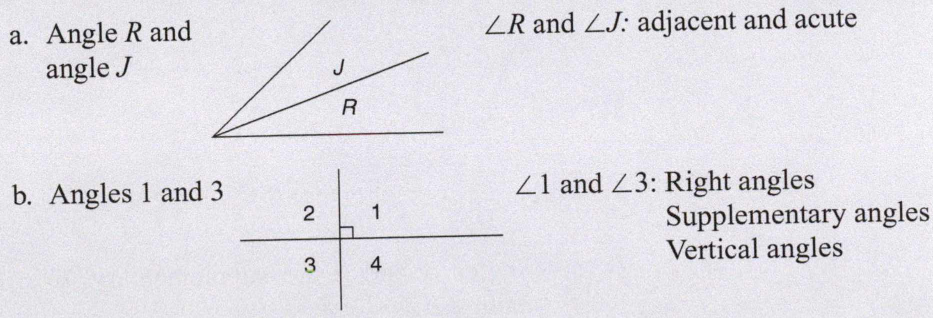

a. Angle R and angle J — $\angle R$ and $\angle J$: adjacent and acute

b. Angles 1 and 3 — $\angle 1$ and $\angle 3$: Right angles
Supplementary angles
Vertical angles

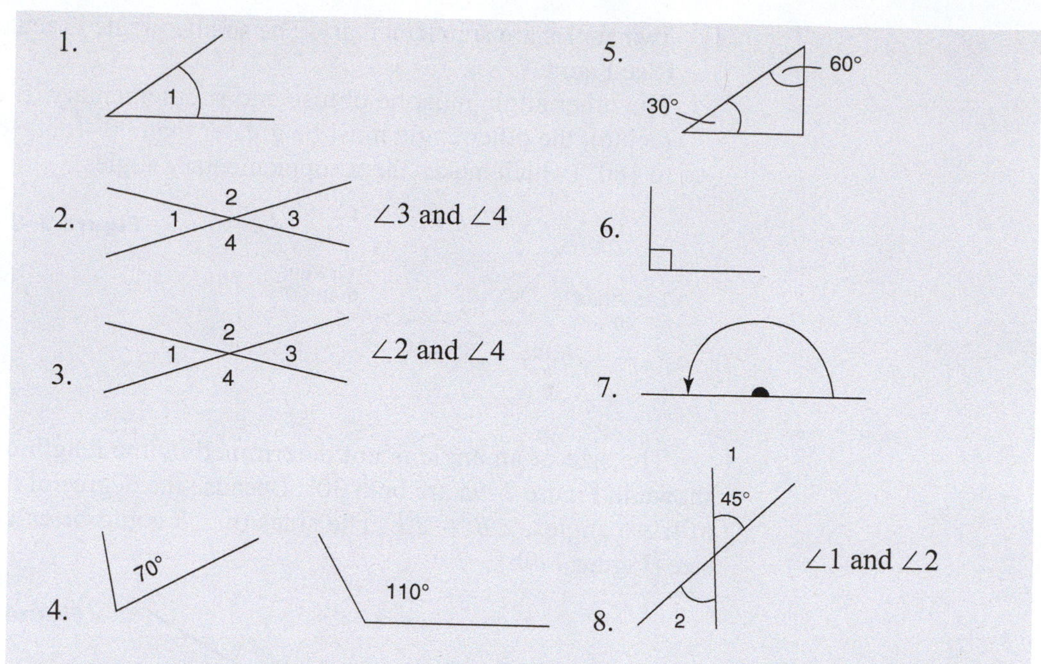

9. Two angles are adjacent and supplementary. One of the angles is 115°. What is the other angle?
10. Two angles are complementary. If one of the angles is 37°, what is the other angle?
11. Two angles are vertical and complementary. What is the size of each angle?
12. Three angles have a total rotation of 142°. If two of the angles are complementary, what is the size of the third angle?
13. Vertical angles 1. and 3. are supplementary. How many degrees are in angles 1, 2, 3, and 4.? (Angles are not drawn as an accurate representation.)

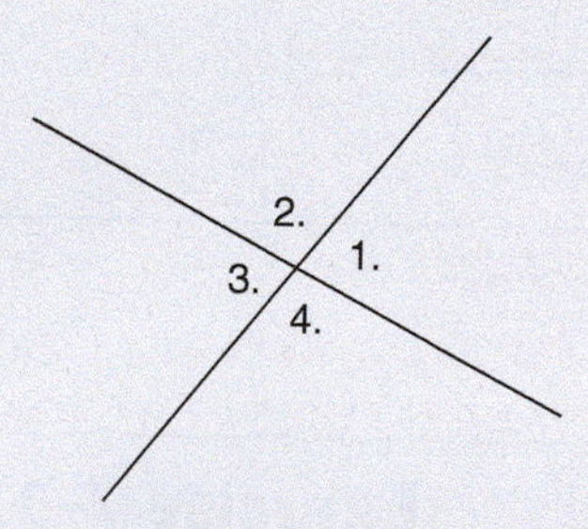

14. Angle *ACB* is 115°. If angles 1. and 3. are complementary, how many Degrees are in angle 2?

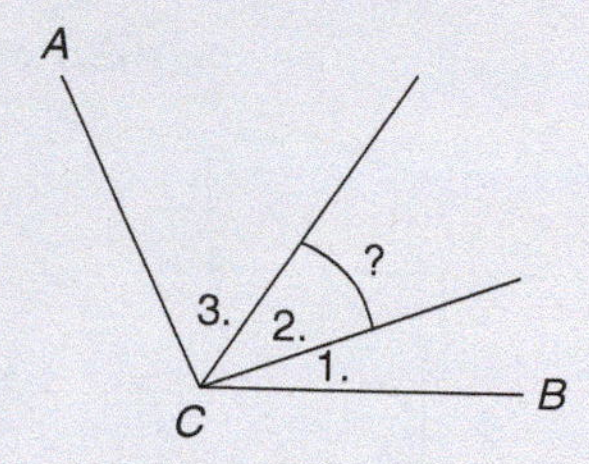

15. Angles *A* and *B* are supplementary. If ∠*A* is obtuse, what do we know about ∠*B*?

7.2 Congruent and Similar Triangles

Triangles are an important tool for the carpenter. They are the most rigid of all shapes and are therefore frequently used in the corners of buildings. (Think of the corner braces often used to keep the walls of a house plumb.) Most roofs are triangular-shaped because triangles are strong and rigid (and shed water and snow).

The name *triangle* means "three angles." A triangle is a three-sided, three-angled closed figure (Figure 7–10). Where two sides of a triangle meet, the *vertex* of the angle is formed. *The sum of the angles of every triangle equals 180°.*

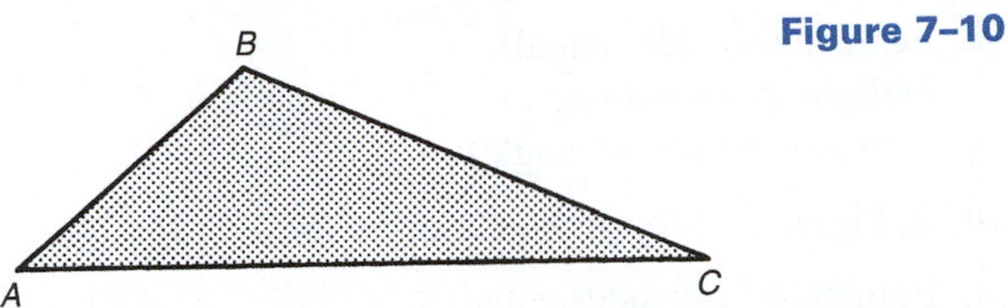

Figure 7–10

Triangles are generally specified by capital letters at their vertices. The triangle shown in Figure 7–11 would be named $\triangle ABC$. The symbol $\triangle$ means triangle. It would be equally correct to name the triangle $\triangle BCA$, $\triangle BAC$, $\triangle CAB$, and so on. The sides of a triangle are identified with the lowercase letter corresponding to the capital letter associated with the opposite angle. Note that $\angle A$ in Figure 7–11 is formed by sides b and c. Therefore, side a is the *opposite* side.

Triangles can be classified by angles and by sides.

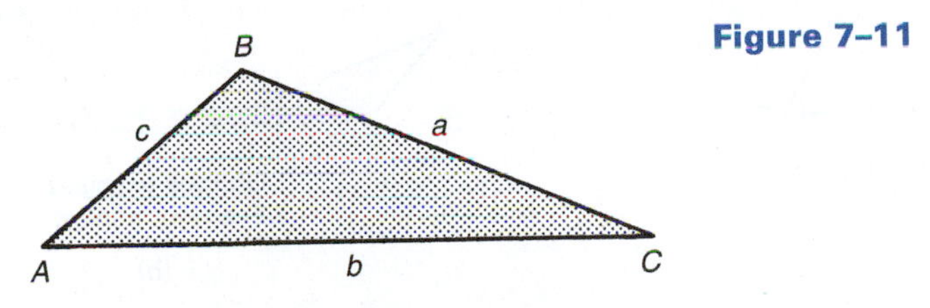

Figure 7–11

Classification by Sides

1. *Scalene triangle.* All three sides are unequal (Figure 7–12a).
2. *Isosceles triangle.* Two sides are equal (Figure 7–12b). The unequal side is called the *base.* The two angles opposite the equal sides are also equal. They are called *base angles.*
3. *Equilateral triangle.* All three sides are equal (Figure 7–12c). All three angles are also equal. Since the sum of the three angles is 180°, each angle in an equilateral triangle is 60°.

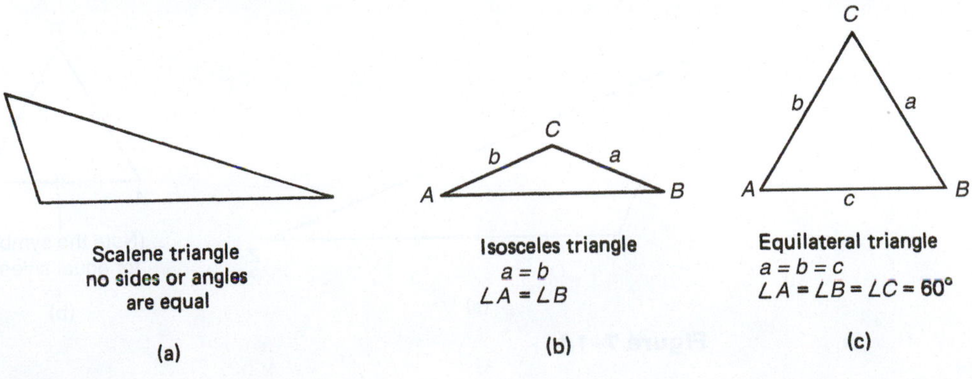

Figure 7–12

Classification by Angles

1. *Right triangle.* Any triangle containing a right (90°) angle (Figure 7–13a).
2. *Oblique triangle.* Any triangle that does not contain a right angle (Figure 7–13b).
3. *Obtuse triangle.* Any triangle that contains an obtuse angle (Figure 7–13c).
4. *Acute triangle.* Any triangle with three acute angles (Figure 7–13d). (*Every* triangle has at least two acute angles.)

A triangle may have more than one classification. Referring to Figure 7–14a. $\triangle XYZ$ can be classified as:

1. Scalene (no sides equal)
2. Oblique (no right angle)
3. Obtuse (one obtuse angle)

$\triangle ABC$ in Figure 7–14b can be classified as:

1. Equilateral (all sides equal)
2. Oblique (no right angle)
3. Acute (all angles acute)

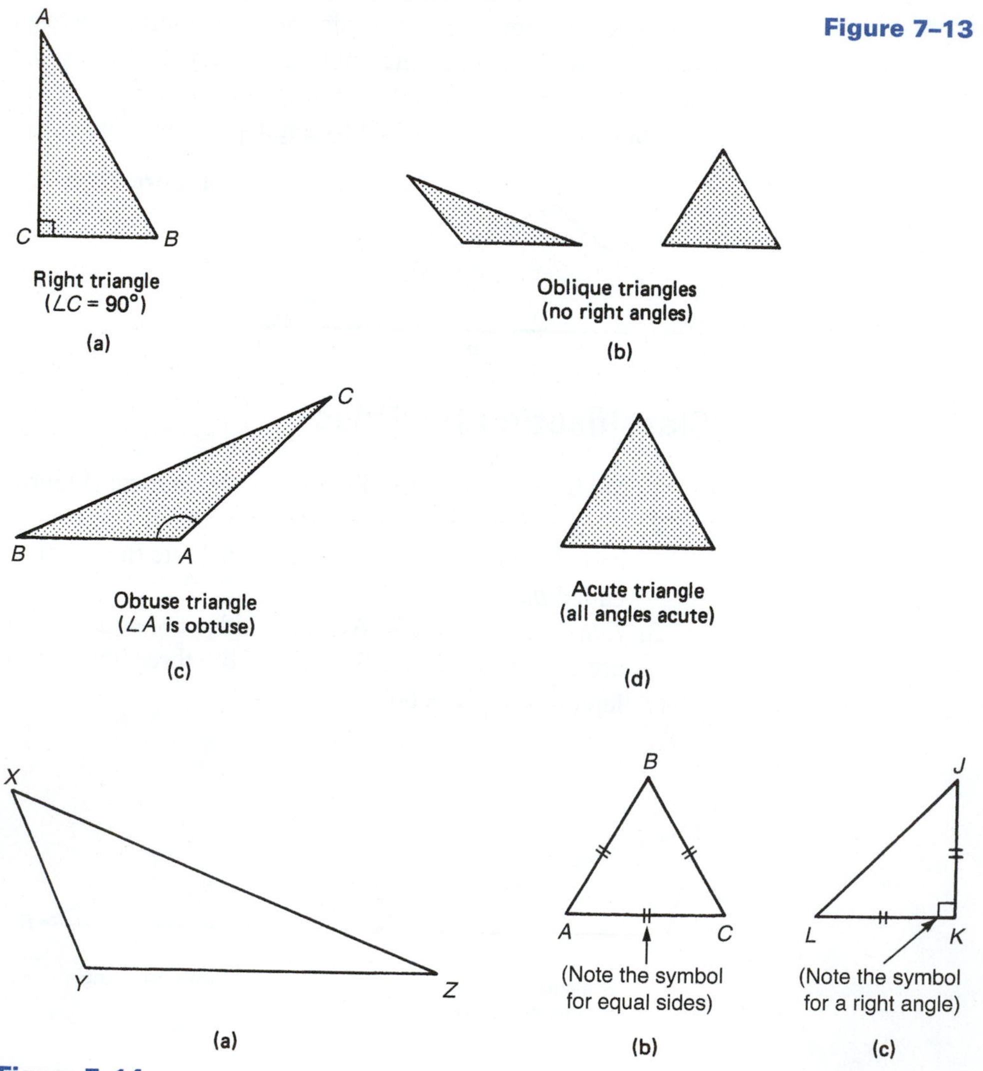

Figure 7–13

Figure 7–14

$\triangle JKL$ in Figure 7–14c can be classified as:

1. Isosceles (two sides equal)
2. Right ($\angle K$ is a right angle)

If two triangles are exactly identical in shape and size, they are said to be *congruent.* $\angle ABC$ in Figure 7–15 could fit exactly on top of $\triangle DEF$ Therefore, $\triangle ABC \cong \triangle DEF$. The symbol $\cong$ means "is congruent to."

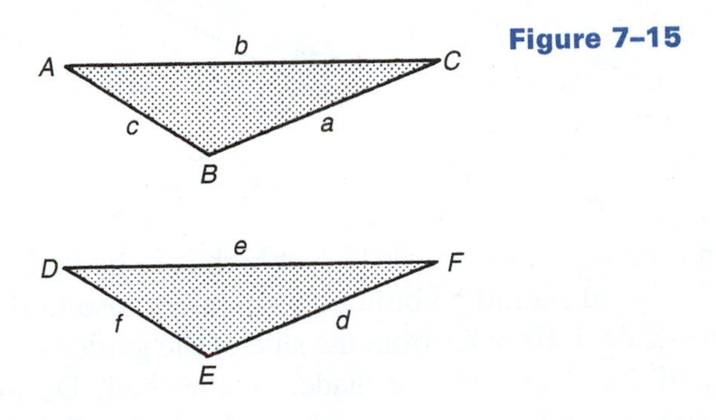

Figure 7–15

In Figure 7–15

$$\angle A = \angle D \qquad a = d$$
$$\angle B = \angle E \qquad b = e$$
$$\angle C = \angle F \qquad c = f$$

In congruent triangles, all corresponding sides and corresponding angles are equal.

If two triangles have the same shape but not the same size, they are said to be *similar* (see Figure 7–16). In order to have the same shape, the angles in one triangle must equal the corresponding angles in the other triangle. All equilateral triangles are similar. In Figure 7–16, $\triangle ABC \sim \triangle XYZ$. The symbol for similar triangles is $\sim$. *In similar triangles, the ratios of the corresponding sides are equal.* This property makes it possible to solve for missing sides of similar triangles.

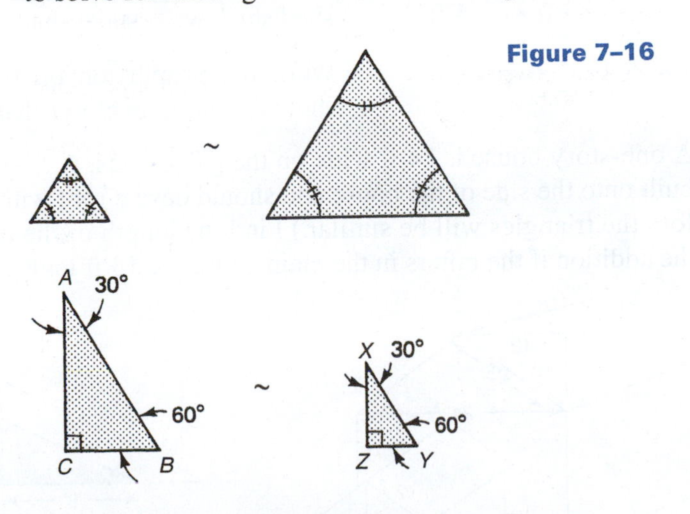

Figure 7–16

Examples

A. Given that $\triangle ABC \sim \triangle XYZ$, what is the length of side y?

$$\frac{4}{10} = \frac{y}{16}$$

$$y = 6.4$$

Since the ratios of the corresponding sides are equal, y can be solved as a proportion.

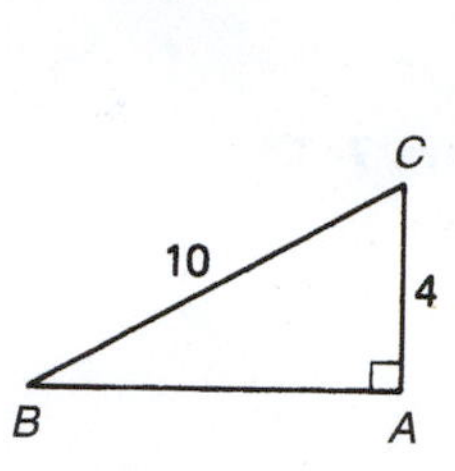

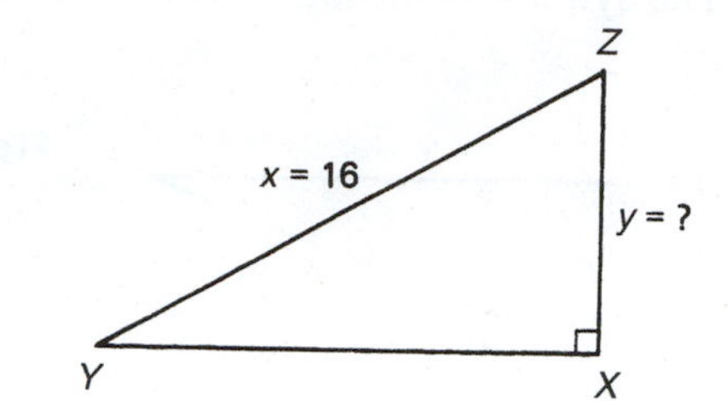

B. A garden shed with a roof peak 15 feet high is to be built on the sunny side of the garden. The shed should be built conveniently close to the garden, but not close enough to shade it. How far from the side of the garden should the shed be built so that none of the garden will be shaded by the shed? During the summer growing season, the longest shadow cast by a stake 4 ft high is 7.2 ft.

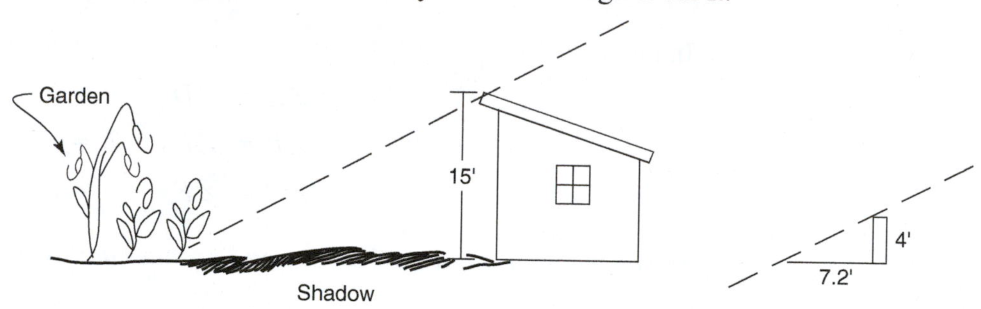

$$\frac{4 \text{ ft}}{7.2 \text{ ft}} = \frac{15 \text{ ft}}{X \text{ ft}}$$

1. Since the triangles are similar, the ratio of the height of the shed to its shadow is the same as the ratio of the height of the stake to its shadow.

then divide by 4 ft
multiply

$$\frac{4 \text{ ft}}{7.2 \text{ ft}} = \frac{15 \text{ ft}}{X \text{ ft}}$$

$$\frac{\text{shed}}{\text{shed shadow}} = \frac{\text{stake}}{\text{stake shadow}}$$

$7.2 \times 15 \div 4 = 27$

2. Work the proportion on the calculator. The shed should be built no closer than 27 ft to the garden.

C. A one-story house is 24 ft wide on the gable ends. A 16-ft wide addition is being built onto the side of the house and should have a roof with the same pitch. (Therefore the triangles will be similar.) Find the length of the roof rafters to be used in the addition if the rafters in the main house are 13 ft long.

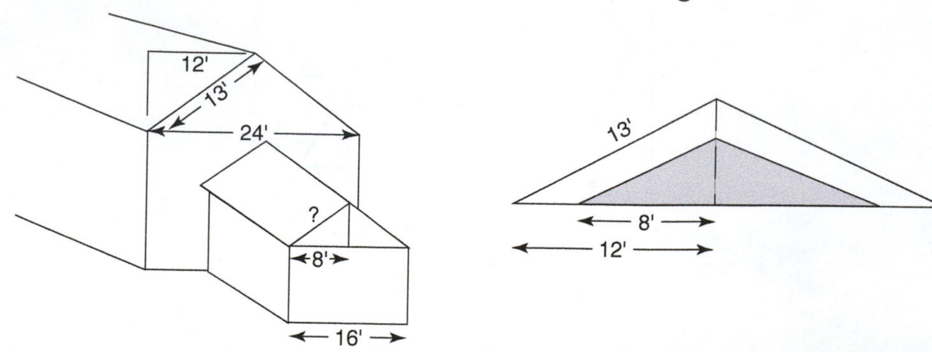

$$\frac{12 \text{ ft (house's run)}}{13 \text{ ft (house's rafters)}} = \frac{8 \text{ ft (addition run)}}{X \text{ ft (addition rafters)}}$$

1. Set up proportion between the house and the addition.

rafters on addition $= 8.\overline{6}$ ft $\rightarrow$ 8 ft 8 in.

2. Solve the proportion and convert decimal feet to feet and inches.

D. A ramp is to be built with vertical braces that form similar triangles of three different sizes (see diagram). If the braces are 5 ft apart, find the height of the two inside braces.

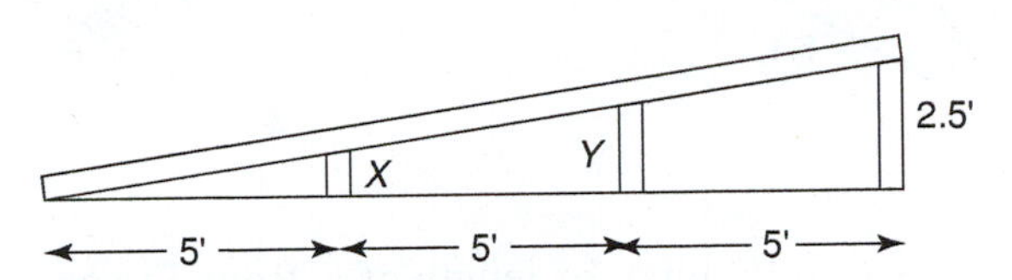

Find length X and length Y.

$$\frac{2.5 \text{ ft}}{15 \text{ ft}} = \frac{X \text{ ft}}{5 \text{ ft}} \rightarrow 0.8\overline{3} \text{ ft}$$

1. Set up the proportion for the similar triangles and solve for X.

$0.8\overline{3}$ ft $\rightarrow$ 10 in.

2. The first brace is 10 inches.

$$\frac{2.5 \text{ ft}}{15 \text{ ft}} = \frac{Y \text{ ft}}{10 \text{ ft}} \rightarrow 1.\overline{6} \text{ ft}$$

3. Set up the second proportion and solve for Y.

$1.\overline{6}$ ft $\rightarrow$ 1 ft 8 in.

4. The second brace is 1 ft 8 in. high or 20 inches. Note that the second brace is twice as far from the vertex as the first brace, and the second brace is also twice as high as the first brace. This proportion will always hold true for similar triangles.

Exercise 7–2

1. $\triangle ABC \cong \triangle XYZ$. Find x, y, and z and $\angle X$, $\angle Y$, and $\angle Z$.

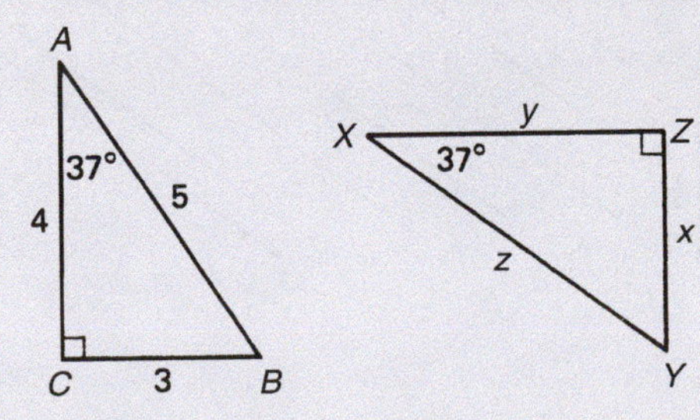

2. $\triangle ABC \sim \triangle JKL$. Find the lengths of sides j and l. Round to one decimal place.

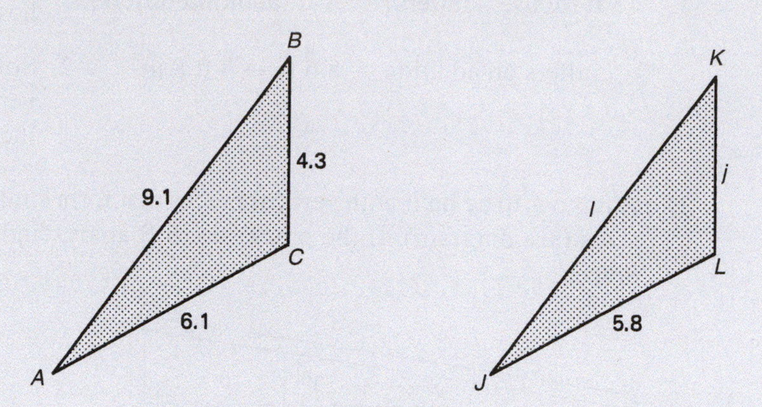

3. $\triangle ABC \sim \triangle ADE$. Find the length of x. Round to one decimal place.

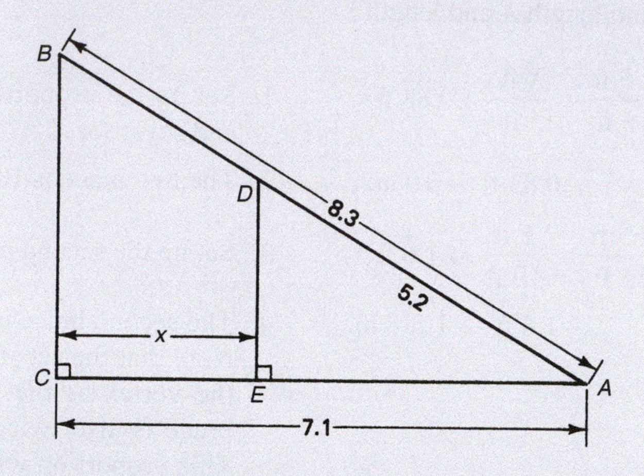

4. What is the height of the tree? Round to the nearest foot.

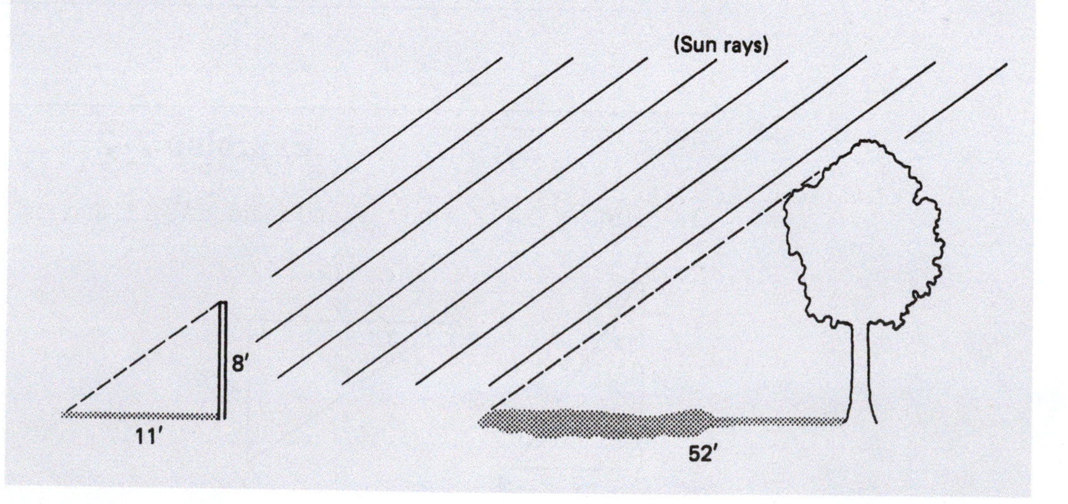

5. Find the height of the lightning rod from the ground.

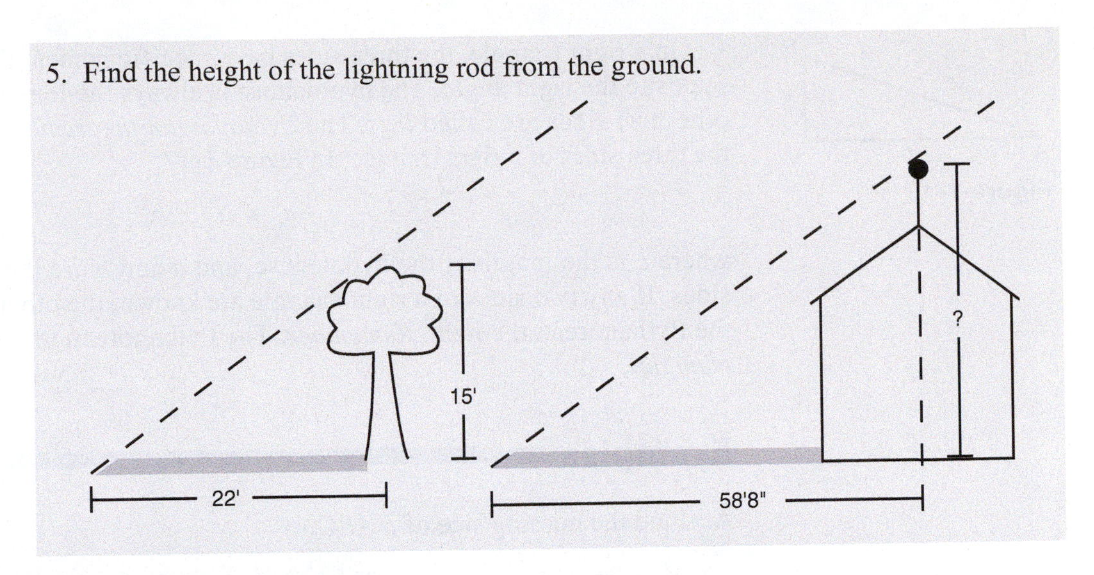

7.3 The Pythagorean Theorem

Before introducing the Pythagorean theorem, a brief discussion of squares and square roots of numbers is important. 5^2 is read "5 squared" and means 5 times 5. Similarly, 4^2 means 4×4 and 18.72^2 means 18.72×18.72. Any number squared is multiplied by itself. $\sqrt{16}$ is read "the square root of 16" and indicates the number that when multiplied by itself equals 16. $\sqrt{16} = 4$ since 4×4 or $4^2 = 16$. $\sqrt{25} = 5$ since 5×5 or $5^2 = 25$. $\sqrt{2} \doteq 1.414$ because $1.414^2 \doteq 2$. $\sqrt{3} \doteq 1.73$ because $1.73^2 \doteq 3$. Although the student is expected to use a calculator to find square roots, it is useful to be able to recognize the first 10 common squares and square roots.

$1 = \sqrt{1}$	$6 = \sqrt{36}$	$1^2 = 1$	$6^2 = 36$
$2 = \sqrt{4}$	$7 = \sqrt{49}$	$2^2 = 4$	$7^2 = 49$
$3 = \sqrt{9}$	$8 = \sqrt{64}$	$3^2 = 9$	$8^2 = 64$
$4 = \sqrt{16}$	$9 = \sqrt{81}$	$4^2 = 16$	$9^2 = 81$
$5 = \sqrt{25}$	$10 = \sqrt{100}$	$5^2 = 25$	$10^2 = 100$

Exercise 7–3A

Using the square and square root keys on a calculator, determine the following answers, rounding to two decimal places where necessary.

1. 8.24^2
2. 15^2
3. 47.81^2
4. 0.564^2
5. 9.3^2
6. 16.25^2
7. 42^2
8. 10^2
9. 5.6^2
10. 1^2
11. $\sqrt{25}$
12. $\sqrt{16}$
13. $\sqrt{1}$
14. $\sqrt{4}$
15. $\sqrt{81}$
16. $\sqrt{9.36}$
17. $\sqrt{17.5}$
18. $\sqrt{10}$
19. $\sqrt{600}$
20. $\sqrt{851}$

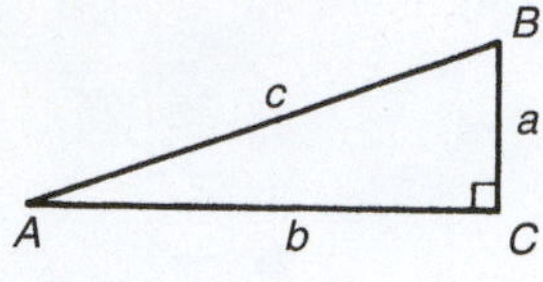

Figure 7–17

In a right triangle, the three sides have specific names. The *hypotenuse* is the side opposite the right angle. The hypotenuse is always the longest of the three sides. The other two sides are called *legs.* The *Pythagorean theorem* states a relationship among the three sides of a right triangle. In Figure 7–17,

$$c^2 = a^2 + b^2$$

where c is the length of the hypotenuse, and a and b are the lengths of the other two sides. If any two sides of a right triangle are known, the other side can be found using the Pythagorean theorem. *Remember:* The Pythagorean theorem applies only to *right triangles.*

Examples

A. Find the missing side of $\triangle ABC$.

$c^2 = a^2 + b^2$ — 1. The hypotenuse c is the unknown.

$= 4^2 + 3^2$ — 2. Substitute the known values.

$= 16 + 9$ — 3. a and b must be squared before adding them together.

$= 25$ — 4. Take the square root of 25 to find the value of c.

$c = \sqrt{25}$

$c = 5$ — 5. Note that c is the longest side. The hypotenuse is *always* the longest side of a right triangle.

B. Find the length of side a in $\triangle ABC$.

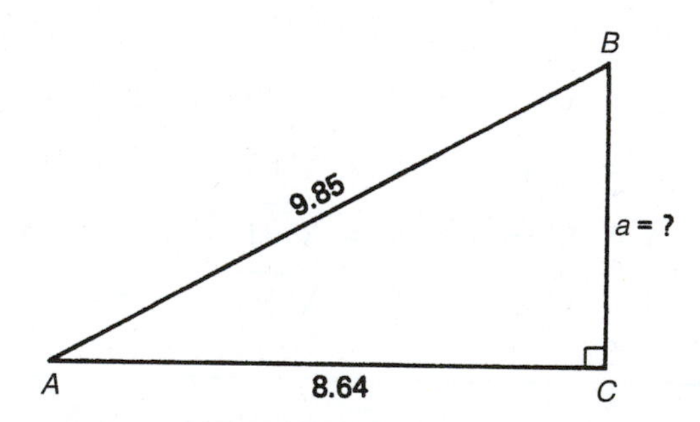

$c^2 = a^2 + b^2$ — 1. In this example the hypotenuse is given.

$c^2 - b^2 = a^2$ — 2. Whenever the hypotenuse is given, the square of the known side is *subtracted* from the square of the hypotenuse.

$9.85^2 - 8.64^2 = a^2$

$97.02 - 74.65 = a^2$ — 3. Solve for a. Remember to square c and b before subtracting.

$22.37 = a^2$ — 4. Take the square root to find a.

$a = \sqrt{22.37}$

$= 4.73$ — 5. Note that both legs are smaller than the hypotenuse.

C. Floor joists are approximately 16″ apart. Find the length of bridging attached to the top and bottom of adjacent 2″ × 12″ joists, as shown.

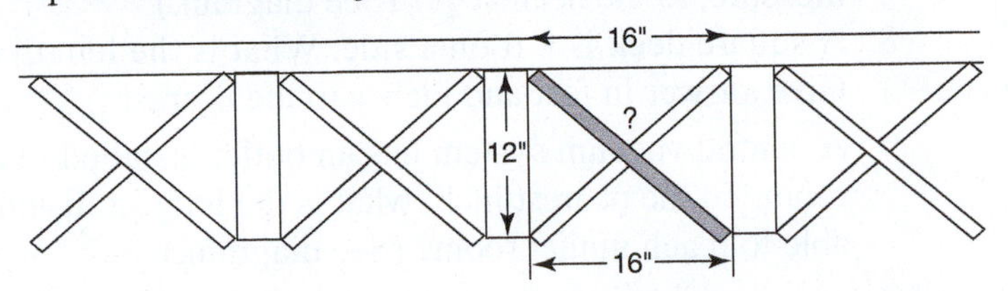

$c^2 = a^2 + b^2$	1. The length of bridging is the hypotenuse.
$c^2 = 12^2 + 16^2$	2. Substitute given values.
$c^2 = 400$	
$c = 20''$	3. Take the square root to find the approximate length of the bridging.

D. A rectangular foundation has sides that measure 36 ft long and 28 ft wide. To check that the foundation is a true rectangle, what would the diagonals measure?

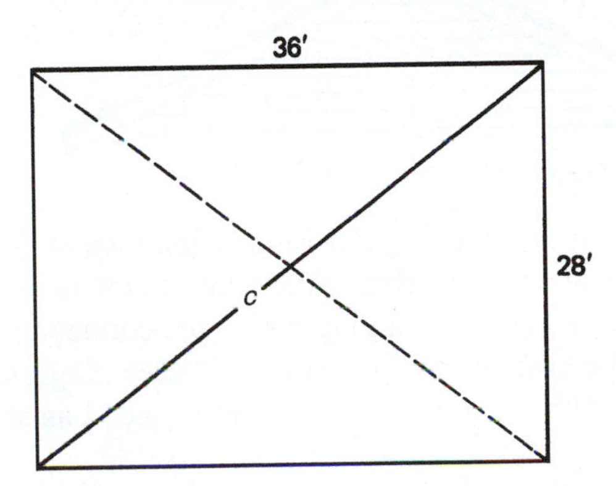

1. The diagonal, labeled c, is the hypotenuse of the two identical right triangles.

$c^2 = 36^2 + 28^2$	2. Find c^2 and take the square root.
$c = 45.607$ ft	
$45.607 \text{ ft} = 45 \text{ ft } 7\frac{1}{4} \text{ in.}$	3. Convert to feet and inches to the nearest $\frac{1}{4}''$.
	4. The diagonals should be equal and should measure 45 ft $7\frac{1}{4}$ in. if the foundation is truly rectangular.

Exercise 7–3B

Complete the table for right triangle $\triangle ABC$. $\angle C$ is the right angle.

	Side a	Side b	Side c
1.		16.14	18.25
2.	2.23	1.65	
3.	8.92	5.36	
4.	62.17	14.59	
5.	8.35	6.23	
6.	19.35		27.33

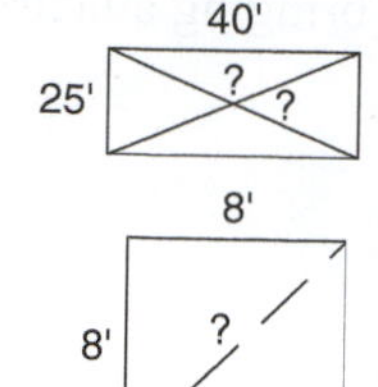

7. A rectangular foundation measures 25 ft × 40 ft. What should the diagonals measure, to the nearest $\frac{1}{4}$″? (See diagram.)
8. A square deck is 8 ft on a side. What is the length of the diagonal of the deck? Give answer in feet and inches to the nearest $\frac{1}{16}$″ (See diagram.)
9. A central vacuum system has an outlet installed in one corner of a 12 ft × 18 ft room. To the nearest inch, what is the longest distance the vacuum hose must be able to reach in that room? (See diagram.)

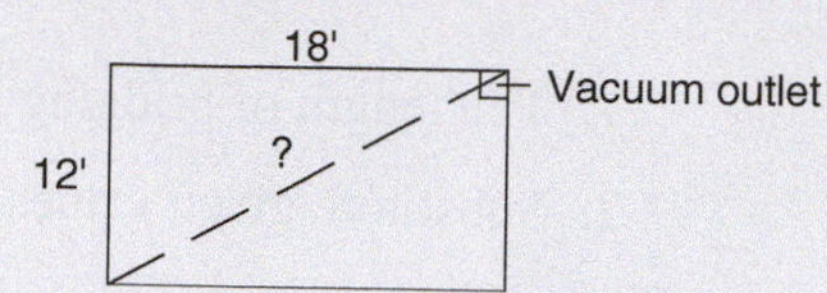

10. A roof has a rise of 6 ft 5 in. and a run of 12 ft 8 in. To the nearest $\frac{1}{4}$″, find the length of the roof rafters, as shown. Disregard any overhang.

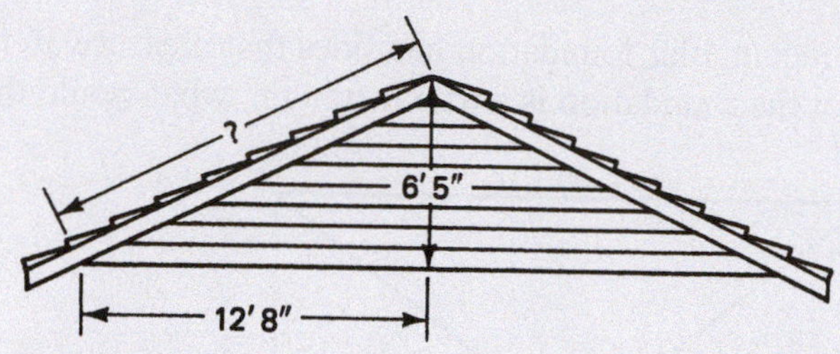

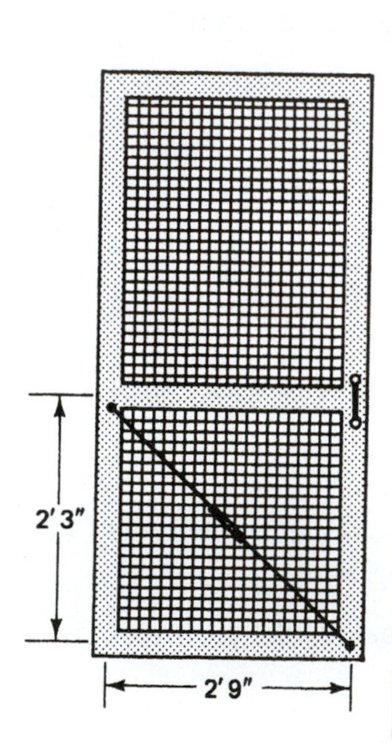

11. The diagonal of a rectangular foundation measures 51 ft $1\frac{1}{4}$ in. If the length of the foundation is 44 ft, what is the width to the nearest $\frac{1}{4}$″?
12. A screen door is to have a diagonal brace connected as shown. What should the length of the brace be to the nearest $\frac{1}{8}$″? (See diagram.)
13. A small shed has floor joists sized and spaced as shown. What is the length of the bridging?

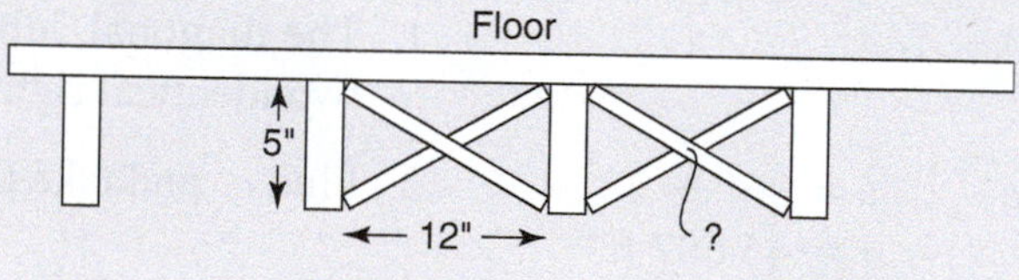

14. A roof has a rise of 4 feet and a run of 7 ft 6 in. The roof rafters have an 18 in. overhang. Determine the length of the rafters. (*Hint:* Calculate the rafter length without the overhang, and then add on the overhang.)

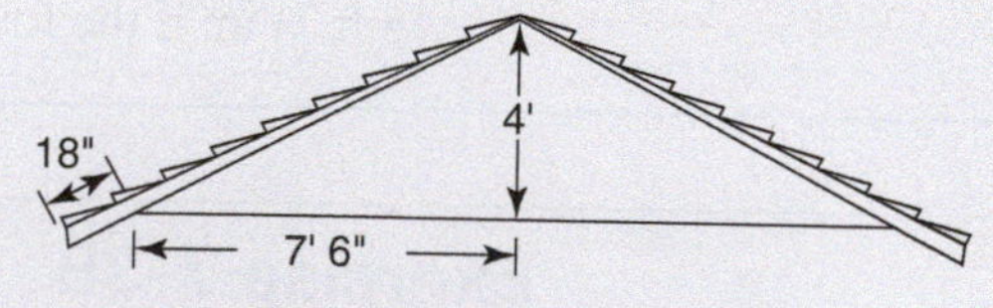

15. A garden gate has dimensions of 3 ft 3 in. wide by 3 ft 9 in. high. Find the minimum length board that can be used as a diagonal brace as shown to the nearest $\frac{1}{8}$ inch.

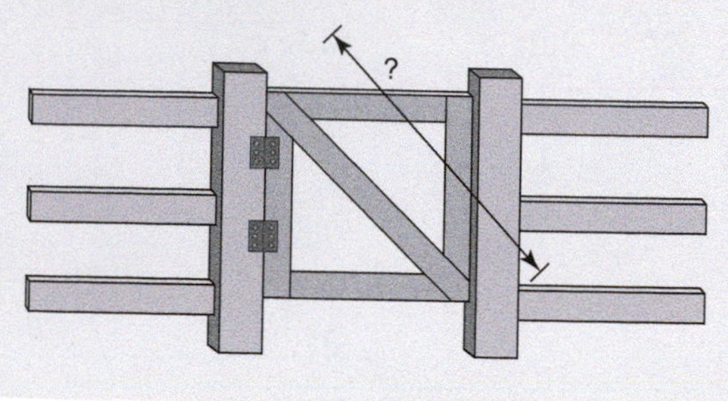

16. A shed roof has rafters that are 19 ft long. That includes a one-foot overhang on each end. If the shed is 15 feet wide, how much taller is the high end than the low end of the roof?

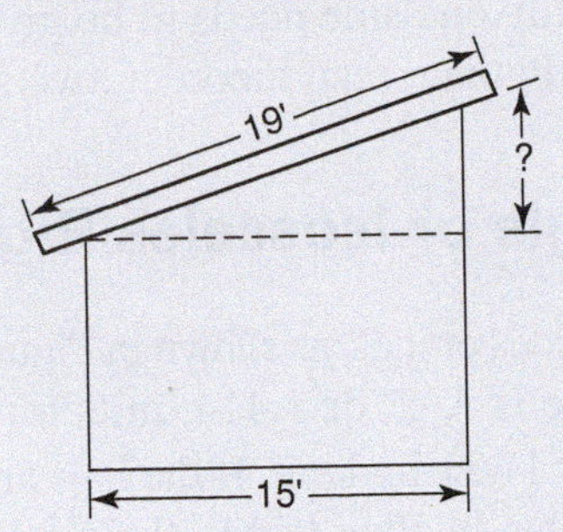

17. Find the length of AD (the hypotenuse of the smaller triangle). This problem requires an understanding of both the Pythagorean theorem and similar triangles.

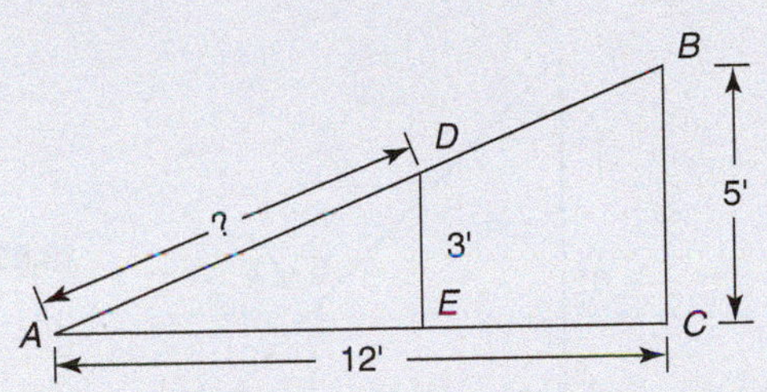

18. A garden building has a floor that is 10 ft square (i.e., 10 ft on each side). The walls are 8 ft high. Each side of the roof is also 10 ft. square (no overhang). Find the height of the roof to the nearest $\frac{1}{2}''$.

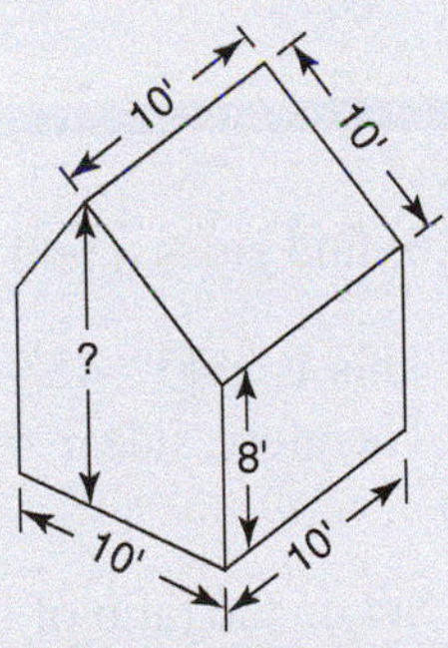

19. Which is longer: the diagonal of a square 20 feet on a side, or the diagonal of a 15 ft × 25 ft rectangle? Both of these have the same perimeter (distance around).

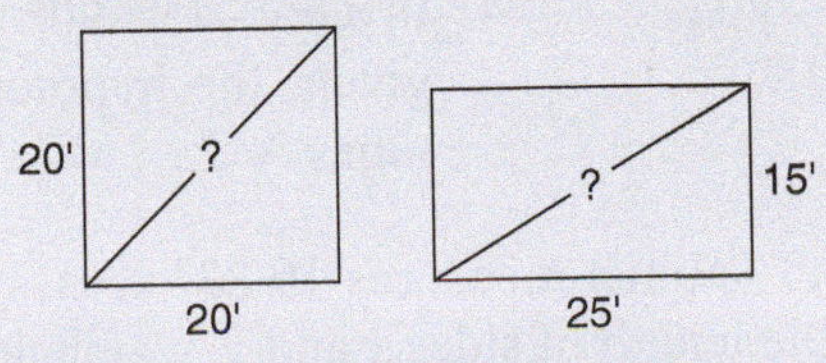

20. What is the diagonal length of a 4 ft × 8 ft sheet of plywood? Give answer to the nearest $\frac{1}{16}$ inch.

7.4 Special Right Triangles

There are two right triangles that are classified as "special right triangles." If the ratios of their sides are known, only one side needs to be given in order to find the remaining two sides. (To solve by the Pythagorean theorem, two sides must be known.)

45°–45°–90° Triangle or Isosceles Right Triangle

This triangle has two equal sides or legs, as shown in Figure 7–18. If the sides (legs) are each 1 unit long, the hypotenuse is $\sqrt{2}$, or 1.414 units long. If the legs are 8 in. long, the hypotenuse is $8 \cdot \sqrt{2}$ in., or 11.31 in. long. If the legs are 19.65 ft long, the hypotenuse is $19.65 \cdot \sqrt{2}$ ft, or 27.79 ft long. In other words, the sides are always in the ratio $1 : 1 : \sqrt{2}$, where $\sqrt{2}$ represents the hypotenuse. Multiply the length of the shorter (equal) sides by $\sqrt{2}$ to find the hypotenuse.

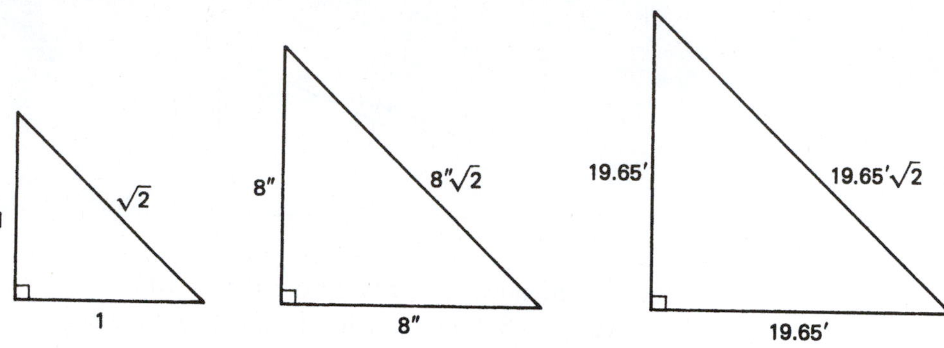

Figure 7–18

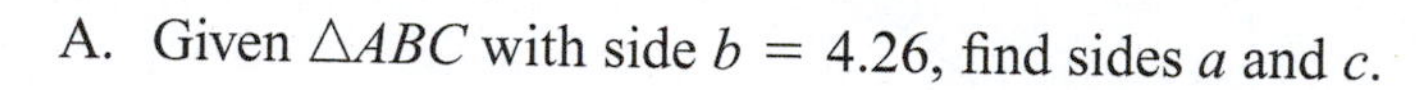

Examples

A. Given $\triangle ABC$ with side $b = 4.26$, find sides a and c.

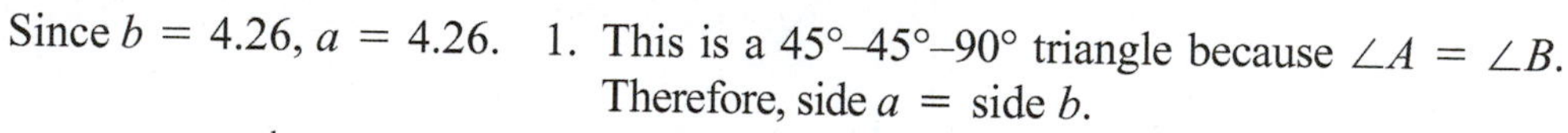

Since $b = 4.26$, $a = 4.26$.

1. This is a 45°–45°–90° triangle because $\angle A = \angle B$. Therefore, side a = side b.

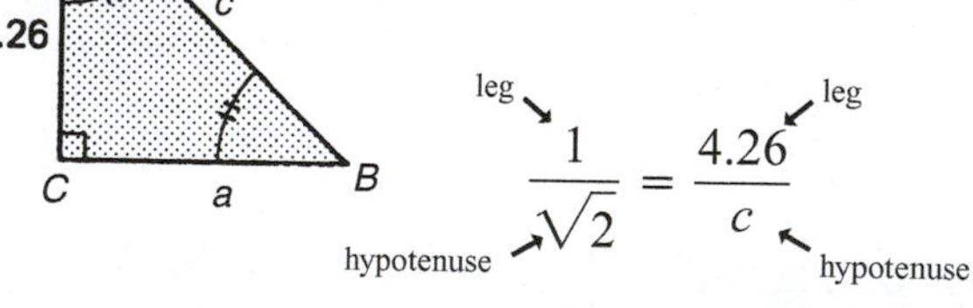

$$\frac{1}{\sqrt{2}} = \frac{4.26}{c}$$

2. Set up the ratio of leg/hypotenuse = leg/hypotenuse.

$$c = 4.26\sqrt{2} = 6.02$$

3. Cross-multiply and solve for c.

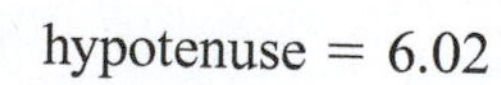

hypotenuse = 6.02

4. In a 45°–45°–90° triangle, if the length of the leg is given, the hypotenuse will always be that length times $\sqrt{2}$.

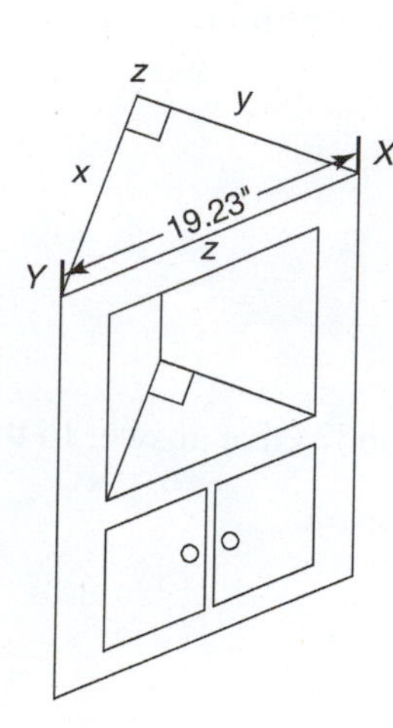

B. A built-in corner cabinet measures 19.23″ across the front (the hypotenuse of the triangle). Find the length of sides x and y, which are equal.

$z = 19.23$ in.

1. This time the hypotenuse is the given value in the 45°–45°–90° triangle.

$$\frac{1}{\sqrt{2}} = \frac{x}{19.23 \text{ in.}}$$

2. Set up the proportion and cross-multiply.

$\sqrt{2} \cdot x = 19.23$ in.	3. Solve for x by dividing 19.23 by $\sqrt{2}$.
$x = 19.23 \div \sqrt{2} = 13.60$ in.	4. Whenever the hypotenuse is known, divide by $\sqrt{2}$ to find the length of the legs in a 45°–45°–90° triangle.
$x = 13.60$ in., $y = 13.60$ in.	5. The legs x and y are equal.

30°–60°–90° Triangle

This is another very useful right triangle. The ratio of its sides are $1 : \sqrt{3} : 2$ (Figure 7–19). The smallest side, opposite the 30° angle, is 1. The side opposite the 60° angle is $\sqrt{3}$ or 1.73. The longest side, the hypotenuse, is 2. Any 30°–60°–90° triangle will have its sides in this same ratio. For instance, if the shortest side of a 30°–60°–90° triangle is 8 inches, the longer leg will be $8 \cdot \sqrt{3}$, or 13.86 inches, and the hypotenuse will be $8 \cdot 2$, or 16 inches.

Figure 7–19

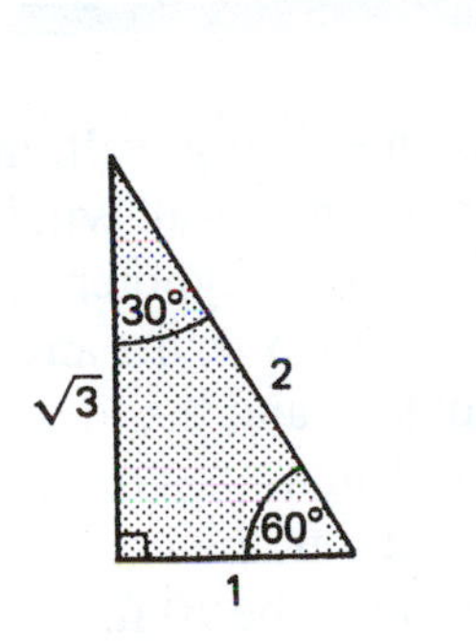

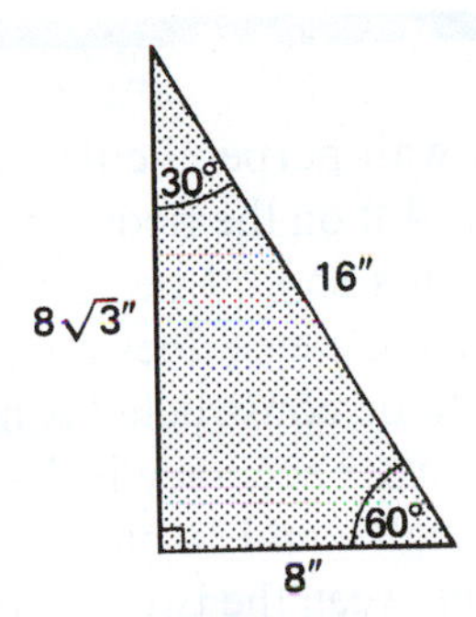

Examples

C. A shed roof has a run of 18.35 ft and a slope of 30°. The other angles are 60° and 90°, as shown. Find the height of the shed roof (side b) and the length of the roof rafters (side c). 30°–60°–90° triangles have sides of ratio $1 : \sqrt{3} : 2$.

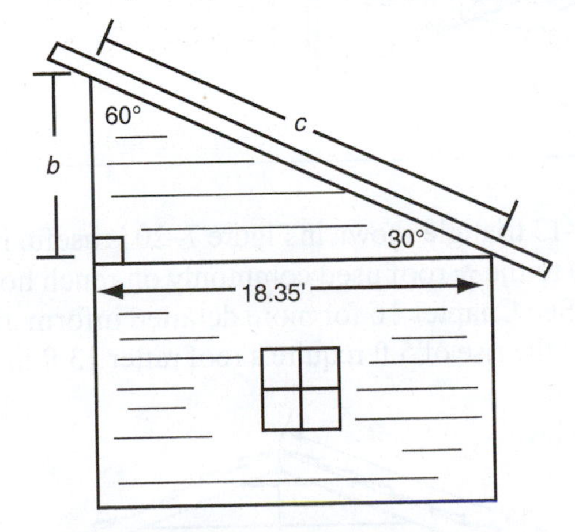

$\dfrac{1}{\sqrt{3}} = \dfrac{b}{18.35}$	1. The ratio of $b:a$ is $1: \sqrt{3}$.
$\sqrt{3} \cdot b = 18.35$ $b = 10.59$ ft	2. Cross-multiply and divide by $\sqrt{3}$ to solve for b.

$$\frac{1}{2} = \frac{10.59}{c}$$

3. When the smallest side is found, use it to find the remaining side. The hypotenuse is always twice the smallest side.

$$c = 21.19 \text{ ft}$$

4. Cross-multiply to solve for c.

The ratios of the sides of all 30°–60°–90° triangles are equal (because they are similar triangles).

3–4–5 and 5–12–13 Triangles

Two other right triangles used commonly in building construction are the 3–4–5 triangle and the 5–12–13 triangle. The 3–4–5 triangle has legs of 3 and 4 units and the hypotenuse is 5 units. This triangle is used commonly to check for right angles.

Example

D. To lay out a wall perpendicular to an existing wall, measure 3 ft on the existing wall, measure 4 ft on the floor where the new wall will be, and the distance between these two points should be 5 ft apart (the hypotenuse of the right triangle thus formed). If the distance does not measure 5 ft, the angle between the existing wall and the wall to be constructed is not 90°, and the location of the new wall must be adjusted. If greater accuracy is desired, multiples of the 3–4–5 triangle can be used. The existing wall could be measured at 6 ft, the location of the new wall at 8 ft, and the distance between the two points should be 10 ft.

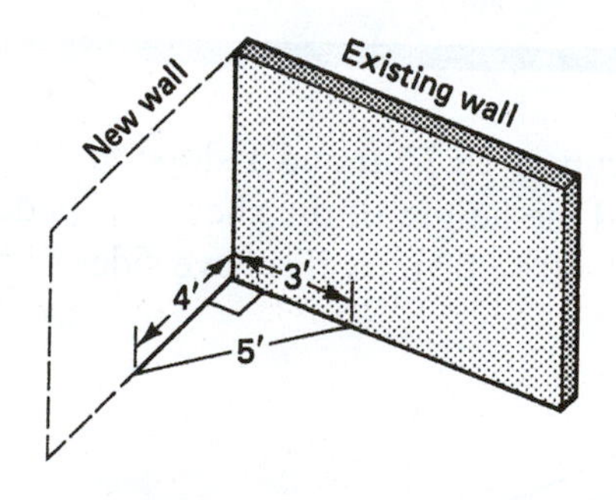

The 5–12–13 triangle shown in Figure 7–20 is useful in building construction because it corresponds to the $\frac{5}{12}$ roof used commonly on ranch houses. ($\frac{5}{12}$ refers to the unit rise, not the pitch. See Chapter 16 for more detailed information on unit rise and pitch.) The run of 12 ft and the rise of 5 ft require a roof rafter 13 ft long (disregarding overhang).

Figure 7–20

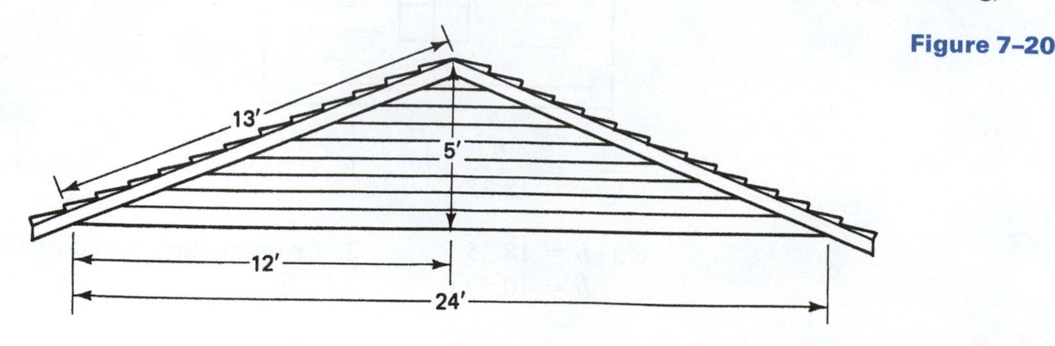

Exercise 7–4

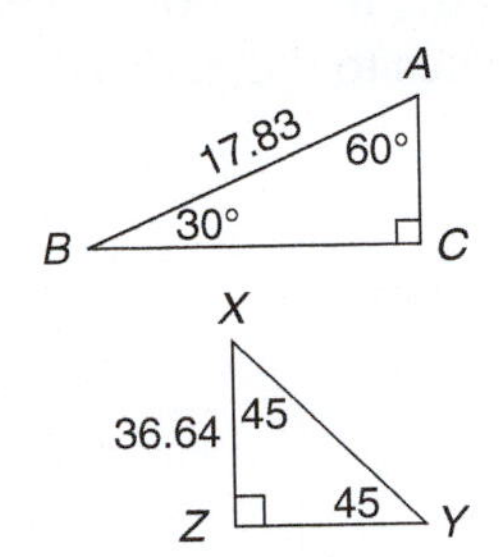

1. $\triangle ABC$: $A = 60°$ $a =$
   $B = 30°$ $b =$
   $C = 90°$ $c = 17.83$
2. $\triangle XYZ$: $X = 45°$ $x =$
   $Y = 45°$ $y = 36.64$
   $Z = 90°$ $z =$
3. $\triangle JKL$: $J = 30°$ $j =$
   $K = 60°$ $k = 46.23$
   $L = 90°$ $l =$
4. $\triangle RST$: $R = 30°$ $r = 41.75$
   $S = 60°$ $s =$
   $T = 90°$ $t =$
5. $\triangle XYZ$: $X = 45°$ $x =$
   $Y = 45°$ $y =$
   $Z = 90°$ $z = 17.85$

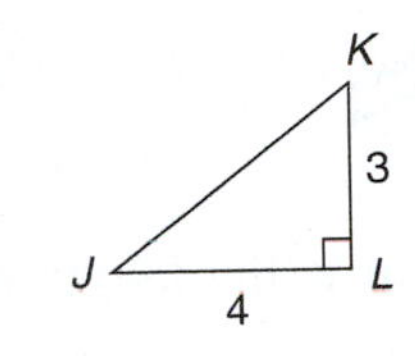

6. Right triangle JKL with $\angle L = 90°$: $j = 3$
   $k = 4$
   $l =$
7. Right triangle ABC with $\angle C = 90°$: $a = 9$ ft
   $b = 12$ ft
   $c =$

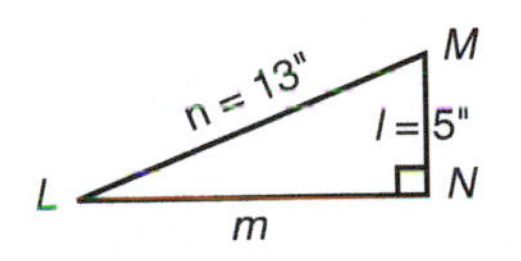

8. Right triangle LMN with $\angle N = 90°$: $l = 5$ in.
   $m =$
   $n = 13$ in.
9. Right triangle NOP with $\angle P = 90°$: $n = 10$ ft $o = 24$ ft
   $p =$
10. Right triangle XYZ with $\angle Z = 90°$: $x = 6$ ft $y = 8$ ft
    $z =$
11. A shed roof has a slope of 30°. If the run is 12 ft, what is the length of the rafters? Ignore overhang. Give the answer to the nearest $\frac{1}{8}''$.

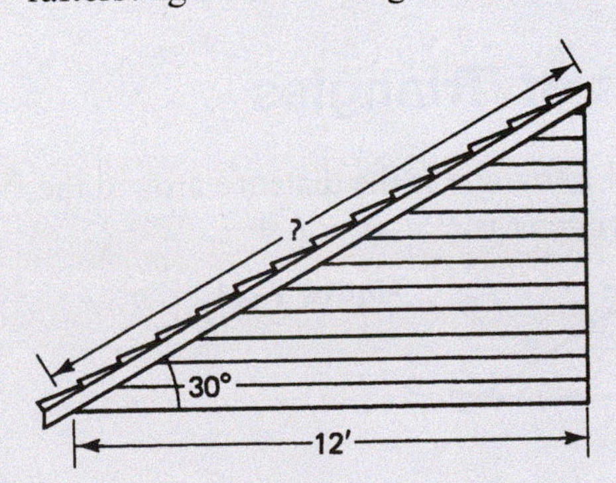

12. A square room measures 15 ft 3 in. by 15 ft 3 in. What is the length in feet and inches of the diagonal? (*Hint:* Change feet and inches to decimal feet or to inches before finding the diagonal, then change back to feet and inches.) Give the answer in feet and inches to the nearest $\frac{1}{16}''$.

13. Stairs sloped at an angle of 30° must reach 10 ft 8 in. to the next floor. What is the total run of the stairs? Give the answer in feet and inches to the nearest $\frac{1}{16}$″.

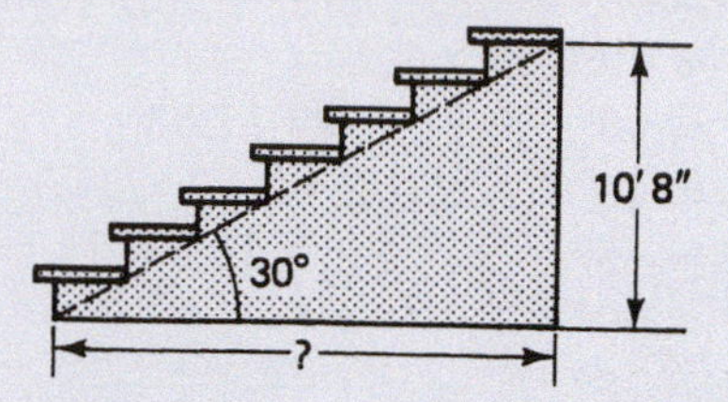

14. The roof shown has a 30° slope and a height of 5 ft $9\frac{1}{4}$ in. Determine the run.

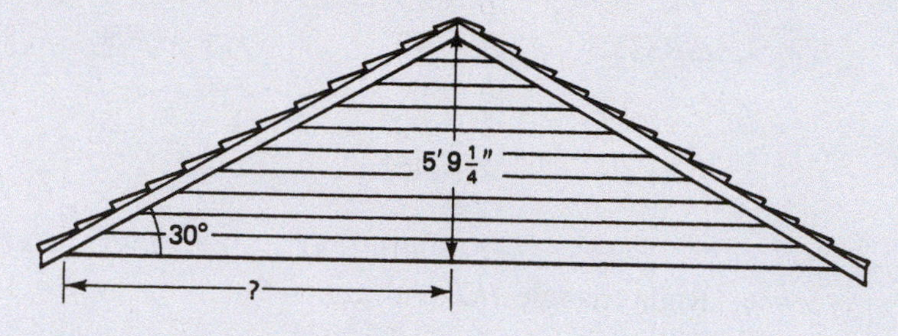

15. A square foundation has diagonals of 33 ft $11\frac{1}{4}$ in. What is the length of the sides? Round to the nearest whole inch.

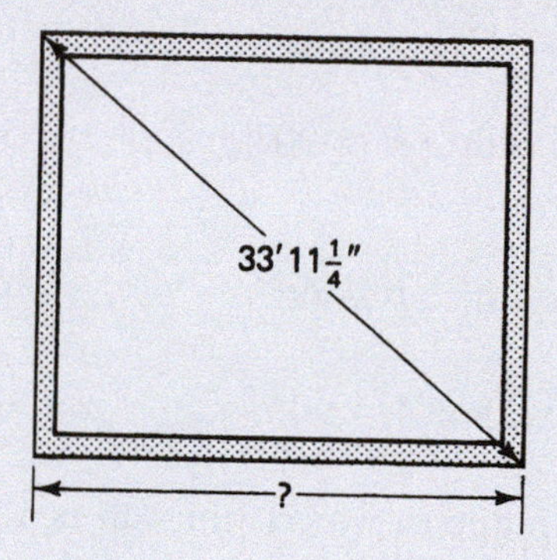

7.5 Perimeters and Areas of Triangles

The *perimeter* of a triangle is the distance around the figure. For $\triangle ABC$ shown in Figure 7–21, the perimeter is $P = a + b + c$.

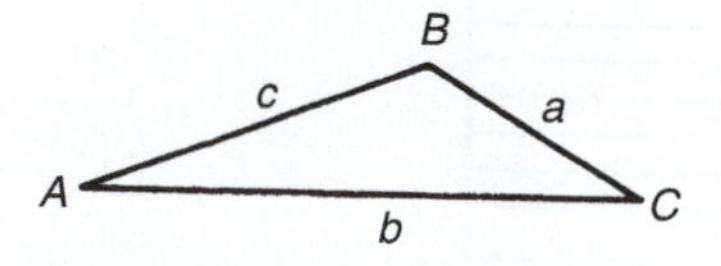

Figure 7–21

Examples

A. Find the perimeter of a triangle with sides of 8, 5, and 7 inches.

$P = a + b + c$	1. The perimeter is the sum of the three sides.
$= 8$ in. $+ 5$ in. $+ 7$ in. $= 20$ in.	2. The perimeter of the triangle = 20 inches.

B. A triangular-shaped window has sides of 22.8″ and 31.6″. The hypotenuse was not measured. The perimeter of the window must be known in order to purchase trim boards. Find the perimeter of the window.

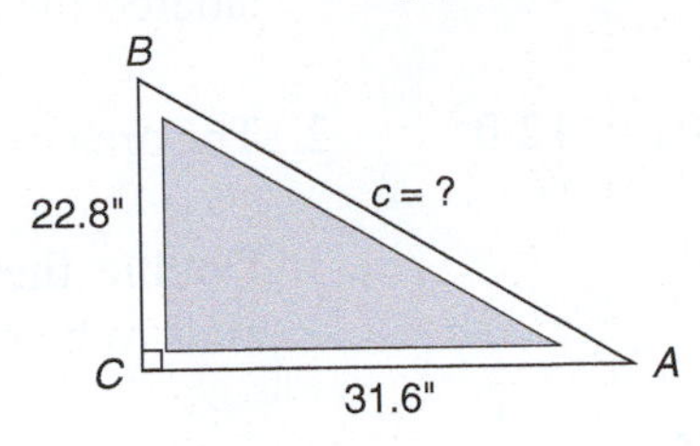

$c^2 = 22.8^2 + 31.6^2$ $c = 39.0''$	1. Since $\triangle ABC$ is a right triangle, side c can be found by the Pythagorean theorem.
$P = 22.8 + 31.6 + 39.0$ $= 93.4''$	2. The perimeter is 93.4 inches. The amount of trim needed will be 93.4″ plus an allowance for cutting waste.

The area of a triangle is given by

$$A = \tfrac{1}{2}bh$$

This is the formula for the area of a triangle where A is the area, b is the base, and h the height of the triangle.

Any of the three sides can be considered the base, but the height must always be perpendicular to the base, as shown in Figure 7–22. In a right triangle, the base and height are the legs of the right triangle. In an oblique triangle, the base is one of the sides, but the height is not.

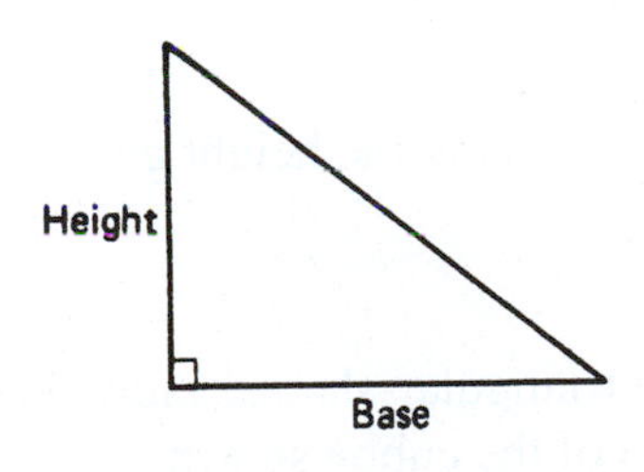

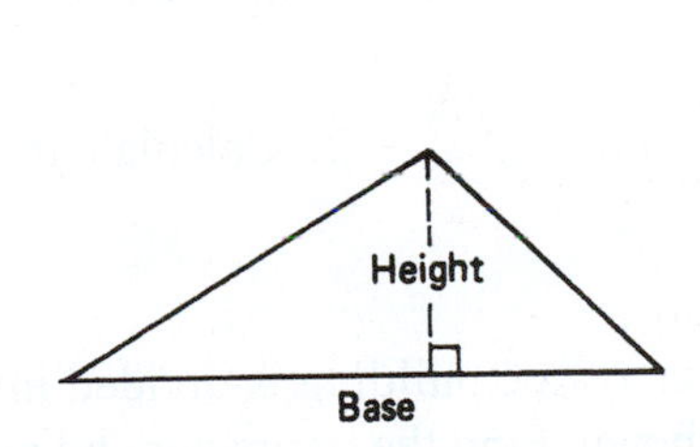

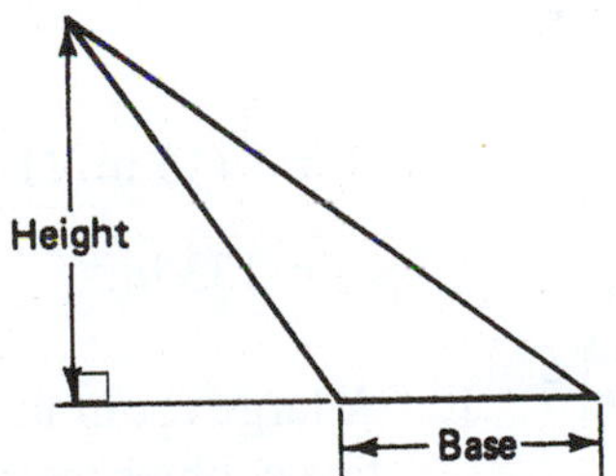

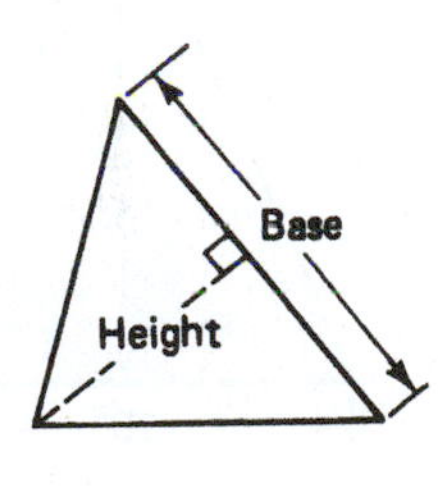

Figure 7–22

Examples

C. A front porch is 8 ft wide and the roof is 3 ft high. The ends of the roof are to be sided as shown. Find the area to be sided.

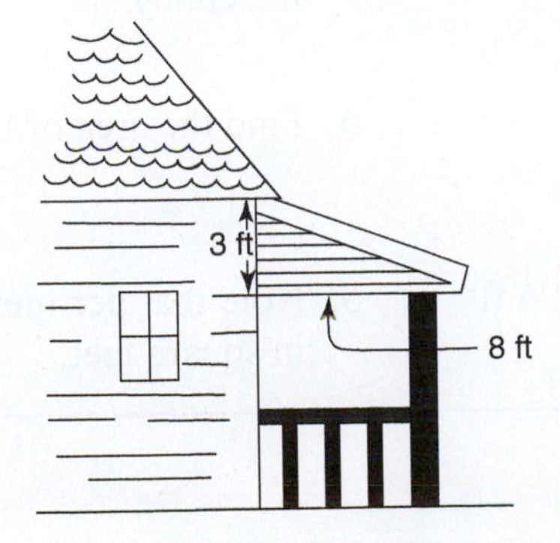

$A = \frac{1}{2}bh$ — 1. Since the triangle-shaped ends are right triangles, the two sides given can be considered the base and height.

$A = \frac{1}{2}(3 \text{ ft})(8 \text{ ft}) = 12 \text{ ft}^2$ — 2. The area of each end is 12 ft^2.

$2(12 \text{ ft}^2) = 24 \text{ ft}^2$ — 3. Double the amount since there are two ends to be sided.

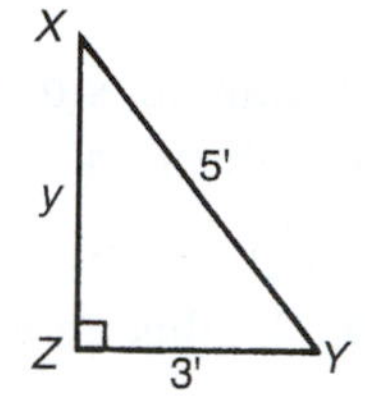

D. Find the area of $\triangle XYZ$.

1. Since this is a right triangle, sides x and y can be considered to be the base and height.

$y^2 = 5^2 - 3^2 = 16 \text{ ft}^2$
$y = 4 \text{ ft}$

2. Using the Pythagorean theorem, solve for y.

$A = \frac{1}{2}bh$
$= \frac{1}{2}(3 \text{ ft})(4 \text{ ft})$

3. Solve for the area using the values 3 ft and 4 ft for the base and height.

$A = 6 \text{ ft}^2$

4. Area is measured in square feet.

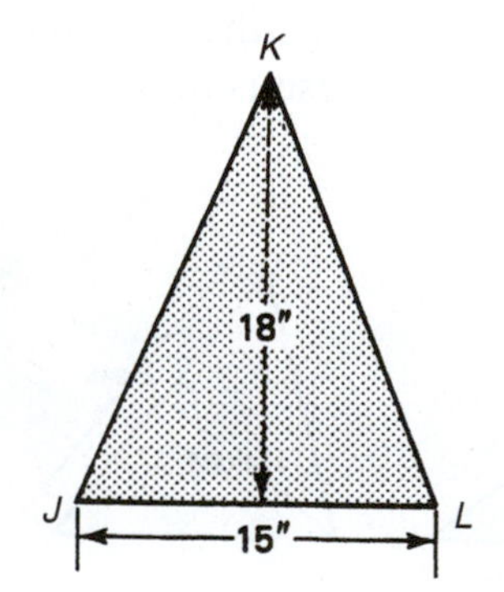

E. Find the area of $\triangle JKL$.

1. Since this is not a right triangle, the altitude is not one of the sides.

$A = \frac{1}{2}(15 \text{ in.})(18 \text{ in.})$
$= 133 \text{ in.}^2$

2. Calculate the area using the height given.

F. A large room in an office building is divided into triangular-shaped individual office cubbies, as shown. Find the perimeter and area of the cubby shown.

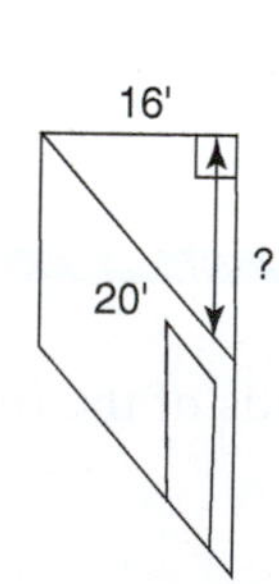

$a^2 = 20^2 - 16^2$

1. Since this is a right triangle, the base of the triangle can be found by the Pythagorean theorem.

$a = 12 \text{ ft}$

2. The base is 12 ft.

$P = 16 \text{ ft} + 20 \text{ ft} + 12 \text{ ft}$
$= 48 \text{ ft}$

3. The perimeter is the sum of the three sides of the cubby.

$A = \frac{1}{2}(12 \text{ ft})(16 \text{ ft})$
$= 96 \text{ ft}^2$

4. Find the area of the cubby.

$P = 48 \text{ ft}; \quad A = 96 \text{ ft}^2$

5. Note that perimeter is in linear feet and area is in square feet.

Area of a Triangle by Hero's Formula

If not enough information is given to find the height of an oblique triangle but all three sides are known, the area can be computed by means of *Hero's formula:*

$$A = \sqrt{s(s-a)(s-b)(s-c)}$$

in which s is $\frac{1}{2}$ the perimeter and a, b, and c are the sides.

Example

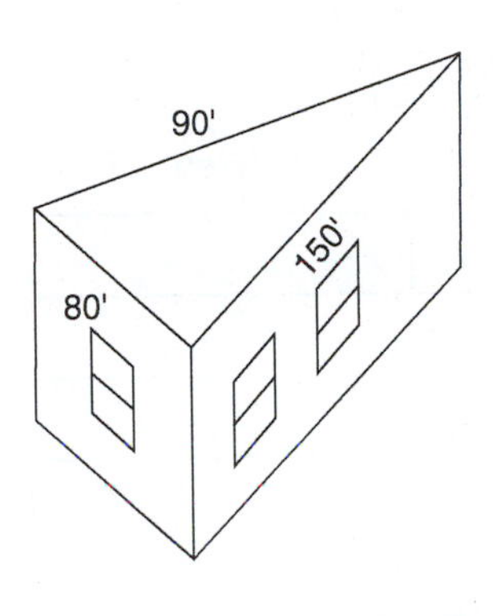

G. A triangular-shaped building has sides of 80 ft, 90 ft, and 150 ft. Find the area of the building.

$P = 80 \text{ ft} + 90 \text{ ft} + 150 \text{ ft} = 320 \text{ ft}$

1. Find the perimeter of the triangle.

$s = \frac{1}{2}(320) = 160 \text{ ft}$

2. s equals one-half the perimeter.

$$\begin{aligned} A &= \sqrt{s(s-a)(s-b)(s-c)} \\ &= \sqrt{160(160-80)(160-90)(160-150)} \end{aligned}$$

3. $(s - a)$ is the difference between s and side a.

$$\begin{aligned} &= \sqrt{160(80)(70)(10)} \\ &= \sqrt{8{,}690{,}000} \end{aligned}$$

4. Multiply and take the square root.

$A = 2993 \text{ ft}^2$

5. The area of the building is 2993 ft^2.

Since the formula $A = \frac{1}{2}bh$ is simpler, it is generally used whenever possible. In this case, there was not enough information to use the simpler formula.

Exercise 7–5

Given right triangle ABC, find the missing side and the perimeter. Assume that $\angle C$ is the right angle.

	Side *a*	Side *b*	Side *c*	Perimeter
1.	8.26	4.11		
2.	3.58	6.24		
3.	4.12	6.85		
4.	3.00		5.00	
5.		5.00	13.00	

Given right triangle ABC, find the missing side and the area. ∠C is the right angle.

	Side *a*	Side *b*	Side *c*	Area
6.	8.21	4.15		
7.		4	5	
8.	6.62	5.17		
9.	3.2	6.4		
10.		12	13	

Given the oblique triangle ABC, use Hero's formula to find the area.

	Side *a*	Side *b*	Side *c*	Area
11.	8.26	4.17	9.23	
12.	27.0	35.6	31.8	
13.	8.3	8.3	8.3	
14.	7.4	5.2	7.4	
15.	5	12	15	

16. Find the area of the triangle.

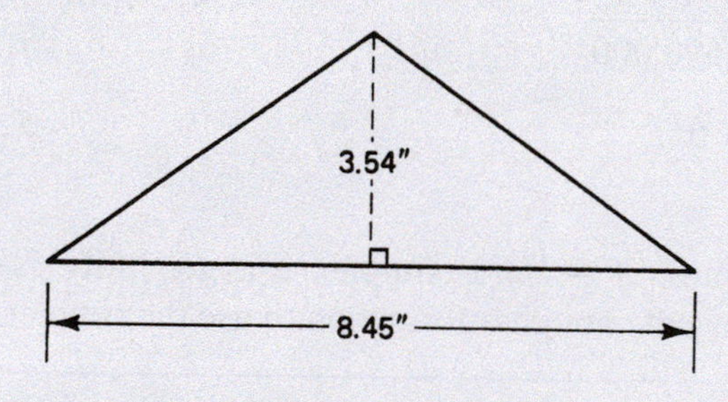

17. Find the area of the triangle.

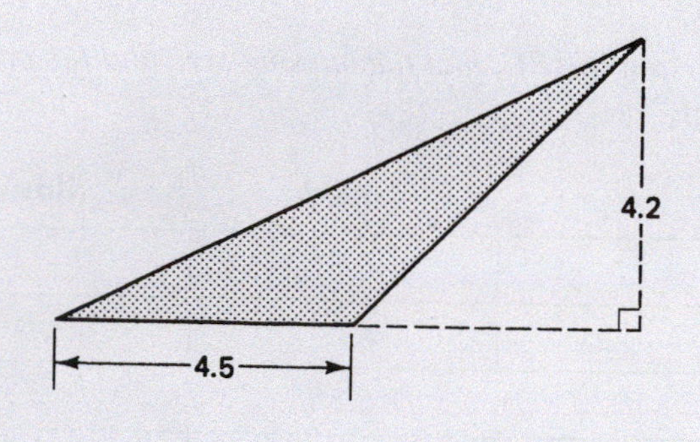

18. Given the 45°–45°–90° triangle shown, find the area of the triangle.

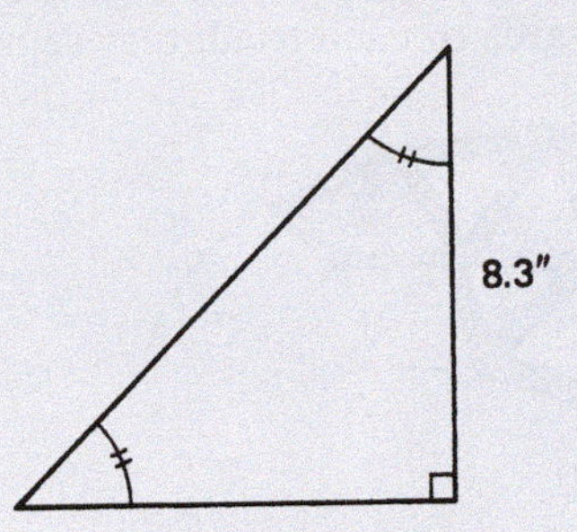

19. A triangular-shaped piece of land is bordered by a road on one side, a stone wall on the second side, and a barbed wire fence on the third side. If the lot borders the road for 428 ft, the stone wall for 1000 ft, and the barbed wire fence for 862 ft, what size is the lot in acres? (1 acre = 43,560 ft^2)
20. A house with cathedral ceilings has windows in the gable ends that are right triangles. Find the area in square feet of the glass in each window.

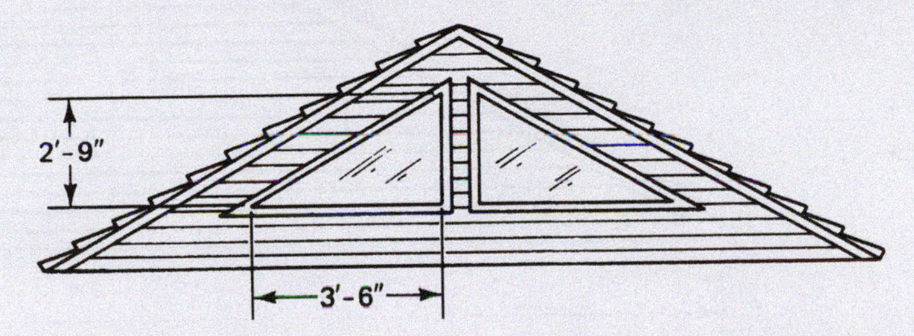

21. In exercise 20, find the amount of trim needed to go around each window. Add 2 ft of trim per window for cutting waste. Give the answer to the nearest foot.
22. A passive solar collector is attached to the side of a house as shown. The collector is 4 ft deep and 5 ft 3 in. high. To the nearest $\frac{1}{16}$″, how long is the exposed glass front in the collector? The frame is 1″ wide.

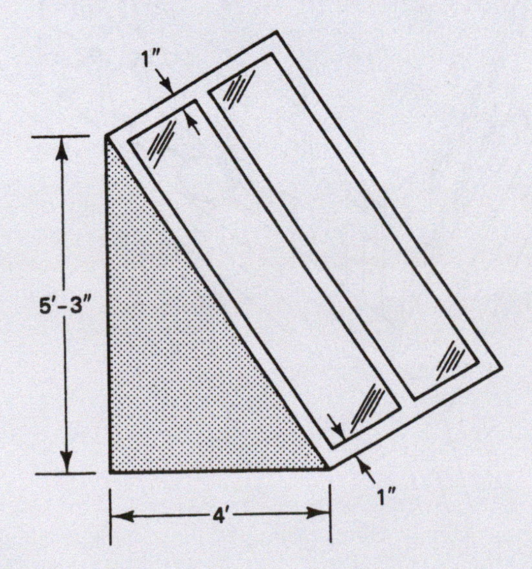

23. The triangular supports in a geodesic dome house are equilateral triangles 4 ft on a side. What is the area of each triangular segment of the roof?

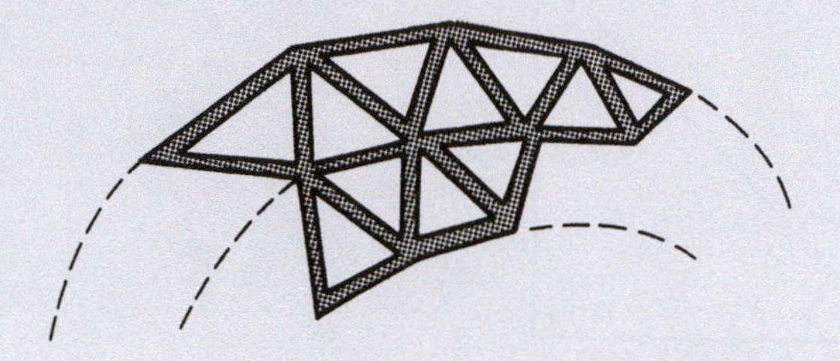

24. The roof shown is designed for a solar hot-water collector. What is the area of the end shown?

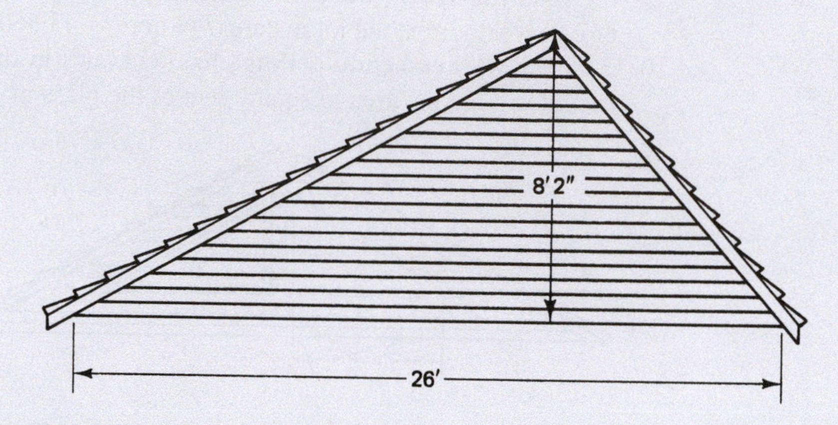

25. A triangular building on a city lot has the dimensions shown. What is the area in square feet of each floor of the building?

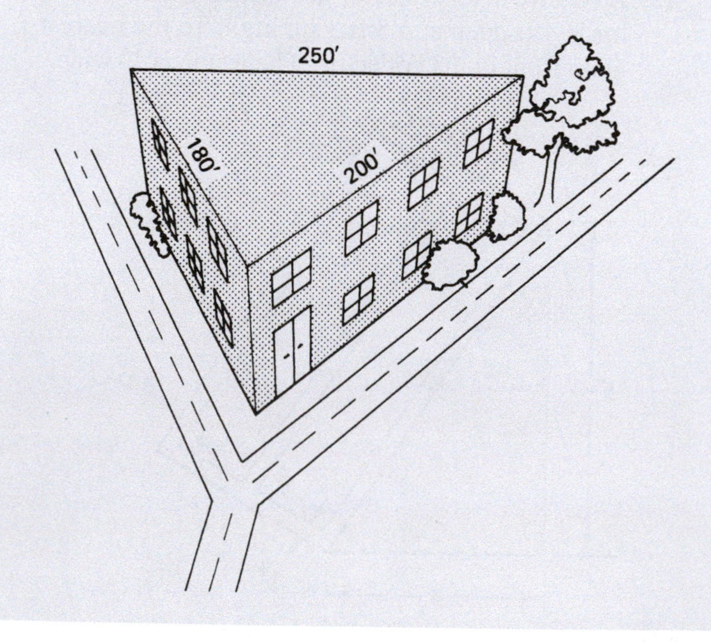

26. An A-frame cottage is 32 ft 8 in. tall and 20 ft wide. If the ends are to be sided with cedar shingles, how many square feet, to the nearest whole foot, are to be sided on each end? Ignore openings for doors and windows.

27. The gable ends of a house are solar reflective glass. Find the combined length of all trim boards (shaded boards in the figure) if the symmetrical roof has a rise of 8 ft, and a run of 15 ft. Ignore waste. Don't forget to include the vertical center board. (*Remember:* run is $\frac{1}{2}$ the width of the house.)

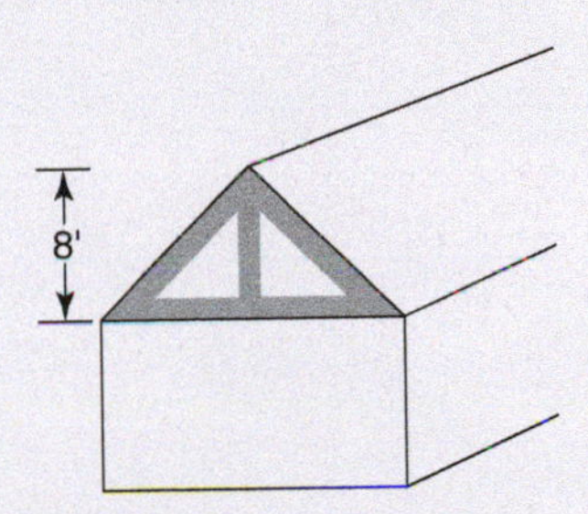

28. Find the approximate number of square feet of solar reflective glass in the gable end in problem 27 above.
29. A solar panel on a recreational vehicle can be adjusted to 3 positions: horizontal, 30° with the roof, and 60° with the roof. What is the distance from the top edge of the panel to the horizontal for the 30° and 60° settings? The dimensions of the panel are 2 ft 3 in. by 4 ft 5 in.

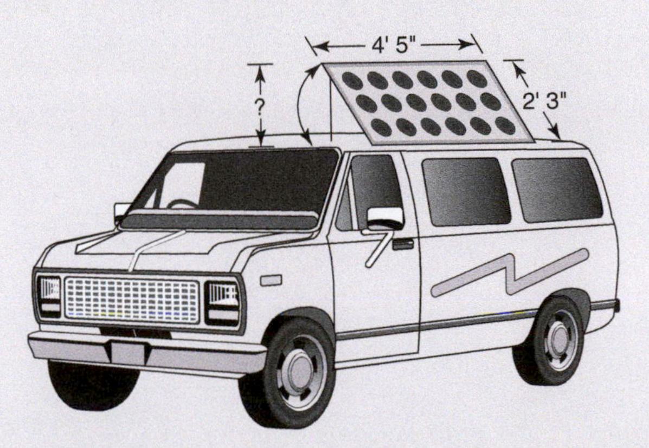

30. Pipes running up the corner of a room are to be hidden behind a false wall. The wall is to be positioned 2 ft. 2 in. from the corner of the room. Find the width of the wall.

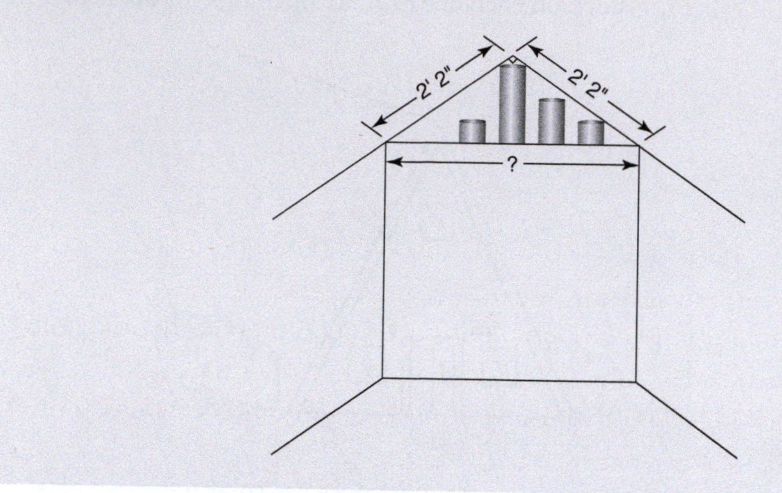

CHAPTER 8

Areas and Perimeters

Did You Know

Student volunteers at the University of North Carolina at Ashville changed 13,000 light bulbs to compact fluorescent bulbs? This is estimated to save $500,000 and 5 million kilowatt hours of electricity over the life of the bulbs. It will also reduce the amount of carbon released in the atmosphere by 3,700 tons.

OBJECTIVES

Upon completing this chapter, the student will be able to:

1. Find the perimeter and area of a rectangle.
2. Find the perimeter and area of a square.
3. Find the perimeter and area of a parallelogram.
4. Find the perimeter and area of a trapezoid.
5. Solve problems involving perimeters and areas of the figures noted above.
6. Find the circumference of a circle.
7. Find the area of a circle.
8. Solve problems involving circles, semicircles, and quarter circles.
9. Find the area of a regular octagon given one of the following: the diameter across the angles, the diameter across the flats, the length of each side, or the perimeter.
10. Given the area of a regular octagon, find any of the following: the diameter across the angles, the diameter across the flats, the length of each side, or the perimeter.
11. Determine the area of composite figures.
12. Determine the perimeter of composite figures.

8.1 Quadrilaterals

A *quadrilateral* is a closed, four-sided figure. *Quad* means "four," and *lateral* means "side."

Parallel lines are lines that are the same distance apart the entire length of the lines, as shown in Figure 8–1. No matter how far they are extended, they will never meet or cross.

Parallel lines are in the same *plane.* Think of a plane as a two-dimensional flat surface such as a floor or a sheet of paper.

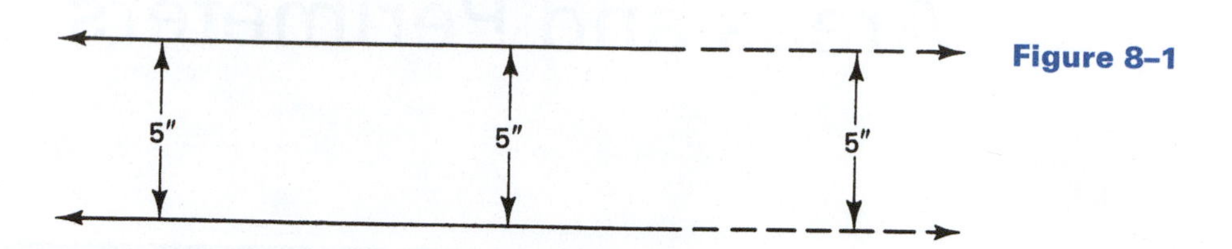

Figure 8–1

A *parallelogram* is a quadrilateral with opposite sides equal and parallel. In parallelogram $ABCD$ in Figure 8–2, sides AB and DC are opposite sides. $AB = DC$ and $AB \parallel DC$ (the symbol $\parallel$ means "parallel to"). Opposite sides AD and BC are also equal and parallel to each other. Opposite angles are also equal. Therefore, $\angle A = \angle C$ and $\angle B = \angle D$. The sum of $\angle A + \angle B + \angle C + \angle D = 360°$. The four angles of any four-sided figure total 360°.

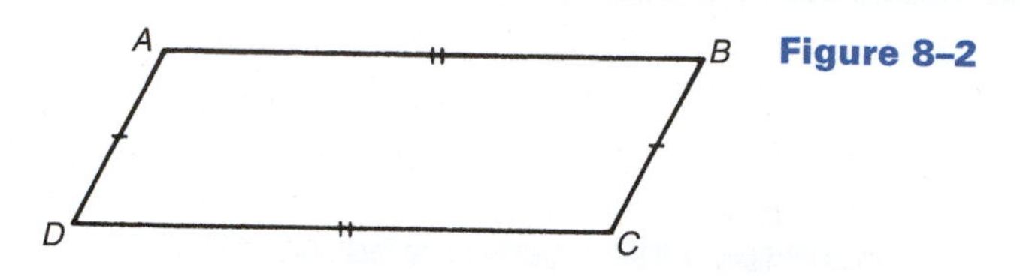

Figure 8–2

A *rectangle* is a parallelogram with four right angles (Figure 8–3). Any parallelogram that has at least one right angle must be a rectangle. Why?

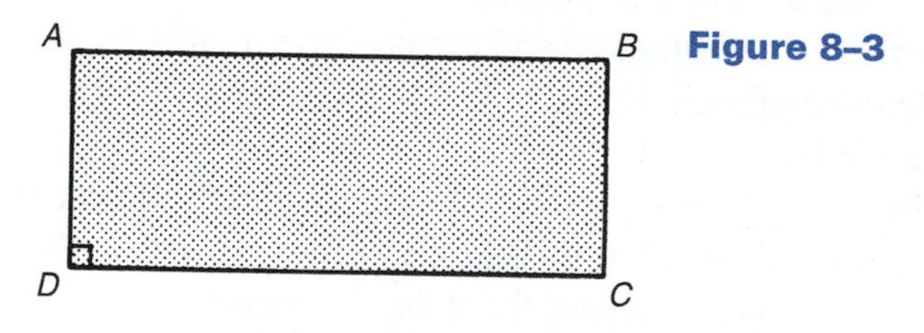

Figure 8–3

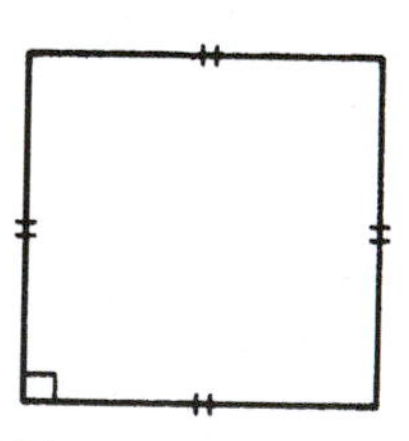

Figure 8–4

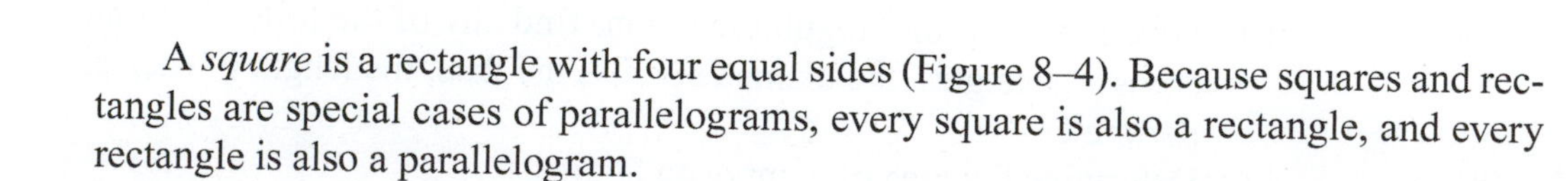

A *square* is a rectangle with four equal sides (Figure 8–4). Because squares and rectangles are special cases of parallelograms, every square is also a rectangle, and every rectangle is also a parallelogram.

Perimeter and Area of Rectangles

The *perimeter* of a rectangle is the distance around it. Therefore, the perimeter is the sum of all four sides. The formula for perimeter is usually written in either of two forms:

$$P = 2(L + W) \quad \text{or} \quad P = 2L + 2W$$

where P is the perimeter, L the length, and W the width.

Figure	Area	Perimeter	Diagram
Rectangle	$A = L \times W$	$P = 2(L + W)$ or $P = 2L + 2W$	W, L
Square	$A = S^2$	$A = 4S$	S
Parallelogram	$A = b \times h$	$P = 2a + 2b$	b, h, a
Trapezoid	$A = \frac{1}{2}h(B + b)$	$P = a + b + c + B$	b, a, h, c, B

Examples

A. Before ordering baseboard trim, the perimeter of a 12 ft × 18 ft room must be found. What is the perimeter of the room?

$P = 2(L + W)$ 1. This is frequently the simpler formula to use.

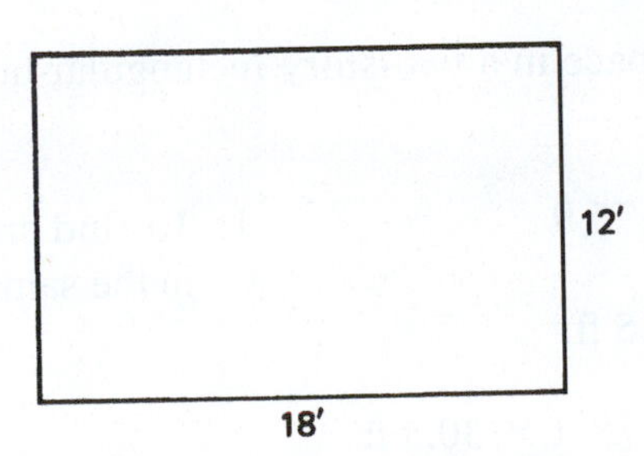

$= 2(12 \text{ ft} + 18 \text{ ft})$ 2. Add the length and width in the parentheses. The length plus the width represent halfway around the room.

$= 2(30 \text{ ft})$ 3. Multiply the sum of the length and width by 2 to find the entire distance around.

$= 60 \text{ ft}$

B. The perimeter of a deck is 48 ft. If the length is 16 ft, what is the width?

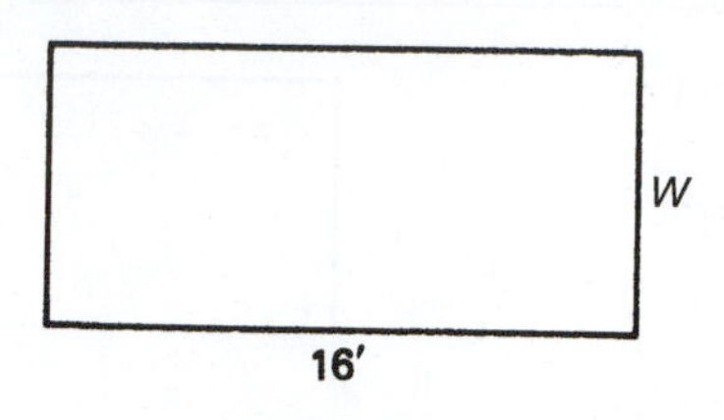

$P = 2(L + W)$ $48 \text{ ft} = 2(16 \text{ ft} + W)$	1. If the entire distance around is $2(L + W)$, then halfway around the deck will be $(L + W)$.
$24 \text{ ft} = 16 \text{ ft} + W$	2. Halfway around (half the perimeter) is half of 48 ft.
$24 \text{ ft} - 16 \text{ ft} = W$	3. Subtract the length from half the perimeter.
$W = 8 \text{ ft}$	4. The width is 8 ft.

Area is measured in square units. The area of a house is the amount of floor space it has. The area of a rectangle is found by the formula

$$A = L \times W$$

where A is the area, L the length, and W the width.

Examples

C. Find the floor space of a rectangular house that is 26 ft wide by 30 ft long.

$A = L \times W$	1. Area of a rectangle equals length times width.
$= 26 \text{ ft} \times 30 \text{ ft}$	2. Substitute known values into the formula and solve for area.
$= 780 \text{ ft}^2$	3. Area is in square units.

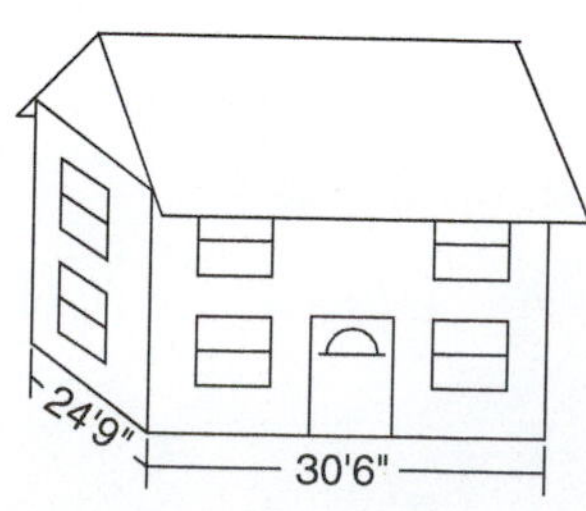

D. Find the floor space in a two-story rectangular house that measures 24 ft 9 in. by 30 ft 6 in.

$24 \text{ ft } 9 \text{ in.} = 24.75 \text{ ft}$ $30 \text{ ft } 6 \text{ in.} = 30.5 \text{ ft}$	1. To find area the length and width must be in the same units.
$A = 24.75 \text{ ft} \times 30.5 \text{ ft}$ $= 754.875 \text{ ft}^2$	2. Find the area of each floor.
$\text{Total area} = 2 \times 754.875 \text{ ft}^2$	3. Multiply by 2 to find the floor space on both floors.
$= 1510 \text{ ft}^2$	4. Round to the nearest square foot.

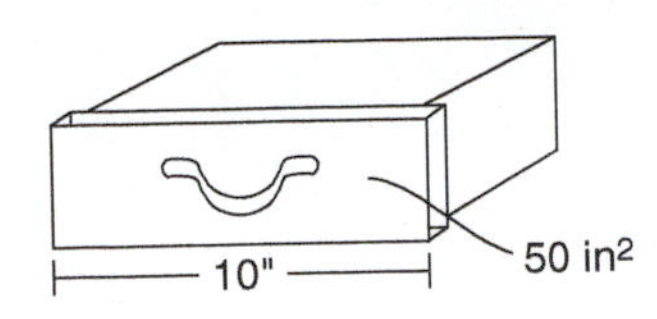

E. A rectangular drawer front has a length of 10″ and an area of 50 in.^{2}. What is its perimeter?

$A = L \times W$ — 1. Substitute the given values in the formula for area of a rectangle.

$50 \text{ in.}^2 = 10 \text{ in.} \cdot W$

$\dfrac{50 \text{ in.}^2}{10 \text{ in.}} = W$

$W = 5 \text{ in.}$ — 2. Rearrange and solve for W.

$P = 2L + 2W$

$= 2(10 \text{ in.}) + 2(5 \text{ in.})$ — 3. Substitute length and width into the formula for perimeter.

$= 30 \text{ in.}$ — 4. Solve for P.

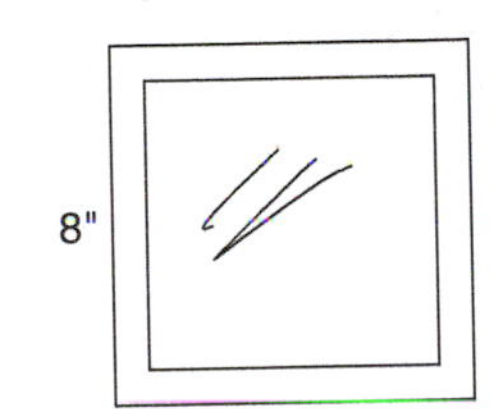

F. Find the perimeter of a square window if one side is 8 inches.

$P = 4S$ — 1. Since the length and width are equal on a square, the perimeter is four times the length of one side.

$P = 4(8 \text{ in.})$

$= 32 \text{ in.}$ — 2. Substitute the known side and multiply.

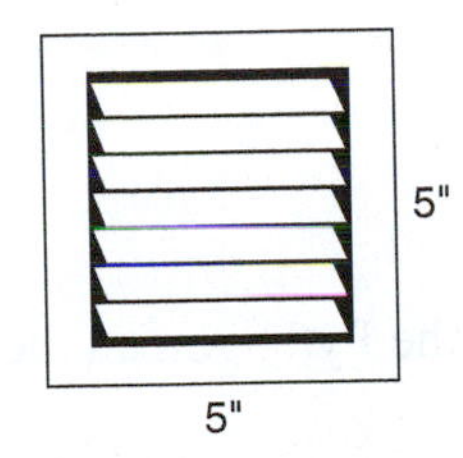

G. Find the area of a square heating vent with sides of 5 inches.

$A = S^2$ — 1. Since the length and width are the same on a square, the area is any side squared.

$A = 5^2$

$= 25 \text{ in.}^2$ — 2. A square 5 inches on a side has an area of 25 in.^{2}.

H. A guest house has a square floor with 576 ft^2 of area. What is the linear distance around the house?

$A = S^2$ — 1. If the area of a square is given, the length of any side can be found.

$576 \text{ ft}^2 = S^2$

$\sqrt{576} = S$

$S = 24 \text{ ft}$ — 2. The length of the side of a square is the square root of its area.

$P = 4S$

$= 4(24 \text{ ft})$ — 3. Knowing the length of a side, the perimeter (distance around) can be found.

$= 96 \text{ ft}$ — 4. The distance around the house is 96 ft.

I. Find the area of the parallelogram shown.

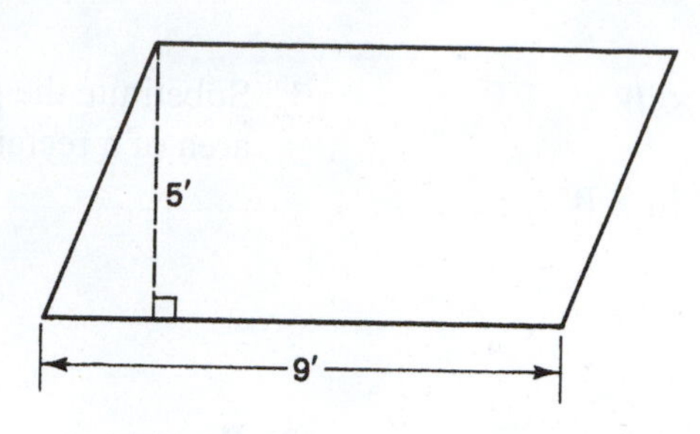

$$A \square = b \times h$$

The area of a parallelogram is the height times the base. *Remember:* The height is always perpendicular to the base.

$A = b \times h$ 1. Substitute given quantities into the formula.

$= 9 \text{ ft} \times 5 \text{ ft}$

$= 45 \text{ ft}^2$ 2. Area is 45 ft^2.

J. Find the area of the parallelogram shown.

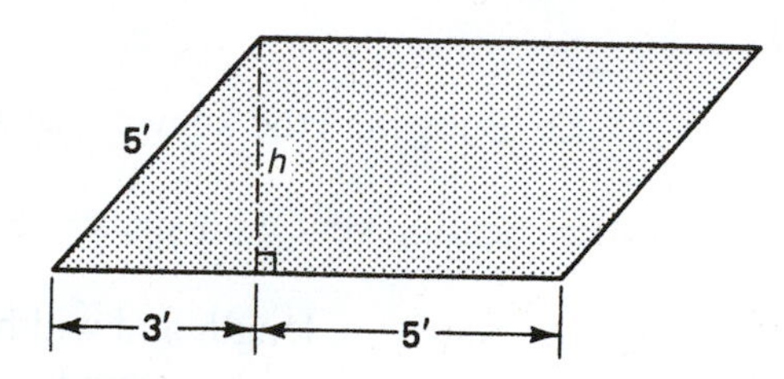

$h^2 = 5^2 - 3^2$ 1. Find the height by using the Pythagorean theorem.

$h = 4 \text{ ft}$

$b = 3 \text{ ft} + 5 \text{ ft} = 8 \text{ ft}$ 2. Determine the base.

$A = 4 \text{ ft} \times 8 \text{ ft} = 32 \text{ ft}^2$ 3. Find the area by the formula.

Perimeter and Area of Trapezoids

A *trapezoid* is a four-sided figure with *two sides parallel* and *unequal*. Several types of trapezoids are shown in Figure 8–5. The two sides that are parallel and unequal are called the *bases*. The longer base is usually represented by B and the shorter base by b. The perimeter of a trapezoid is the sum of its four sides. The area of a trapezoid is given by the formula

$$A = \frac{1}{2}h(B + b)$$

where A is the area, h the height, B the longer base, and b the shorter base.

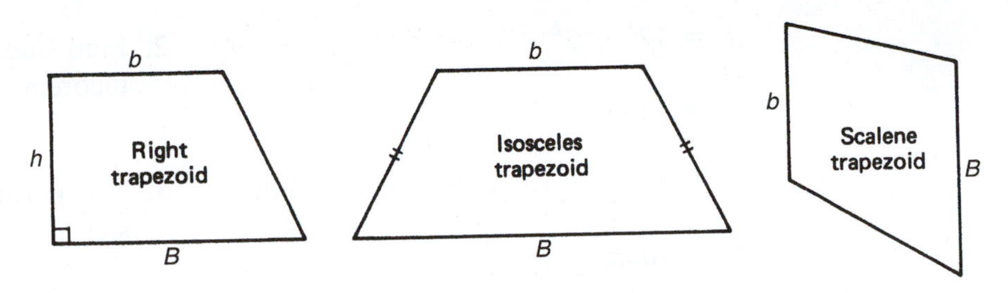

Figure 8–5

Examples

K. A trapezoidal window has the dimensions shown. Find the area of the window.

$$A = \frac{1}{2}h(B + b)$$

In a right trapezoid the height h is the side perpendicular to the bases.

$A = \frac{1}{2}(6 \text{ ft})(10 \text{ ft} + 16 \text{ ft})$ — 1. Substitute the known values into the formula.

$= \frac{1}{2}(6 \text{ ft})(26 \text{ ft})$ — 2. Perform the addition inside the parentheses first.

$= 78 \text{ ft}^2$ — 3. Multiply to determine the area of the window in square feet.

L. Find the perimeter of the following trapezoid.

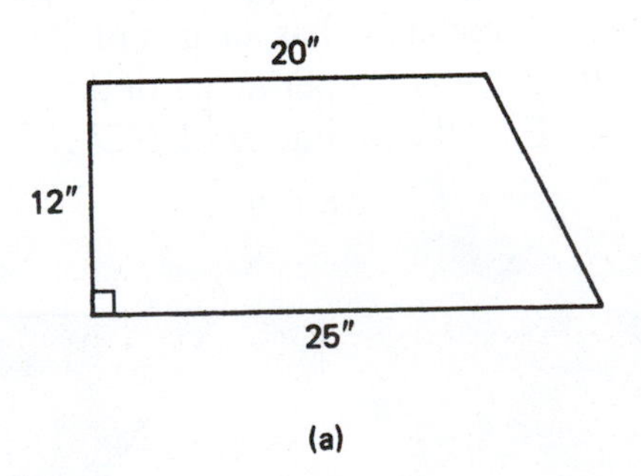

(a)

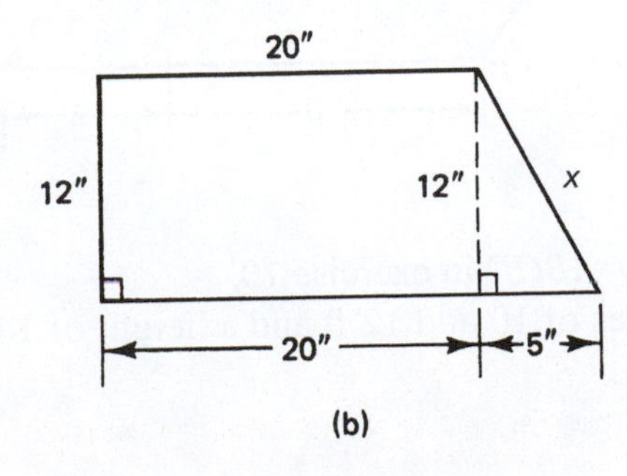

(b)

1. Draw in the height, forming a right triangle and a rectangle. Label the sides with the information given.

$x^2 = 12^2 + 5^2$ $x = 13$ in.	2. Find side x using the Pythagorean theorem.
$P = 25$ in. + 12 in. + 20 in. + 13 in. $= 70$ in.	3. The perimeter is the sum of the sides.

Exercise 8–1

The student may find it useful to draw and label a diagram when one is not given.

1. Find the perimeter of a rectangle that is 8 in. long and 5 in. wide.
2. Find the area of the rectangle in exercise 1.
3. Find the perimeter of ▱ *ABCD.*

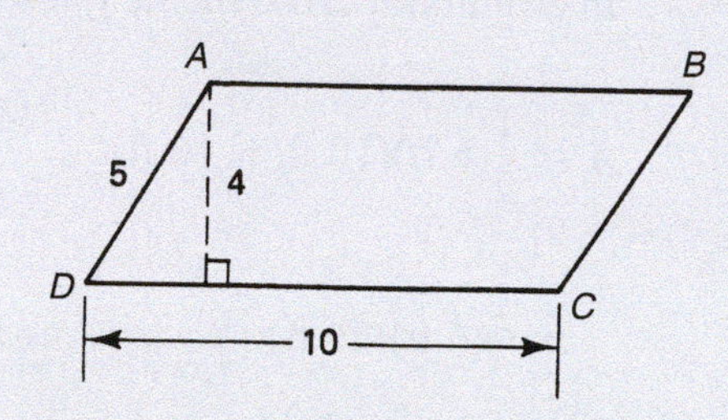

4. Find the area of ▱ *ABCD* in exercise 3.
5. Find the area of a square which is 3 ft on a side.
6. Find the perimeter of the square in exercise 5.
7. A square has a perimeter of 20 ft. What is its area?
8. A rectangle has an area of 200 ft^2 and a width of 8 ft. What is its length?
9. What is the perimeter of the rectangle in exercise 8?
10. Find the perimeter of ▱ *ABCD.*

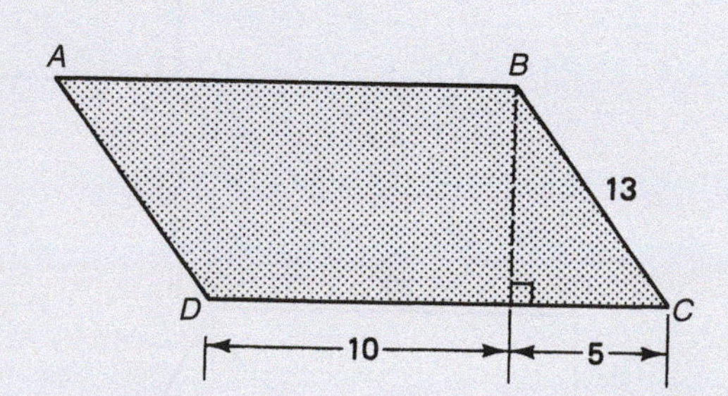

11. Find the area of ▱ *ABCD* in exercise 10.
12. A trapezoid has bases of 10 and 12 ft and a height of 14 ft. Find the area of the trapezoid.

Complete the table for the rectangles described.

	Length	Width	Perimeter	Area
13.	10 in.		28 in.	
14.	16 ft			192 ft^2
15.	15 ft 3 in.	8 ft 6 in.		
16.		26 ft	112 ft	

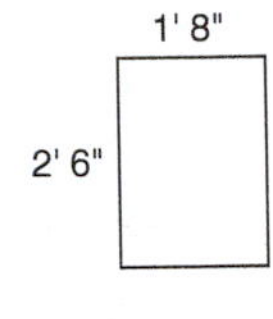

17. A rectangular window calls for a rough opening of 2 ft 6 in. by 1 ft 8 in. What is the area, rounded to the nearest square foot, of the R.O. (rough opening)? (*Hint:* Do not round until the final answer is obtained.) See diagram.

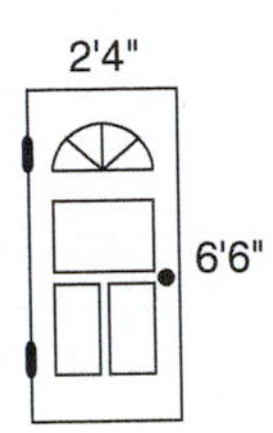

18. A door is 6 ft 6 in. high by 2 ft 4 in. wide. What is the area, in square inches, of the door? See diagram.
19. A square table is to be covered with plastic laminate costing $2.95 per square foot. If the perimeter of the table is 12 ft, what is the cost of the laminate to cover it?
20. A room 12 ft × 16 ft 6 in. is to be carpeted. Assuming no waste, how many square feet of carpeting are needed?
21. If carpeting costs $40.95 per square yard installed, what will it cost to carpet the room in exercise 20? (*Hint:* Change square feet in problem 20 to square yards.)

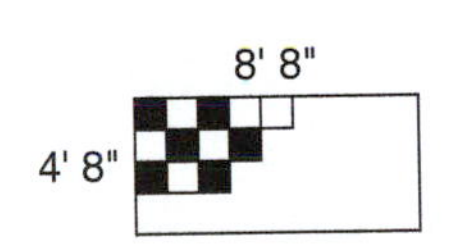

22. Floor tiles 1 ft × 1 ft are to be installed on a kitchen floor measuring 8 ft 8 in. × 4 ft 8 in. How many tiles are needed? Add six tiles for waste and proper fitting. See diagram.

23. Sills are to be placed on top of a rectangular foundation that is 26 ft wide and has an area of 780 ft^2. Ignoring waste, how many lineal feet of 2 × 6 lumber will be needed for the sill? See diagram.
24. A straight driveway 15 ft wide and 93 ft long is to be hot-topped at a cost of $17.50 per square yard. What will it cost to hot-top the driveway?
25. A square gazebo is 48 ft around. What is the area of the gazebo?

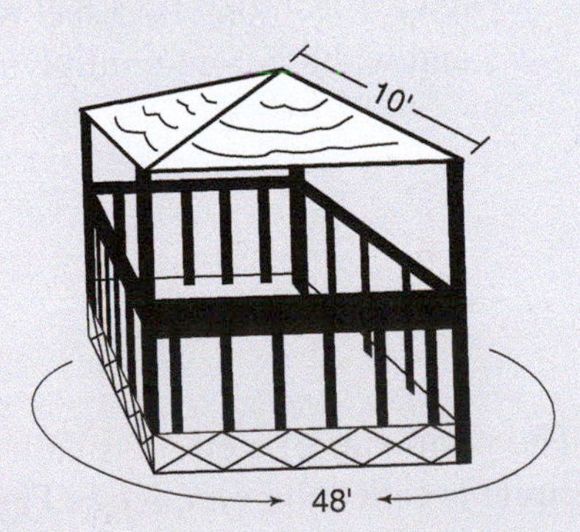

26. Find the area of the roof on the gazebo in problem 25, if each roof ridge is 10 ft long, as shown.

27. Pipes are concealed in a corner alcove as shown. If the wall is 8 ft high, find the area of the panel concealing the pipes.

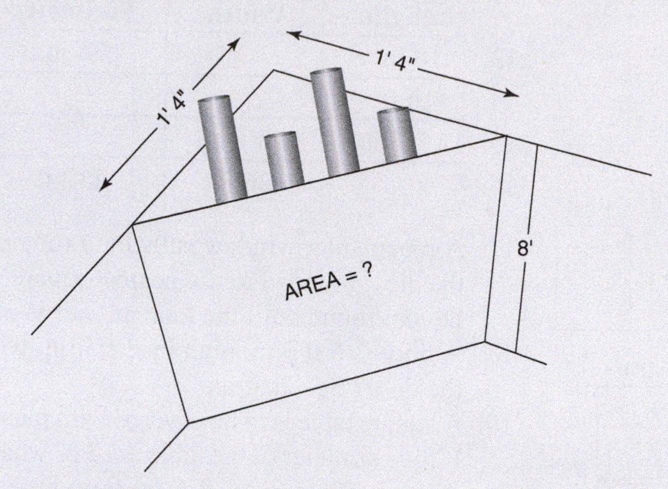

28. A window is an isosceles trapezoid, as shown. Find the perimeter of the window.

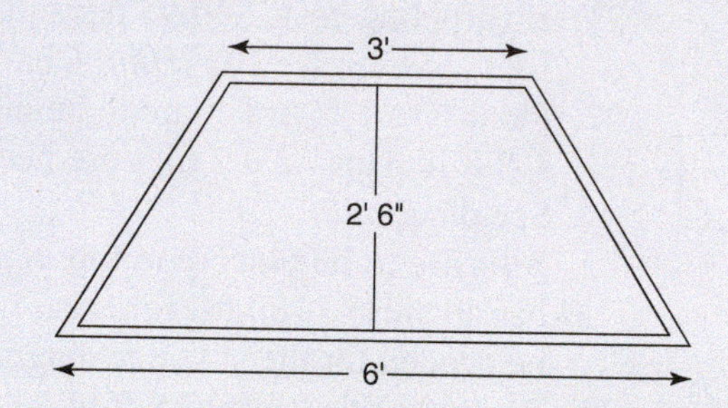

29. Find the area of the window in 28. above.
30. Square floor tiles are 9 inches on a side. How many are needed to cover an alcove 3 feet long by 5 feet wide? (Round up to a whole number of tiles for length and width before multiplying.)

8.2 Circles

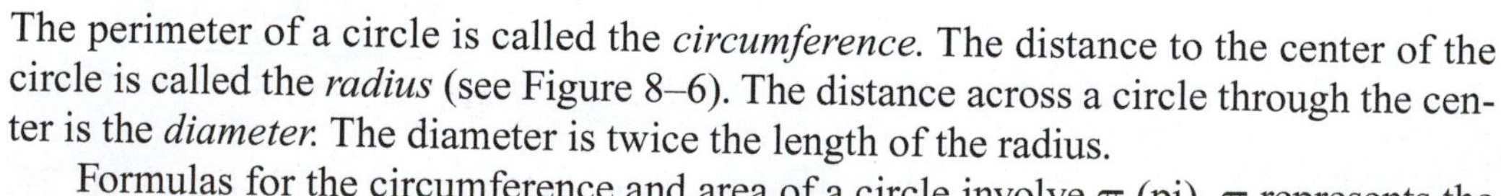

The perimeter of a circle is called the *circumference.* The distance to the center of the circle is called the *radius* (see Figure 8–6). The distance across a circle through the center is the *diameter.* The diameter is twice the length of the radius.

Formulas for the circumference and area of a circle involve π (pi). π represents the ratio of the circumference of any circle to its diameter, and it is always the same regardless of the size of the circle. π is approximately $\frac{22}{7}$ or 3.14. Many calculators have a π key because it is a value that is used so frequently. The area of a circle is found by the formula

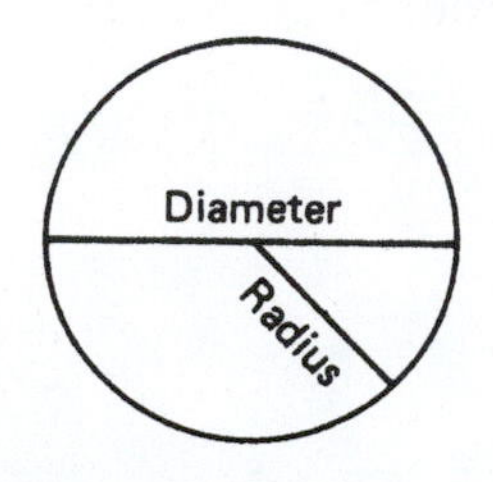

Figure 8–6

$$A = \pi r^2$$

where A is the area and r is the radius (π is always approximately 3.14).

Example

A. Find the area of a circle with a radius of 10 inches.

$A = \pi r^2$ — 1. Formula for the area of a circle.

$= \pi(10)^2$ — 2. Be sure to square the radius before multiplying by π.

$= \pi(100)$

$A = 314 \text{ in.}^2$ — 3. The area is in square inches.

The circumference of a circle is found by the formula

$$C = \pi d \quad \text{or} \quad C = 2\pi r$$

where C is the circumference, d the diameter, and r the radius.

Examples

B. A circular garden is to be walled in with a concrete wall. Find the outside circumference of the wall if the outside diameter of the circle is 20 feet.

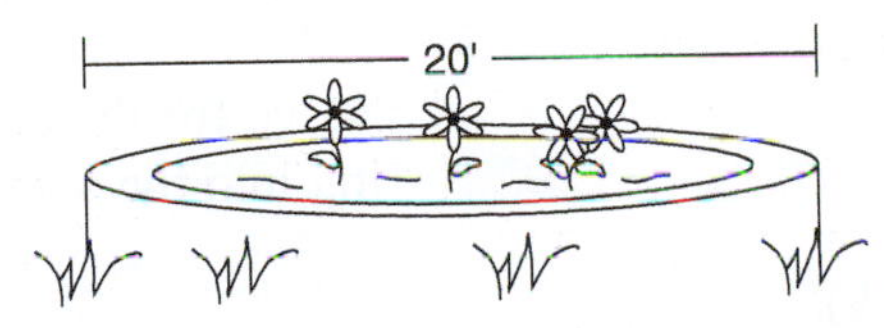

$C = \pi d$ — 1. Substitute the given values into the formula.

$= \pi(20 \text{ ft})$

$C = 62.8 \text{ ft}$ — 2. The circumferences of the garden is in linear feet.

C. A circular deck around an above-ground swimming pool has a circumference of 80 feet. Find the radius of the deck.

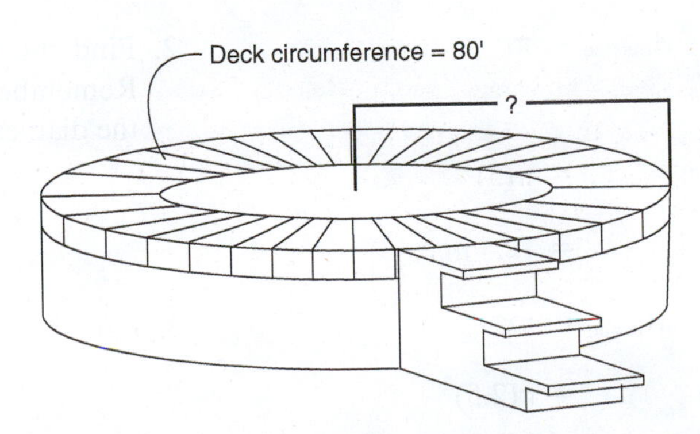

$C = \pi d$ — 1. Substitute the known values into the formula.

$80 \text{ ft} = \pi d$

$\frac{80 \text{ ft}}{\pi} = d$ — 2. Rearrange and solve for d. Use $\pi = 3.14$ if a calculator with a π key is not available.

$d = 25.5 \text{ ft}$ — 3. Since the diameter is twice the radius, divide by 2 to find r.

$r = 25.5 \text{ ft} \div 2 = 12.7 \text{ ft}$ — 4. Radius of the deck = 12.7 ft.

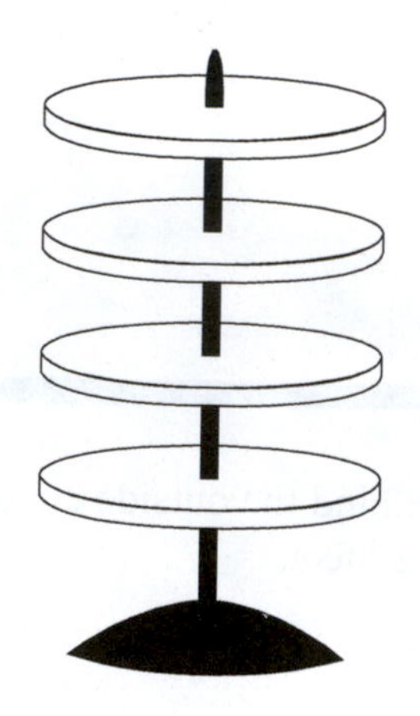

D. A series of rotating circular shelves are to be built to display watches and jewelry in a department store. Find the area of each shelf if the circumference is 100 inches.

$C = \pi d$ — 1. Substitute into the formula for circumference.

$100 \text{ in.} = \pi d$ — 2. Divide 100 by π to solve for d.

$31.8 \text{ in.} = d$

$r = \frac{31.8 \text{ in.}}{2} = 15.9 \text{ in.}$ — 3. Solve for the radius ($\frac{1}{2}$ the diameter).

$A = \pi r^2$ — 4. Solve for the area of each shelf by substituting into the area formula.

$= \pi(15.9)^2$

$= 795.8 \text{ in.}^2$

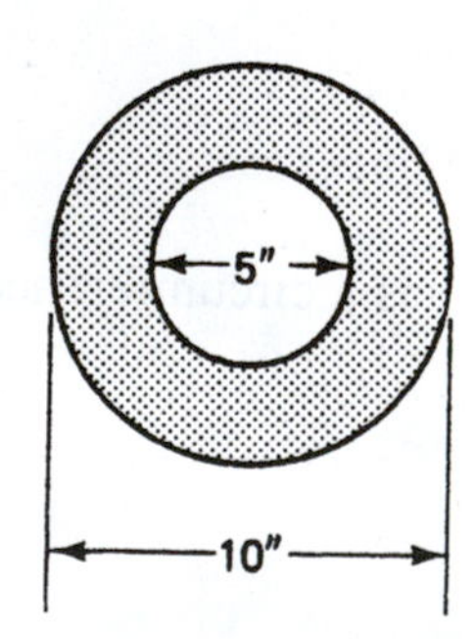

E. Find the area of the ring shown (the shaded area).

$\text{Area of ring} = A_{\text{outside}} - A_{\text{inside}}$ — 1. To find the area of the ring, find the area of the inside circle and subtract it from the area of the outside circle.

$A_{\text{outside}} = \pi r^2$ — 2. Find the area of the outside circle. Remember to square the radius, not the diameter.

$= \pi(5)^2$

$= 78.5 \text{ in.}^2$

$A_{\text{inside}} = \pi r^2$ — 3. Find the area of the inside circle.

$= \pi(2.5)^2$

$= 19.6 \text{ in.}^2$

$A_{\text{ring}} = 78.5 \text{ in.}^2 - 19.6 \text{ in.}^2$ — 4. The area of the ring is the difference between the areas of the outside and inside circles.

$A_{\text{ring}} = 58.9 \text{ in.}^2$

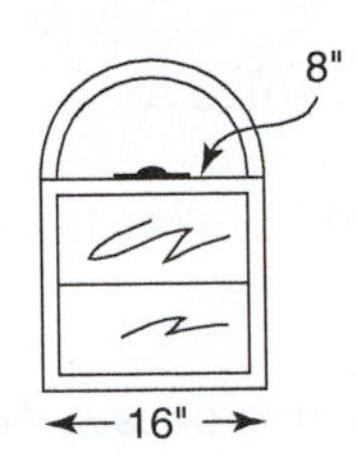

F. A semicircular arch is to be installed above a window. Find the area and perimeter of the arch.

$$A = \pi(8)^2$$

$$= 201 \text{ in.}^2$$

1. This is the area of a circle with radius = 8 inches.

$$\text{Area of semicircle} = 100.5 \text{ in.}^2$$

2. The area of the semicircle will be $\frac{1}{2}$ the area of the circle.

$$C = \pi d$$

$$= \pi(16) = 50.3 \text{ in.}$$

3. Find the circumference and divide by 2.

$$\frac{1}{2} \cdot \text{circumference} = 25.1 \text{ in.}$$

$$P = \frac{1}{2}C + d$$

$$= 25.1 \text{ in.} + 16 \text{ in.}$$

$$= 41.1 \text{ in.}$$

4. The perimeter equals half the circumference plus the diameter of the circle.

G. Two drain pipes measuring $1\frac{1}{2}$ inches I.D. (inside diameter) flow into one main pipe by means of a "Y" connector. What should be the minimum diameter of the main pipe to accommodate the flow from the two smaller pipes? (The cross-sectional area of the main pipe must be as large or larger than the total cross-sectional area of the two pipes flowing into it.)

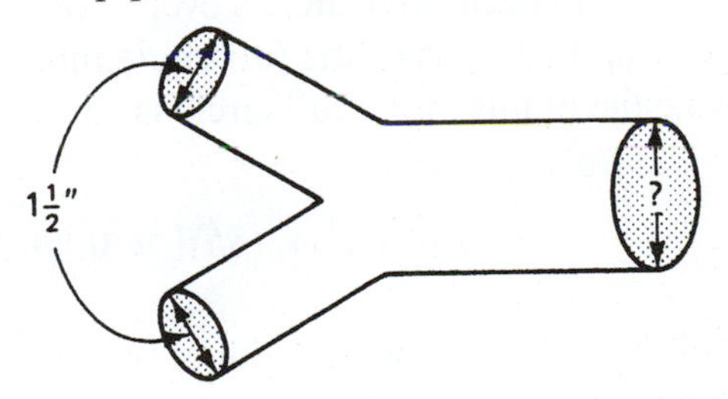

$$A = \pi r^2$$

$$= \pi(0.75 \text{ in.})^2$$

$$= 1.77 \text{ in.}^2$$

1. Find the cross-sectional area of each of the smaller pipes.

$$A_{\text{total}} = 2 \times 1.77 \text{ in.}^2$$

$$= 3.53 \text{ in.}^2$$

2. This is the minimum cross-sectional area the main pipe can have.

$$A = \pi r^2$$

$$3.53 = \pi r^2$$

3. Substitute the area into the formula.

$$r^2 = \frac{3.53}{\pi} = 1.125 \text{ in.}^2$$

4. Rearrange and solve for r^2.

$r = \sqrt{1.125} = 1.06$ in.	5. Take the square root to solve for r.
$d = 2.12$ in.	6. Find the diameter and convert to the nearest $\frac{1}{8}''$. Round up.
$= 2\frac{1}{8}$ in. I.D.	7. Note that doubling the area does *not* double the diameter.

H. Find the area of the landing for the circular staircase shown.

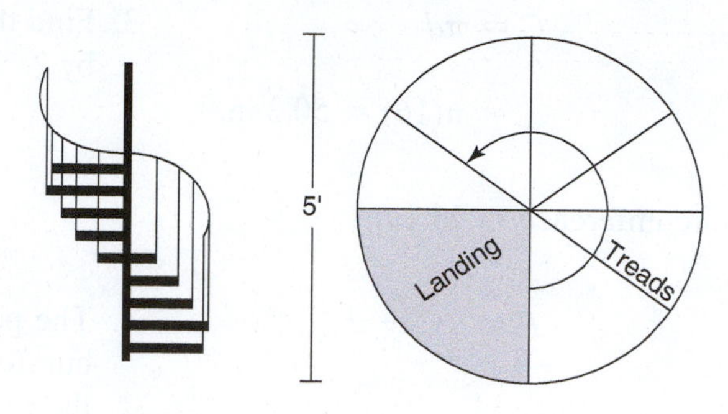

$r = 5 \div 2$	1. Determine the radius.
$A = \pi(2.5)^2 = 19.63\ \text{ft}^2$	2. Find the area of the circle.
$\frac{1}{4} \times 19.63 = 4.9\ \text{ft}^2$	3. Area of the landing is $\frac{1}{4}$ of the area of the circle.

Carpenters often build decorative arches over doors and windows. In order to make the pattern for the arch, the radius of the full circle must be known. A formula that can be used to determine the radius of the full circle is as follows:

$$r = [w^2 \div (8h)] + 0.5h$$

where r = radius of the full circle, w = width of the door or window, and h = height of the arch above the door. Study example I to determine one method for finding the radius of the arch.

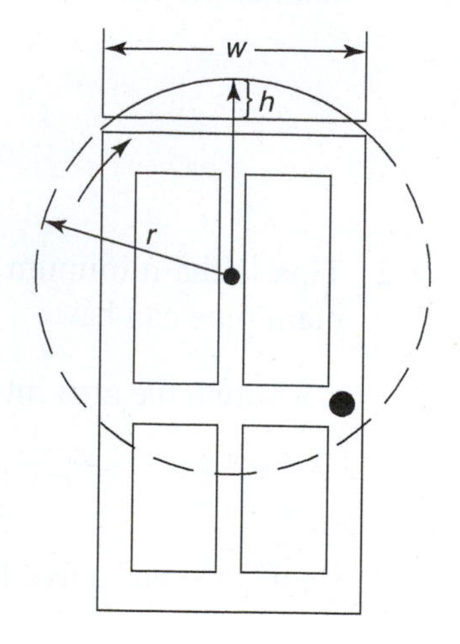

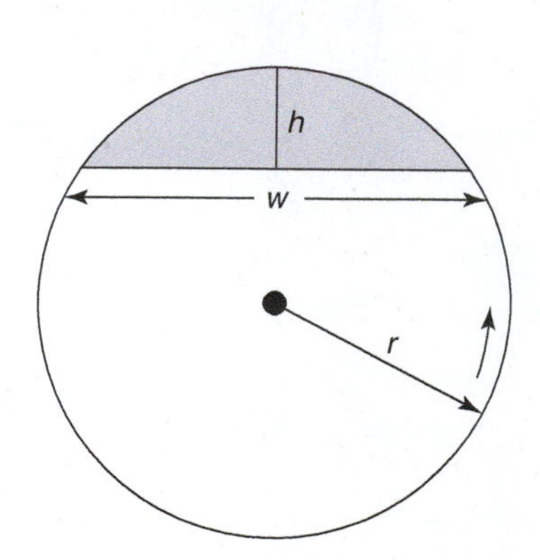

Example

I. A curved arch over a window is to be constructed. The radius of the complete circle must be determined in order to draw the pattern for the arch. The arch is to have a height of 7″ and a width of 36″. Find the radius of the complete circle.

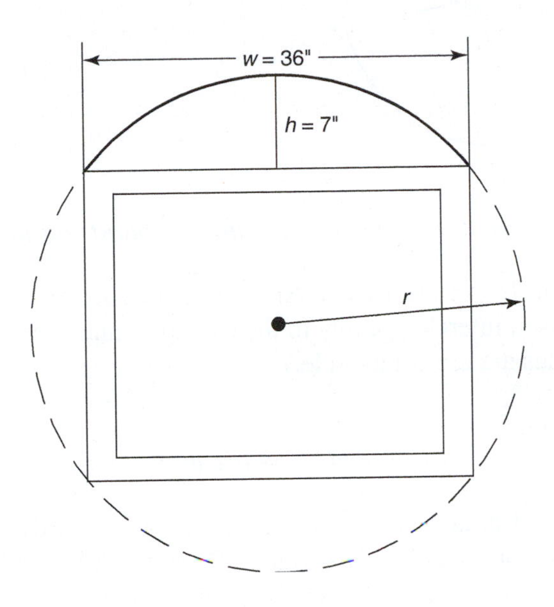

$r = [w^2 \div (8h)] + 0.5h$	1. Use the formula to determine the radius of the circle.
$r = [36^2 \div (8 \times 7)] + 0.5(7)$	2. Substitute the appropriate values into the formula.
$r = [1296 \div 56] + 3.5$	3. Be sure to square w and multiply $8 \times h$ before dividing.
$r = 23.14 + 3.5 = 26.64''$	4. The radius of the circle is about 26.6 inches.
$26.64'' \rightarrow 26\frac{5''}{8}$	5. Convert to the equivalent fraction (an approximation).

A more straightforward method for finding the radius of the arch does not use the formula employed in example I. In order to use this method, several definitions and principles must be understood.

Definitions

A chord is a line segment that joins two points on the circle.

In the diagram (see page 172), AB is a chord, as is CD. EF is a specific type of chord, the diameter of a circle. All diameters are chords, but not all chords are diameters.

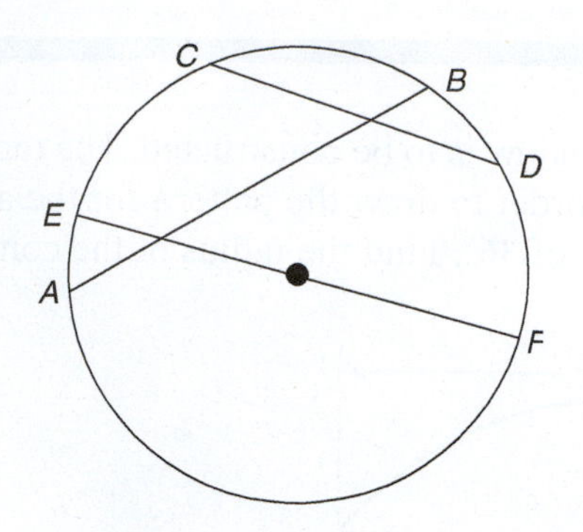

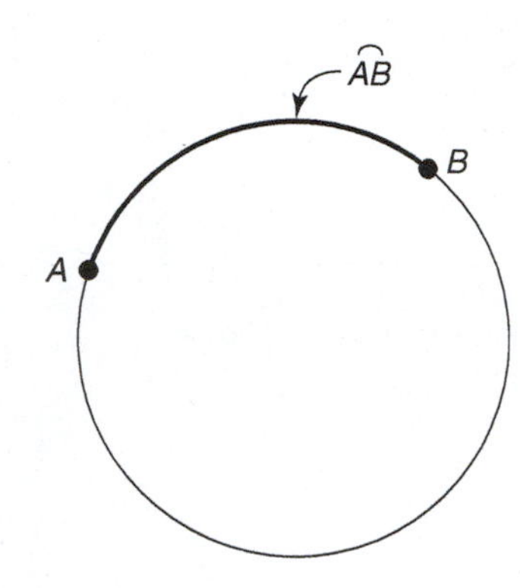

An arc is a part of a circle between any two points on the circle.

The symbol $\overset{\frown}{AB}$ means arc AB. An arch (construction term) and an arc (mathematical term) are used interchangeably in the examples and problems in this book. (Technically not all arches are arcs of a circle.)

A segment is the area formed by an arc and a chord.

The darkened area between $\overset{\frown}{AB}$ (arc) and AB (chord) is a segment. A curved arch above a door or window is a segment. Do not confuse a line segment (meaning part of a line) with a segment of a circle (part of a circle). One is a distance and the other one an area.

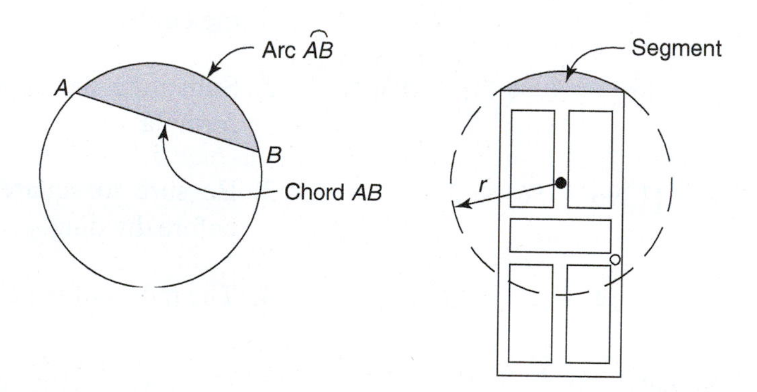

To bisect a chord (or line or arc, etc.) is to divide it into equal segments.

Chord AB is divided into equal segments a and b.

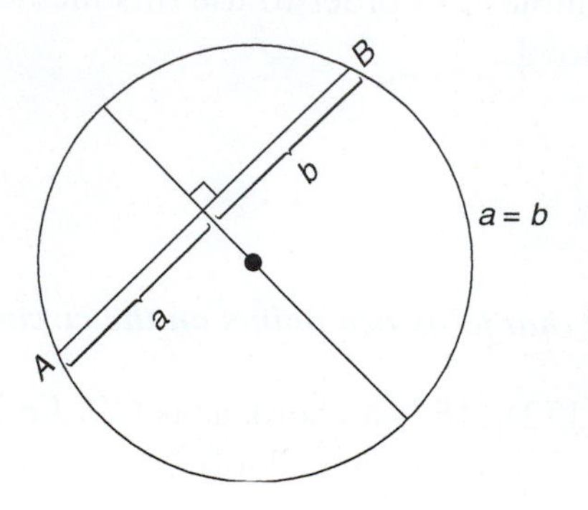

Principles

a. A diameter (a chord which goes through the center of a circle) that intersects a chord at right angles bisects the chord (divides it into equal segments).

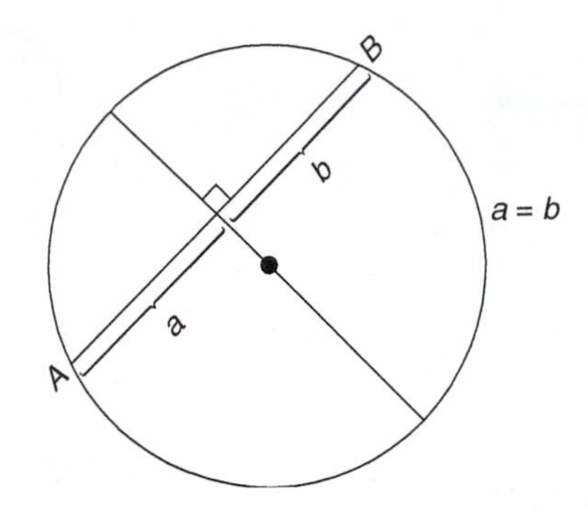

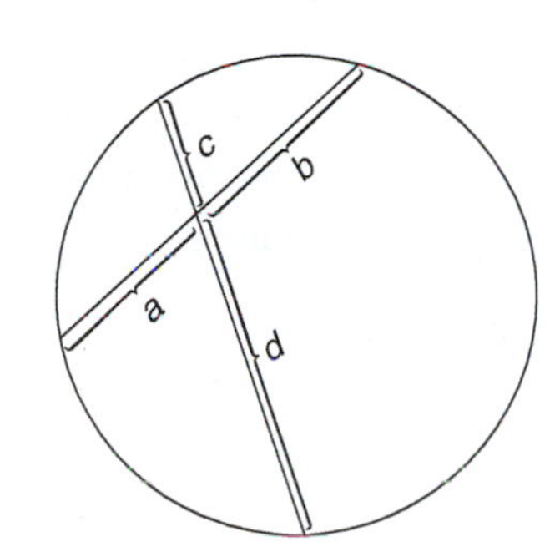

b. If two chords intersect inside a circle, the product of the two segments of one chord equals the product of the two segments of the other chord.

Example 1: $a \times b = c \times d$ or $ab = cd$

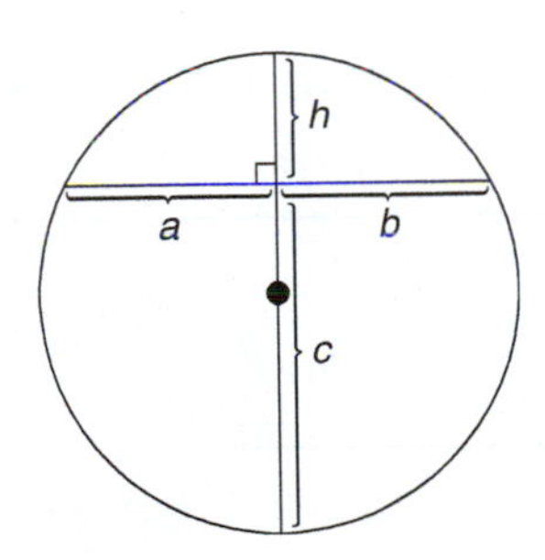

Example 2: $a \times b = h \times c$

In this example, $a = b$ (principle a, above). Therefore, $hc = a^2$.

Examples

J. Using the definitions and principles just stated, determine the radius of the circle containing the segment shown.

$a = 18''$	1. a will be $\frac{1}{2}$ of 36 inches since the segment is bisected by the diameter (principle a).
$a^2 = hb$ $18^2 = 6b$	2. Application of principle b.
$324 = 6b$ $b = 54''$	3. Square a before dividing by 6.
$6'' + 54'' = 60''$	4. $b + h =$ diameter of the circle.
$r = 30''$	5. Radius is $\frac{1}{2}$ the diameter.

K. A curved arch tops a 6′ sliding glass door. The arch is 6 in. high and 72 in. wide. In order to draw the pattern for the arch, the radius of the full circle must be determined. Find the radius of the segment's circle.

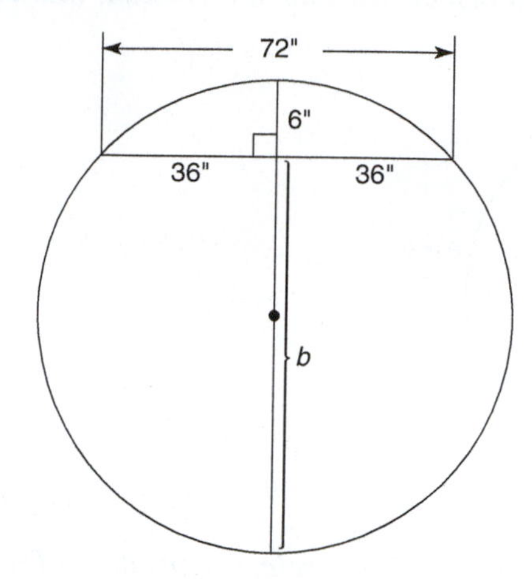

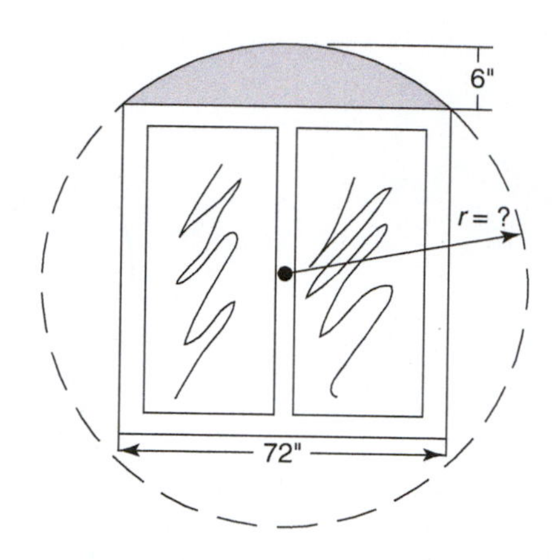

$36^2 = 6b$	1. The product of the line segments of one chord equal the product of the segments of the other chord.
$1296 = 6b$ $b = 216''$	2. Square 36 before dividing by 6.
$b + h = \text{diameter}$ $216'' + 6'' = 222''$	3. The diameter of the circle is 222 inches.
$r = 111''$ $r = 9 \text{ ft } 3 \text{ in.}$	4. The radius is $\frac{1}{2}$ the diameter. Convert to feet and inches.

Exercise 8–2

Complete the table for the circles described.

	Radius	Diameter	Circumference	Area
1.	5 in.			
2.		12 ft		
3.			82.6 ft	
4.		10.35 in.		
5.	8 in.			
6.		2.6 in.		
7.			31.4 in.	
8.	$6\frac{1}{4}$ in.			
9.				50.265 in.^{2}

10. A concrete storm pipe has an I.D. of 5 ft and an outside diameter of 7 ft. What is the cross-sectional area of the concrete walls?

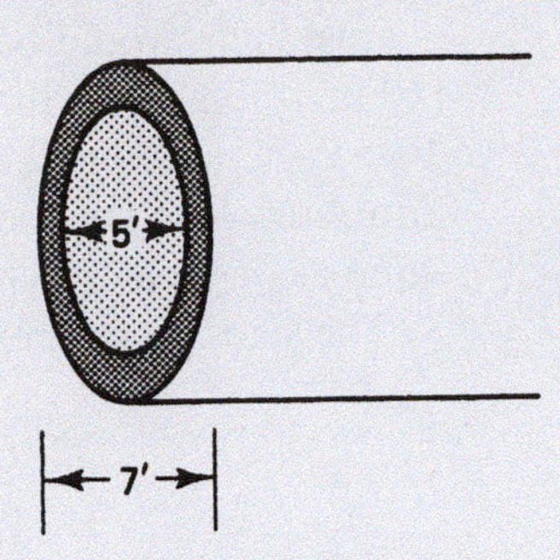

11. A window is a quarter circle with a radius of 1 ft 4 in. How many square inches of glass are in the window?

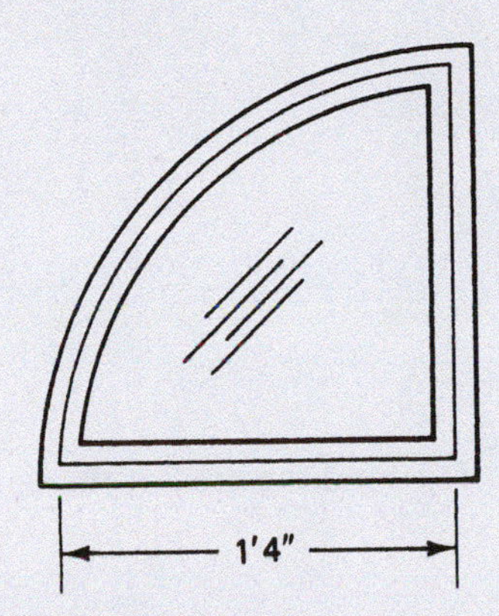

12. To the nearest $\frac{1}{8}''$, what is the perimeter of the window in exercise 11?
13. A walkway is made of circular concrete slabs 18 inches in diameter. What is the area of each slab?

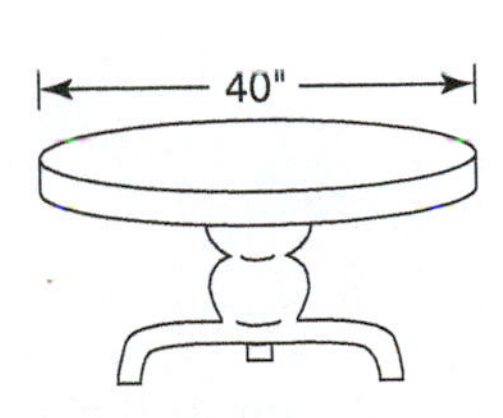

14. A circular dining room table with a diameter of 40 inches is to be covered with woodgrain plastic laminate. How many square feet of laminate are needed? Round up to the next whole number of square feet. See diagram.
15. The table in exercise 14 is to have the edge trimmed with a thin strip of laminate. How long should the laminate strip be? Round up to the next whole number of feet.
16. A tree being sawed into lumber measures 32 inches around. To the nearest inch, what is the diameter of the tree?
17. A tapered newel post has a diameter of $3\frac{1}{8}''$ at the top and $4\frac{1}{2}''$ at the bottom. To the nearest square inch, what is the difference in the cross-sectional area at the top and the bottom of the post?

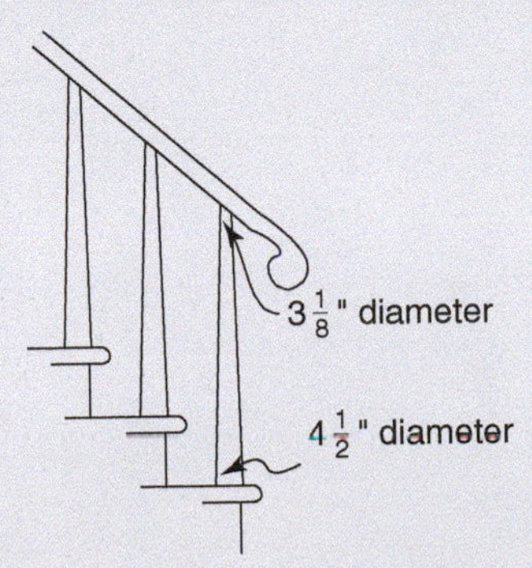

18. A wooden dowel has a diameter of $\frac{3}{4}''$. What is its cross-sectional area?
19. If a 10-in. pizza feeds one medium-hungry carpenter, what size pizza, to the nearest inch, would be needed to split between two medium-hungry carpenters? (10 in. refers to the diameter.)
20. Two $1\frac{1}{2}$-in. and one 2-in. drain pipes merge into a main drain pipe. What is the minimum size the main pipe can be to have a cross-sectional area that is the same as or greater than the three pipes draining into it? Round up to the nearest $\frac{1}{4}''$. The sizes given are diameters. (Refer to example G.)

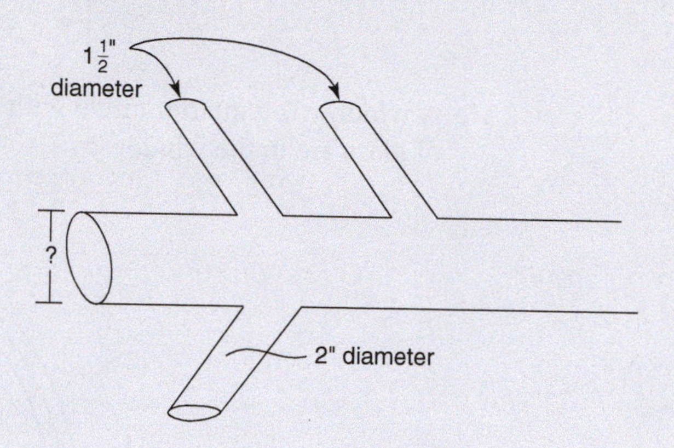

21. Find the area of the landing for the circular staircase shown.

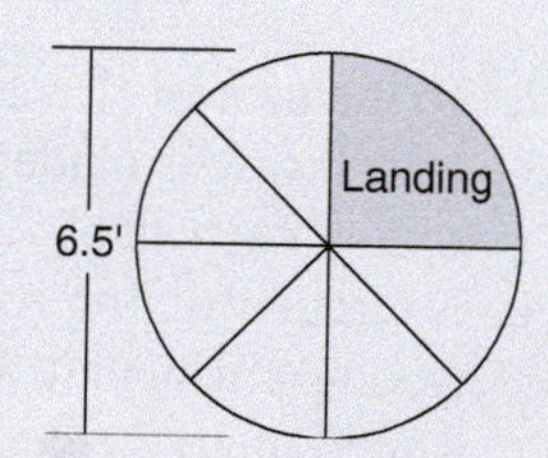

22. A 36″-wide door is to be topped with a curved archway as shown. If the arch is to have a height of 5 inches, find the radius of the circle needed to draw the pattern. Use the formula to determine the radius.

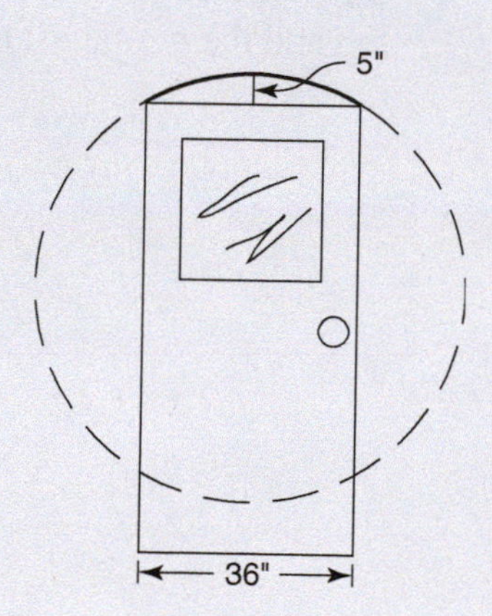

23. A triple-sliding glass door is 108″ wide. Determine the radius needed to draw the pattern for a curved arch that has a height of 10 inches. Convert your answer to feet and inches to the nearest $\frac{1}{16}$″. Use the principle of intersecting chords to determine the radius of the arch.

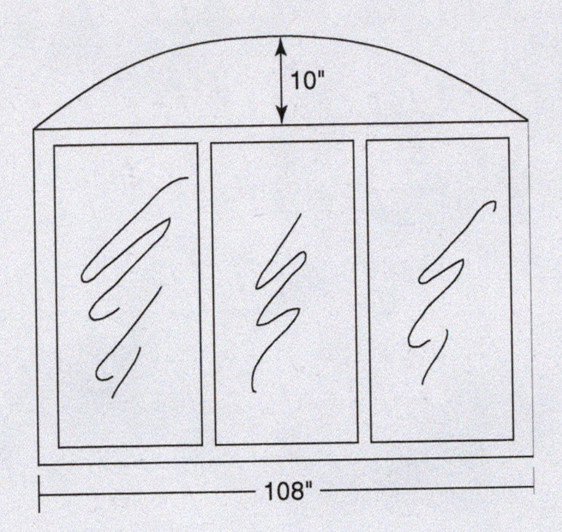

24. The arch over a 34″-wide door is to have a height of 6 inches. Find the radius of the arc's circle.

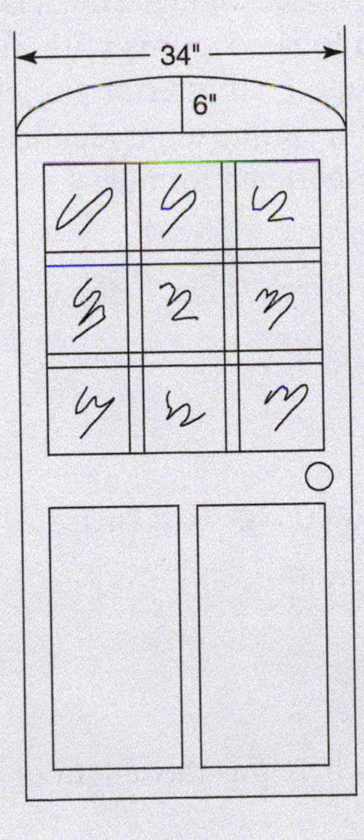

25. A window is a semicircle above a door 3 feet wide. If the door is 6 ft 8 in. tall, what is the height from the floor to the top of the semicircular window?

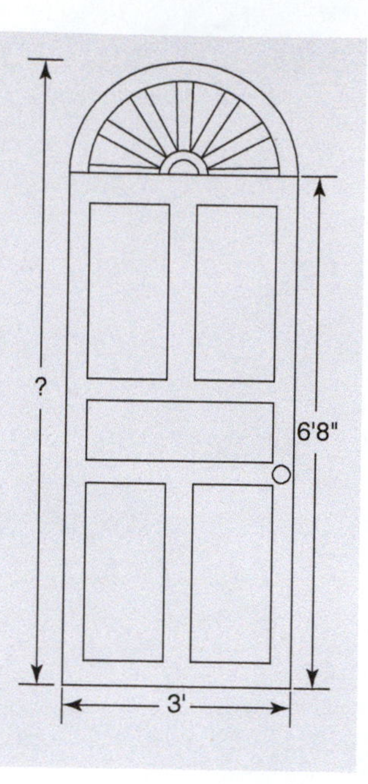

8.3 Octagons

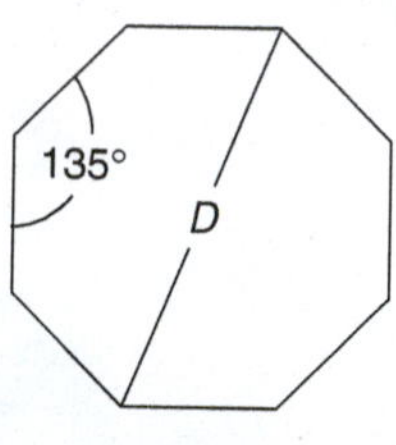

An *octagon* is any eight-sided closed figure. The octagons used most often in building construction are *regular* octagons—figures with all eight sides equal and all eight angles equal. In a regular octagon, all interior angles are 135°. The area of a regular octagon can be found by breaking it into eight identical isosceles triangles; however, we will use one of the following "shortcut" formulas.

$A = 0.7071\ D^2$, where D is the diameter across the angles.

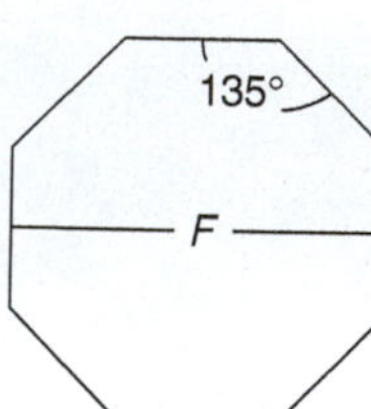

$A = 0.8284\ F^2$, where F is the distance across the flats.

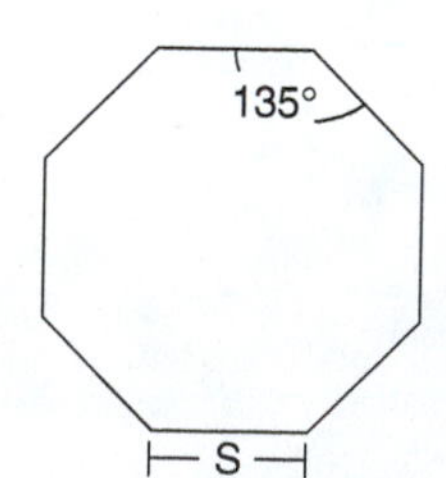

$A = 4.8275\ S^2$, where S is the length of one side.

All three formulas work equally well. Which one is used depends on the information one starts with.

Examples

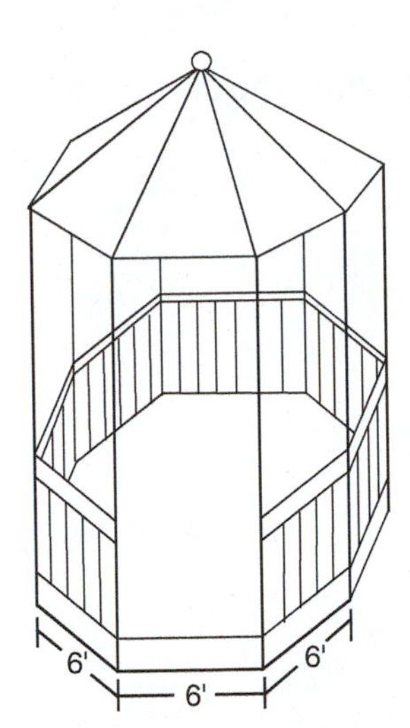

A. Find the area of an octagonal gazebo if the builder wishes to use 6-ft lengths of 2 × 6 lumber for footings.

$A = 4.8275\ S^2$	1. Determine the correct formula to use.
$A = 4.8275\ (6)^2$	2. Square 6 before multiplying by 4.8275.
$A = 173.79\ \text{ft}^2$	3. Calculate the area.
$A = 174\ \text{ft}^2$	4. Round to the nearest whole number of square feet.

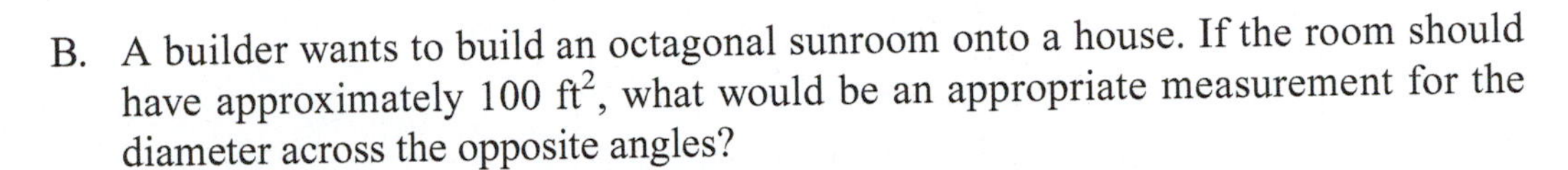

B. A builder wants to build an octagonal sunroom onto a house. If the room should have approximately 100 ft², what would be an appropriate measurement for the diameter across the opposite angles?

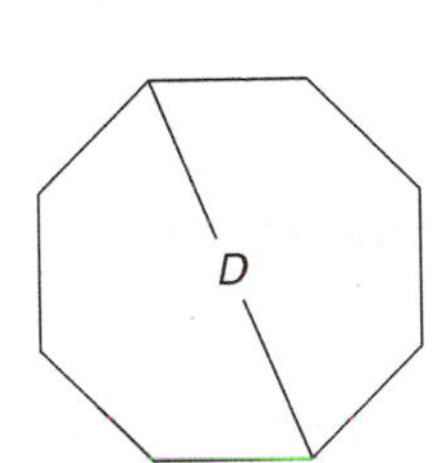

100 square feet

$A = 0.7071\ D^2$	1. Determine the correct formula to use.
$100 = 0.7071\ D^2$	2. Substitute known values into the formula.
$\frac{100}{0.7071} = D^2$	3. Rearrange to solve for D^2.
$D^2 = 141.4;\quad D = 11.89\ \text{ft}$	4. Solve for D.
$D = 11.89 \rightarrow 12\ \text{ft}$	5. A 12-foot diameter across the angles would provide approximately 100 square feet of area.

C. A carpenter is designing a gazebo for a park that will have a 38-ft distance across the flats. Calculate the lengths of the sides.

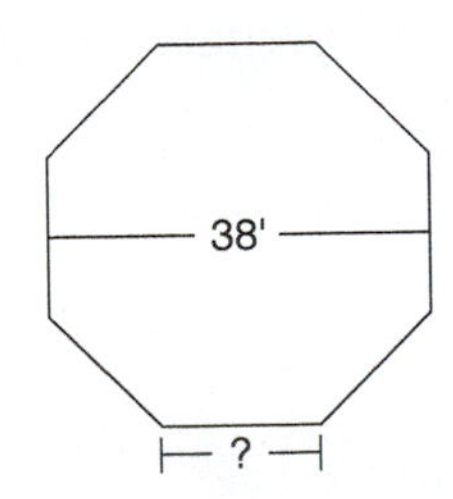

$A = 0.8284\ F^2$	1. Choose the appropriate formula.
$A = 0.8284\ (38)^2$	2. Calculate the area.
$1196\ \text{ft}^2 = 4.8275\ S^2$	3. Substitute the calculated area into the formula that involves the sides S.
$\frac{1196}{4.8275} = S^2$	4. Rearrange and solve for S^2.
$S^2 = 247.8\ \text{ft}^2,\ S = 15.74\ \text{ft}$	5. Solve for S.
Sides require 16 ft lengths	6. The sides will need to be almost 16 ft long in order to have a 38-ft span across the flats.

Exercise 8–3

1. A regular octagon has a diameter across the angles of 15 ft. Find the area of the octagon.

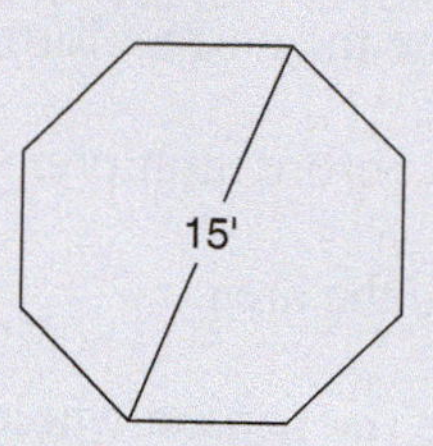

2. An octagonal birdhouse is built with 1-ft sides. What is the area of the birdhouse?

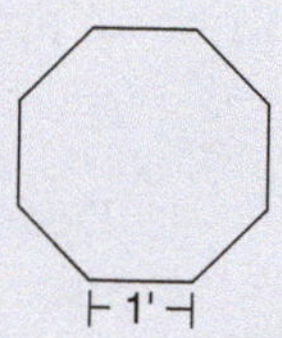

3. A regular octagon has an area of 162 ft^2. Find the distance across the flats. (Round to the nearest whole number.)
4. A regular octagon has a diameter across the angles of 26 ft. Find the length of one side. Round to the nearer whole foot.

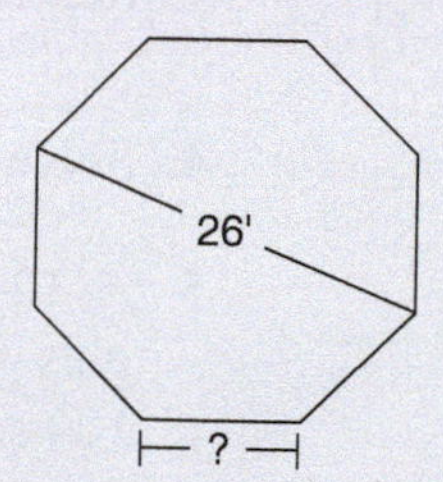

5. A regular octagon has an area of approximately 483 ft^2. Find the perimeter of the octagon. (*Hint:* Find one side first.)

8.4 Composite Figures

To determine the areas and perimeters of composite shapes, divide them into sections that can be determined individually.

Examples

A. Here are three different methods to find the area of the house plan shown.

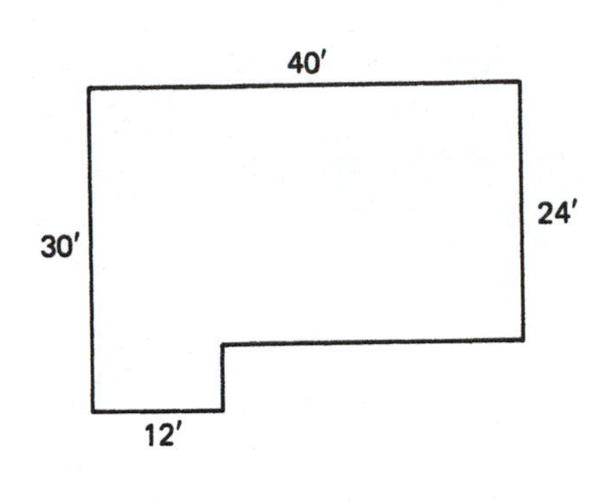

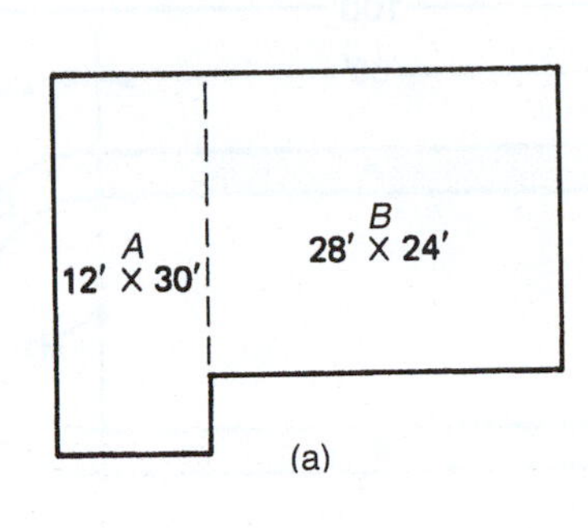

(a)

$$\begin{aligned} A &= 30 \text{ ft} \times 12 \text{ ft} = 360 \text{ ft}^2 \\ B &= 24 \text{ ft} \times 28 \text{ ft} = \underline{672 \text{ ft}^2} \\ \text{Area of house:} & \qquad 1032 \text{ ft}^2 \end{aligned}$$

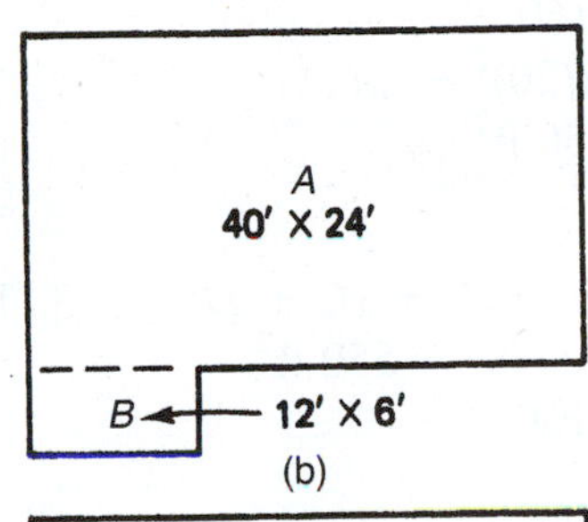

(b)

$$\begin{aligned} A &= 24 \text{ ft} \times 40 \text{ ft} = 960 \text{ ft}^2 \\ B &= \ \ 6 \text{ ft} \times 12 \text{ ft} = \underline{\ \ 72 \text{ ft}^2} \\ \text{Area of house:} & \qquad 1032 \text{ ft}^2 \end{aligned}$$

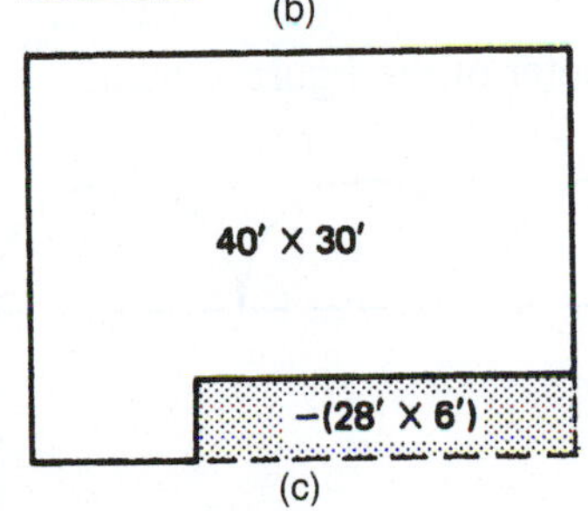

(c)

$$\begin{aligned} 40 \text{ ft} \times 30 \text{ ft} &= \ \ 1200 \text{ ft}^2 \\ -(28 \text{ ft} \times 6 \text{ ft}) &= \underline{-168 \text{ ft}^2} \\ \text{Area of house:} & \qquad 1032 \text{ ft}^2 \end{aligned}$$

In the third case, the area of the expanded rectangle is found and the portion that is not on the blueprint is then subtracted.

B. The 5-ft-wide track shown below is to be paved. How many square feet are to be paved? Assume that the ends are semicircles.

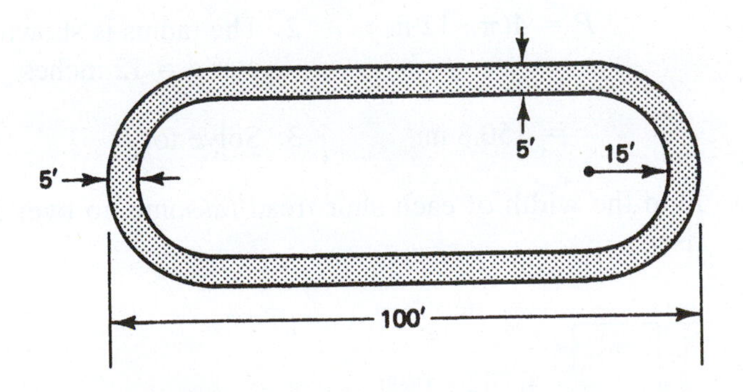

1. Areas A and B are rectangles 60 ft × 5 ft. Study how the length of 60 ft was determined.

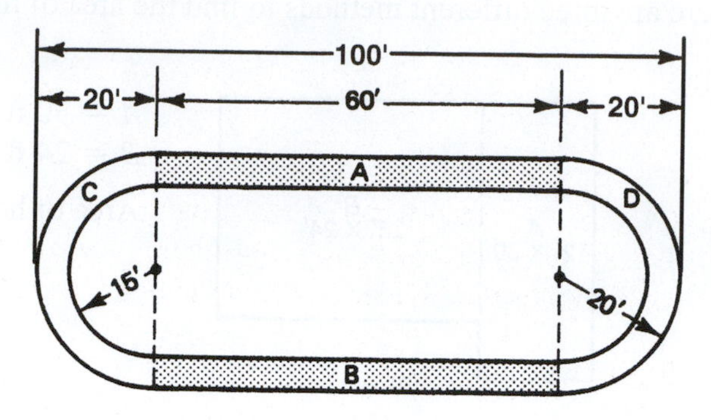

$$A + B = 2(60 \times 5) = 600 \text{ ft}^2$$

$$\text{Area}_{C+D} = \pi(20)^2 - \pi(15)^2 = 550 \text{ ft}^2$$

2. The two ends combined form a ring with outside radius = 20 ft and inside radius = 15 ft.

$$\text{Total area} = (A + B) + (C + D) = 600 \text{ ft}^2 + 550 \text{ ft}^2 = 1150 \text{ ft}^2$$

3. Find the sum of all four areas.

C. Find the perimeter of the figure shown.

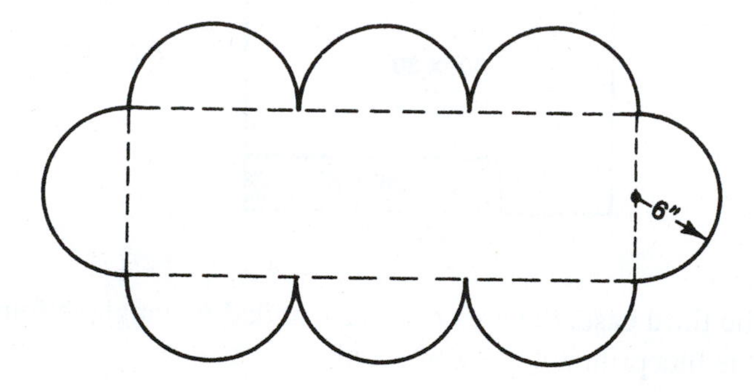

Perimeter $= 4(\pi d)$

1. The perimeter is the equivalent of the circumferences of four circles.

$P = 4(\pi \cdot 12 \text{ in.})$

2. The radius is shown as 6 inches; hence the diameter $d = 12$ inches.

$= 150.8$ in.

3. Solve for P.

D. Find the width of each stair tread (assume no overhang) and the height of each riser.

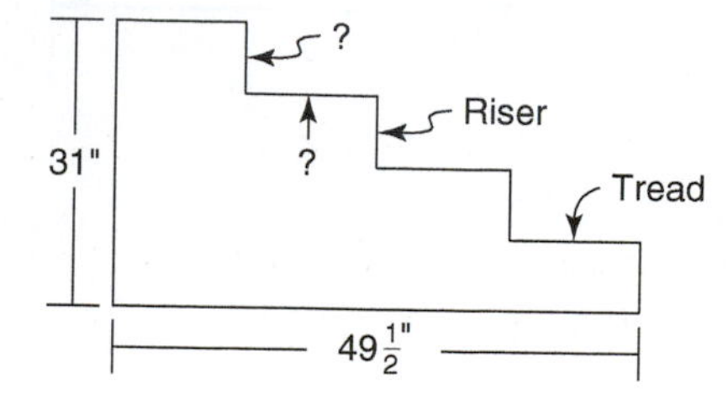

$31'' \div 4 = 7\frac{3''}{4}$ 1. To find the height of each riser, divide the total rise by the number of risers.

$49\frac{1''}{2} \div 4 = 12\frac{3''}{8}$ 2. Divide the total run by the number of treads.

E. A custom-made window is a semicircle positioned atop a rectangle, as shown. (This is known as a Norman window.) (a) Find the area of the glass (ignore the divider) and (b) find the total number of feet of trim needed. The trim will be the perimeter of the figure plus the divider between the two window sections.

r

6'

7' 6" (7.5')

a. $7.5 \text{ ft} - 6 \text{ ft} = 1.5 \text{ ft}$ 1. The radius of the semicircle is 1.5 ft.

$A = \pi r^2 \rightarrow \pi(1.5^2) = 7.1 \text{ ft}^2$ 2. The area of an entire circle is 7.1 ft^2.

$A = 7.1 \div 2 = 3.5 \text{ ft}^2$ 3. The area of the semicircle is 3.5 ft^2.

$D = 2 \times 1.5 \text{ ft} = 3 \text{ ft}$ 4. The diameter of the semicircle is twice the radius, and the diameter equals the width of the window.

$A_{rectangle} = 3 \text{ ft} \times 6 \text{ ft} = 18 \text{ ft}^2$ 5. The area of the rectangular window.

$18 \text{ ft}^2 + 3.5 \text{ ft}^2 = 21.5 \text{ ft}^2$ 6. The total area of the window is the sum of the rectangular and semicircular window sections.

b. $C = \pi D \rightarrow \pi(3 \text{ ft}) = 9.4 \text{ ft}$ 1. This is the circumference of the entire circle.

$9.4 \div 2 = 4.7 \text{ ft}$ 2. This is the perimeter of the top of the semicircle.

$P_{rectangle} = 2L + 2W$ 3. This is the formula for the area of a rectangle.

$P_{rectangle} = 2(6 \text{ ft}) + 2(3 \text{ ft}) = 18 \text{ ft}$ 4. The perimeter of the rectangle is 18 ft.

$18 \text{ ft} + 4.7 \text{ ft} = 22.7 \text{ ft} \rightarrow 23 \text{ ft}$ 5. This is the amount of trim needed (ignoring waste for cutting and fitting).

Note: It is a coincidence that both the area of the rectangle and the perimeter of the rectangle equal 18. These numbers are usually different. Even in this case, they are *not* the same. The area equals 18 *square* feet and the perimeter equals 18 *linear* feet. Linear feet and square feet are not interchangeable.

F. The degrees in a regular polygon (a multisided figure) can be found by the following formula: Degrees = 180°(N – 2), where N is the number of sides (and the number of angles) of the figure. Since all angles (and sides) are equal in a regular polygon, the number of degrees in each angle is $\frac{180°(N - 2)}{N}$.

For a regular hexagon (6-sided figure), the number of degrees in each angle can therefore be found by the formula $\frac{180°(6 - 2)}{6} = 30°(4) = 120°$.

Find the perimeter and area of the regular hexagon shown below:

$P = 6 \times 10'' = 60''$

1. Each side is 10 inches; therefore the perimeter of the hexagon is 60 inches

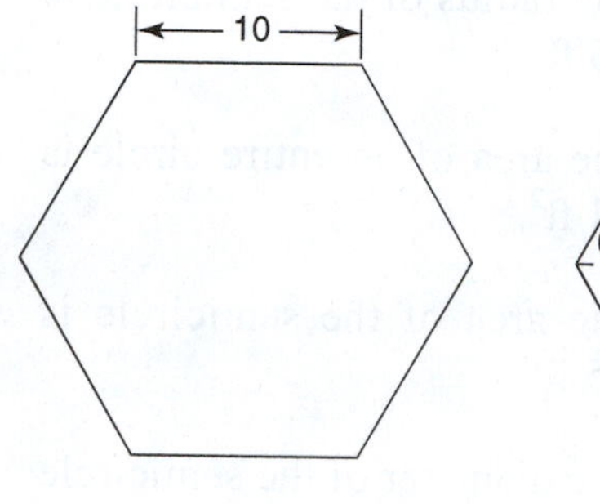

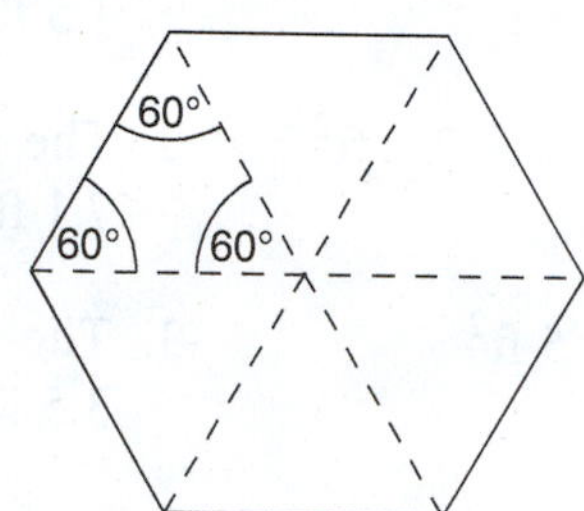

2. Divide the hexagon into 6 congruent equilateral triangles

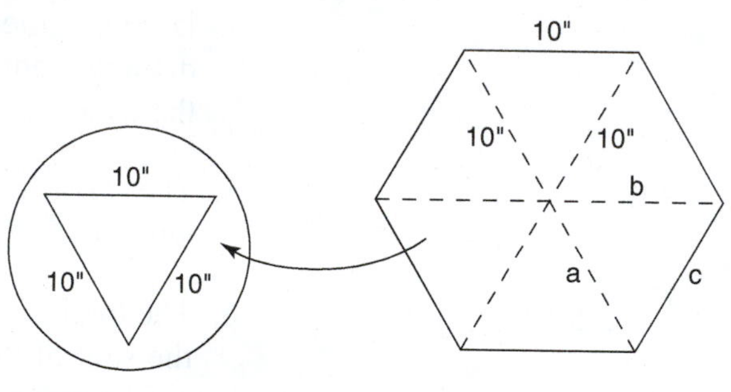

3. Each side of each equilateral triangle is 10 inches, as shown, and each angle is 60°. (Why?)

$$A_\Delta = \sqrt{s(s - a)(s - b)(s - c)}$$
$$A_\Delta = \sqrt{15(15 - 10)(15 - 10)(15 - 10)}$$
$$A_\Delta = \sqrt{1875} \doteq 43.3 \text{ in}^2$$

4. Use Hero's formula to find the area of one triangle $S = \frac{1}{2}P$ (perimeter of triangle), and $a = b = c =$ sides of triangle

$$43.3 \text{ in}^2 \times 6 \doteq 260 \text{ in}^2$$
$$A_{⬡} \doteq 260 \text{ in}^2$$

5. Multiply by 6 to find the area of the hexagon.

G. Find the perimeter and area of the regular hexagon shown below. This time we will utilize the special properties of the 30°–60°–90° triangles. Recall that triangles with those angles have corresponding sides in the ratio of $1:\sqrt{3}:2$

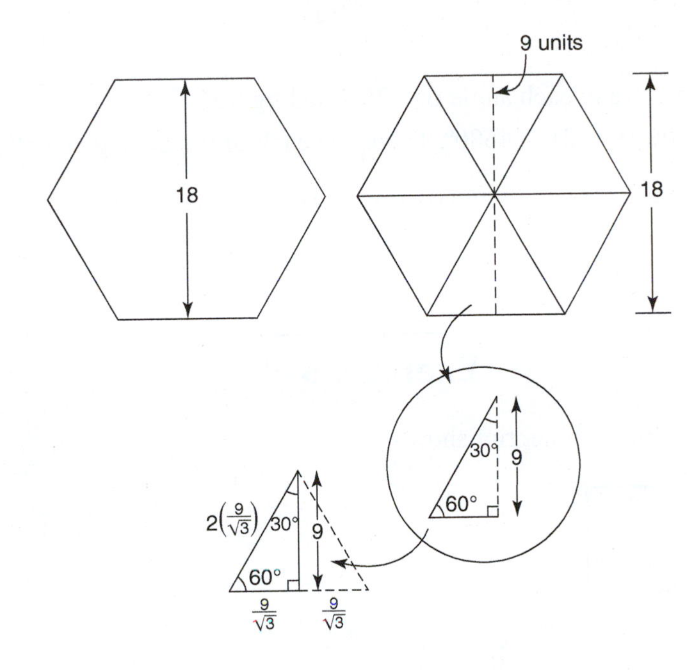

1. The distance across the flats is 18. Therefore the height of each of the 6 equilateral triangles is 9.

2. Using the special 30°–60°–90° triangle relationship: sides are in the ratio of $1:\sqrt{3}:2$

3. Sides are $\dfrac{9}{\sqrt{13}} : 9 : 2\left(\dfrac{9}{\sqrt{3}}\right)$

4. Sides $\doteq$ 5.196, 9, and 10.39 for each 30–60–90° triangle

$A_\Delta = \frac{1}{2}bh;\ A_\Delta = \dfrac{1}{2}\left(2\dfrac{9}{\sqrt{3}}\right)$

$A_\Delta \doteq 46.76$ sq. units

5. Approximate area of each equilateral triangle (this will be twice the size of the 30–60–90° triangle calculated above)

$A_{\hexagon} \doteq 6(46.76) \doteq 281$ sq. units

6. Approximate area of 6 triangles = approximate area of hexagon

$P_{\hexagon} \doteq 6\left(2\left(\dfrac{9}{\sqrt{3}}\right)\right) \doteq 62$ units

7. Approximate perimeter = 6 × one side

H. **Important concepts:**

1. In every polygon the number of sides equals the number of angles. (A 5-sided figure has 5 angles, an 11-sided figure has 11 angles, etc.)
2. All angles in a regular polygon will be equal. Therefore, a 5-sided figure (called a pentagon) will have 5 angles, and each angle will be $\frac{1}{5}$ the sum of all angles in the pentagon.
3. The total number of degrees in all angles in a polygon can be found by the following formula: $\text{Degrees}_{\text{total}} = 180°(\text{N} - 2)$ where N is the number of sides (or angles).
4. The number of degrees in each angle of a regular polygon can be found by the formula: $\text{Degrees}_{\text{each angle}} = \dfrac{180°(\text{N} - 2)}{\text{N}}$.

We can see that a regular 10-sided figure will have the following degrees in each angle: $\text{Degrees}_{\text{total}} = 180°(10 - 2) = 180(8) = 1440°$. Therefore, each of the 10 angles will have:

$$\frac{1440}{10} = 144°$$

How many degrees are in each angle of a 28-sided figure?

$\text{Degrees}_{\text{total}} = 180°(28 - 2) = 4680°$. Therefore each of the 28 angles will have

$$\text{Degrees}_{\text{each}} = \frac{180°(28 - 2)}{28} = \frac{4680}{28} = 167°$$

Exercise 8–4

1. Find the area of the blueprint shown.

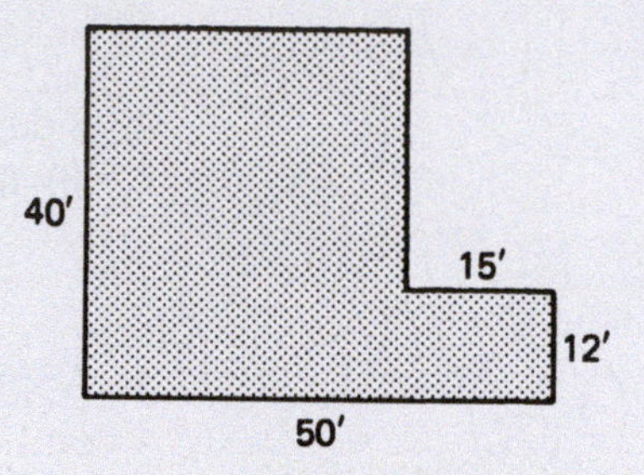

2. Find the perimeter of the figure in exercise 1.
3. Find the area of the track.

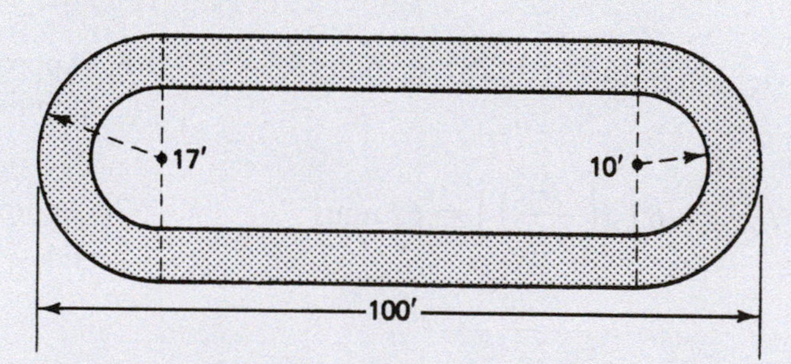

4. Find the outside perimeter of the track in exercise 3.
5. Find the inside perimeter of the track in exercise 3.
6. The square in the center of the figure has an area of 100 in^2. Find the area of the entire figure. (*Hint:* The sides of the square are the diameters of the semicircles.)

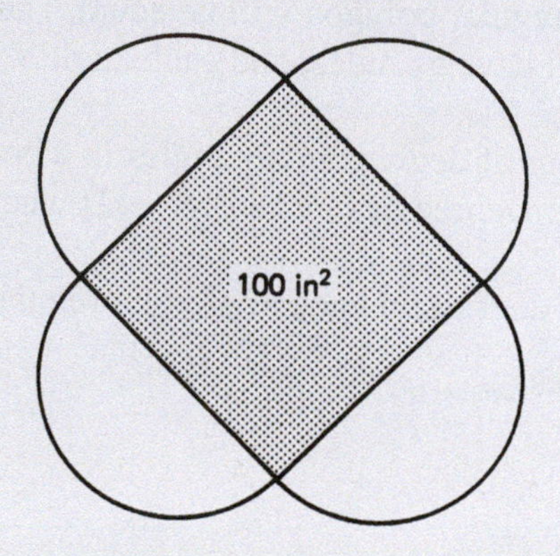

7. Find the perimeter of the figure shown in exercise 6.
8. Find the shaded area and the perimeter of the square.

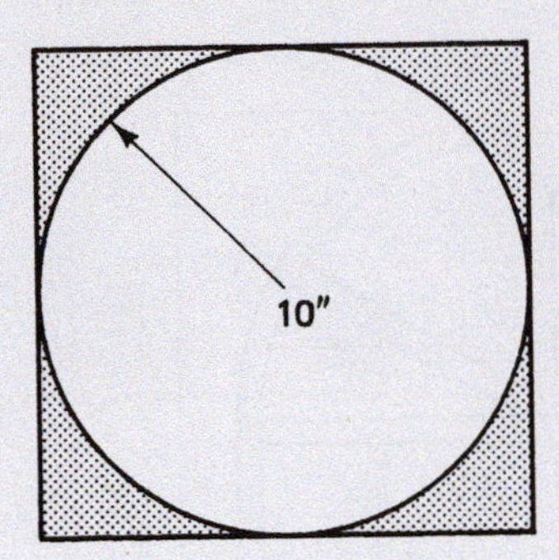

Find the perimeter and area of the following figures.

9.

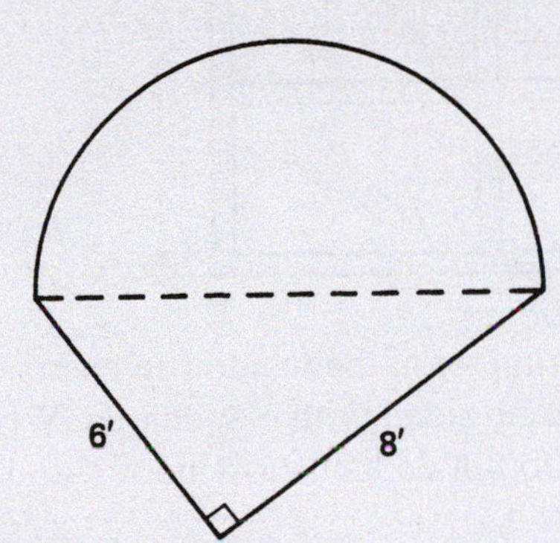

10.

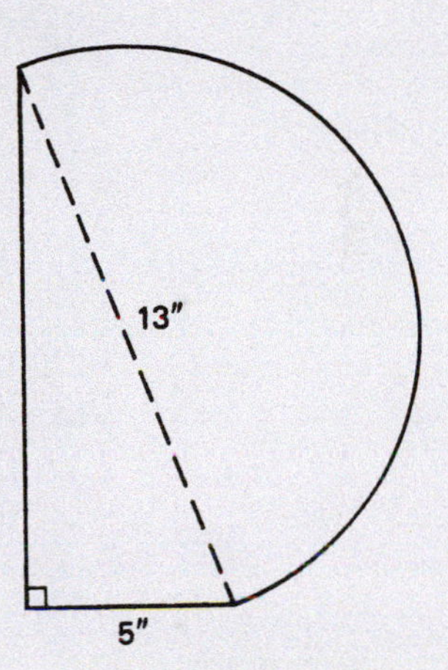

11.

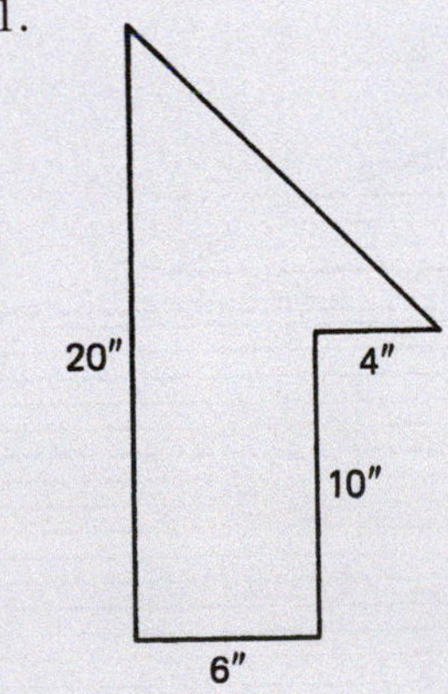

12. A semicircular window is positioned atop a rectangular window as shown. Determine the total window area (glass and dividers).

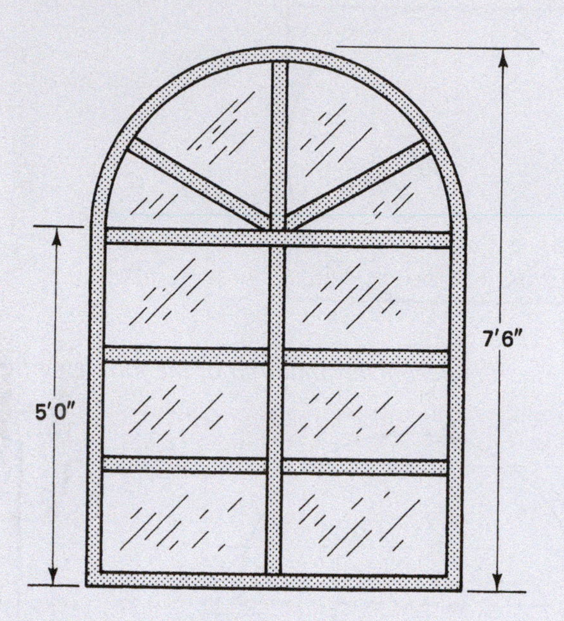

13. Determine the perimeter of the window in exercise 12.
14. A new house is to be sided with clapboards. What is the area in square feet of the end shown? Do not deduct for windows and doors. (*Hint:* Divide into two trapezoids, as shown.)

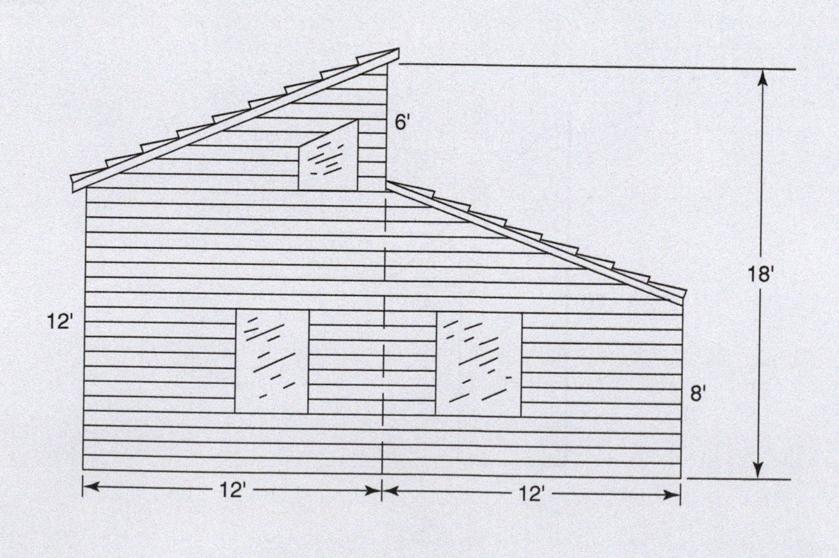

15. A tabletop for a corporation conference room is shown. Find the area of the tabletop.

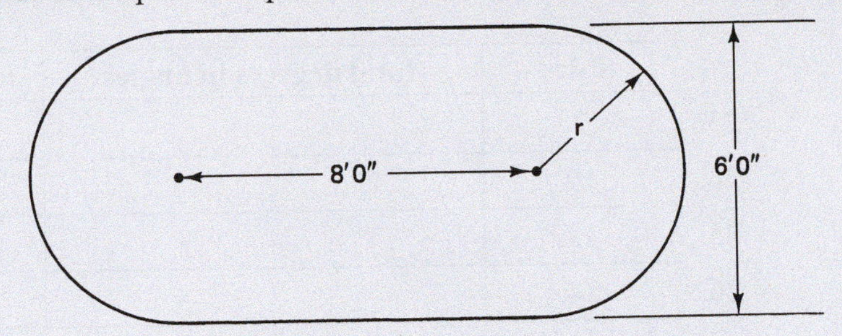

16. Find the perimeter of the tabletop in exercise 15.
17. Find the area of the cross-section of the steel I-beam shown.

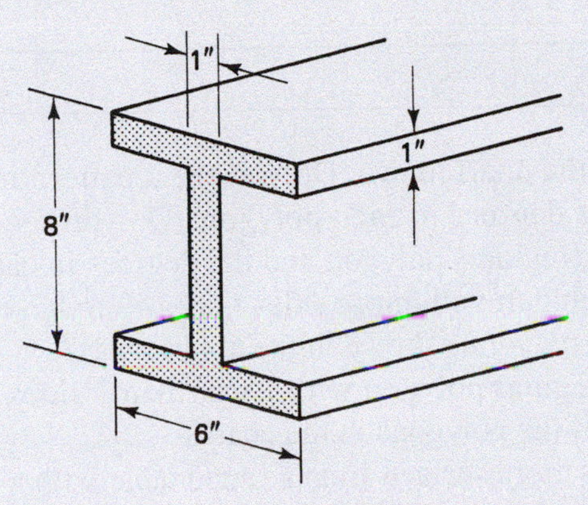

18. Find the perimeter of the cross-section of the I-beam in exercise 17.
19. A rectangular pool has a semicircular spa attached to one side. Find the total area of the pool and spa.

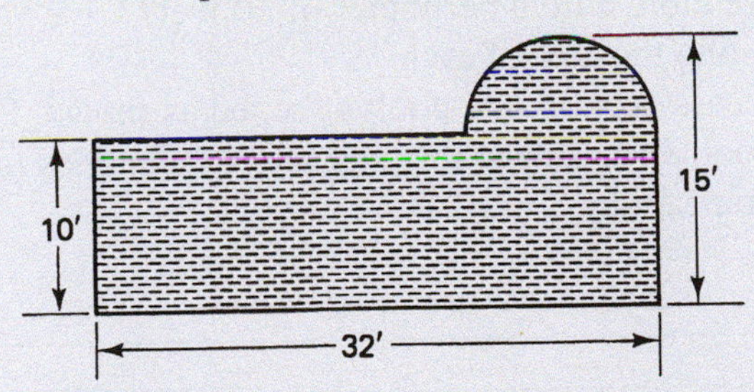

20. Find the perimeter of the pool and spa in exercise 19.
21. A square window 4 feet on a side is topped by a semicircular window. Find the total perimeter of the window.

22. Find the total area of the window in exercise 21 above.

The polygons in the table below are all regular polygons.

	Sides	Total degrees in angles	Degrees in each angle
23.	7		
24.	19		
25.	28		
26.	34		
27.	55		
28.	80		
29.	101		
30.	140		

31. Examine the chart above. Do you see a pattern in the number of sides and the number of degrees in each polygon? Do you see a pattern with the number of sides of a (regular) polygon and the degrees in each angle? Would you expect a regular polygon with more sides than 140 to have smaller or larger angles than any of the polygons shown in the above chart?
32. Would a regular polygon with fewer than 7 sides have smaller or larger angles than any of the polygons in the chart?
33. Which has more surface area: a round table with a diameter of 4 feet, or a square table 4 feet on a side? Give your answer before calculating. Then find the area of each.
34. Which has more surface area: a round table with a circumference of 20 feet, or a square table with a perimeter of 20 feet? Give your answer before calculating. Then find the area of each.
35. The cross-section of a pool is shaped as shown. Find the cross-sectional area of the pool. *Hint:* divide the pool into 2 rectangles (deep end and shallow end) and one trapezoid.

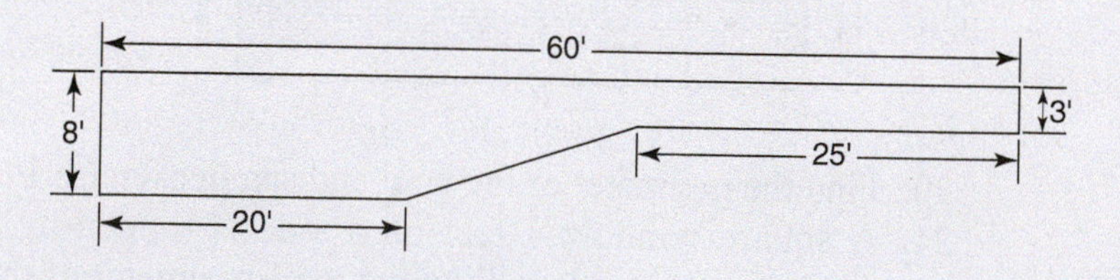

CHAPTER 9

Volume and Surface Area of Solids

Did You Know

A goal of the Building Technology Program (BTP) is a net-zero building (one that produces as much energy as it consumes).

OBJECTIVES

Upon completing this chapter, the student will be able to:

1. Find the volume of rectangular solids.
2. Find the LSA (lateral surface area) and TSA (total surface area) of rectangular solids.
3. Find the volume of right prisms.
4. Find the LSA and TSA of right prisms.
5. Find the volume of right cylinders.
6. Find the LSA and TSA of right cylinders.
7. Find the volume of cones.
8. Find the LSA of cones.
9. Find the volume of spheres.
10. Find the surface area of spheres.
11. Solve problems involving volume and surface area of spheres.
12. Find the volume of composite shapes.
13. Find the surface area of composite shapes.
14. Solve problems involving volume and surface area of composite shapes.

9.1 Rectangular Solids and Prisms

Rectangular solids are three-dimensional figures with six sides, all of which are rectangles (or squares). The adjoining sides of a rectangular solid are perpendicular to each other. Figure 9–1 shows examples of rectangular solids.

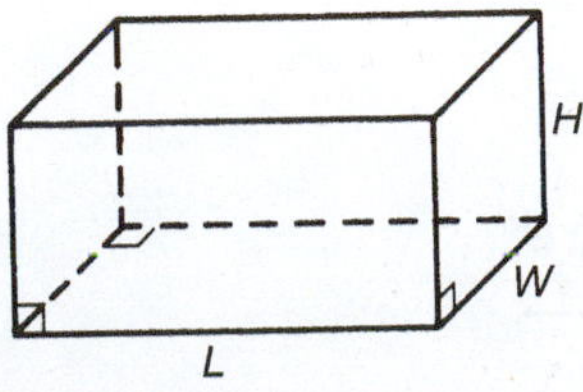

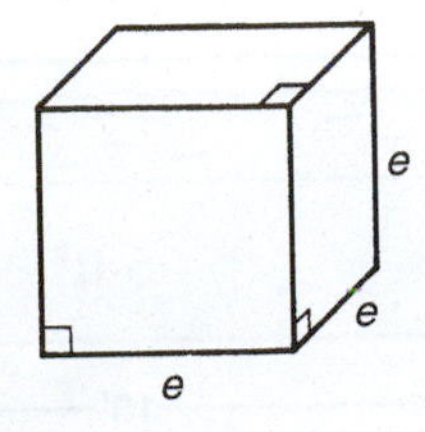

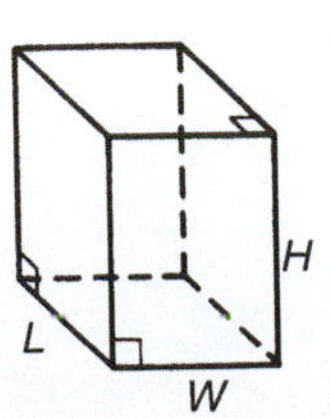

Figure 9–1

The volume of a rectangular solid is found by the formula

$$V = LWH$$

where V is the volume, L the length, W the width, and H the height. When no operation symbol is shown, multiplication is implied. Hence L, W, and H are multiplied together.

Examples

A. Find the volume of air to be heated in a room that measures 12 ft × 18 ft with 8-ft-high ceilings.

$V = LWH$	1. The area of the room ($L \times W$) times the height equals the volume.
$= 18 \text{ ft} \times 12 \text{ ft} \times 8 \text{ ft}$	2. Substitute given values into the formula.
$= 1728 \text{ ft}^3$	3. Volume is measured in cubic units.

B. What is the volume of a cube that is 10 inches on an edge?

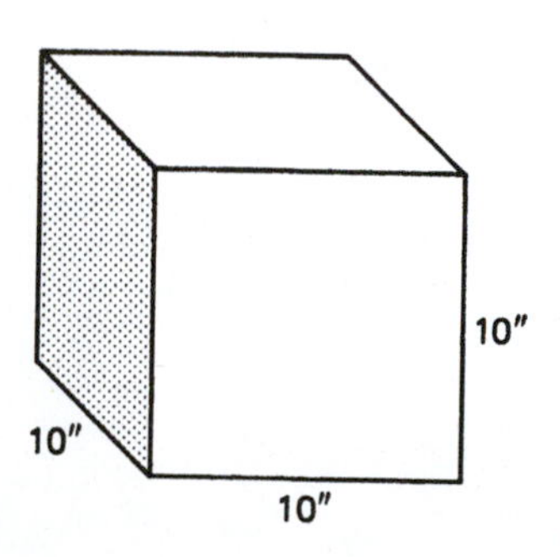

$V = e^3$	1. A cube is a rectangular solid with length, width, and height equal. Therefore, $V = LWH = e^3$, where e stands for any edge.
$= (10 \text{ in.})^3$	2. Substitute known values into the formula.
$= 1000 \text{ in.}^3$	3. in.^3 is the abbreviation for cubic inches.

C. The volume of a rectangular watering trough is 75 ft^3. If the trough is 2 ft 6 in. deep and 10 ft long, how wide is it?

$V = LWH$ $75 \text{ ft}^3 = 10 \text{ ft} \times W \times 2.5 \text{ ft}$	1. Change 2 ft 6 in. to 2.5 ft and substitute known values into the formula.
$= 25 \text{ ft}^2 \times W$ $\dfrac{75 \text{ ft}^3}{25 \text{ ft}^2} = W$	2. Simplify and rearrange formula.
$W = 3 \text{ ft}$	3. Divide to obtain the width. Note that cubic feet ÷ square feet yield (linear) feet.

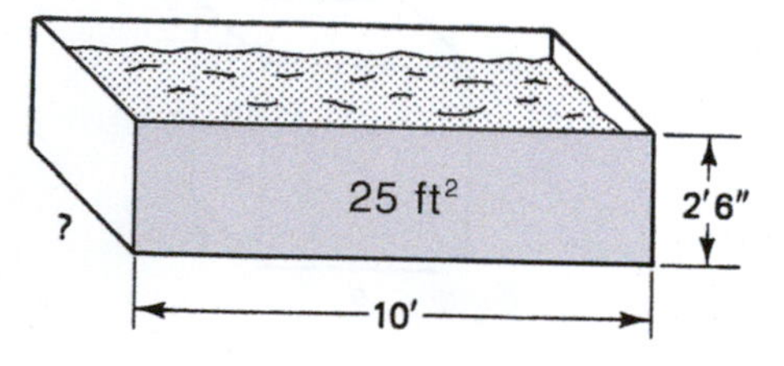

D. A concrete foundation 8 ft deep measures 24 ft × 40 ft (outside dimensions) and has walls 9 in. thick. How many cubic yards of concrete is there in the foundation walls?

$$V_{\text{outside}} = (40 \text{ ft})(24 \text{ ft})(8 \text{ ft}) = 7680 \text{ ft}^3$$

1. Find the volume of the outside dimensions of the walls.

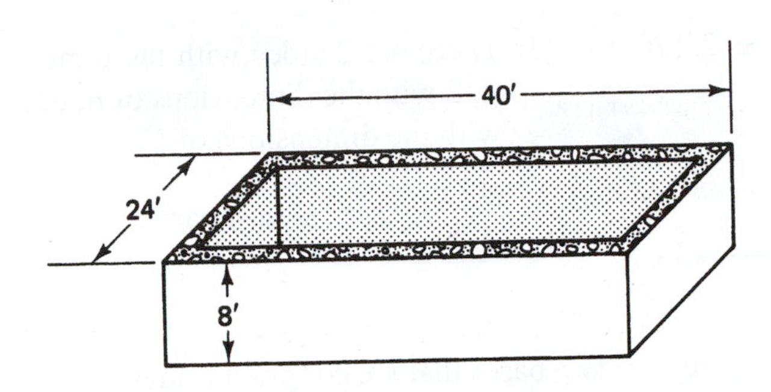

$$40 \text{ ft} - 1.5 \text{ ft} = 38.5 \text{ ft}$$
$$24 \text{ ft} - 1.5 \text{ ft} = 22.5 \text{ ft}$$

2. Find the inside dimensions of the foundation by subtracting twice the thickness of the walls from the outside dimensions. (9-in. walls on both sides = 18 in. = 1.5 ft)

$$V_{\text{inside}} = (38.5 \text{ ft})(22.5 \text{ ft})(8 \text{ ft}) = 6930 \text{ ft}^3$$

3. Find the volume of the inside dimensions of the walls.

$$V_{\text{walls}} = 7680 \text{ ft}^3 - 6930 \text{ ft}^3$$

4. The difference in the volume of the outside and inside dimensions is the volume of the walls in cubic feet.

$$= \frac{750 \text{ ft}^3}{1} \times \frac{1 \text{ yd}^3}{27 \text{ ft}^3}$$

5. Convert cubic feet to cubic yards.

$$= 27.8 \text{ yd}^3$$

6. Approximately 28 yd^3 of concrete are required for the foundation walls. For another approach, see Chapter 13.

The *total surface area* (TSA) is the sum of the areas of all the sides of a rectangular solid.

Example

E. A craftsman is making a 12 in. × 8 in. × 10 in. jewelry box that will be covered with a rosewood veneer. Find the total surface area to be covered.

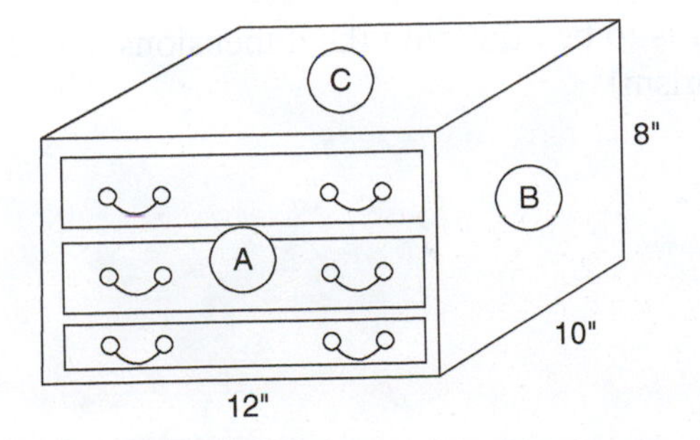

$TSA = 2A + 2B + 2C$	1. The TSA is twice the area of sides A, B, and C.
$A = 12\text{ in.} \times 8\text{ in.} = 96\text{ in.}^2$ $B = 10\text{ in.} \times 8\text{ in.} = 80\text{ in.}^2$ $C = 10\text{ in.} \times 12\text{ in.} = 120\text{ in.}^2$	2. Find the area of each side A, B, and C.
$TSA = 2(96) + 2(80) + 2(120)$	3. There are 2 sides with the dimensions of A, 2 with the dimensions of B, and 2 sides with the dimensions of C.
$TSA = 592\text{ in.}^2$	4. Surface area is in square units.

A right *prism* is a solid figure with two bases that are congruent (identical) polygons and faces that are rectangles. The faces of a right prism are always perpendicular to the bases. Figure 9–2 shows several examples of right prisms. (Note that a rectangular solid is a special case of a prism. In this book, only right prisms are discussed; therefore, any reference to "prism" will mean "right prism.")

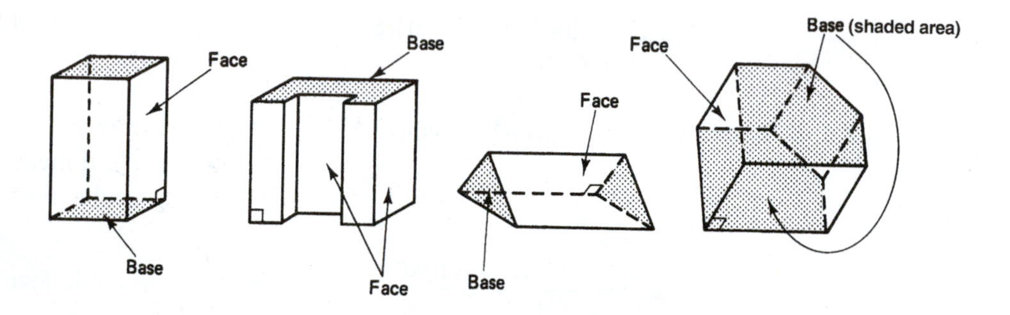

Figure 9–2

The volume of any right prism is found by the same formula, regardless of its shape:

$$V = B \times h$$

where V is the volume of the prism, B the area of the base, and h the height of the prism. (Height is the dimension perpendicular to the base.)

Example

F. A small barn is to be built with the dimensions shown. Find the volume of the barn (which is a prism).

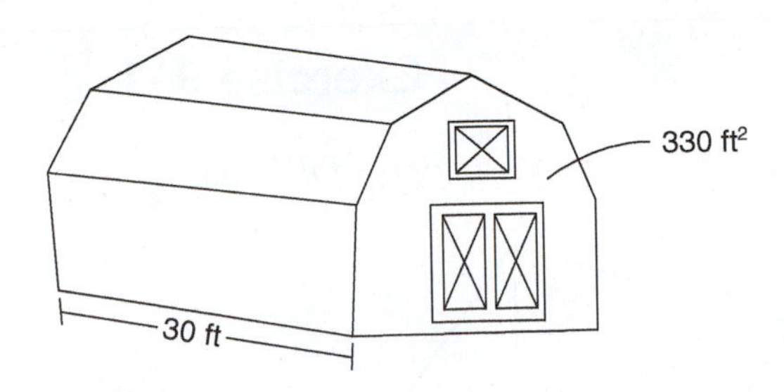

$V = B \times h$ $330 \text{ ft}^2 \times 30 \text{ ft} = 9900 \text{ ft}^3$	1. The bases are the congruent polygons; the height is perpendicular to the base.
$V = 9900 \text{ ft}^3$	2. Volume is in cubic units.

The LSA of a prism is the area of all the *faces*. The TSA is the area of all the *faces* plus the *two bases*.

Example

G. Find the LSA and TSA of the prism shown.

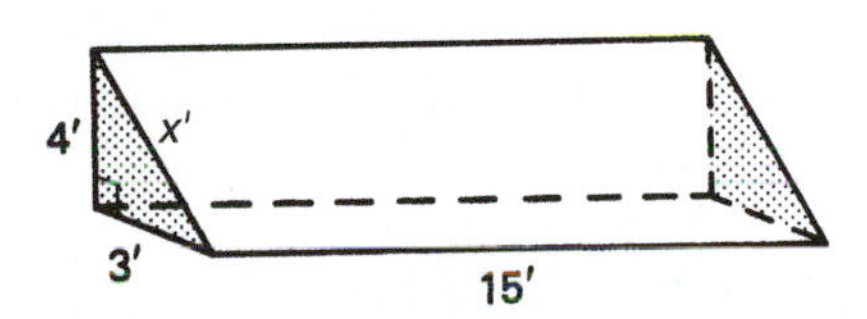

$x^2 = 3^2 + 4^2$ $= 5 \text{ ft}$	1. Find the unknown side of the base by the Pythagorean theorem.
$\text{LSA} = 3(15) + 4(15) + 5(15)$	2. Find the area of each rectangular face.
$= 180 \text{ ft}^2$	3. The LSA is the sum of the area of the three faces.
$\text{Area}_{\text{base}} = \frac{1}{2}bh$ $= \frac{1}{2}(4)(3) = 6 \text{ ft}^2$	4. Find the area of each base.
$\text{TSA} = 2(6 \text{ ft}^2) + 180 \text{ ft}^2$ $= 192 \text{ ft}^2$	5. The area of the two bases plus the LSA equals the total surface area.

Shortcut method for finding LSA:

$3 \text{ ft} + 4 \text{ ft} + 5 \text{ ft} = 12 \text{ ft}$	1. Find the perimeter of the base.
$12 \text{ ft} \times 15 \text{ ft} = 180 \text{ ft}^2$	2. Multiply perimeter × height. Note that the LSA is the same using either method.

Exercise 9–1

Find the LSA, TSA, and volume of the following figures.

1.

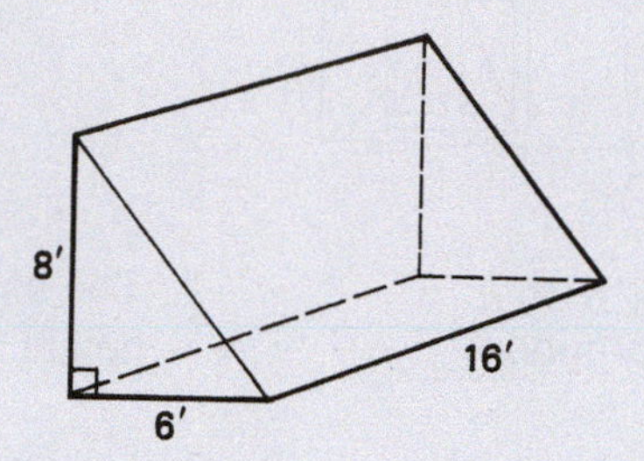

2.

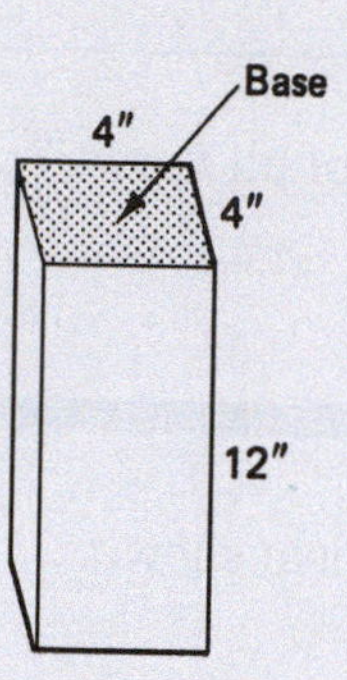

3.

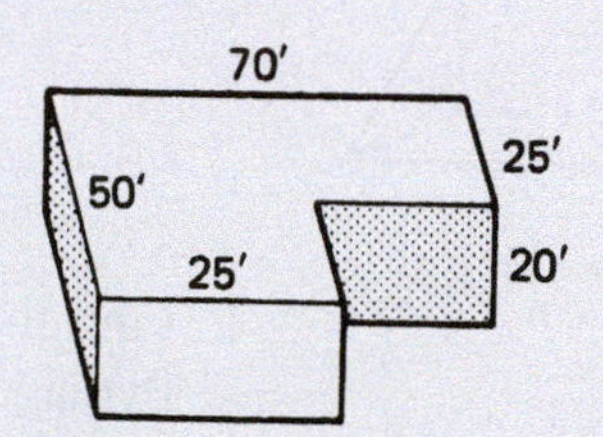

4.

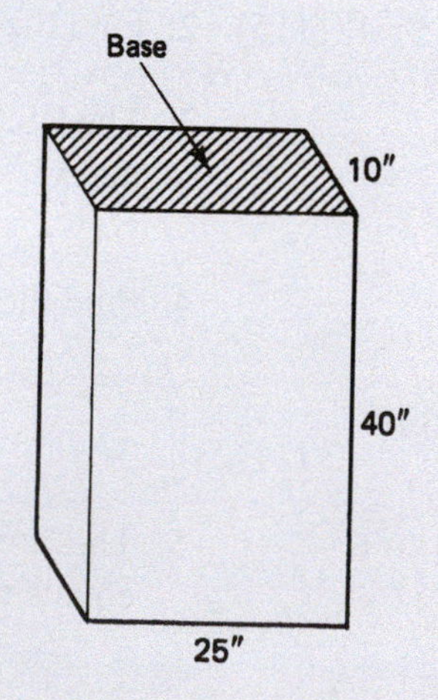

5. A foundation is to have outside measurements of 28 ft × 36 ft. If the foundation is to be 7 ft high and the walls 9 in. thick, how many cubic yards of concrete are needed for the foundation walls? Round up to a whole number of cubic yards. (*Hint:* Refer to example D.)
6. The outside of the foundation walls in exercise 5 is to be waterproofed. What is the surface area of the exterior walls?

7. A roof has the dimensions shown. How many cubic feet of air space are there in the attic?

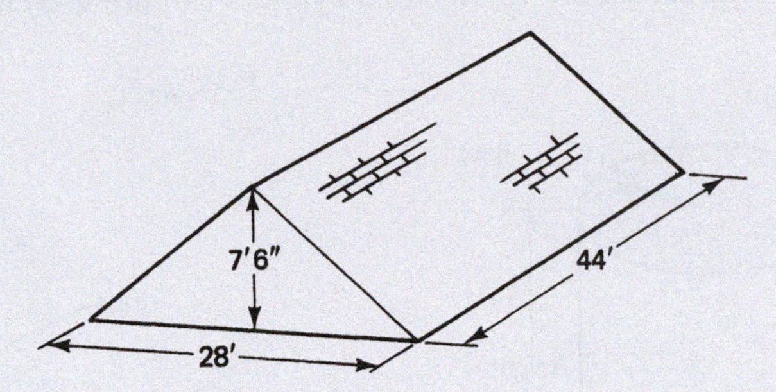

8. A small shed, open in the front, is to be completely shingled on the outside with cedar shingles. Assuming minimal waste, how many bundles of shingles are required to shingle the three sides and roof of the shed? One bundle covers 25 ft^2. (Note that the two ends are trapezoids.)

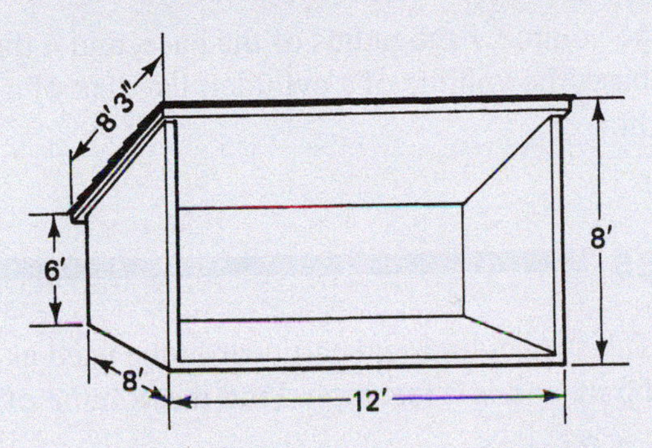

9. If the shed in exercise 8 is enclosed in front, what volume of air does the shed enclose?
10. An office building has the dimensions shown. Find (a) the volume of the building and (b) the lateral surface area of the building.

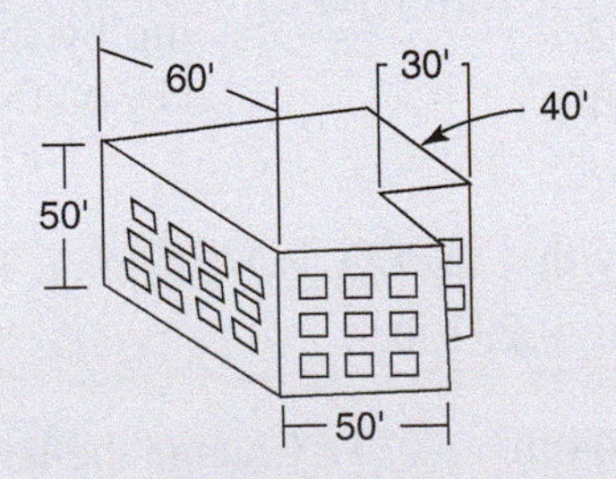

The right prism shown has right-triangle bases with the given dimensions. The volume of the prism is 1080 cubic inches. Find the following:

11. Find the perimeter of each base.
12. Find the area of each base.
13. Find the height of the prism.
14. Find the lateral surface area of the prism.
15. Find the total surface area of the prism.

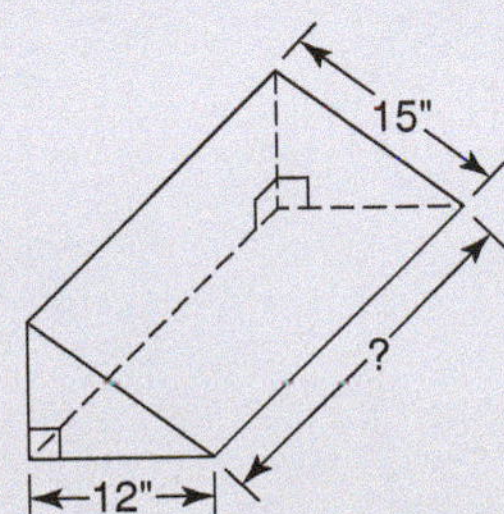

9.2 Cylinders and Cones

The formula for the volume of a cylinder (Figure 9–3) is

$$V = \pi r^2 h$$

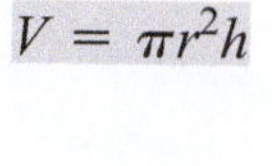

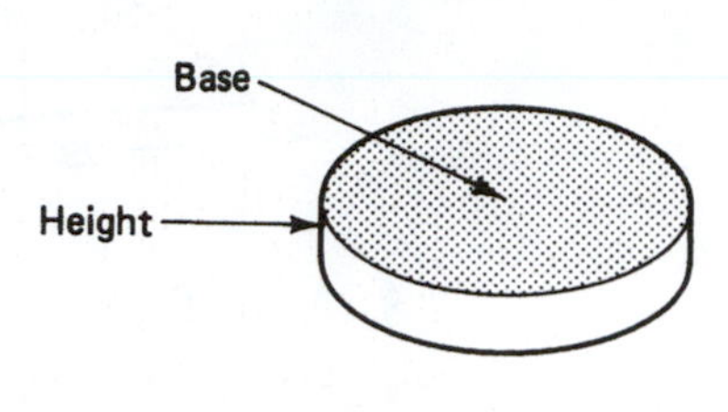

Figure 9–3

where V is the volume, r the radius of the base, and h the height. Since πr^2 is the area of the circular base, the volume of a cylinder, like that of a prism, is actually the area of the base × height.

Examples

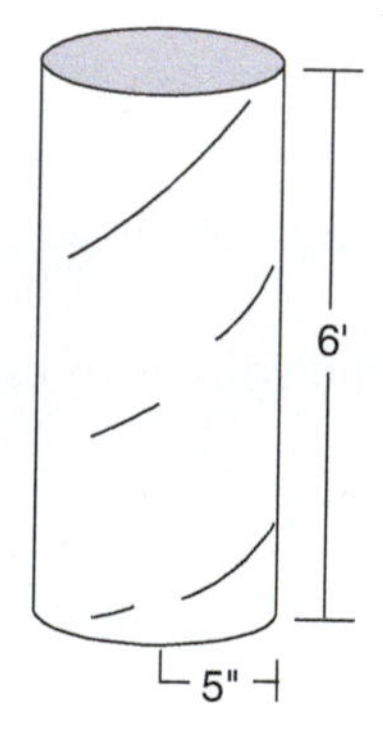

A. A Sono Tube® (a cylindrical cardboard tube used as a form for concrete posts) has a radius of 5 in. and is 6 feet high. Find the volume of concrete needed to fill the tube.

$\frac{5 \text{ in.}}{1} \times \frac{1 \text{ ft}}{12 \text{ in.}} = \frac{5}{12} \text{ ft}$ — 1. Change radius to feet.

$V = \pi r^2 h$ — 2. Substitute the given values into the formula for the volume of a cylinder. Be sure to square the radius before multiplying by the height and by pi. (Use the π key on the calculator.)

$V = \pi\left(\frac{5}{12} \text{ ft}\right)^2 (6 \text{ ft})$

$V = 3.3 \text{ ft}^3$

B. A copper water pipe with $\frac{3}{4}$ in. I.D. is 87 ft long. How many gallons of water are there in the pipe? (1 gal = 231 in.^{3})

$87 \text{ ft} \times 12 \text{ in./ft} = 1044 \text{ in.}$ — 1. Change the length to inches.

$V = \pi r^2 h$ — 2. A pipe is a cylinder. Be sure to square the radius, not the diameter.

$= \pi(0.375 \text{ in.})^2(1044 \text{ in.})$

$= 461 \text{ in.}^3$

$= \frac{461 \text{ in.}^3}{1} \times \frac{1 \text{ gal}}{231 \text{ in.}^3}$ — 3. Convert cubic inches to gallons.

$= 2 \text{ gal}$ — 4. The pipe contains approximately 2 gal of water.

The LSA of a cylinder (think of the label on a tin can) can be found by the formula

$$\text{LSA} = 2\pi rh \text{ or LSA} = \pi dh$$

If the label is cut and removed as shown in Figure 9–4, the rectangle thus formed has a length of $2\pi r$ or πd (the circumference of the cylinder) and a width of h (the height of the cylinder).

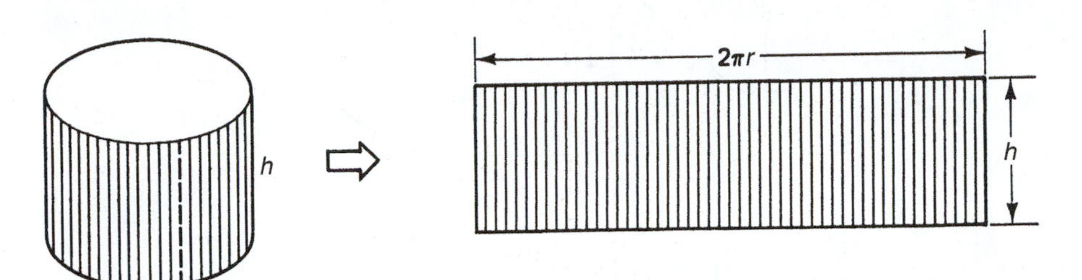

Figure 9–4

The TSA is the entire surface area of the cylinder.

$$\text{TSA} = 2\pi rh + 2\pi r^2$$

If the formula looks complicated, just remember it as the lateral surface area plus the area of the two bases.

$$\text{TSA} = \underbrace{2\pi rh}_{\text{LSA}} + \underbrace{2\pi r^2}_{\text{bases}}$$

Examples

C. Find the LSA and the TSA of a cylinder with a height of 10 inches and a diameter of 8 inches.

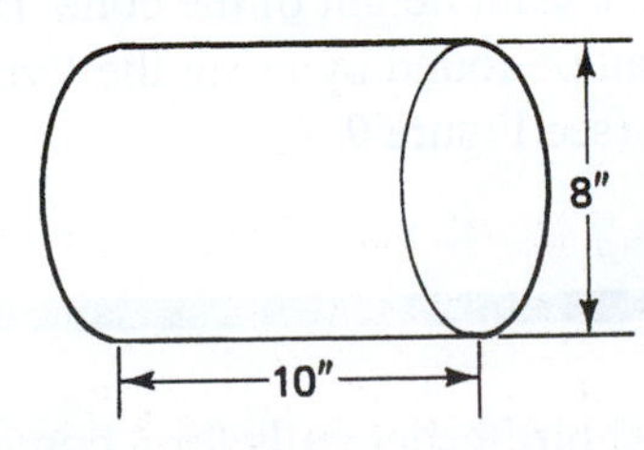

$$\begin{aligned} \text{LSA} &= 2\pi rh \\ &= 2\pi(4 \text{ in.})(10 \text{ in.}) \\ &= 251 \text{ in.}^2 \end{aligned}$$

1. Substitute known values into the formula for LSA.

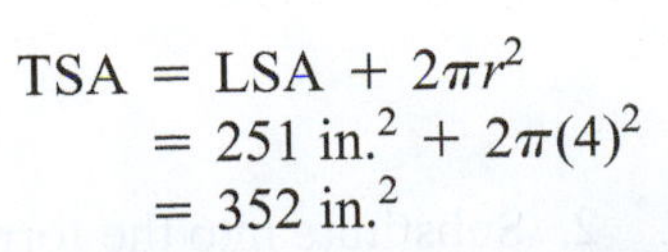

$$\begin{aligned} \text{TSA} &= \text{LSA} + 2\pi r^2 \\ &= 251 \text{ in.}^2 + 2\pi(4)^2 \\ &= 352 \text{ in.}^2 \end{aligned}$$

2. Find the area of the bases and add to the LSA.

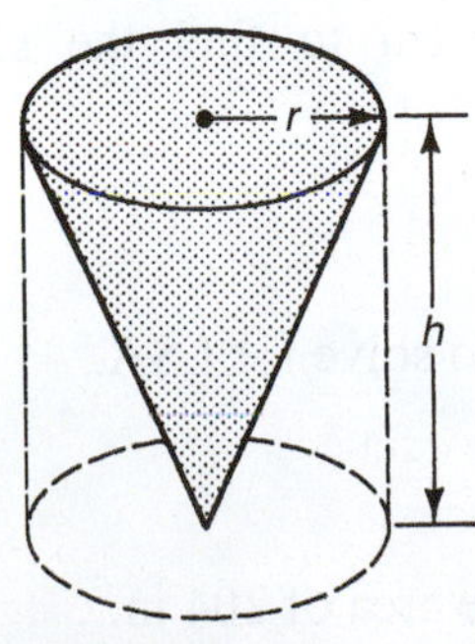

Figure 9–5

The volume of a cone (Figure 9–5) is $\frac{1}{3}$ that of a cylinder with the same base and height.

$$V = \frac{1}{3}\pi r^2 h$$

Therefore, the formula for the volume of a cone is identical to the formula for a cylinder except that it is multiplied by $\frac{1}{3}$.

D. A conical (cone-shaped) pile of sand is to be used in a concrete mix. Find the approximate number of cubic yards of sand in the pile with the dimensions shown.

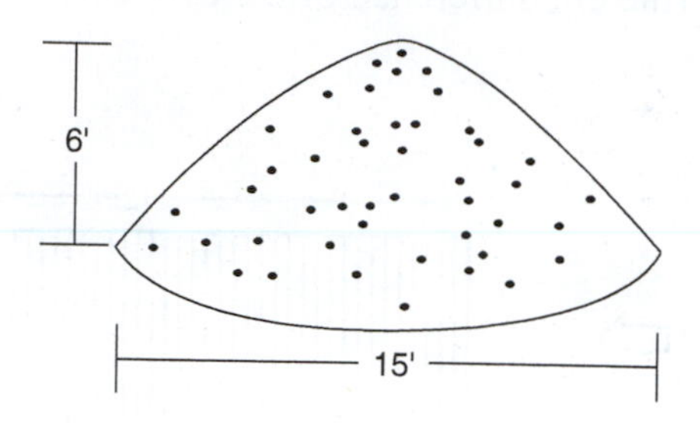

$V = \frac{1}{3}\pi r^2 h$ 1. Formula for the volume of a cone.

$= \frac{1}{3}\pi(7.5 \text{ ft})^2(6 \text{ ft})$ 2. Substitute and solve for volume.

$= 353.4 \text{ ft}^3$

$= \frac{353.4 \text{ ft}^3}{1} \times \frac{1 \text{ yd}^3}{27 \text{ ft}^3}$ 3. Convert cubic feet to cubic yards.

$= 13.1 \text{ yd}^3$

The LSA of a cone is found by the formula

$$\text{LSA} = \pi rs$$

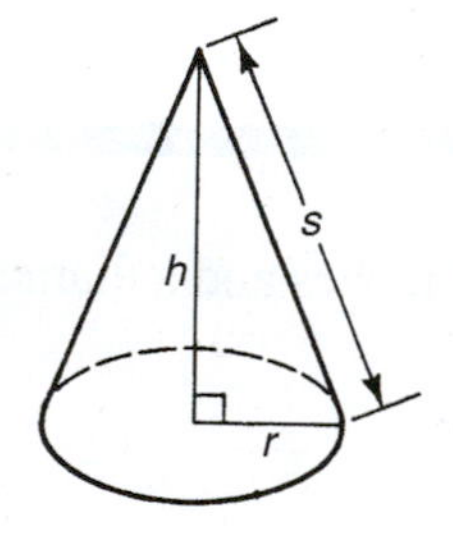

Figure 9–6

where r is the radius and s is the slant height of the cone. If the height h and the radius r are given, the slant height s can be found by using the Pythagorean theorem, where the slant height is the hypotenuse (see Figure 9–6).

Examples

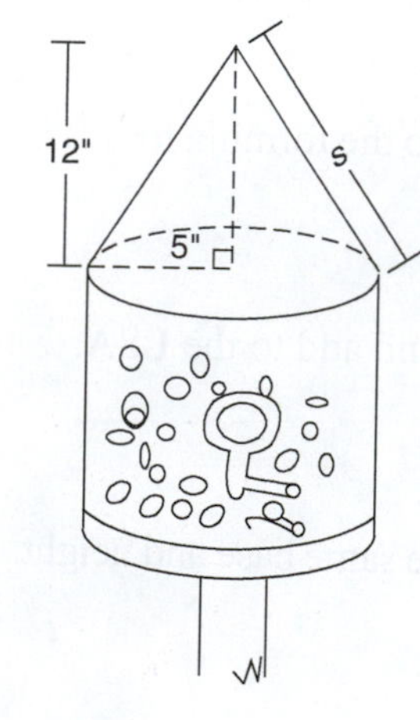

E. The design for a cylindrical birdfeeder calls for a conical roof. Find the LSA of the roof as shown.

$s^2 = 5^2 + 12^2$ 1. Use the Pythagorean theorem to find the slant height.

$s = 13 \text{ in.}$

$\text{LSA} = \pi rs$ 2. Substitute into the formula to solve for LSA.

$= \pi(5 \text{ in.})(13 \text{ in.})$

$= 204 \text{ in.}^2$ 3. The roof has a lateral surface area of 204 in.^{2}.

F. A 5-gal bucket of joint compound has an average diameter of $10\frac{1}{2}$ inches. Find the height of the bucket if there is 1 inch of space at the top of the bucket. (One gal = 231 in.^{3}.)

$5 \times 231 \text{ in.}^3 = 1155 \text{ in.}^3$	1. Find the volume in cubic inches of 5 gallons.
$V = \pi r^2 h$	2. Volume of a cylinder (bucket).
$1155 \text{ in.}^3 = \pi(5.25 \text{ in.})^2(h)$	3. Substitute known values into the formula.
$1155 \text{ in.}^3 = 86.6 \text{ in.}^2(h)$	4. Multiply values on the right side.
$1155 \text{ in.}^3 \div 86.6 \text{ in.}^2 = h$	5. Divide to find the height.
$h = 13.3 \text{ in.} + 1 \text{ in.} = 14.3 \text{ in.}$	6. Height of the bucket is 1 inch more than the calculated height.

Exercise 9–2

Complete the table for the cylinders described.

	Radius	Diameter	Circum-ference	Area of Base	Height	Volume	LSA	TSA
1.	8 in.				5 in.			
2.		12 in.			8 in.			
3.			15.7 in.		5 in.			
4.	$3\frac{1}{4}$ in.				8 in.			
5.		1 ft 10 in.			1 ft 3 in.			

6. A cone used as an oil funnel has a radius of 5 inches and a height of 10 inches. What volume of oil could it hold if the opening at the bottom is plugged?
7. A cone and a cylinder have the same height and the same size base. If the cone has a volume of 125 in.^{3}, what is the volume of the cylinder?
8. A 1-in I.D. water pipe has a 55-ft run. How many gallons of water are in the pipe? (1 gal $\doteq$ 231 in.^{3}) (See example B.)
9. A drilled well has a diameter of 6 inches. The water standing in the well is 200 ft deep. How many gallons of water are standing in the well? (A drilled well is a cylinder.) (1 ft^3 $\doteq$ 7.5 gal.)
10. A conical (cone-shaped) pile of sand is being mixed with cement and gravel to make concrete. If the pile of sand is 5 ft high and 7 ft across at its base, how many cubic feet of sand are in the pile?
11. What is the lateral surface area of the pile of sand in exercise 10.
12. A 1-gal paint can has a radius of approximately $3\frac{1}{4}$ inches. What is the height of the can? Round to the nearest inch. (1 gal $\doteq$ 231 in.^{3}) (See diagram.)

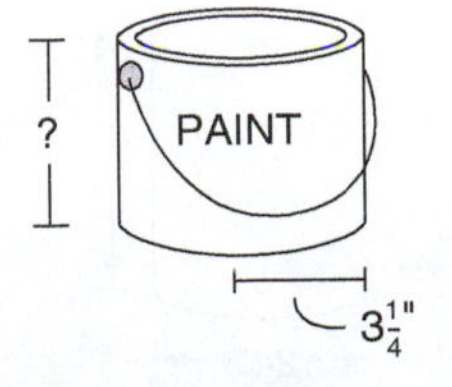

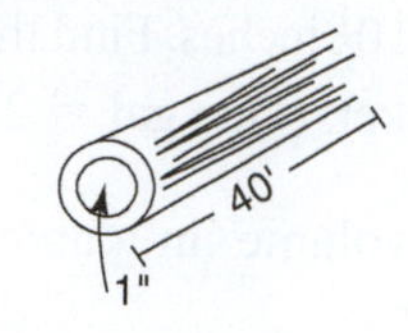

13. Water pipes with an outside diameter (O.D.) of 1 inch are to be insulated with a thin sheet of foam. Assuming no waste or overlap, how many square inches of foam are needed to insulate pipes with a total length of 40 ft? Round to the nearest square inch.

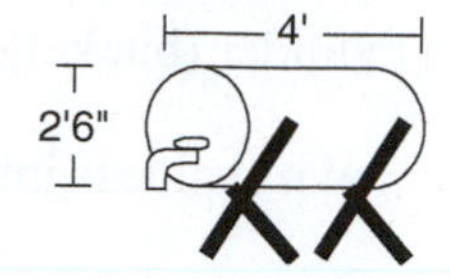

14. A cylindrical drum used for storing kerosene has a diameter of 2 ft 6 in. and is 4 ft long. What is the total surface area of the drum? (See diagram.)
15. A tent (without a floor) is to be constructed in the form of a tepee (cone-shaped) 12 ft high and 10 ft wide at the base. Ignoring openings, determine the number of square feet of canvas needed to construct the tent. Add 20% for waste and round to the nearest whole square foot.

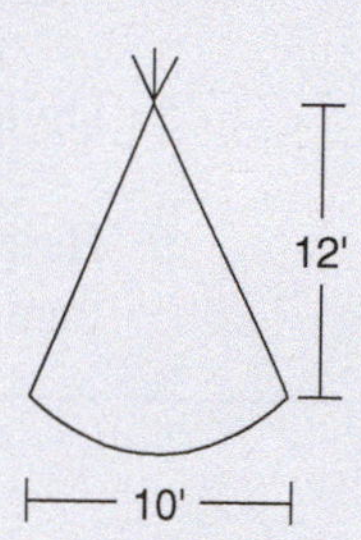

16. A water pipe has an O.D. of $\frac{3}{4}''$ and an I.D. of $\frac{1}{2}''$. If a 98-foot length pipe must be drained, find the number of gallons of water that must be drained. A gallon of water $\doteq$ 231 in.^{3}. Give answer rounded to the nearest gallon.
17. The above water pipe is to be insulated with a sheet of thin foam wrapped around the outside. Find the lateral surface area of the pipe. Give answer in square feet, rounded up to a whole number of square feet. 1 square foot = 144 square inches.
18. A grain dispenser is a cylinder atop a cone, as shown. Find the volume of grain that can be stored, assuming the cylinder can be filled to the top. Round to the nearest cubic foot.
19. Find the LSA of the cone in problem 18. Round to the nearest square foot.
20. Find the total surface area of the figure in problem 18. Assume the cylindrical storage area has a flat top on it. Round to the nearest square foot.
21. Paint the figure in problem 18, using a rust-inhibiting paint that covers 220 square feet/gallon. How many gallons are needed for two coats? (Round up to the nearest tenth of a gallon.)
22. A cylindrical storage tank used in conjunction with a solar collector has the dimensions shown. Find the capacity in gallons of the storage tank. Use the conversion 1 gal $\doteq$ 231 in.^{3}
23. Find the total surface area of the above tank.

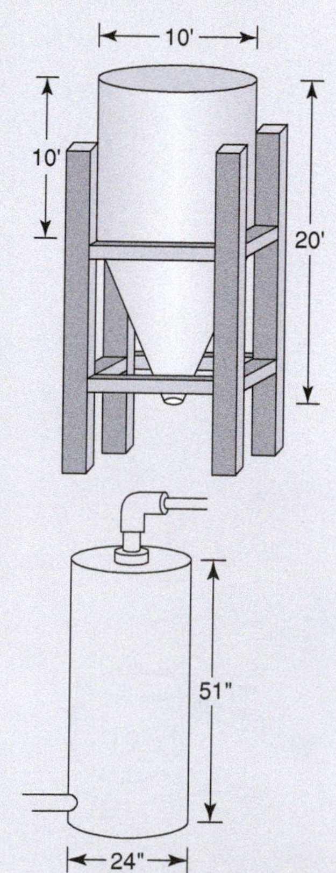

24. A rectangular storage tank used in conjunction with a solar collector has the dimensions shown. Find the capacity in gallons of the storage tank.
25. Find the total surface area of the rectangular tank in problem 24.
26. Which is a more efficient shape? Since both of these tanks have the same storage capacity, the shape with less surface area is the more efficient, for two reasons: it uses less material in the construction, and the smaller surface area allows less heat to escape.

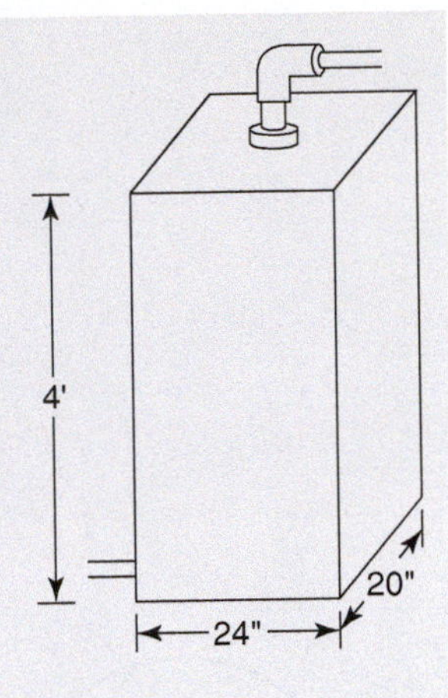

9.3 Spheres

The formula for the volume of a sphere is

$$V = \frac{4}{3}\pi r^3$$

The surface area of a sphere is found by the formula

$$SA = 4\pi r^2$$

Examples

A. An ornate staircase has a spherical cap on the top of the newel post as shown. Find the volume of the spherical cap.

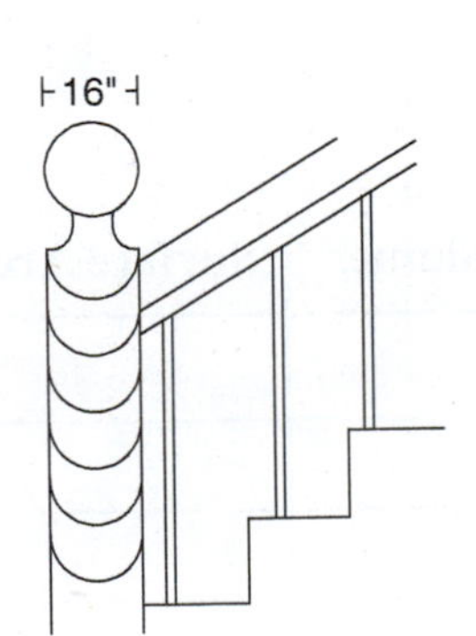

$V = \frac{4}{3}\pi r^3$ 1. Formula for the volume of a sphere.

$= \frac{4}{3}\pi(8 \text{ in.})^3$ 2. Find 8^3 before multiplying by $\frac{4}{3}\pi$. (8^3 on the calculator is 8 [y^x] 3 [≃] (512) or use the [x^3] key if available.)

$= 2145 \text{ in.}^3$ 3. Volume is in cubic units.

B. Find the surface area of a sphere with a diameter of 22 ft.

$SA = 4\pi r^2$ 1. Formula for surface area of a sphere.

$= 4\pi(11 \text{ ft})^2$ 2. Be sure to square radius, not diameter.

$= 1521 \text{ ft}^2$ 3. Surface area is in square units.

C. A spherical water tank has a circumference of 62 ft. (Circumference of a sphere is the distance around at its widest point.) Find the number of gallons of water it can hold. ($1 \text{ ft}^3 = 7.5 \text{ gal}$)

$C = \pi d$	1. Use the formula for the circumference of a circle or a sphere.
$62 = \pi d$ $d = 19.74 \text{ ft}$	2. Find the diameter of the sphere.
$r = 9.87 \text{ ft}$	3. The values for diameter and radius shown here are rounded. When using a calculator, it is best not to round until the final answer is obtained.
$V = \frac{4}{3}\pi(9.87 \text{ ft})^3$	4. Substitute the radius into the formula for volume.
$= 4024.6 \text{ ft}^3$	5. To find $(9.87)^3$ on a calculator with a [yˣ] key: 9.87 [yˣ] 3 [=].
$= \frac{4024.6 \text{ ft}^3}{1} \cdot \frac{7.5 \text{ gal}}{1 \text{ ft}^3}$ $= 30{,}185 \text{ gal}$	6. Convert cubic feet to gallons.

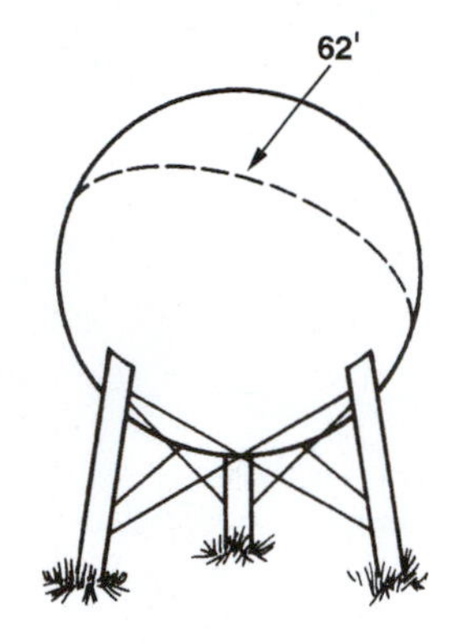

D. The spherical water tank in example C is to be painted. Find the number of gallons of paint required for two coats if each gallon covers 250 ft².

$SA = 4\pi r^2$	1. Formula for the surface area of a sphere.
$SA = 4\pi(9.87 \text{ ft})^2 = 1224 \text{ ft}^2$	2. Find the surface area using the radius determined in example C.
$1224 \text{ ft}^2 \div 250 \text{ ft}^2/\text{gal} = 4.9 \text{ gal}$	3. Find the number of gallons of paint to cover the surface of the spherical water tank.
$4.9 \text{ gal/coat} \times 2 \text{ coats} = 9.8 \text{ gal}$	4. 10 gallons of paint are needed for two coats of paint on the spherical water tank.

Exercise 9–3

Complete the table for the spheres described.

	Radius	Diameter	Circumference	Volume	Surface Area
1.	8 in.				
2.	5 ft				
3.		16 ft			
4.	3 ft 4 in.				
5.			22 ft		
6.	15 in.				
7.	8.6 in.				
8.		12 ft 6 in.			

9.4 Composite Shapes

To find the volume and surface area of composite shapes, divide the form into parts that can be determined individually.

Examples

A. A propane gas storage tank is a cylinder with hemispheric ends. (A hemisphere is half a sphere.) Find the volume of the tank. The tank is 20 ft long and has a diameter of 5 ft.

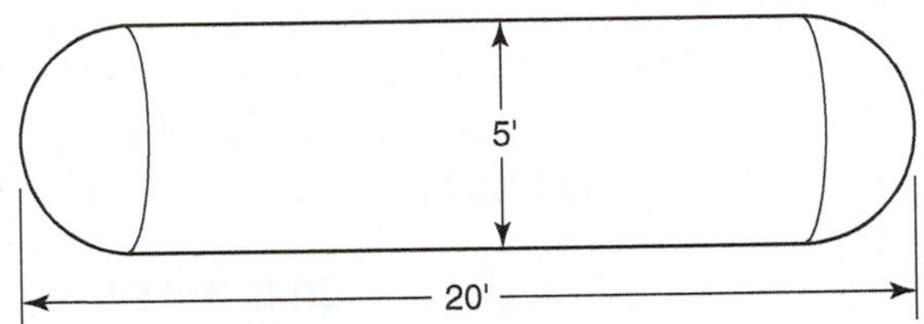

$V_{\text{sphere}} = \frac{4}{3}\pi r^3$	1. The volume of the two hemispheres = the volume of a sphere with radius 2.5 ft.
$= \frac{4}{3}\pi(2.5 \text{ ft})^3$	
$= 65.4 \text{ ft}^3$	2. Substitute and solve for volume.
$V_{\text{cylinder}} = \pi r^2 h$	3. Solve for the volume of the cylinder. Why is the height of the cylinder 15 ft?
$= \pi(2.5 \text{ ft})^2(15 \text{ ft})$	
$= 294.5 \text{ ft}^3$	4. The volume of the cylindrical portion of the tank.
$V_{\text{tank}} = 65.4 \text{ ft}^3 + 294.5 \text{ ft}^3$	5. Find the total volume of the tank.
$= 360 \text{ ft}^3$	6. Round to the nearest cubic foot.

B. The gas tank in example A is to be painted with rust-inhibiting paint. How many square feet are to be painted?

$\text{SA}_{\text{sphere}} = 4\pi r^2$	1. Find the surface area for the two hemispheres.
$= 4\pi(2.5 \text{ ft})^2$	
$= 78.5 \text{ ft}^2$	
$\text{LSA}_{\text{cylinder}} = 2\pi rh$	2. Find the LSA of the cylindrical middle part of the tank.
$= 2\pi(2.5 \text{ ft})(15 \text{ ft})$	
$= 235.6 \text{ ft}^2$	
$\text{TSA}_{\text{tank}} = 78.5 \text{ ft}^2 + 235.6 \text{ ft}^2$	3. Find the surface area of the entire figure.
$= 314 \text{ ft}^2$	

C. A container shaped as shown is filled with sand. If sand weighs 98 lb/ft^3, what is the weight of the sand in the container? Consider the thickness of the container to be negligible. (See diagram on page 206.)

$V_{\text{cone}} = \frac{1}{3}\pi r^2 h$	1. Find the volume of the conical (cone-shaped) top.

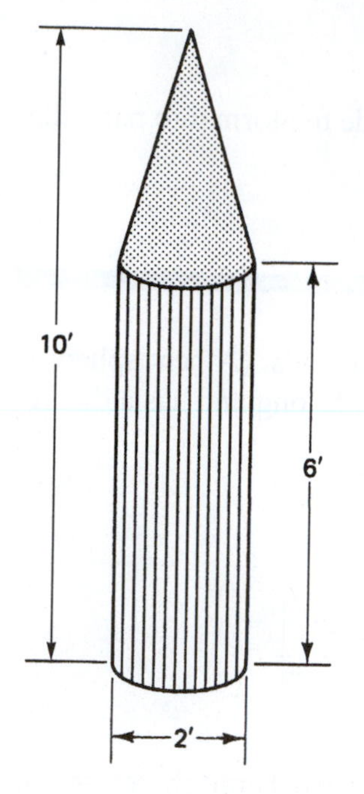

$$= (\tfrac{1}{3})\pi(1\text{ ft})^2(4\text{ ft})$$
$$= 4.2\text{ ft}^3$$

$V_{cylinder} = \pi r^2 h$ 2. Find the volume of the cylindrical bottom.

$$= \pi(1\text{ ft})^2 \cdot (6\text{ ft})$$
$$= 18.8\text{ ft}^3$$

$V_{total} = 4.2\text{ft}^3 + 18.8\text{ft}^3 = 23\text{ ft}^3$ 3. Find the total volume.

$\text{Weight} = \dfrac{23\text{ ft}^3}{1} \times \dfrac{98\text{ lb}}{1\text{ ft}^3}$ 4. Use unit conversion to find weight.

$= 2250\text{ lb}$ 5. Approximate weight of sand.

D. Find the volume of the barn shown.
The barn is actually a prism with seven-sided bases and rectangular faces. To find the area of the base, divide the ends (base) of the barn into three shapes: rectangle, trapezoid, and triangle.

$A_{rectangle} = 20\text{ ft} \times 15\text{ ft} = 300\text{ ft}^2$ 1. Area of the rectangle $= 300\text{ ft}^2$.

$A_{trapezoid} = \frac{1}{2}(5\text{ ft})(20\text{ ft} + 16\text{ ft}) = 90\text{ ft}^2$ 2. Area of trapezoid $= 90\text{ ft}^2$.

$A_{triangle} = \frac{1}{2}(16\text{ ft})(4\text{ ft}) = 32\text{ ft}^2$ 3. Area of the triangle $= 32\text{ ft}^2$.

$300 + 90 + 32 = 422\text{ ft}^2$ 4. Area of base (barn end) $= 422\text{ ft}^2$.

$422\text{ ft}^2 \times 40\text{ ft} = 16{,}880\text{ ft}^3$ 5. Volume of the barn is $16{,}880\text{ ft}^3$.

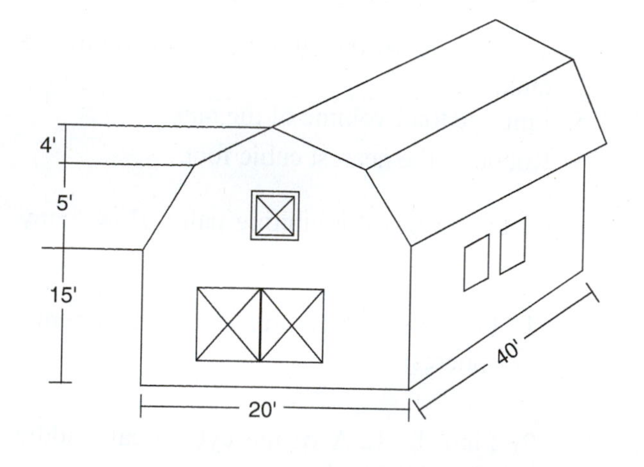

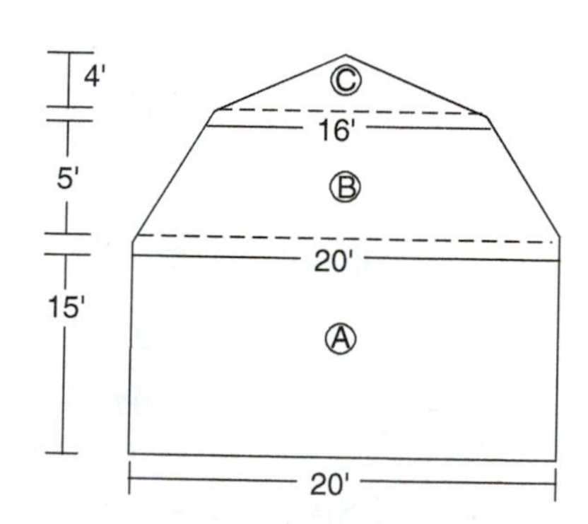

Ⓒ = triangle
Area = $\frac{1}{2}$16 (4)
Ⓑ = trapezoid
Area = $\frac{1}{2}$ (5)(20+16)
Ⓐ = rectangle
Area = 15 × 20

Exercise 9–4

1. Find the volume of the silo shown (see page 207).
2. A small greenhouse is to be installed on the side of a house. Find the number of square feet of glass to be used. The only nonglass side is the end wall against the house. *Note:* Ignore deductions for framing around the glass. See page 207.

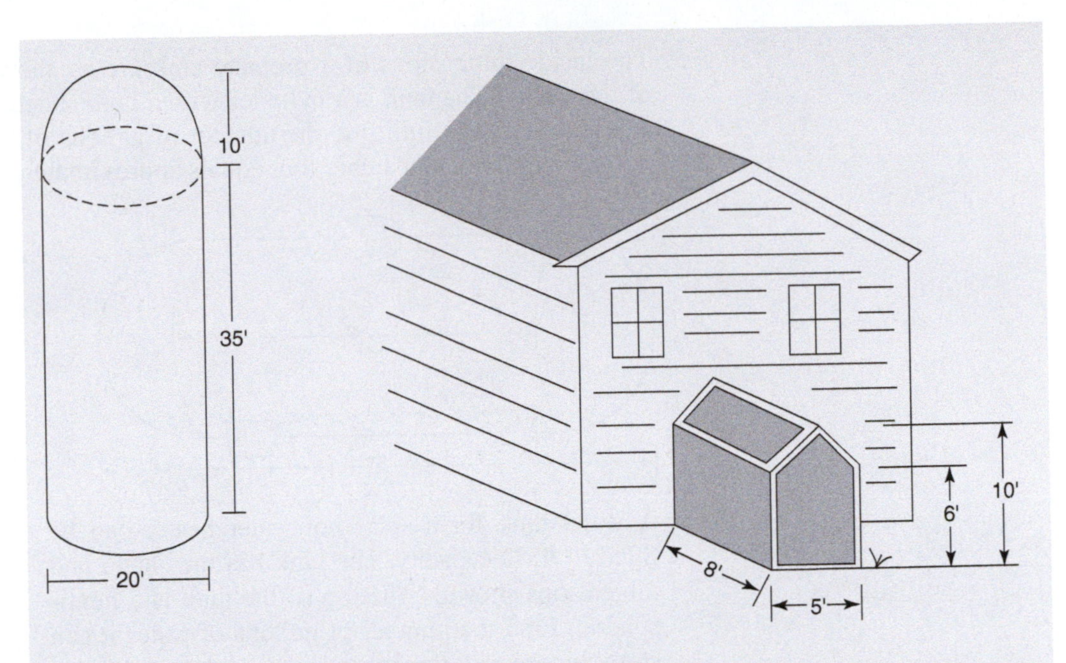

3. A small house is to be sided with clapboards. How many square feet are to be sided? (Don't put clapboards on the roof!) Ignore openings for doors and windows.

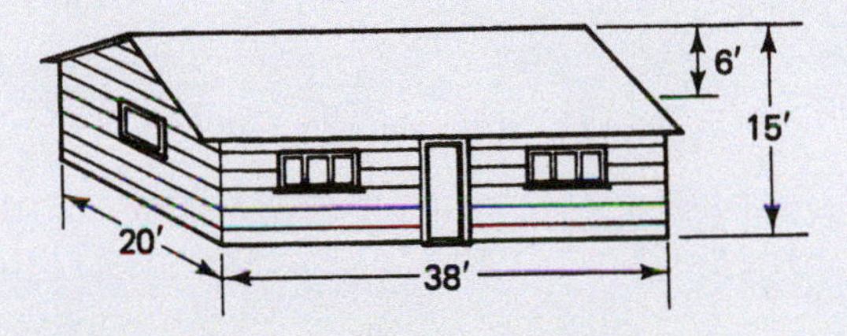

4. A metal silo has the dimensions shown. The silo is to be painted with two coats of metal paint. How many gallons are needed if each gallon of paint covers 350 ft^2? Round up.

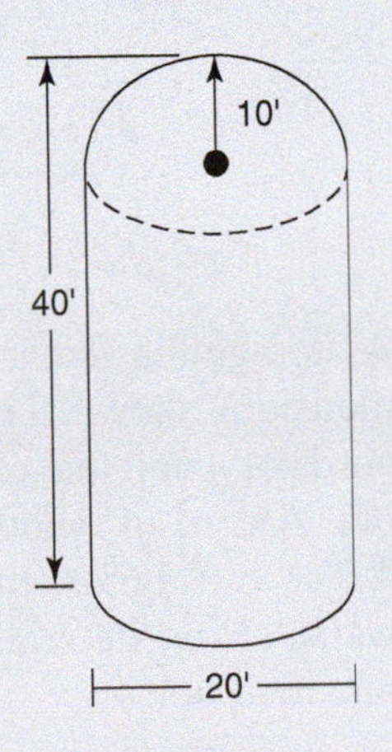

5. The inside dimensions of a propane tank are as shown. (Ignore the thickness of the walls.) The tank is a cylinder with a hemisphere on each end. If the tank can be filled 80% full, find the number of gallons of liquid propane the tank is designed to hold. One cubic foot equals approximately 7.5 gallons.

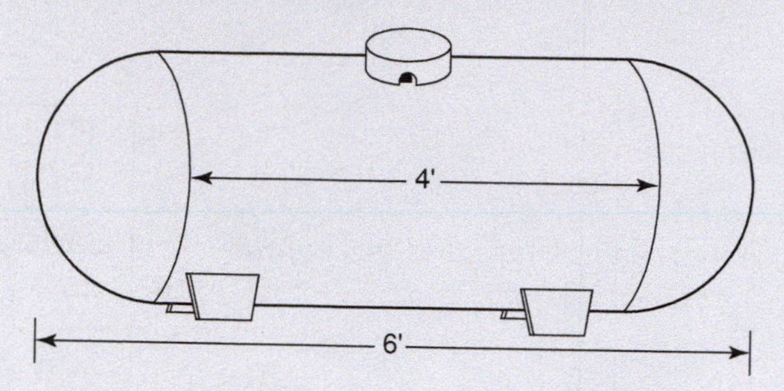

6. A water tank for a solar hot water heater can be filled to 95% capacity. The tank has the shape and dimensions shown. (The top of the tank is a hemisphere.) Find the number of gallons of water it can store. Ignore wall thickness.

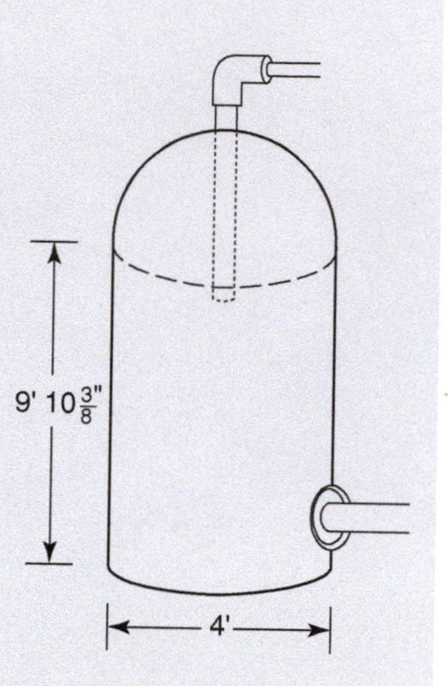

7. A 55-gallon drum (cylinder) is 34-in. high. What is the approximate diameter of the drum? One gallon $\doteq$ 231 cubic inches.

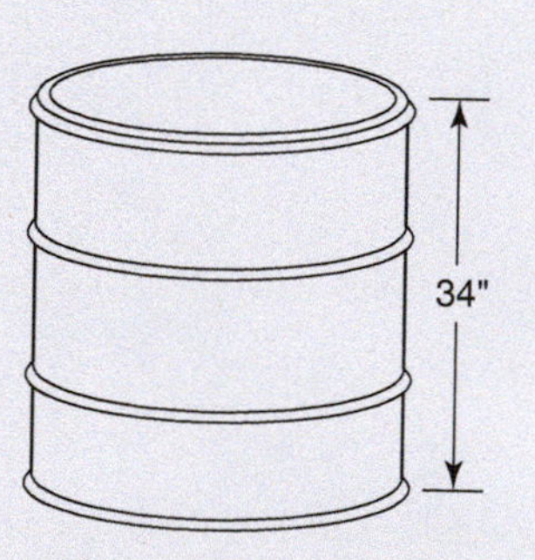

8. A storage bin for firewood has the dimensions shown. Find the number of cords of wood that can be stored in the bin. A cord of wood measures 4 ft × 4 ft × 8 ft. Ignore the thickness of the walls. (*Hint:* determine the volume of one cord first.)

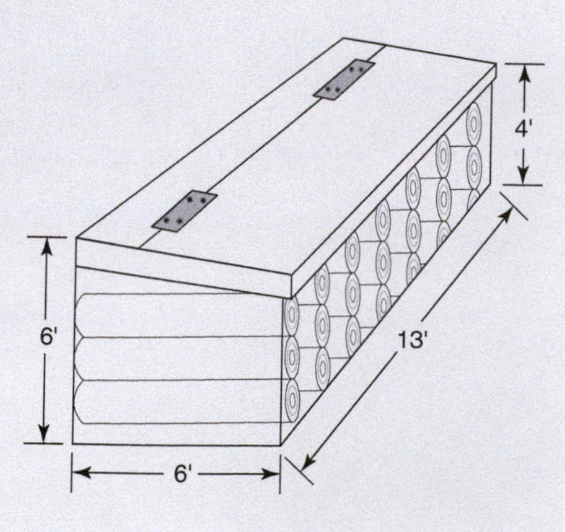

9. A swimming pool has dimensions as shown. Water is filled to 6 inches from the top. Find the number of gallons of water in the pool. One cubic foot $\doteq$ 7.5 gallons.

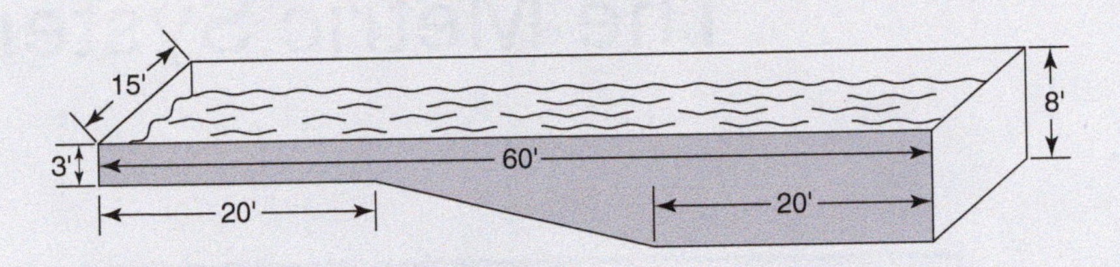

10. A watering trough is a half-cylinder with a half-hemisphere on each end. Find the capacity in gallons, if the trough is filled to the top. Use the conversion 1 gal $\doteq$ 231 in.^{3}.

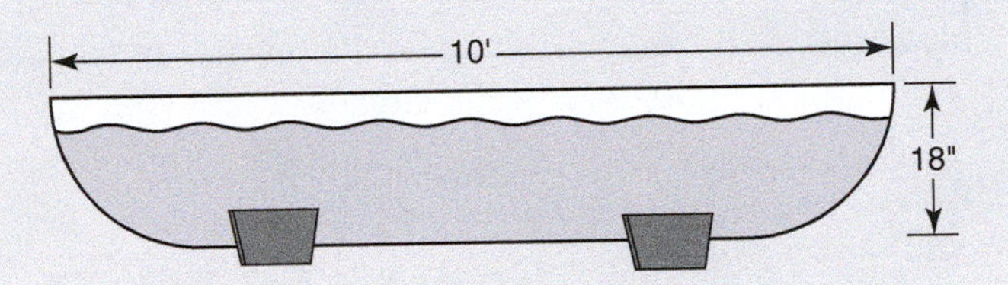

11. The outside of the house illustrated is to be painted. Find the number of square feet to be painted, deducting 10% for window and door openings.

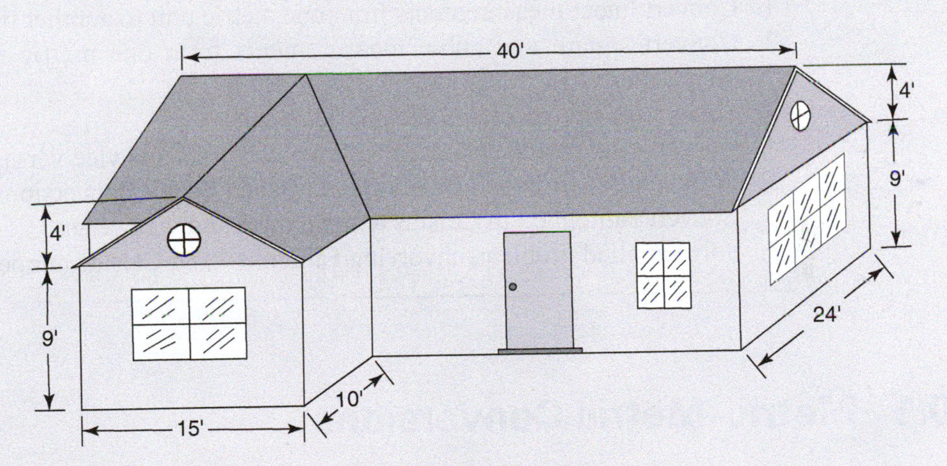

12. What is the square footage of the above house?
13. When using blower door testing to determine air leaks, the volume of the test space must be calculated. Determine the volume of the above house. Ignore wall thickness.

CHAPTER 10

The Metric System

Did You Know

On-Demand water heaters (tankless or instantaneous) can be as much as 50% more efficient than conventional storage water heaters.

OBJECTIVES

Upon completing this chapter, the student will be able to:

1. Convert linear measurements from one metric unit to another metric unit.
2. Convert square and cubic measurements from one metric unit to another metric unit.
3. Solve problems involving the metric system.
4. Convert common metric units to American units and vice versa.
5. Work applied problems involving single and double conversions.
6. Convert Fahrenheit to Celsius temperatures and vice versa.
7. Solve applied problems involving Fahrenheit and Celsius temperatures.

10.1 Metric–Metric Conversions

The United States is the only major industrialized country in the world still saddled with the cumbersome "American" or *customary* system of measurement (formerly called the *English* system). The metric system is in use almost everywhere else. The United States is slowly converting to the metric system, and the conversion will eventually allow a smooth flow of equipment and products from one country to another. Equally or more important, the metric system is a much simpler and more sensible system to use. To illustrate, here is a comparison of some American and metric conversion equivalents. Note that conversions in the metric system involve multiples of 10 and can therefore be calculated mentally in most cases.

1 foot = 12 inches	1 centimeter = 10 millimeters
1 yard = 3 feet	1 decimeter = 10 centimeters
1 rod = $5\frac{1}{2}$ yards	1 meter = 10 decimeters

1 hand = 4 inches	1 decameter = 10 meters
1 fathom = 6 feet	1 hectometer = 10 dekameters
1 chain = 22 yards	1 kilometer = 10 hectometers
1 mile = 5280 feet	1 kilometer = 1000 meters
1 pound = 16 ounces	1 kilogram = 1000 grams
1 gallon = 4 quarts	1 liter = 10 deciliters

Here is an example of the clear superiority of the metric system of measurement:

Convert 2.4 miles to inches and also convert 5.1 kilometers to centimeters. (Converting kilometers to centimeters in the metric system is comparable to converting miles to inches in the American system.)

Metric System of conversion: 5.1 km (kilometers) = 510,000 cm. With a little practice, this problem can be calculated mentally—no pencil or paper needed.

American System of conversion: $\frac{2.4\ \not{mi}}{1} \times \frac{5280\ \not{ft}}{1\ \not{mi}} \times \frac{12\text{ in.}}{1\ \not{ft}} = 152{,}064$ inches. Try working this problem in your head!

Converting from one metric unit to another is accomplished by multiplying or dividing by a power of 10. Recall that multiplying or dividing by 10, 100, 0.01, and so on can be accomplished simply by moving the decimal point a certain number of places to the left or right.

NOTE

In addition to meter, liter, and gram, there are many other units of measurements that conform to the metric prefix system. A few of these include watt, ohm, ampere, bel, byte, and second. Additionally, there are other prefixes both smaller and larger that are not covered in this text. Most individuals have some familiarity with *mega*watts, *giga*bytes, *nano*seconds, and *micro*waves, to name a few.

In the metric system, each larger unit is 10 times as large as the preceding unit. Hence 10 of the smaller units equal 1 unit of the next larger size. There are three major types of measurements in the metric system: *meters* measure distance or length, *liters* measure volume, and *grams* measure mass. (Mass is interchangeable with weight for most purposes.) Meters will be used to illustrate the method of conversion, but liters and grams are converted in the same manner. A meter, the standard unit of length, is slightly longer than a yard (about 39 in.). Units smaller than and larger than a meter are indicated by a prefix:

Name:	*kilo*meter	*hecto*meter	*deka*meter	meter	*deci*meter	*centi*meter	*milli*meter
Abbrev:	km	hm	dam	m	dm	cm	mm
Size:	1000	100	10	1	0.1	0.01	0.001

The prefixes kilo, hecto, and so on always mean the same number, regardless of what type of unit is being measured. *Kilo*meter (km) means 1000 meters, *kilo*gram (kg) means 1000 grams, and *kilo*liter (kl) means 1000 liters. A milligram (mg) is $\frac{1}{1000}$ or 0.001 gram, a deciliter (dl) = $\frac{1}{10}$ or 0.1 liter, and a hectometer (hm) represents 100 meters. To use the metric system, it is necessary to memorize the prefixes, what they represent, and in the proper order from left to right. It is then a simple matter to convert from one metric unit to another.

Examples

A. Convert 58.5 cm to meters.

58.5 cm → 0.585 m

1. Note that meters (m) are *two places to the left* of cm. To convert cm to m, move the decimal point *two places* to the *left.*

58.5 cm = 0.585 m

B. Convert 0.35 km to cm.

0.35 km → 35000 cm

1. Cm are 5 units to the right of km on the chart. To convert *km* to *cm*, move the decimal point *five places* to the *right.*

0.35 km = 35 000 cm

C. Convert 8 cm to dam.

8 cm → 0.008 dam

1. To convert *cm* to *dam* requires a movement of *three places* to the *left.*

8 cm = 0.008 dam

2. A leading zero is shown if there is no whole number to the left of the decimal point.

D. A blueprint indicates that a wall is to be 4250 mm long. What is the length in meters?

4250 mm → 4.250 m

1. Move the decimal place 3 places to the left to convert millimeters to meters.

4.250 m = 4.25 m

2. Drop trailing zeros after the decimal point.

E. A microwave oven is rated at 1100 watts. What is this in kilowatts?

1100 W → 1.100 kW

1. Move the decimal place 3 places to the left to convert watts to kilowatts.

1.100 kW = 1.1 kW

2. Drop trailing zeros after the decimal point.

Examples A–E are shown in the following chart. Using a chart can be very helpful when learning to do metric conversions.

Name:	kilo*unit*	hecto*unit*	deca*unit*	unit	deci*unit*	centi*unit*	milli*unit*
Abbrev:	k_	h_	da_	_	d_	c_	m_
Size:	1000	100	10	1	0.1	0.01	0.001
A				0.585 m ←	———	58.5 cm	
B	0.35 km ———	———	———	———	———	→ 35 000 cm	
C			0.008 dam ←	———	———	8 cm	
D				4.25 m ←	———	———	4250 mm
E	1.100 kW ←	———	———	1100 W			

Example

F. A nail weighs about 10.5 grams. Approximately how many nails are in a 5-kilogram box?

5 kg → 5000 g	1. Both weights must be in the same unit.
5000 g ÷ 10.5 g = 476.19	2. Arithmetic operations are performed the same way whether the units are metric or American.
476 (or approximately 480) nails	3. Be sure to round to a whole number of nails!

Exercises 10–1A

Study completed examples a, b, and c, and complete the table.

	km	hm	dam	m	dm	cm	mm
a.	8.2 km ———	———	→ 820 dam				
b.		0.053 hm ←	———	5.3 m			
c.			2.5 dam ———	———	———	———	→ 25 000 mm
1.	? km			47 m			
2.				47 m		? cm	
3.			0.83 dam			? cm	
4.				? m		380 cm	
5.					0.25 dm		? mm
6.					? dm	360 cm	
7.	0.048 km					? cm	
8.		? hm		800 m			
9.					46 dm		? mm
10.				? m		820 cm	

Using exactly the same method, it follows that 27 kg = 27 000 g, and 8000 ml = 8l. (Note that spaces are used in place of commas in the metric system.) Convert the following values.

11. 18 dg = ________ mg

12. 48 000 cl = ________ kl

13. 0.0058 kg = ________ dg

14. 0.5 liter = ________ kl

15. 5000 mg = ________ g

To convert square measure the number of places the decimal point is moved is *doubled*. To convert cubic measure, the number of places the decimal point is moved is *tripled*. Study the following examples to determine the correct method.

Examples

G. Convert 83 000 cm^2 (square centimeters) to square meters.

83 000 cm = ? m

1. First determine the direction and number of places the decimal point would be moved for the linear conversion. In this case, the decimal point would be moved *two* places to the *left*.

83 000 cm^2 = 8.3000 m^2

2. Double the number of places moved in square conversion, so that the decimal point is moved *four* places to the *left* instead of 2.

83 00 cm^2 = 8.3 m^2

3. These are equivalent square units.

H. Convert 80 000 cc to cubic meters. (Cubic centimeters or cm^3 is often written as cc.)

80 000 cm = ? m

1. First determine the direction and number of places the decimal point would be moved for the linear conversion. In this case, the decimal point would be moved *two* places to the *left*.

80 000 cc. = 0.080000 m^3

2. Triple the number of places the decimal point is moved for cubic measure. Therefore, the decimal point is moved *six* places to the *left*.

80 000 cc = 0.08 m^3

3. These are equivalent cubic units.

I. Change 5 000 000 dm^3 to km^3.

5 000 000 dm = ? km

1. Determine the direction and number of places the decimal point would be moved for the linear conversion. In this case, the decimal point would be moved *four* places to the *left*.

$5\ 000\ 000\ dm^3 = .000\ 005\ 000\ 000.\ km^3$	2. Triple the number of places the decimal point is moved for cubic measure; therefore, the decimal point is moved *12* places to the *left.*
$5\ 000\ 000\ dm^3 = 0.000\ 005\ km^3$	3. These are equivalent cubic units.

J. A house has floor space equal to $1\ 200\ 000\ cm^2$. How many square meters is this?

$1\ 200\ 000 \rightarrow 120.0000.$	1. Double the number of places the decimal point is moved. (In this case, 4 instead of 2.)
$1\ 200\ 000\ cm^2 = 120\ m^2$	2. Always drop the trailing zeros after the decimal point.

K. A rectangular warehouse is 20 meters by 35 meters by 8 meters high. Find the volume of the warehouse in cubic decameters. Use two different conversion methods.

a. $20\ m \times 35\ m \times 8\ m = 5600\ m^3$ $5600\ m^3 = 5.6\ dam^3$	1. Find the volume in meters. 2. Triple the number of places the decimal point is moved (in this case, 3 places instead of 1) to convert from cubic meters to cubic decameters.
b. $20\ m = 2\ dam$ $35\ m = 3.5\ dam$ $8\ m = 0.8\ dam$ $2\ dam \times 3.5\ dam \times 0.8\ dam = 5.6\ dam^3$	1. Convert meters to decameters before finding the volume. 2. Note that results are the same.

Exercises 10–1B

Using the prefixes for reference, find the equivalences.

km hm dam m dm cm mm

a. $0.0005\ km^2$ = ___5___ dam^2

b. $20\ 000\ mm^2$ = ___0.02___ m^2

1. 38 000 cc = ________ m^3

2. $0.005\ m^2$ = ________ hm^2

3. $4200\ hm^3$ = ________ km^3

4. $0.05\ mm^2$ = ________ cm^2

5. $82\ 000\ mm^2$ = ________ m^2

6. $0.05\ m^2 =$ __________ cm^2

7. 48 000 cc = __________ dm^3

8. 0.00008 cc = __________ mm^3

9. $41\ m^2 =$ __________ cm^2

10. $18\ 000\ cm^2 =$ __________ m^2

11. Find the number of cubic millimeters in $3.2\ cm^3$ (often abbreviated as cc).

12. A house is 15 000 mm long by 8000 mm wide. (Most metric blueprints are shown in millimeters.) Find the area of the house in square meters.

13. A 4200 cc (cubic centimeter) engine is equivalent to how many liters? (*Hint:* 1 cubic centimeter is exactly equal to 1 milliliter.)

14. A small cabin is 3000 mm wide and 5000 mm long. Find the area in square meters. Mentally calculate your answer!

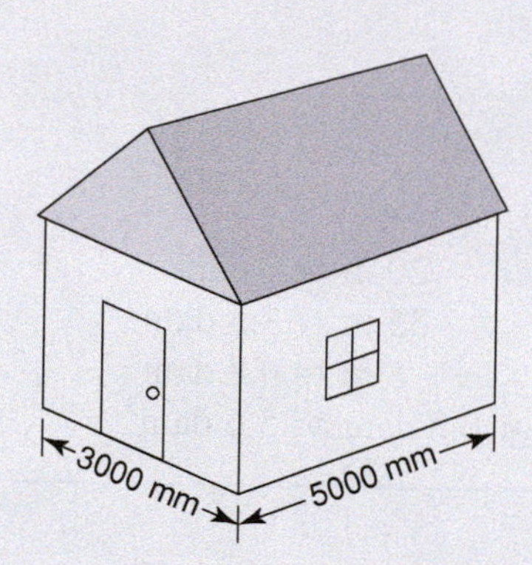

15. A storage barn has the dimensions shown. Find the volume of the barn in cubic meters. (*Hint:* Convert all units to meters before calculating.)

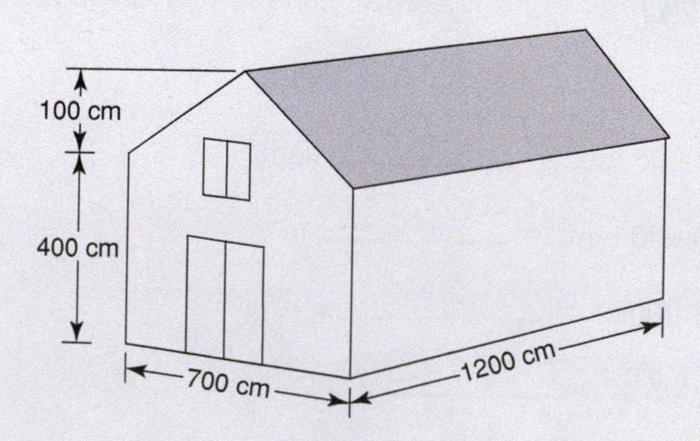

16. The storage barn in problem 15 is to be painted. Doors at each end will be painted also. Determine the number of square meters to be painted. (Don't paint the roof!)

17. Each bucket of paint covers 25 square meters and costs $38.50. Determine the cost of the paint for two coats of paint. (Remember to round up to the higher number of buckets before calculating price.)

18. A 1-liter can has a base with an area of 50 cm^2 (square centimeters). Find the height of the can in centimeters. (*Hint:* 1 liter = 1000 ml = 1000 cc)

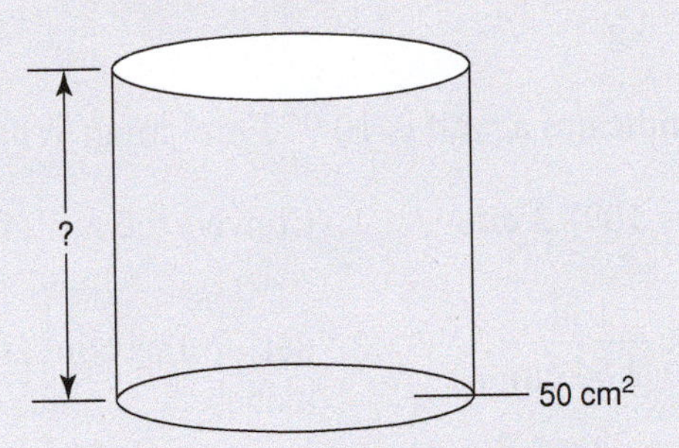

19. A cistern is to be used in a solar water-heating system. In the metric system, one milliliter (1 ml) of water (which is also 1 cc of water) weighs 1 gram. Find the weight in kilograms of the water contained in a rectangular cistern with inside dimensions of 1500 mm by 2000 mm by 3000 mm. Assume the cistern is filled to the top. *Hint:* change all dimensions to cm before multiplying, and the equivalent volume will be in cubic centimeters. (1 cc = 1 ml = 1 gram.)

20. In the metric system, land is measured in hectares (instead of acres). One hectare is one square hectometer. Find the number of hectares in a rectangular parcel of land measuring 200 m by 185 m. (*Hint:* change dimensions to hectometers before multiplying.)

10.2 Metric–American Conversions

As long as the American system is still in use, conversions from one system to the other will be necessary. The most common conversions are listed here, and it is worthwhile to memorize these. One inch is exactly equal to 2.54 cm. All other conversions are approximate.

1 inch	= 2.54 cm
1 meter	≐ 3.28 ft
1 mile	≐ 1.61 km
1 liter	≐ 1.06 qt
1 kilogram	≐ 2.2 lb
1 ounce	≐ 28 g

To convert from one system to another, use the unit conversion method discussed in Chapter 4.

Examples

A. What is the weight in kilograms of a 50-lb box of nails?

$\frac{50\text{ lb}}{1} \times \frac{1\text{ kg}}{2.2\text{ lb}}$ 1. Set up the unit ratio 1 kg = 2.2 lb so that pounds cancel.

$\frac{50}{1} \times \frac{1 \text{ kg}}{2.2}$ 2. Cancel units and perform the division (since 2.2 is in the denominator).

$50 \text{ lb} \doteq 22.7 \text{ kg}$

B. A blueprint indicates a wall is 10 973 mm long. What is the length in feet?

$10\,973 \text{ mm} = 1097.3 \text{ cm}$ 1. Convert the millimeters to centimeters.

$\frac{1097.3 \text{ cm}}{1} \times \frac{1 \text{ in.}}{2.54 \text{ cm}}$ 2. Set up the conversion between centimeters and inches.

$1097.3 \text{ cm} = 432 \text{ in.}$ 3. Perform the division.

$\frac{432 \text{ in.}}{1} \times \frac{1 \text{ ft}}{12 \text{ in.}}$ 4. Convert inches to feet.

$10\,973 \text{ mm} = 36 \text{ ft}$ 5. The wall is 36 ft long.

C. Gas at a certain station in Canada costs $\$1.40^4$ (American) per liter. What is the equivalent price per gallon?

$\frac{\$1.40^4}{1 \text{ liter}} \times \frac{?}{?} = \frac{\$\,?}{\text{gal}}$ 1. Set up the given and desired units.

$\frac{\$1.40^4}{1 \cancel{\text{liter}}} \times \frac{1 \cancel{\text{liter}}}{1.06 \cancel{\text{qt}}} \times \frac{4 \cancel{\text{qt}}}{1 \text{ gal}}$ 2. Since the conversion factor for liters to gallons is not given, a two-step conversion is necessary. Convert liters to quarts and quarts to gallons, canceling the unwanted units of liters and quarts.

$\frac{\$1.40^4}{1} \times \frac{1}{1.06} \times \frac{4}{1 \text{ gal}} = \frac{\$5.29^8}{\text{gal}}$ 3. When only the desired units are left, perform the multiplication and division.

$\frac{\$1.40^4}{\text{liter}} = \frac{\$5.29^8}{\text{gal}}$ 4. Most gas stations price to the mil. (A mil is $\frac{1}{10}$ cent or \$0.001.)

D. A carpenter can buy identical nails in a 50-lb box at one hardware store and a 20-kg box at another store for the same price. Which is the better buy?

$20 \text{ kg} \times 2.2 \text{ lb/kg}$ 1. Determine the number of pounds equivalent to 20 kg.

$20 \text{ kg} = 44 \text{ lb}$ 2. The 50-lb box contains more nails for the same price.

E. A built-up girder has a width of 11 inches and a depth of $7\frac{1}{2}$ inches. Find the cross-sectional area in square centimeters.

$11 \text{ inches} = 27.94 \text{ cm}$ 1. Convert the dimensions from inches to centimeters.

$7\frac{1}{2} \text{ inches} = 19.05 \text{ cm}$

$27.94 \text{ cm} \times 19.05 \text{ cm} = 532.257 \text{ cm}^2$

2. Find the area in square centimeters.

$532.257 \text{ cm}^2 \rightarrow 532 \text{ cm}^2$

3. Round appropriately.

It is important here to recognize why solutions sometimes vary slightly, even though the problems are worked using a correct procedure and accepted conversion factors. Because the majority of the metric–American conversions are approximations rather than exact conversions, variations in the calculated values can occur, sometimes quite significantly. In particular, inaccuracies can multiply when measurements are multiplied. If accuracy and precision are required to the hundredth place, for example, use the most accurate conversion factors, even if it involves extra steps. In particular, when one is using a calculator, it is easy to calculate answers to many decimal places, but the solution can have no greater accuracy and/or precision than the conversion factors used to make the conversions. Study the examples below:

A. A surveyor measures a driveway to be 45.82 m long. What is this in feet, measured to the hundredth of a foot? (Most surveying done in the American or customary system measures to the hundredth of a foot.)

$$\frac{45.82 \cancel{\text{m}}}{1} \times \frac{3.28 \text{ ft}}{1 \cancel{\text{m}}}$$

1. This is a simple, one-step conversion, using an *approximate* conversion factor.

$$\frac{45.82}{1} \times \frac{3.28 \text{ ft}}{1}$$

2. Cancel units and perform the multiplication.

$45.82 \text{ m} \doteq 150.29 \text{ ft.}$

3. This is an *approximate* equivalent.

B. A surveyor measures a driveway to be 45.82 m long. What is this in feet, measured to the hundredth of a foot? This is the same problem as A above, but the calculations will be done using exact equivalents instead of approximations.

$45.82 \text{ m} = 4582 \text{ cm}$

1. 100 cm = 1 m is an *exact* equivalent, and one that can be done mentally.

$$\frac{4582 \cancel{\text{cm}}}{1} \cdot \frac{1 \cancel{\text{in.}}}{2.54 \cancel{\text{cm}}} \cdot \frac{1 \cancel{\text{ft}}}{12 \cancel{\text{in.}}}$$

2. For measuring length, 1 inch = 2.54 cm is an *exact* conversion between the metric and customary (American) systems, and 1 foot = 12 inches is an exact customary-to-customary conversion. Unlike example A above, no *approximate* conversions are used in this problem.

$$\frac{4582}{1} \cdot \frac{1}{2.54} \cdot \frac{1 \text{ feet}}{12}$$

3. Cancel like units, leaving only feet.

150.33 feet

4. Length of driveway, measured to the hundredths of a foot.

150.29 ft ≐ 150. ft
150.33 ft ≐ 150. ft

5. In most applications, saying the driveway is 150 feet long would be sufficient. Either conversion factor would work, and example A is simpler to calculate.

A: 150.29 ft (not correct)
B: 150.33 ft (correct)
Either method: rounding to 150 ft (correct)

6. Both A and B above are calculated to the hundredths place, but only B is *accurate* in the hundredths place. Using either method, it is correct to say the driveway = 150 ft (rounded to the nearest whole foot); using method B, it is correct to say the driveway = 150.33 ft (rounded to the nearest hundredth of a foot). It is NOT correct to use Method A and say the driveway = 150.29 ft. Never include more decimal places in the solution than the method of calculation can accurately figure!

C. Plans call for a 5-kiloliter cistern connected to solar panels to be installed on a platform. The platform must be able to support 50% more than the weight of the water in the cistern. The 50% allowance is a safety factor. Will a platform designed to hold 10 tons be sufficient?

5 kl = 5000 liters

1. This is an exact equivalent (most metric–metric conversions are exact equivalent).

5000 liters = 5000 kg

2. The metric system is conveniently designed such that 1 liter of water weighs 1 kilogram.

5000 kg + 2500 kg = 7500 kg

3. 7500 kg provides the 50% safety allowance for the platform.

$$\frac{7500\ \cancel{kg}}{1} \times \frac{2.2\ \cancel{lbs}}{1\ \cancel{kg}} \times \frac{1\ \text{Ton}}{2000\ \cancel{lbs}}$$

4. Find the weight in tons equal to 7500 kg using the unit conversion method.

8.25 Tons

5. Ten tons will provide a stronger-than-necessary platform. (This is a good thing!)

10T > 8.25T

6. The equivalent 1 kg ≐ 2.2 lbs is an approximation. In this example, the approximation is quite sufficient, because the platform is designed to hold 1.75 tons (3500 lbs) more than necessary.

Exercise 10–2

1. A building lot has 38.21 m of road frontage. Using the conversion factor 1 m ≐ 3.28 ft, find the frontage in feet. Round to the nearest hundredths of a foot. (Land deeds are generally recorded in decimal feet to two decimal places.)
2. Calculate the above frontage by changing meters to cm, cm to inches, and inches to feet. Is the answer the same? Why or why not? Which is the accurate answer?

3. A builder's van has to travel 37 km to a job site. What is this distance in miles?
4. How many grams does a 22-oz framing hammer weigh?
5. A radial arm saw table weighs 38 kg. What is its weight in pounds?
6. The stiles on a raised door panel measure $21\frac{5}{16}$ inches. What is this in millimeters?
7. A 12-oz can of soda weighs approximately how many grams?
8. How much thicker is a $\frac{1}{4}''$ sheet of plywood than a 5-mm sheet? Give the answer in millimeters, to the nearest whole mm.
9. A pickup truck gets 18 mpg (miles/gallons). Find the equivalent fuel efficiency in km/liter (kilometers/liters).
10. A blueprint shows the following dimensions on a small ranch house. Find the approximate area in square feet.

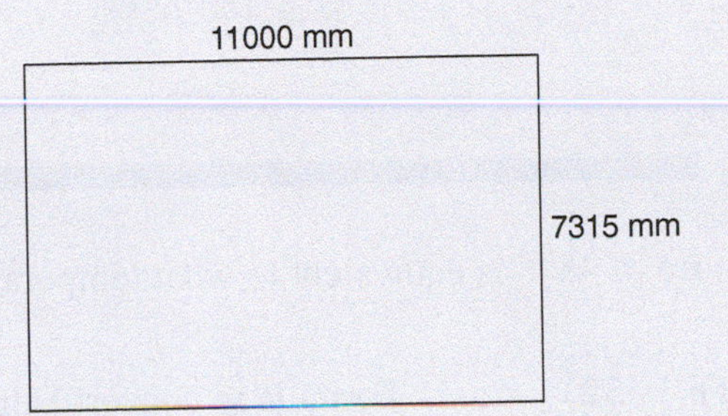

11. A newel post on a stairway measures 32 inches. Find the height in millimeters.
12. Gasoline costs $\$1.59^9$ (American dollars) per liter at a Canadian gas station. Find the equivalent price per gallon.
13. A blueprint for an oversized garage calls for a tall garage door to be installed at the driveway. The garage door was made in Canada and is 3750 mm high. Will a Recreational Vehicle 11 ft 10 in. high fit through the doorway? If so, what will be the clearance in inches?
14. A project requires $\frac{3}{8}''$ plywood. What thickness is required in mm?
15. A 10-foot long $6'' \times 6''$ beam made of dry, seasoned white oak is to be used in a timber-framed house. The beam weighs approximately 12 pounds per lineal foot. Find the weight in kilograms of the beam.
16. A storage shed measuring 2743 mm $\times$ 3353 mm is to be built on an RV site. First change the dimensions from mm to meters, and then find the dimensions of the shed in feet. Use 1 meter $\doteq$ 3.28 ft as the conversion factor.
17. Calculate the dimensions of the shed by first converting the mm to cm, the cm to inches, and the inches to feet.
18. Of the two measurements 16 and 17 above, which is more accurate? If the measurements are to be measured precise to the nearest $\frac{1}{4}''$, does it matter which conversion method is used?
19. The frontage on a lot is to be recorded in the county court house as decimal feet to the nearest hundredth foot. What frontage will be recorded if a metric tape measures 48.29 m, and 1 m $\doteq$ 3.28 feet is used as the conversion factor?
20. On the lot in problem 9, what frontage will be recorded if the meters are first converted to cm, the cm to inches, and the inches to decimal feet? Which measurement is more accurate? For such a small variation, does it actually matter?

10.3 Temperature Conversions

The Celsius scale is used to measure temperature in the metric system. Water freezes at 0°C and boils at 100°C under standard conditions. The Fahrenheit scale, used in the American system, sets the freezing point of water at 32°F and the boiling point at 212°F.

To convert from Fahrenheit to Celsius, use the formula:

$$C = \frac{5}{9}(F - 32)$$

To convert from Celsius to Fahrenheit, use the formula:

$$F = \frac{9}{5}C + 32$$

Examples

A. A temperature of 48°F is equivalent to what temperature on the Celsius scale?

$C = \frac{5}{9}(F - 32)$ 1. Formula to convert Fahrenheit to Celsius.

$= \frac{5}{9}(48 - 32)$ 2. Substitute Fahrenheit temperature into formula.

$= \frac{5}{9}(16)$ 3. Perform operation in parentheses first.

$= 8.9°C$ 4. Multiply by $\frac{5}{9}$.

$48°F = 8.9°C$ 5. These are approximately equivalent.

B. The thermostat in a house is set at 20°C. What temperature is this on the Fahrenheit scale?

$F = \frac{9}{5}C + 32$ 1. Formula to convert Celsius to Fahrenheit.

$= \frac{9}{5} \cdot 20 + 32$ 2. Substitute Celsius temperature.

$= 36 + 32$ 3. Since there are no parentheses, perform the multiplication first.

$= 68°F$ 4. Perform the addition.

$20°C = 68°F$ 5. These are exactly equivalent.

C. A certain adhesive should be used between the temperatures of 8°C and 35°C. What are the equivalent Fahrenheit temperatures?

$F = \frac{9}{5}C + 32$ 1. Use the formula to convert Celsius to Fahrenheit.

$F = \frac{9}{5}(8) + 32$ 2. Substitute into the formula.

$F = 14.4 + 32$ 3. Perform multiplication first.

$F = 46.4$, or 46°F 4. Round to a whole number of degrees.

$F = \frac{9}{5}(35) + 32$ 5. Substitute the other temperature.

F = 95°F 6. Perform the calculations.

The adhesive may be used between 46°F and 95°F.

Exercise 10–3

1. Find the Celsius equivalent of the following Fahrenheit temperatures:
   a. 48°F
   b. 32°F
   c. 68°F
   d. 52°F
   e. 98.6°F
   f. 80°F
2. Find the Fahrenheit equivalent of the following Celsius temperatures:
   a. 35°C
   b. 22°C
   c. 10°C
   d. 15°C
   e. 0°C
   f. 30°C
3. The temperature in Death Valley sometimes registers 45°C. What is this in Fahrenheit degrees?
4. On a certain day Dallas had a temperature of 77°F while Toronto registered 25°C. Which city was warmer?
5. A paint primer must be applied at temperatures between 10°C and 35°C. What is the temperature range in Fahrenheit degrees?
6. Which is colder: −40°F or −40°C?
7. Glue must be applied between 45°F and 95°F. What is that temperature range in degrees Celsius?
8. On the Fahrenheit scale water freezes at 32° and boils at 212°. On the Celsius scale, water freezes at 0° and boils at 100°. Which is larger: a temperature increase of 1 degree Fahrenheit or 1 degree Celsius? Does one of these scales make more sense than the other? Why or why not?
9. The coldest temperature ever recorded on earth is −89.2°C in Antarctica on July 21, 1983 (*Remember:* July is winter in the southern hemisphere). What is that in degrees Fahrenheit? Brrrrr!
10. During a storm, the temperature dropped from 8°C to −5°C in one hour in Halifax, Nova Scotia. During that same hour, the temperature rose from 45°F to 68°F in Miami, Florida. Which city experienced the greater change in temperature?

REVIEW EXERCISES

Perform the following conversions.

1. 5 lb/m = __________ lb/ft
2. A roof is 14.8 m long. What is its equivalent length in feet and inches, to the nearest $\frac{1}{4}''$?
3. A wood lathe weighs 462 kg. How much does the lathe weigh in pounds?
4. Which is the better buy: nails at $1.25 per pound or $2.85 per kilogram?
5. A carpenter has only a metric rule available. If he is framing a wall with studs 16 inches o.c., what length, in millimeters, should he measure on the metric rule? (In Canada, all building layout dimensions are in millimeters.)
6. A finish board is 2.6 m long. To the nearest $\frac{1}{4}''$, how long is the board in feet and inches?
7. 2×6 framing lumber weighs 2.1 lb per lineal foot. What is the weight, in kilograms, of an 8-ft-long 2×6?
8. Convert 47°C to its equivalent Fahrenheit temperature.
9. Convert 85°F to its equivalent Celsius temperature.
10. The normal body temperature in a healthy human being is 98.6°F. What is this on the Celsius scale?
11. A sheet of plywood is $18'' \times 24''$. What is the area in square centimeters?
12. For best results, a certain urethane finish should not be applied at a temperature below 55°F. What is this on the Celsius scale?
13. The highest winds ever recorded on earth were clocked at 231 mph on Mt. Washington, New Hampshire on April 12, 1934. What is this in km/hr? (These are the highest officially recognized winds in the United States that are not associated with a tornado and were measured on the surface of the earth.)
14. A board is 3 ft $7\frac{1}{4}$ inches long. What is this in millimeters?
15. A material used in solar collection panels expands 0.01 mm/m (0.01 millimeter per meter length) per degree Celsius. Find the expansion of a panel 12 feet long for a temperature change from 35°F to 139°F. Give answer to the nearest $\frac{1}{32}''$. (This problem requires multiple steps and conversions.)

CHAPTER 11

Board Measure

Did You Know

The number of board feet (bf) of lumber contained within a tree can be estimated by the following procedure. (1) Find the DBH (diameter at breast height) in feet. (2) Calculate the area ($A = 3.14\ R^2$, where R is the radius in feet). (3) Multiply this area by an estimate of the tree's height in feet; (4) Divide by 4 to compensate for the taper. (5) Multiply this result by 12 board feet per cubic foot of tree.

OBJECTIVES

Upon completing this chapter, the student will be able to:

1. Calculate the number of board feet for any size and quantity of lumber.
2. Determine the number of linear feet of lumber needed for a certain job, given the number of board feet.

11.1 Finding the Number of Board Feet

One Board Feet (bf) is equal to 144 cubic inches (144 in.^{3}) of wood.

Several examples of 1 bf are shown in the figure.
To find the number of bf in a piece of lumber,

1. Calculate its volume in cubic inches.
2. Divide by 144.

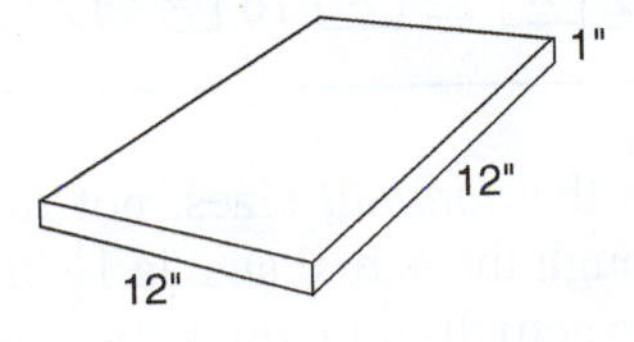

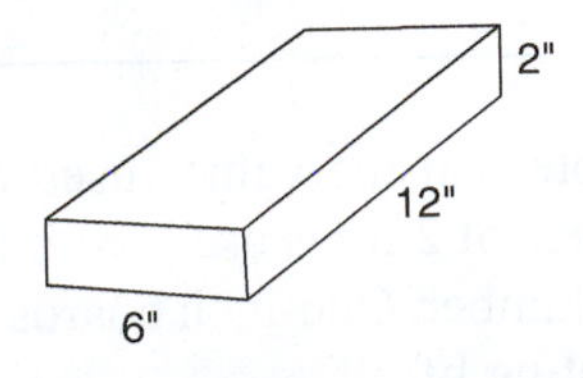

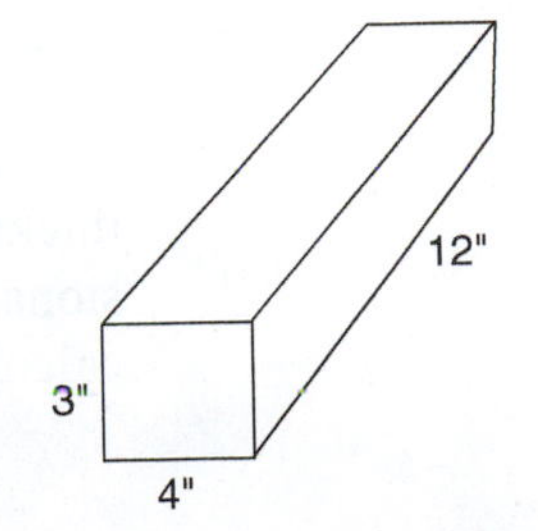

Example

A. How many board feet are in a 2 × 12 that is 16 feet long?
Volume = length × width × height. The dimensions must be in inches in order to get a result in cubic inches.

$16 \text{ ft} \times 12 = 192 \text{ in.}$	Multiply 16 feet by 12 to get inches.
$2 \text{ in.} \times 12 \text{ in.} \times 192 \text{ in.} = 4608 \text{ in.}^3$	Multiply to get volume.
$4608 \div 144 = 32 \text{ bf}$	Divide by 144 to get bf.

There are 32 bf in the piece of lumber.

An Easier Method

The bf formula is

$$\text{bf} = \frac{T \times W \times L}{12}$$

where T is the thickness in inches, W the width in inches, and L the length in feet.

Notice that when using this formula the length does not have to be converted to inches. That has been taken care of in the formula, since the divisor is 12 instead of 144. So instead of multiplying the length in feet by 12 and then dividing by 144, one 12 in the numerator "cancels" one 12 in the denominator.

Example

B. Find the number of bf in the same 16-ft long 2 × 12 used in example A, this time by using the bf formula.

$\text{bf} = \dfrac{2 \times 12 \times 16}{12}$	Substitute in the proper values.
$\text{bf} = \dfrac{384}{12}$	Multiply the terms in the numerator.
$\text{bf} = 32$	Divide 384 by 12.

Calculator solution: Enter 2 [×] 12 [×] 16 [÷] 12 [=] to get 32.

Note that the values used are the *nominal* sizes, not the *actual* sizes. That is, the thickness of 2 in. is used, even though the actual size is $1\frac{1}{2}$ in. This applies to all dimensional lumber. One-inch boards are actually $\frac{3}{4}$ in. thick, but the 1-in. dimension is used in calculating bf.

Example

C. Find the number of bf in 200 pieces of 2×6, each 8 ft long. Use the bf formula.

$$\text{bf} = 200 \times \frac{2 \times 6 \times 8}{12} = 1600$$

Notice that the factor of 200 has been inserted into the formula to account for the 200 pieces.

Calculator solution: Enter 200 [×] 2 [×] 6 [×] 8 [÷] 12 [=] to get 1600.

11.2 Finding Linear Feet When the Number of Board Feet Is Known

COMMENT

When an area is to be covered with 1-in.-thick boards, each square foot requires 1 bf. When an area is to be covered with 2-in. lumber (such as 2×6s), each square foot requires 2 bf.

Example

A. Suppose you need to order the lumber needed to cover a deck measuring 14 ft by 20 ft with 1×8 boards. How many linear feet of 1×8 boards would you order? First, calculate the area to be covered.

$$A = 20 \times 14 = 280 \text{ ft}^2$$

Second, use the bf formula with bf = 280 (since 1 sq ft = 1 bf for 1-in. boards).

$$280 = \frac{1 \times 8 \times L}{12}$$ Substitute in the proper values.

$$280 \times 12 = 8 \times L$$ Multiply both sides by 12.

$$L = \frac{280 \times 12}{8}$$ Divide by 8.

$$L = 420 \text{ ft}$$

You may wish to solve the formula for L first. The formula would become

$$L = \frac{bf \times 12}{T \times W}$$

So it takes 420 linear feet of 1×8 boards to cover 280 ft^2. Since a 1×8 is actually closer to $7\frac{1}{2}$ in. wide, some error is present in this calculation. If the boards are to be spaced, some of that error will be reduced, depending on the size of the spacing. It is recommended that 10% to 15% be added for waste.

Another consideration is the length of the boards to be ordered. This will depend on whether the boards are to be laid with the 20-ft length or the 14-ft length.

Five-quarter decking is a material often used to cover a deck. The term "five-quarter" or $\frac{5}{4}$ refers to the thickness. It is rough sawn at $1\frac{1}{4}$ in., but milling reduces that to 1 full inch of thickness. This makes for a lightweight, but sturdy, material for decking.

Example

B. A deck is to be covered with $\frac{5}{4}$ by 6 spruce decking material. The deck measures 10 ft by 16 ft. How many linear feet of material are needed?

There are two steps to the solution. First, find the number of bf required to cover the area with $\frac{5}{4}$ decking. Second, find the number of linear feet of $\frac{5}{4}$ material needed to provide this number of bf.

Using the bf formula,

$$\text{bf} = \frac{1.25 \times (10 \times 12) \times 16}{12} = 200$$

The thickness is 1.25 in., and the 10-ft width is multiplied by 12 because the width has to be in inches.

So 200 bf of material are required.

Now, using the 200 bf, we find the length of material needed. Using the bf formula again, but solved for L,

$$L = \frac{\text{bf} \times 12}{T \times W}$$

$$L = \frac{200 \times 12}{1.25 \times 6}$$

Substitute in the proper values.

$$L = 320 \text{ ft}$$

It takes 320 linear feet of $\frac{5}{4} \times 6$ decking to cover the area.

Once again, the carpenter will decide what lengths will provide the most efficient use of the lumber. He will need to consider the spacing of the floor joists and the direction of laying.

REVIEW EXERCISES

1. The most commonly used sizes in framing lumber are 2 × 4, 2 × 6, 2 × 8, 2 × 10, and 2 × 12. For each of these sizes, calculate the number of bf in 1 linear foot of material. Notice the sequence formed by these answers. This is an easy-to-remember sequence and will be useful in estimating amounts and prices of lumber. Do you see how? (See "Hint" in exercise 2a below.)
2. Use your results from exercise 1 to calculate the number of bf in each of the following quantities of lumber.
   a. 120 pieces of 8-ft-long 2 × 4. (*Hint:* Find the total length; then multiply by $\frac{2}{3}$.)
   b. 84 pieces of 14-ft-long 2 × 10.
   c. 26 pieces of 12-ft-long 2 × 6.
   d. 33 pieces of 10-ft-long 2 × 8.
   e. 15 pieces of 16-ft-long 2 × 12.
3. Find the number of bf in the following quantities of lumber:
   a. 535 pieces of 8-ft-long 2 × 4.
   b. 96 pieces of 14-ft-long 2 × 10.
   c. 14 pieces of 12-ft-long 2 × 6.
   d. 76 pieces of 10-ft-long 2 × 8.
   e. 18 pieces of 14-ft-long 2 × 12.
4. How many linear feet of 1 × 8 boards are needed to cover an area of subflooring 26 ft wide by 42 ft long? (*Hint:* Use the bf formula with L as the unknown.)

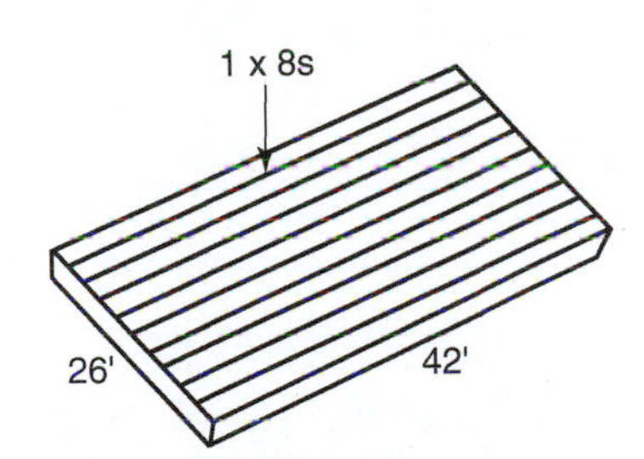

5. A deck measuring 8 ft by 4 ft is to be covered with 2 × 6s. Find the number of linear feet needed.

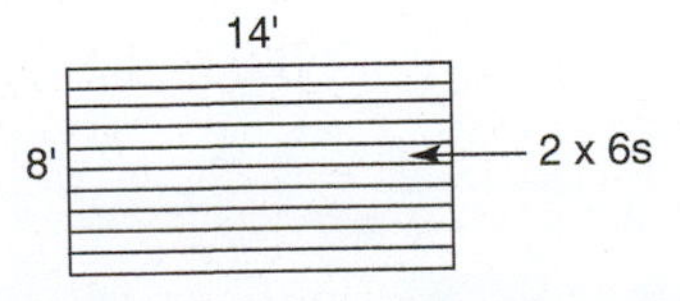

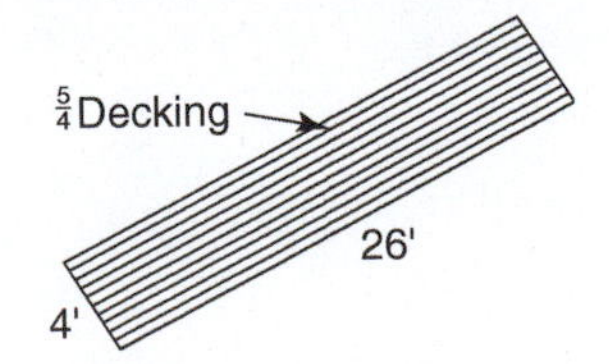

6. $\frac{5}{4}$ decking, 4 in. wide will be used to cover a porch 4 ft wide by 26 ft long. How many bf are needed?
7. Find the number of linear feet of $\frac{5}{4}$ by 4 decking needed to cover the deck described in exercise 6.
8. The roof of a house contains 1120 ft^2. How many linear feet of 1 × 8 roofers are needed to cover this roof? (A "roofer" is a tongue-and-grooved board sometimes used to board-in an area.)
9. How many sheets of $\frac{5}{8}$-in. CDX plywood would be needed to cover the roof in exercise 8? (Plywood measures 4 ft by 8 ft.)

10. A deck measuring 10 ft by 16 ft is to be covered with 2 × 6s. How many linear feet of 2 × 6s are needed?
11. It has been estimated that the corner boards and fascia trim for a house require the following amounts of No. 2 pine boards:
    1 × 4: 8/10s (that is, 8 pieces, each 10 ft long)
    1 × 6: 6/8s; 10/12s; 2/14s
    1 × 8: 8/10s
    Find the number of bf in the materials list.
12. A bundle of 1 × 10 boards measures 4 ft high, 6 ft wide, and 12 ft long.
    a. Estimate the number of linear feet of boards in this bundle.
    b. Estimate the number of bf in this bundle.
13. A cord of firewood is stacked 4 ft high, 4 ft wide, and 8 ft long. Approximately how many bf does this represent?
14. A shed roof is 10 ft wide and 12 ft long.
    a. How many bf of 1 in. boards are needed to cover this roof?
    b. How many linear feet of 1 × 6 boards are needed to cover this roof?
15. If 16 sheets of plywood are needed to cover an area, how many bf of lumber are needed to do the same job?

CHAPTER 12

Lumber Pricing

> **Did You Know**
>
> Advanced framing techniques can reduce the number of board feet (bf) in a house by 11% to 19%. This includes using roof trusses, floor trusses, laminated beams, and structural insulated panels.

OBJECTIVES

Upon completing this chapter, the student will be able to:

1. Calculate the cost of lumber based on a price quoted per board foot.
2. Calculate the cost of lumber based on a price quoted per linear foot.
3. Convert a price per board foot to either a price per piece or a price per linear foot.
4. Convert a price per linear foot to either a price per board foot or a price per piece.
5. Convert a price per piece to either a price per board foot or a price per linear foot.

12.1 Pricing Systems

Lumber is priced in a variety of ways:

- By the piece, such as a 2 × 4-8 feet long, offered for $2.23.
- By the linear foot, such as 2 × 4s, offered for 28 cents per linear foot.
- By the board foot, such as 2 × 4s offered for 51 cents per board foot.
- By 1000 board feet (mbf), such as $498 mbf.

When receiving a quote from a lumber yard, it is important to know which of these methods is being used. If you receive a quote of, say, "32 cents," you need to know if that is a price per linear foot, per board foot, or some other quantity.

It is also important that a buyer know how to convert from one method of pricing to another.

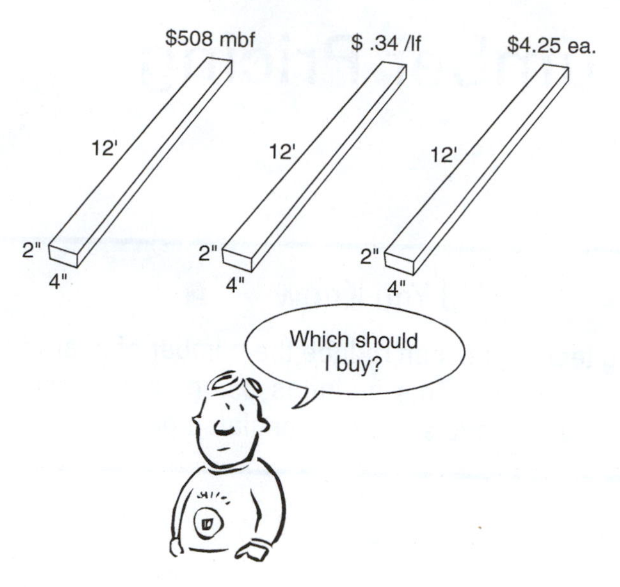

12.2 Price Conversions

First review the sections of the book that dealt with "rates" and "unit conversions."

In brief, a rate such as miles per gallon (mpg) may be expressed as a fraction, mi/gal, and a fraction can be converted into a decimal by dividing the numerator by the denominator, which gives the rate. For example, the fraction $\frac{3}{4}$ becomes 0.75 when 3 is divided by 4. Likewise, to determine miles per gallon after traveling 340 miles on 20 gallons of gasoline, think of mpg as mi/gal, which leads to 340/20, which divides out to 17 mi/gal.

If you need to find a price per foot, think of it as price/feet and *divide the price by the length.* For example, a 2×4 that is 8 feet long and selling for $2.24 would have a price per foot of $\$2.24 \div 8 = 0.28$, or 28 cents per foot.

"Price per board foot" may be abbreviated as "$bf." This unit suggests that *price is divided by bf.* $45 for 100 bf becomes $\frac{\$45}{100}$, or $.45 per board foot.

A price per linear foot may be abbreviated as "$/lf." This unit suggests that the *price is divided by the number of linear feet.* $2.40 for an 8-ft-long board becomes $\frac{\$2.40}{8}$, or $.30 per foot.

Example

A. If 2×4s are quoted at 52 cents per board foot, find the cost of a piece 16 ft long.

Solution:

$$\frac{2 \times 4 \times 16}{12} = 10.667 \quad \text{First, find the number of bf.}$$

$$0.52 \times 10.667 = \$5.55 \quad \text{Second, multiply by the price.}$$

When using a calculator to solve any problem, it is easiest to set up the entire problem on paper before using the calculator. In this way it is easy to check your steps should an error be discovered. Furthermore, the calculations can be done in a continuous sequence. The setup for example A would look like this:

Calculator solution: 0.52 [×] 2 [×] 4 [×] 16 [÷] 12 [=] to get $5.55

Example

B. Find the cost of 20 pieces of 2 × 8, each 12 ft long, if the price quoted is $512 mbf (that is, $512 per 1000 bf).
First, convert the quoted price to a price per board foot.

$512 mbf = $0.512 per board foot	Since "mbf" means per 1000 bf, divide the quote by 1000, getting the price per board foot. (This is easily accomplished by moving the decimal point three places to the left.)
$0.512 \times \dfrac{20 \times 2 \times 8 \times 12}{12} = \$163.84$	Multiply this price by the number of bf.

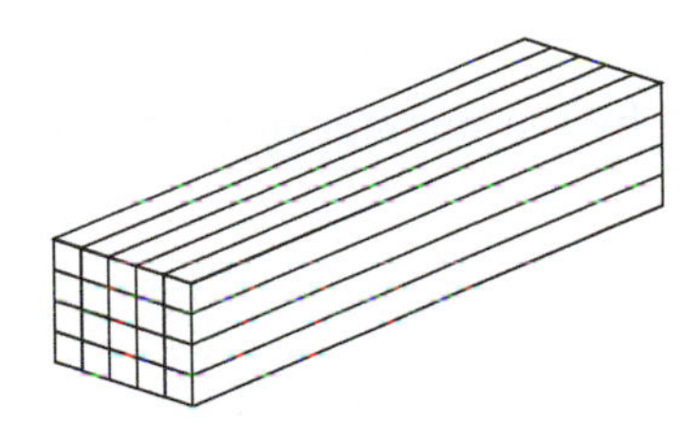

Calculator solution: 0.512 [×] 20 [×] 2 [×] 8 [×] 12 [÷] 12 [=] to get $163.84

Converting a Price Per Linear Foot to a Price Per Board Foot

Example

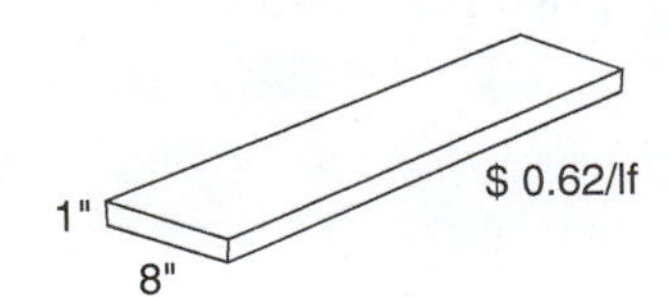

C. If 1 × 8 pine boards are quoted at $1.26 per linear foot, what is the price per board foot?
Since we want the price per board foot, find the number of bf in 1 foot of a 1 × 8.

$\dfrac{1 \times 8 \times 1}{12} = \dfrac{2}{3}$ board foot, or 0.667 Use the board foot formula.

"Price per board foot" is price/bf, so divide the price of $1.26 by the number of bf, namely, 0.667.

$1.26 \div 0.6667 = \$1.89$

Therefore, a price of $1.26 per linear foot of 1 × 8 is equivalent to a price of $1.89 per board foot.

REMINDER

A general rule for determining the number of decimal places to use for accuracy is use two more places than are desired in the answer; then round.

Converting a Price Per Piece to a Price Per Board Foot

Example

D. A 6-ft-long 1 × 8 pine board is priced at $7.52. Find the price per board foot.

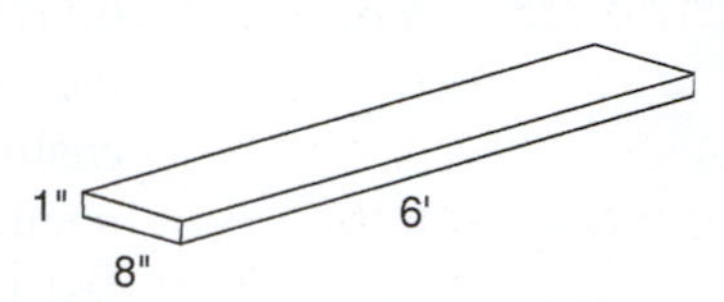

First, find the number of bf.

$$\frac{1 \times 8 \times 6}{12} = 4 \text{ bf}$$ Use the bf formula.

Second, divide the price by the number of bf. (*Remember:* price per board foot is price/bf.)

$$\frac{\$7.52}{4} = \$1.88$$

Therefore, the price is $1.88 per board foot. Notice that this could also be quoted as $1.880 mbf.

REVIEW EXERCISES

1. Eight-feet-long 2 × 4s are offered at $2.26 each. Find the price per board foot.

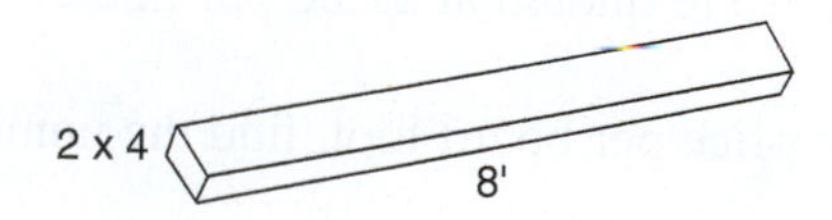

2. 1 × 3 strapping is offered at $16\frac{1}{2}$ cents per linear foot. Find the price per board foot.
3. If 2 × 8s are quoted at $518 mbf, what is the price per linear foot?
4. A piece of 1 × 10 pine 10 ft long is marked $28.10. Find the price per board foot.

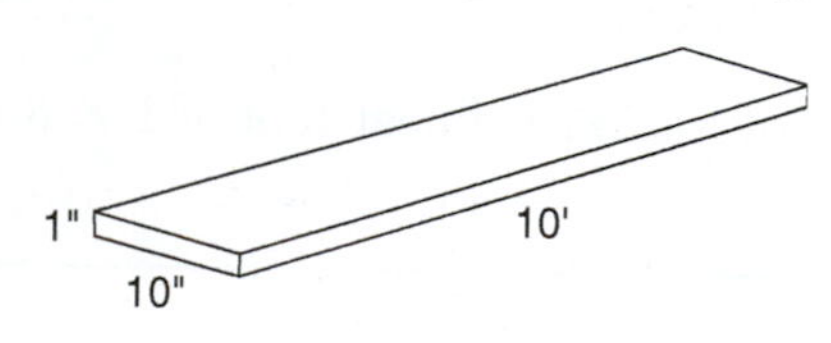

5. Knothole Lumber Company advertises the following prices for lumber. All quotes are prices per board foot.
   2 × 4s: 42 cents up to and including 12 ft; 51 cents for over 12-ft lengths
   2 × 6s: same as 2 × 4s
   2 × 8s: 54 cents for 8-ft lengths; all others 59 cents
   2 × 10s: 56 cents up to and including 12-ft lengths; all others 63 cents
   2 × 12s: 59 cents for 8- and 10-ft lengths; all others 67 cents
   Find the total price for the following list of materials.
   a. 2 × 4s: 325/8s; 44/12s
   b. 2 × 6s: 84/8s; 26/12s; 12/14s
   c. 2 × 8s: 6/14s
   d. 2 × 10s: 8/10s; 64/14s
   e. 2 × 12s: 3/8s; 9/12s; 6/16s
   f. Total price for materials (a) to (e)
   (*Note:* The term "325/8s" and similar terms as used in describing the material means the "number of pieces/lengths.")
6. A rectangular building measures 26 ft × 44 ft. Compare the materials cost of subflooring this area using $\frac{1}{2}$-in. CDX plywood at $12.97 per sheet to the cost using No. 4 grade boards priced at $365 mbf. (Whole sheets of 4 ft × 8 ft plywood will be purchased; add 10% to the board requirements for waste.)
7. 1 × 3 strapping is quoted at $14\frac{1}{2}$ cents per linear foot. Find the price of 256 linear feet.
8. Find the board foot price of the strapping in exercise 7.
9. 1 × 6 select pine is offered at $3,750 mbf by one supplier, while another quotes $1.825 per linear foot. Which is less expensive and how much less?

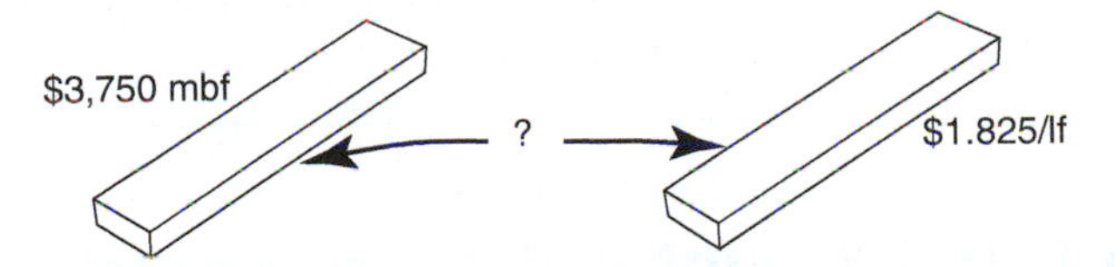

10. A newspaper advertisement offered 2 × 10s at the following prices per piece:
    - 8 ft: $7.53
    - 10 ft: $9.45
    - 12 ft: $11.88
    - 14 ft: $13.42
    - 16 ft: $16.57

    a. Find the price per board foot in each case.
    b. Is the price per board foot the same for all lengths?
    c. Suggest a reason why the price of lumber increases with length.
11. Five-quarter by 6 cedar decking is priced at $.96 per board foot.
    a. What is the price per linear foot?
    b. What is the cost of 24 pieces each 12 ft long?
    c. How many pieces of the 12 footers are needed to cover a deck measuring 10 ft by 16 ft?
    d. What is the cost of the material in part c?
12. In each example below, which costs less:
    a. 8-ft 2 × 4s at $1.35 each or at $.25 per board foot?
    b. 10-ft 2 × 8s at $81 per linear foot or at $8.40 each?
    c. 12-ft 2 × 10s at $1.19 per linear foot or at $.70 per board foot?
    d. 1 × 8 boards at $530 mbf or at $.34 per linear foot?
    e. 2 × 6s at $.58 per linear foot or at $.60 per board foot?

CHAPTER 13

Footings, Foundations, and Slabs

Did You Know

Insulated concrete forms may be substituted for wood, saving trees, but for each ton of Portland cement manufactured, an equal amount of carbon monoxide is released into the atmosphere.

OBJECTIVES

Upon completing this chapter, the student will be able to:

1. Calculate the amount of components of concrete.
2. Calculate the amount of concrete required for a foundation footing.
3. Calculate the amount of concrete required for a foundation wall.
4. Calculate the amount of concrete required for a slab.

13.1 Calculating the Amount of Components in a Concrete Mixture

Concrete is a mixture of cement, sand, and aggregate, the latter being some sort of gravel. Concrete used for masonry (bricks and blocks) is composed of cement and sand. The strength of concrete, after it has properly cured, depends on several factors, including the ratio of cement to sand to aggregate, the amount of water and air entrainment, and curing conditions. In addition, there are different types of cement that affect the strength of the resulting concrete. These are called Type I, Type II, etc., to Type V. For more information on concrete, see www.cement.org.

Our objective here is to be able to calculate the amounts of various components of the concrete mix. A typical mix ratio of cement to sand to aggregate is 1:2:3 (one part cement to two parts sand, to three parts aggregate). For example, six cubic yards of concrete might be comprised of one cubic yard of cement, two cubic yards of sand, and three cubic yards of aggregate. In another case, one cubic yard (27 ft^3) of concrete might be made up of $4\frac{1}{2}$ ft^3 of cement, 9 ft^3 of sand, and $13\frac{1}{2}$ ft^3 of aggregate.

The 1:2:3 ratio is by no means stated here as the best ratio for concrete. Concrete that is well designed will more often use a mix prescribed for its particular use. The range of ingredients in a typical mix is 10 to 15% cement, 60 to 75% aggregate (including sand), 15 to 20% water, and entrained air of 5 to 8%.

Example

For 12 yds^3 of concrete, using the ratio of one part cement to two parts sand to three parts aggregate, how many cubic yards of each of these components are used?

Solution:

The 1 : 2 : 3 ratio suggests that there are 1 + 2 + 3 parts, or a total of 6 parts. Dividing the 12 yds^3 by 6 tells us that the basic unit is 2 yds^3. So in quantity, the ratio becomes 2 yds^3 to 4 yds^3 to 6 yds^3. Thus there will be 2 yds^3 of cement, 4 yds^3 of sand, and 6 yds^3 of aggregate.

Example

If 5 yds^3 of concrete are needed and the specifications call for 12% cement, 20% sand, and 55% aggregate, find the number of cubic yards of each ingredient. (Note that the total is not 100%—the balance being air entrainment and water.

Solution:

Find the percent of 5 yds^3 that each component represents. This answer will be in yds^3.

Cement: 12% of 5 = 0.6 yds^3
Sand: 20% of 5 = 1 yd^3
Aggregate: 55% of 5 = 2.75 yds^3

Notice that this does not total 5 yds^3, just as the percentages given did not total 100%.

13.2 Calculating the Amount of Concrete Needed

Concrete is measured in cubic yards (yd^3). The unit indicates a volume is being measured. To calculate a volume in cubic yards, all three dimensions have to be in the same units. They can be yards or feet or inches, but they all must be the same. Since a foundation or a footing is usually measured partly in inches and partly in feet, one unit must be converted to the other. Feet is the unit recommended. Therefore, a footing 8 inches high and 20 inches wide should be thought of as $\frac{8}{12}$ or 0.667 ft high and $\frac{20}{12}$ or 1.667 ft wide. A foundation that is 10 inches thick and 7 feet 6 inches high would be converted to 0.833 ft thick and 7.5 ft high.

When calculating a volume with all of the dimensions in feet, the result will be a volume in cubic feet (ft^3). The cubic feet must be converted into cubic yards. There are 27 cubic feet in 1 cubic yard, so

$$ft^3 \div 27 = yd^3$$

RULE

To calculate the amount of concrete needed for any rectangular solid (slab, footing, foundation),

1. Convert all measurements to feet,
2. Calculate the volume in ft^3,
3. Divide by 27 to get yd^3.

13.3 Foundations

There are two ways to calculate the number of cubic yards of concrete needed for the foundation shown in Figure 13–1.

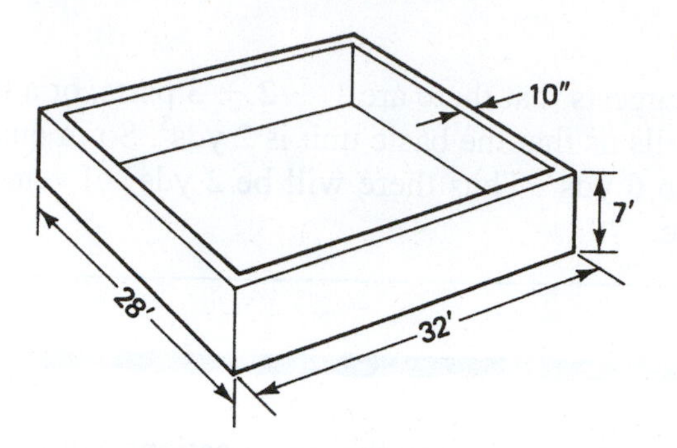

Figure 13–1

The Exact Method

To get the exact solution, separate the foundation into four rectangular solids. The front and rear sections are identical, as are the left and right sections (see Figure 13–2).

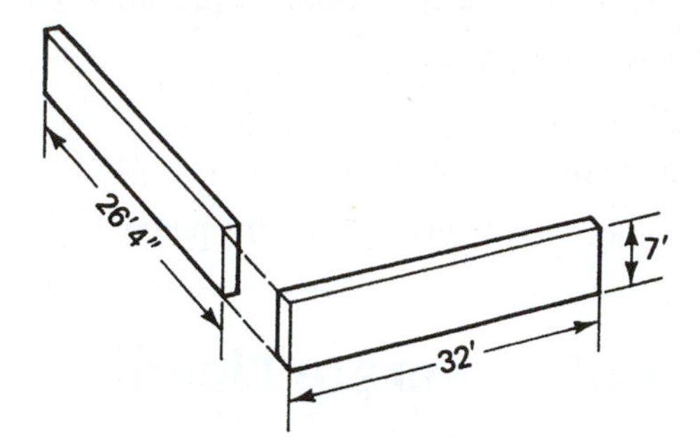

Figure 13–2

The dimensions of these sections are 32 ft long by 7 ft high and 10 in. thick (Figure 13–2). The volume of the front and rear sections may be calculated as follows:

$$V = 32 \times 7 \times \frac{10}{12}$$
$$= 186.67 \text{ ft}^3$$

(Notice that the 10-in. dimension is divided by 12, converting it to feet.) Double this figure to take both sections into account.

$$2 \times 186.67 = 373.34 \text{ ft}^3$$

Now, for the side walls, notice that the length must be reduced by twice the wall thickness—a total of 20 in. The length then becomes

$$28 \text{ ft} - 20 \text{ in.} = 26 \text{ ft } 4 \text{ in., or } 26.33 \text{ ft}$$

The volume of an end section is

$$V = 26.33 \times 7 \times \frac{10}{12}$$
$$= 153.59 \text{ ft}^3$$

Doubling this to take into account both ends, we have a total volume of

$$2 \times 153.59 = 307.18 \text{ ft}^3$$

The total volume of the front, rear, and two ends is

$$373.33 + 307.18 = 680.51 \text{ ft}^3$$

Now, convert this to cubic yards, the standard unit of measure for concrete. Since there are 27 ft^3 in 1 yd^3, we divide the number of cubic feet by 27.

$$\begin{aligned} V &= 680.51 \div 27 \\ &= 25.20 \text{ yd}^3 \end{aligned}$$

The Approximate Method

This technique imagines the foundation to be "unwrapped" at the corners and considers it as one long rectangular solid (see Figure 13–3). The length of this wall equals the perimeter of the building, that being

$$\begin{aligned} P &= 2(32 + 28) \\ &= 120 \text{ ft} \end{aligned}$$

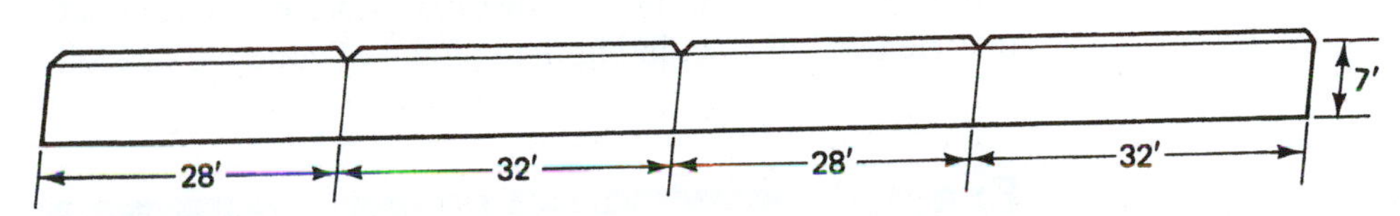

Figure 13–3

The total volume, V, of the foundation is

$$\begin{aligned} V &= \text{perimeter} \times \text{height} \times \text{thickness} \\ &= 120 \times 7 \times \frac{10}{12} \\ &= 700 \text{ ft}^3 \end{aligned}$$

Convert this to cubic yards:

$$\begin{aligned} V &= 700 \div 27 \\ &= 25.93 \text{ yd}^3 \end{aligned}$$

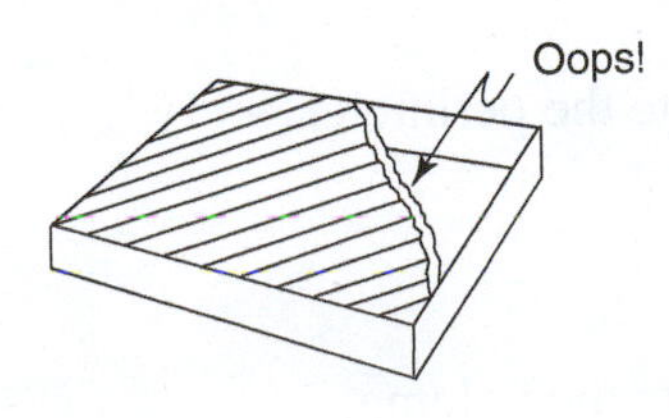

Comparing the results of the two methods, there is a difference of 0.73 yd^3—approximately $\frac{3}{4}$ yd^3. (A study of Figure 13–3 will show the source of this error to be the triangular prisms formed in each corner when the foundation is "unwrapped." The approximation method assumes that these prisms are filled in.) Too much concrete is a better situation than too little.

The approximation method provides an excess, so it is the better method.

Bear in mind, too, that the measuring of concrete at a ready-mix batch plant involves some error due to the variable moisture content of sand and aggregate as it is weighed. The final argument for using the approximation method is its relative simplicity.

SUMMARY

To calculate the concrete needed for a foundation:

1. Convert all dimensions to feet.
2. Calculate the perimeter of the foundation.
3. Use volume = perimeter $\times$ thickness $\times$ wall height to calculate the volume.
4. Divide volume by 27 to get cubic yards.

It should be noted that the size requirements for foundation footings, wall widths, and floor slab thicknesses are dictated by the soil classification, the use and size of the building to be placed on the footings and walls, and so on. This book does not cover the matter of sizing. Students interested in learning about sizing should consult references that address specifications.

13.4 Footings

Calculate the amount of concrete necessary for footings in the same manner as for foundations. The question of accuracy is left to the student as an exercise (see exercise 1). Suffice it to say that the outside dimensions of the foundation that will sit on the footing may be used in calculating the volume of concrete needed.

Example

A. How many cubic yards of concrete are needed for the footing of a house with dimensions 28 ft $\times$ 32 ft? The footing is to be 20 in. wide and 8 in. high.

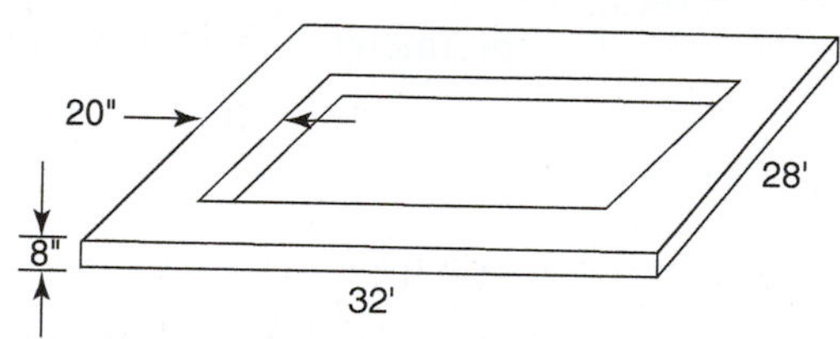

Follow the steps used for a foundation.

1. $20 \div 12 = 1.667$ and $8 \div 12 = 0.667$ — Convert dimensions to feet.

2. $P = 2(28 + 32)$
   $= 120$ ft — Calculate the perimeter.

3. $V = 120 \times \dfrac{20}{12} \times \dfrac{8}{12}$
   $= 133.33\ \text{ft}^3$ — Calculate the volume.

4. $V = 133.33 \div 27$
   $= 4.94\ \text{yd}^3$ — Convert to cubic yards.

13.5 Slabs

Next, consider the slab of concrete that will serve as the cellar floor in the building described in example A. Again, the question of accuracy will be examined in an exercise (see exercise 2). The length and width of the cellar floor may be approximated using the outside dimensions of the building.

Example

A. Find the number of cubic yards of concrete needed for the cellar floor in a house having outside dimensions of 28 ft × 32 ft if the floor is to be 3 in. thick.

1. $3 \text{ in.} = \frac{3}{12} \text{ ft}$ — Convert dimensions to feet.

2. $V = 28 \times 32 \times \frac{3}{12}$ — Calculate the volume.
   $= 224 \text{ ft}^3$

3. $V = 224 \div 27$
   $= 8.30 \text{ yd}^3$ — Convert to cubic yards.

SUMMARY

For footings and slabs,

1. Use the dimensions given by the foundation for perimeter or length and width.
2. Calculate the volume of the solid in cubic feet.
3. Divide by 27 to get cubic yards.

COMMENT

Always round up. Remember that a bit too much concrete is better than too little. Having to wait for another batch truck to deliver concrete to fill in a shortage can result in a cold joint, which can be a weak point. Therefore, when making a calculation, always round up.

REVIEW EXERCISES

1. A typical ration of cement, sand, and gravel is 1:2:3. Using this ratio, determine the volume in cubic feet of the amounts needed for each of the following.
   a. 1 cubic yard of concrete.
   b. 6 cubic yards of concrete.
   c. 20 cubic yards of concrete.

2. 6 ft^3 of cement are used in making a batch of concrete. Using the 1:2:3 ratio of cement to sand to aggregate,
   a. How many ft^3 of sand and how many ft^3 of aggregate are used?
   b. How many yds^3 of concrete can be made?
3. It is estimated that 15 yds^3 of concrete are needed to fill the forms for a slab. Using the 1:2:3 ratio, find the number of yds^3 of cement, sand, and aggregate needed.
4. Find the number of cubic yards needed for each of the following slabs.
   a. 10 ft × 60 ft × 3 in.
   b. 22 ft × 24 ft × 6 in.
   c. 18 ft × 26 ft × 4 in.
   d. 12 ft × 48 ft × 6 in.
   e. 4 ft × 180 ft × 4 in.
5. Calculate the number of cubic yards of concrete needed for the footing of a house measuring 28 ft by 32 ft. The foundation is to be 10 in. thick. The footing is to be 20 in. wide and 8 in. high. Adjust the dimensions of the footing to fit the foundation and calculate the answer accurately. Then compare your results to the results from example A on page 240.
6. Calculate the number of cubic yards of concrete needed for the cellar floor in the 28 ft × 32 ft house using accurate methods. The slab is 3 in. thick. Compare your results with the results from the approximate method example A on page 241.
7. The outlines of several foundations are shown in the figure on page 242. Dimensions are noted, including wall thickness and height. Assume the footings to be 8 in. high and twice the thickness of the foundation wall; the cellar floor is to be 3 in. thick. For each foundation, calculate the amount of concrete needed for the footing, the walls, and the floor. Include a total figure.
8. If concrete costs \$95 per cubic yard, find the cost of the concrete for the footings, the walls, and the floor for each foundation in exercise 3.

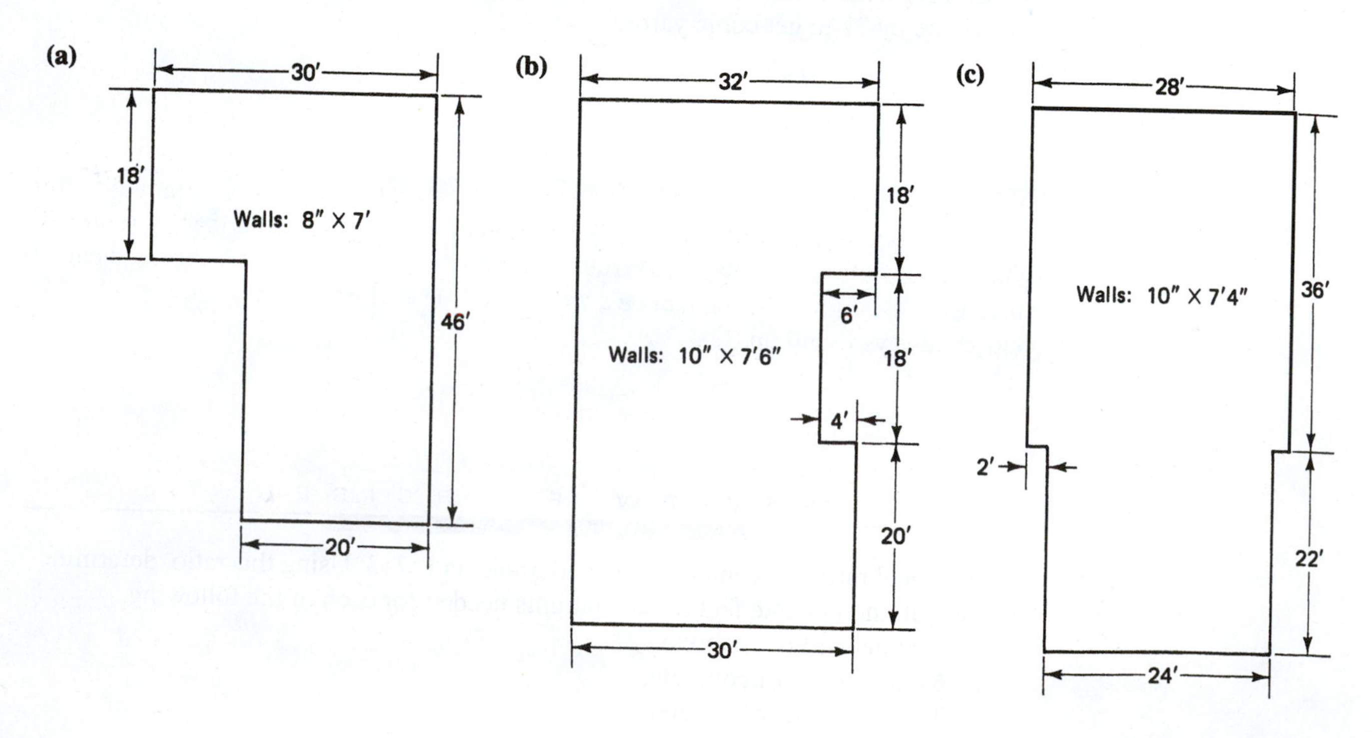

CHAPTER 14

Girders, Sill Plates, Bridging, Floor Joists, and Floor Covering

Did You Know

Stress-skin panels can save 25% to 50% of framing lumber in a house.

OBJECTIVES

Upon completing this chapter, the student will be able to:

1. Determine an economical materials list for a built-up girder.
2. Determine an economical materials list for sill plates.
3. Calculate the number and length of floor joists and floor joist headers for a particular framing plan.
4. Calculate the amount of bridging material needed for a particular framing plan.
5. Calculate the amount of floor covering material needed for a certain area.

14.1 The Girder

A girder may be made of one of several types of materials. The material may be reinforced concrete, steel, solid wood, or built-up laminations of wood. The last type will be considered here.

A Word About Girder Design

The proper size for a girder is dependent on several factors, among them:

- type of lumber making up the girder,
- load to be carried,
- distance between supports,
- distance between other load-bearing girders or walls.

Structural tables are available for sizing a girder. The objective here is to determine the materials required to construct a built-up girder once its specifications have been determined from a proper source.

Supports and Butt Joints

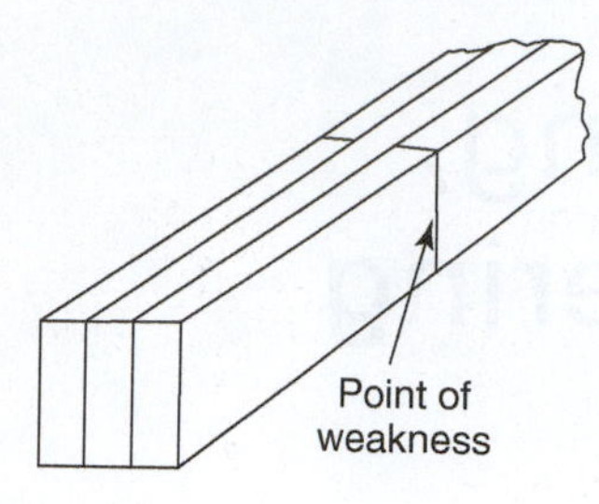

A butt joint in a laminated girder is a point of weakness. Column supports should be placed where the joints occur. In certain situations, space utilization may play a role in positioning column supports. Therefore, both the desired support positions and the butt joint positions are of consideration in constructing a built-up girder.

Another factor that determines support position is the distance required between supports as per design specifications. Two guidelines that should be kept in mind when designing the materials layout for a girder are

1. Butt joints should be supported.
2. Joints should be no closer to each other than 4 ft.

The example that follows assumes a particular distance between supports.

Example

A. Determine the length of materials needed for a built-up girder made of three laminations of 2 × 10s. Support posts are to be placed every 8 ft and the overall length is 40 ft.

- Begin by sketching a plan view of the girder. See Figure 14–1.
- Mark the support positions at 8-ft intervals.
- Sketch the possible butt-joint positions, making note of the support positions. To the extent possible, butt joints should occur at the support post.

It is acceptable to have two butt joints opposite each other, as long as they are separated by a lamination and are supported by a post. In this case, lengths of 8 ft and 16 ft work well. Thus, materials for this girder could be 2 × 10s: 6 pcs. 16 ft; 3 pcs. 8 ft.

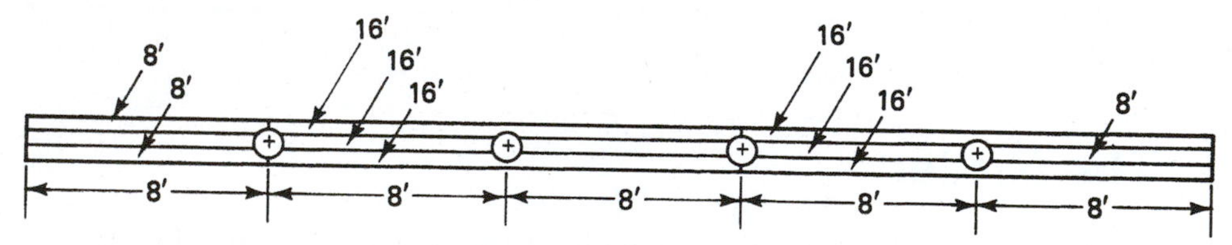

Figure 14–1

Check: Total length of material = 3 times girder length.

In this case, the materials are

$$(6 \times 16) + (3 \times 8) = 96 + 24 = 120 \text{ ft}$$

and three times the grider length is

$$3 \times 40 = 120 \text{ ft}$$

It should be noted that there are other layouts that are equally correct. After determining butt-joint placement, economy is the next consideration. To the extent possible, materials should be ordered in lengths that minimize waste.

14.2 Sill Plates

Sill plates are ordinarily made of a single layer of 2 × 6 or 2 × 8 lumber. It is advisable to place a layer of sill seal, an unfaced fiberglass insulation product, along the perimeter of the foundation. The sill plate is then laid on the sill seal and fastened to the top of the foundation by anchor bolts that have been embedded in the concrete at intervals of approximately 4 ft. The length of the wall segments will dictate the economical lengths of material to order for sill plates.

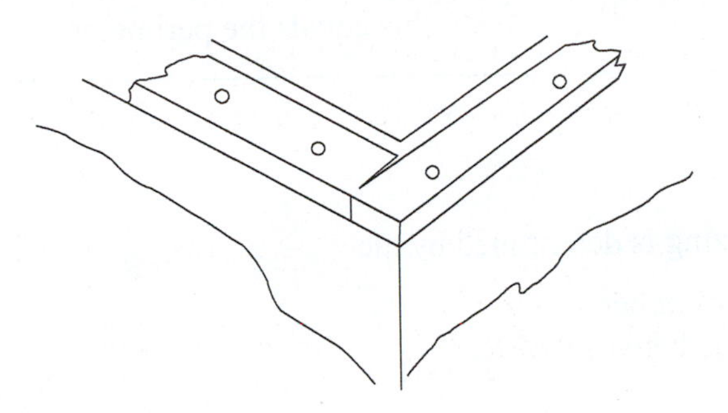

The total length of material needed for sill plates equals the perimeter of the foundation.

Example

A. Figure 14–2 shows a plan view outline of a foundation. Determine an economical order for the 2 × 6 sill plates.

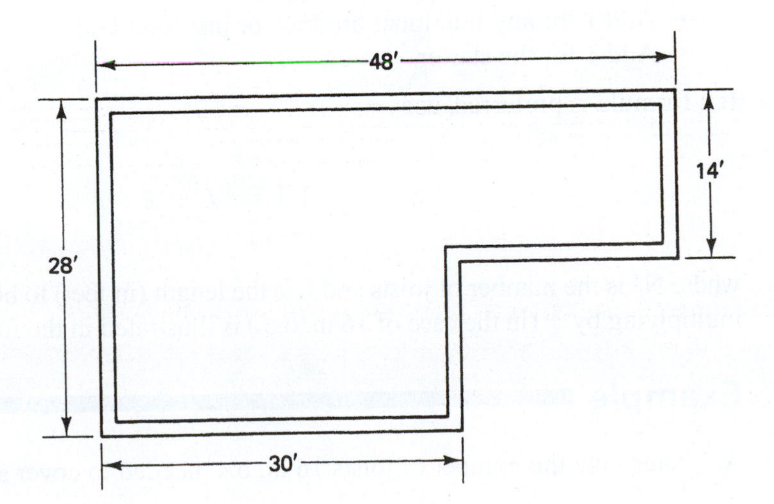

Figure 14–2

The perimeter is 2(28 + 48) = 152 ft. As for an economic layout,

- For the 48-ft-long rear wall use four pieces, 12 ft long. (Three 16s will work, also; however, a premium price may be charged for longer pieces.)
- The left and right ends may be covered with four pieces 14 ft long.
- The 30-ft and 18-ft segments along the front are the equivalent of the rear section and, therefore, may be covered with four 12-ft pieces.

In summary,

2 × 6s: 8/12s; 4/14s   Check: (8 × 12) + (4 × 14) = 96 + 56 = 152 and this equals the perimeter.

14.3 Floor Joists

Floor-joist sizing is determined by the

- type of lumber,
- distance between joists,
- span,
- load the joists are to carry.

The joist size will be indicated on a set of plans prepared by a designer or by structural tables.

To determine the number of joists:

- First, determine the spacing (that is, 16 inches on center, 24 in. on center (o.c.) etc.)
- Second, multiply the distance in feet to be covered with joists by
  - $\frac{3}{4}$ or 0.75 for 16 in. o.c.
  - $\frac{1}{2}$ or 0.5 for 24 in. o.c.
- Add 1 for any fractional product, or just round up.
- Add 1 for the starter.

If a formula is preferred, use

$$NJ = \frac{3}{4}L + 1$$

where NJ is the number of joists and L is the length (in feet) to be covered. The reason for multiplying by $\frac{3}{4}$ (in the case of 16 in. o.c.) is illustrated in the following example.

Example

A. Determine the number of joists 16 in. o.c. needed to cover a distance of 40 ft.

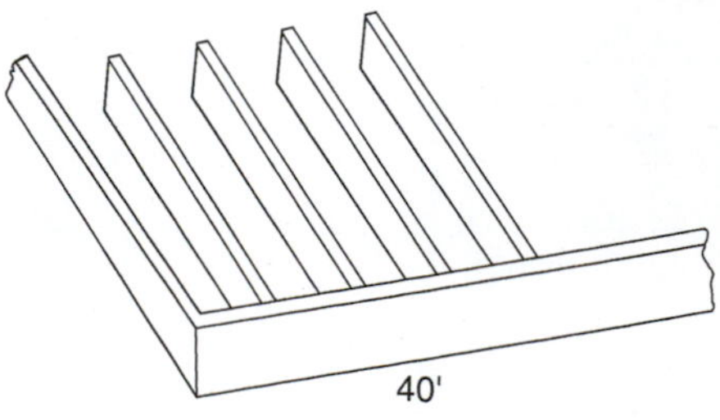

40'

The problem could be solved by first converting 40 ft to inches and then determining the number of 16-in. spaces.

$$40 \times 12 = 480 \text{ in.}$$

$$\frac{480}{16} = 30 \text{ spaces}$$

There will be one more joist than the number of spaces. That is the reason for adding 1 in the formula.

In example A, the length, 40 ft, was multiplied by 12 and then divided by 16. The result would be the same if the length were multiplied by the fraction $\frac{12}{16}$ (or $\frac{3}{4}$, in reduced form). This is the reason for the factor of $\frac{3}{4}$ in the formula.

If joists are to be spaced 12 in. o.c., the number of joists equals the length in feet, plus 1. That is,

$$\text{NJ} = L + 1$$

If joists are to be spaced 24 in. o.c., the number of joists is one-half the length in feet, plus 1. That is,

$$\text{NJ} = L/2 + 1$$

NOTE

When the calculation of the number of joists results in a fraction, the result should be rounded up. A fractional space will require a joist. Other adjustments must be made for end trimmers, stairwells, or other considerations, such as extra joists needed for load bearing.

Example

B. The floor framing for the building outlined in Figure 14–3 calls for 2 × 10s placed 16 in. o.c. How many are needed if a girder is centered as shown? Assume that the joists will be hung from the girder with joist hangers.

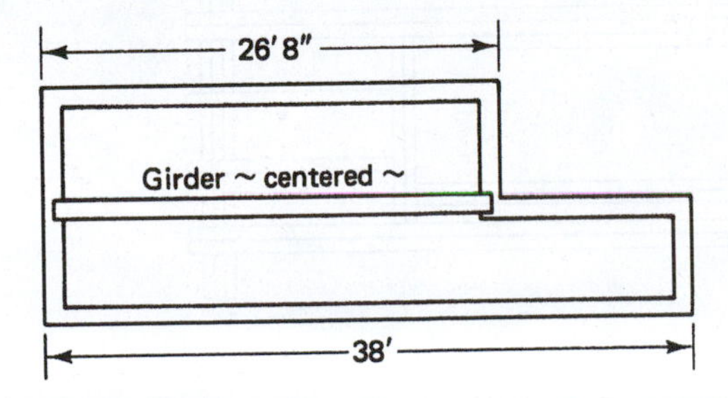

Figure 14–3

Along the front, the length to be covered is 38 ft.

$$
\begin{aligned}
NJ &= \frac{3}{4} \times 38 + 1 && \text{Substitute in the proper values.}\\
&= 28.5 + 1 && \text{Multiply.}\\
&= 29 + 1 && \text{Round up.}\\
&= 30 \text{ joists}
\end{aligned}
$$

Along the rear, the length is 26 ft 8 in., or 26.67 ft.

$$
\begin{aligned}
NJ &= \frac{3}{4} \times 26.67 + 1 && \text{Substitute in the proper values.}\\
&= 20 + 1 && \text{Multiply.}\\
&= 21 \text{ joists}
\end{aligned}
$$

Notice that in both of the calculations, a fractional number of joists is rounded to the next higher whole number. Therefore, a total of 51 floor joists are needed.

Floor-Joist Headers

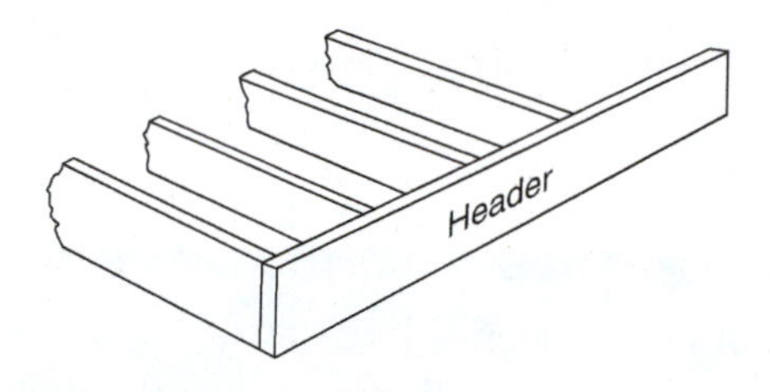

In addition to floor joists, floor-joist headers should be included at this point. Recalling the material lengths used for sill plates along the front and rear will be of assistance. The foundation plan we are using would call for headers along the 38-ft length twice. Two 12-ft pieces and one 14-ft will cover 38 ft with no waste. Therefore, the floor-joist headers will be 4/12s and 2/14s.

Stairwell Framing

Figure 14–4

Figure 14–4 shows the framing for a stairwell. Adjustments to the amount of framing materials needed are made according to the stairwell plan. If the stairwell is conventional and joists are spaced 16 in. o.c., parts of two joists will be eliminated for a 36-in.-wide well. But these should not be eliminated from the count, because the joists along the sides of the well will be doubled. Furthermore, both ends of the stairwell will be double-headed.

In general, two or three floor joists should be added for framing a standard stairwell.

14.4 Bridging

Bridging provides added stiffness between floor joists. It should be applied for spans greater than 8 feet. Bridging may be

- purchased as steel bracing,
- in the form of blocking between adjacent floor joists, using material of the same dimension as the joists,
- made from 1 × 3 strapping.

Both the steel and the strapping are applied in the form of an X between joists.

The following example determines the amount of 1 × 3 strapping needed for a single piece of diagonal bridging between a pair of 2 × 10s placed 16 in. o.c.

Example

A. Calculate the length of material needed to cover the inside diagonal between 2 × 10s (actual width $9\frac{1}{2}$ in.) that have been placed 16 in. o.c. (see Figure 14–5).

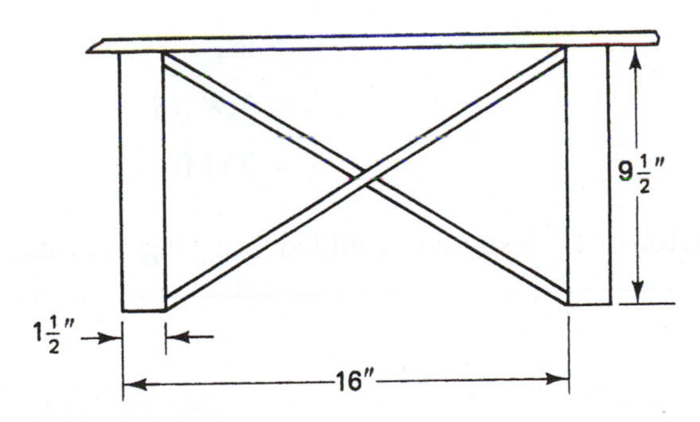

Figure 14–5

The joists are $1\frac{1}{2}$ in. thick. This makes the inside distance between joists $14\frac{1}{2}$ in. The depth of the 2 × 10 is $9\frac{1}{2}$ in. Use the Pythagorean theorem. The diagonal, representing the bridging, is the hypotenuse of a right triangle; the legs are the inside distance between joists and the joist width.

$$L^2 = 9.5^2 + 14.5^2$$

$$L^2 = 90.25 + 210.25$$

$$L^2 = 300.50$$

$$L = \sqrt{300.50}$$

$$L = 17.3 \text{ in.}$$

This example could be repeated for floor joists of other sizes and spacing. However, the cost of 1 × 3 strapping does not warrant expending much effort on precise calculations. The solution of 17.3 in. could be rounded to 18 in.; another 18 in. will make up the second member of the X that forms the bridging between joists. This amounts to a total of 36 in. or 3 ft of material for each space to be bridged.

Since the spacing between floor joists may be 12 in. o.c., 16 in. o.c., 24 in. o.c., and so on, it seems reasonable to establish an estimating factor that will accommodate all these variables. Multiplying the length of feet to be bridged by 3 produces a reasonable estimate of bridging material. That is,

$$B = 3L$$

where L is the length in feet to be bridged and B is the number of feet of material needed.

Example

B. Determine the amount of 1 × 3 strapping needed to bridge the floor joists from the plan shown in Figure 14–2.
Assuming a girder is placed midway between the front wall (30 ft) and the rear wall (48 ft), two rows of bridging are required: one row parallel to the front wall, back 7 ft; another parallel to the rear wall, in 7 ft. The total length to be bridged is

$$30 + 48 = 78 \text{ ft}$$

The material needed is

$$\begin{aligned} B &= 3L \\ &= 3 \times 78 \\ &= 234 \text{ ft} \end{aligned}$$

A waste factor of 10% should be added, bringing the total to 268 feet.

14.5 Floor Covering

The final problem considered in this chapter is the amount of floor covering material needed for a certain area. The typical subflooring material comes in sheets measuring 4 ft × 8 ft, whether the material to be used is plywood of various thicknesses, particleboard, or oriented strand board (OSB). Since 4 ft × 8 ft covers an area of 32 ft^2, the number of sheets needed may be determined by dividing the floor area in square feet by 32 and rounding to the next higher whole number.

For example, a house measuring 34 ft × 26 ft has an area of

$$\begin{aligned} A &= 34 \times 26 \\ &= 884 \text{ ft}^2 \end{aligned}$$

Dividing this area by 32 and rounding up, we get 28 sheets.

$$884 \div 32 = 27.625 \text{ or } 28$$

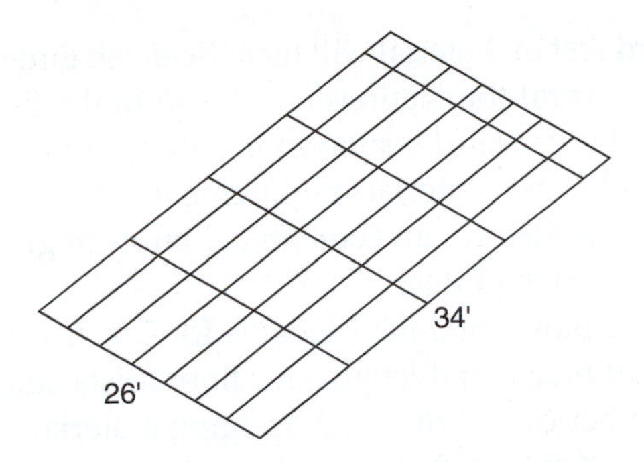

SUMMARY

To find the amount of material needed for floor covering,

1. Find the area to be covered in square feet.
2. Divide by 32.
3. Round up.

REVIEW EXERCISES

1. A rectangular framed floor measures 24 ft by 34 ft. How many sheets of $\frac{1}{4}''$ T&G plywood are needed to cover this floor?
2. The foundation for the 24 ft by 34 ft building is ready for floor framing. Determine the following:
   a. Economical girder payout using 3 laminations of 2 × 10s with support posts no more than 8-ft apart.
   b. Number of linear feet of sill plate material. Suggest economic lengths that will minimize waste.
   c. Number and lengths of 2 × 10 floor joist headers. Suggest economic lengths.
   d. Number and lengths of 2 × 10 floor joists placed 16″ o.c. A girder is placed halfway across the 24-ft width, running 34-ft long.
   e. Number of linear feet of 1 × 3 bridging material needed. Two rows of bridging will be placed halfway between the girder and the exterior walls.
3. A built-up girder is to be constructed of three laminations of 2 × 12s. Steel column supports will be placed every 10 ft. The overall length is 52 ft. Determine the materials needed for an economical layout.

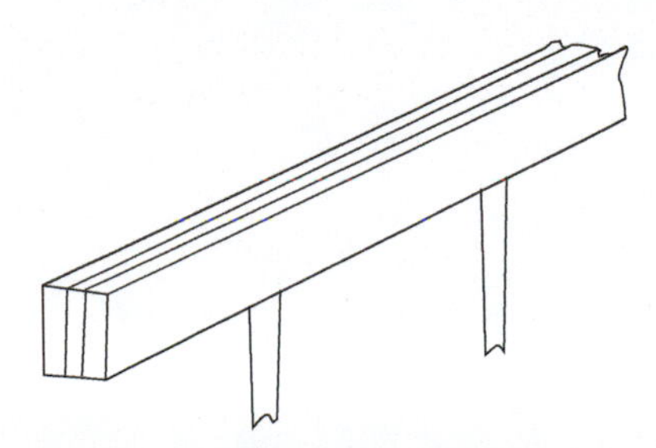

4. How many board feet of material will there be in the girder described in exercise 1?
5. The outlines of several foundations are shown in the figure. Girder positions are shown by dashed lines. Girder support positions are noted. For each foundation, determine the following materials requirements:
   a. The materials needed for an economical built-up girder design (assume three laminations for each girder)
   b. The number of pieces and their lengths for 2 × 6 sill plates
   c. The number of pieces and lengths for floor joists and floor-joist headers
   d. The total number of feet of 1 × 3 bridging material
   e. The number of 4 ft by 8 ft sheets of sheathing needed for subflooring

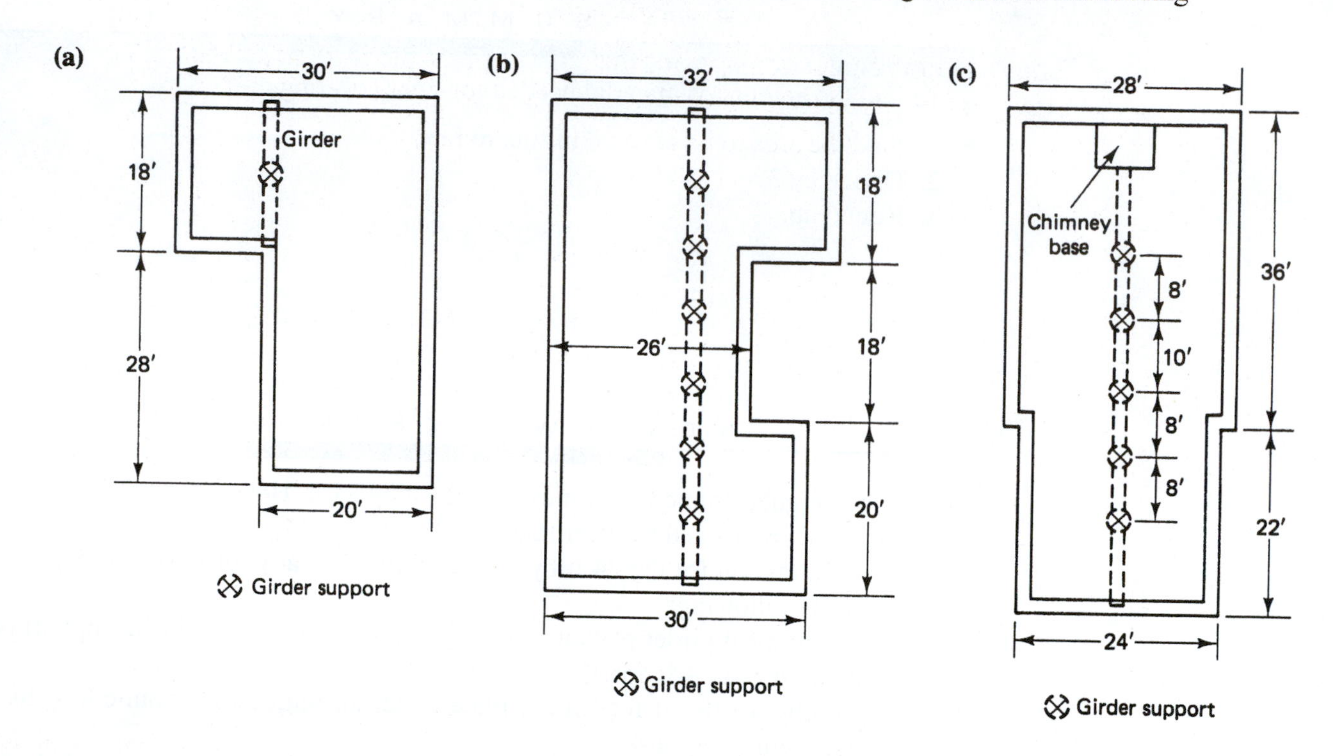

CHAPTER 15

Wall Framing

Did You Know

Today's average new home is approximately twice the size as a 1960s home, has fewer occupants, and needs wood from $\frac{3}{4}$ acre of forest.

OBJECTIVES

Upon completing this chapter, the student will be able to:

1. Calculate the number of studs needed to frame the exterior walls of a structure.
2. Calculate the number of studs needed to frame the interior walls of a structure.
3. Calculate the amount of material needed for shoes and plates.
4. Compare various methods of wall framing with respect to costs and energy efficiency.

A builder must develop accurate and efficient methods for estimating the costs of construction. The accuracy of any estimate of the quantity of materials affects profit, not only in terms of the price quoted to a customer but in terms of having sufficient materials on the job site to maintain the flow of construction.

The most useful measure for determining the materials needed for shoes, plates, and wall studs is the *wall length.* For the exterior walls, this is the *perimeter* of the building.

Consider a simple rectangular structure with foundation size 24 ft × 32 ft (see Figure 15–1). The perimeter of this structure is

$$\begin{aligned} P &= 2(24 + 32) \\ &= 112 \text{ ft} \end{aligned}$$

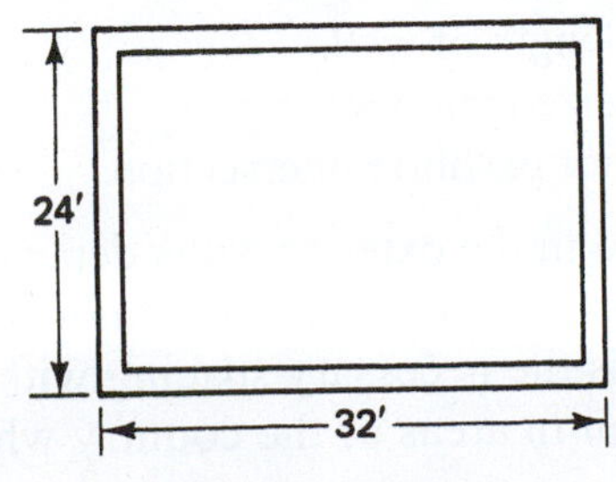

Figure 15–1

15.1 Shoes and Plates—Exterior Walls

In general, materials needed for the shoe and plates, S, is given by

$$S = 3 \times P$$

where P is the perimeter of the structure in feet. The amount of materials needed for the single shoe (also called *sole plate*) and double top plate for an exterior wall is three times the perimeter. In the case of our example,

$$3 \times 112 = 336 \text{ linear feet}$$

When selecting the length of pieces making up this order, both economy and convenience should be considered. Material lengths should be matched to the length of the wall segments to be constructed with the objective of minimizing waste. However, the lengths should also be determined by what the carpenter finds convenient in terms of handling wall segments, which will be prefabricated on the floor and then stood in place. The number of workers on hand is a factor.

The amount of material may be checked by comparing the value of three times the perimeter to the sum of the lengths of all pieces. For the building in Figure 15–1, assume that 12-ft lengths are chosen for the shoe and plate material, with 8-ft lengths to complete the 32-ft segments. This would require 24 pcs. 12 ft and 6 pcs. 8 ft. Checking the total length against the perimeter, the total length is

$$(24 \times 12) + (6 \times 8) = 288 + 48 = 336 \text{ linear feet}$$

The preceding formula yields

$$\begin{aligned} S &= 3 \times P \\ &= 3 \times 112 \\ &= 336 \text{ ft} \end{aligned}$$

15.2 Exterior Wall Studs

An efficient way of counting the number of studs needed is to apply a factor to the perimeter of the structure, then adjust for openings, corners, odd lengths, and interior partition starters.

To determine the number of studs needed for the exterior walls,

1. Use NS $= \frac{3}{4} \times P$ for 16 in. on center (o.c), NS $= \frac{P}{2}$ for 24 in. o.c., where NS is the number of studs and P is the perimeter of the building.
2. Add 2 studs per wall segment.
3. Add 1 stud for each odd length of wall.
4. Add 2 studs per average-sized window.
5. Add 1 stud for each interior partition intersection.

The number of studs needed in the exterior walls depends on the spacing and the number of openings for doors and windows. The common stud spacing in residential construction is 16 in. o.c. This is the necessary spacing with 2 × 4 framing material. 2 × 6 framed walls are common in areas of the country where the benefits of extra insulation are significant. When 2 × 6 framing material is used, sufficient structural strength is gained with 24-in. o.c. stud spacing. However, an argument in favor of 16-in. o.c. spacing can be made in terms of the application of drywall. It is possible to detect slight waves in drywall applied to studs spaced 24 in. o.c.

Example

A. Determine the number of studs needed to frame the exterior walls of the first story of the structure shown in Figure 15–1. The spacing is to be 16 in. o.c.

The method used is the same as that established for counting floor joists spaced 16 in. o.c. (see Chapter 14), with some modification. A basic count is first established by multiplying the perimeter by $\frac{3}{4}$. (Recall that $\frac{3}{4}$ is the reduced fraction formed when a number is multiplied by 12 to convert feet to inches, and then divided by 16 to determine the number of 16-in. spaces.) In Figure 15–1, the perimeter is 112 ft. Letting NS represent the number of studs,

$$NS = \frac{3}{4} \times 112$$ Multiply the perimeter by $\frac{3}{4}$.

$$= 84 \text{ studs}$$ Basic count.

Recall that this method actually counts *spaces,* not *studs.* Therefore, 1 stud must be added for each wall segment to account for the "starter" stud. In addition, 1 stud must be added because of the manner in which corners are usually framed (see Figure 15–2). For each wall segment, then, add 2 studs.

This increases the count for Figure 15–1 by 2 studs for each of the four wall segments, a total of

$$4 \times 2 = 8 \text{ studs}$$ 2 studs are added for each wall segment.

and brings the total count to

$$84 + 8 = 92 \text{ studs}$$ Add the 2 studs for each wall segment to the basic count.

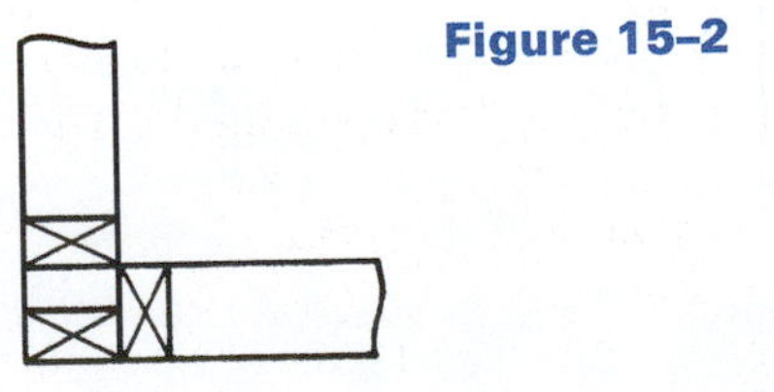

Figure 15–2

(Checking this against the total found by counting the studs needed for each wall segment will prove the accuracy of using the perimeter.)

One other adjustment to the basic count may be necessary when the length of a wall is not evenly divisible by 4. This means that a fractional space exists. This space requires a stud and one should be added.

Example

B. Determine the number of studs needed for a wall 22 ft 6 in. long. Spacing is 16 in. o.c. Use the formula NS $= \frac{3}{4} \times P$ with 22 ft 6 in., or 22.5 ft, in place of P.

$$NS = \frac{3}{4} \times 22.5$$
$$= 16.875$$

This is rounded up to 17 studs; 2 additional studs are needed for the starter and the corner—a total of 19 studs.

The estimator should use the perimeter to establish a basic count; then examine the plan for wall segments that are not divisible by 4, adding 1 stud for each such segment.

When the spacing is 24 in. o.c., the factor applied to the perimeter becomes $\frac{1}{2}$ instead of $\frac{3}{4}$. This is because the number of 2-ft spaces dictates the number of studs. Essentially, the perimeter in feet is divided by 2, the divisor being 2 ft. Adjustments to the basic count are otherwise the same as with other spacing: add 1 for the starter, add 1 for the corner, add 1 for an odd length of wall—this being a wall segment not evenly divisible by 2.

Example

C. Determine the number of studs needed 24 in. o.c. to frame the exterior walls of the building shown in Figure 15–3.

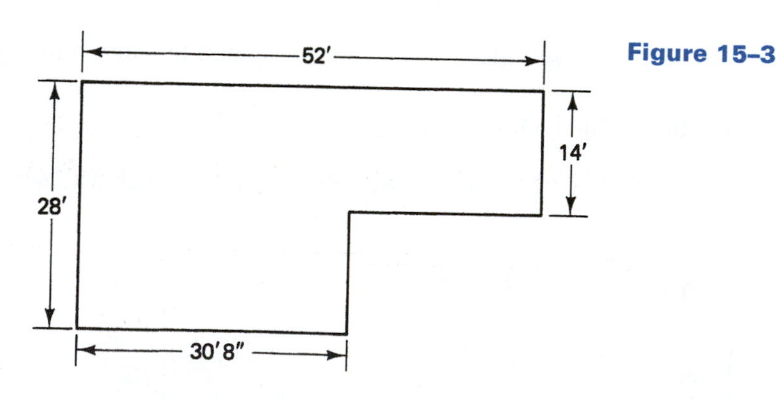

Figure 15–3

$$P = 2(28 + 52)$$ Determine the perimeter.
$$= 2(80)$$
$$= 160 \text{ ft}$$

$$NS = \frac{P}{2}$$ Determine the basic count of the number of studs (NS).
$$= \frac{160}{2}$$
$$= 80 \text{ studs}$$

The adjustments are as follows:

6 wall segments	6 studs
6 corners	6 studs
2 odd lengths (30 ft 8 in. and 21 ft 4 in.)	2 studs
Total adjustments	14 studs
Total studs	94 studs

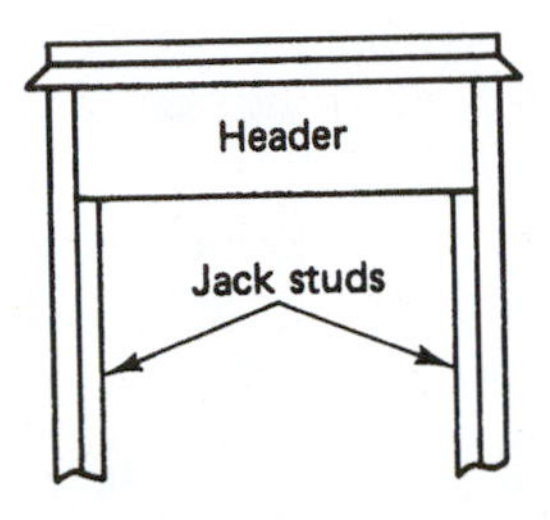

Figure 15–4

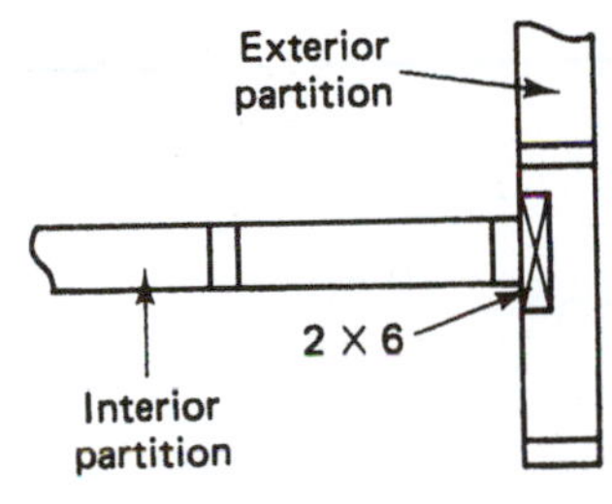

Figure 15–5

Wall openings for doors and windows should also be considered as adjustments to the number of studs. Standard door and window widths require an average opening of approximately 3 ft. Depending on where the openings occur with respect to stud location, two or three studs would normally be eliminated. Stud spacing also has some effect on the number to be eliminated. On the other hand, jack studs will be placed on each side of the opening to support the header (see Figure 15–4). The studs to be eliminated become the jack studs, resulting in no adjustment to the count. (The estimator may find it more economical to replace these studs with 7-ft studs for use as jacks, resulting in less material waste.) For wider openings, such as those caused by sliding patio doors, the estimator may wish to eliminate from 2 to 4 studs, depending on spacing and door width.

The framing under a window does require additional material. For the average-sized window, the addition of 2 studs will ordinarily be sufficient to frame the sill and its supports.

The final adjustment to the number of studs in the exterior walls results from the intersection with interior partitions. Locating a 2 × 6 stud at the centerline of an interior partition serves as a corner for drywall nailing (see Figure 15–5). Therefore, add one 2 × 6 for each intersection with an interior wall.

Example

D. Determine the number of studs and the shoe and plate materials for the exterior walls of the house shown in Figure 15–6. The framing will be done with 2 × 6, 16 in. o.c.

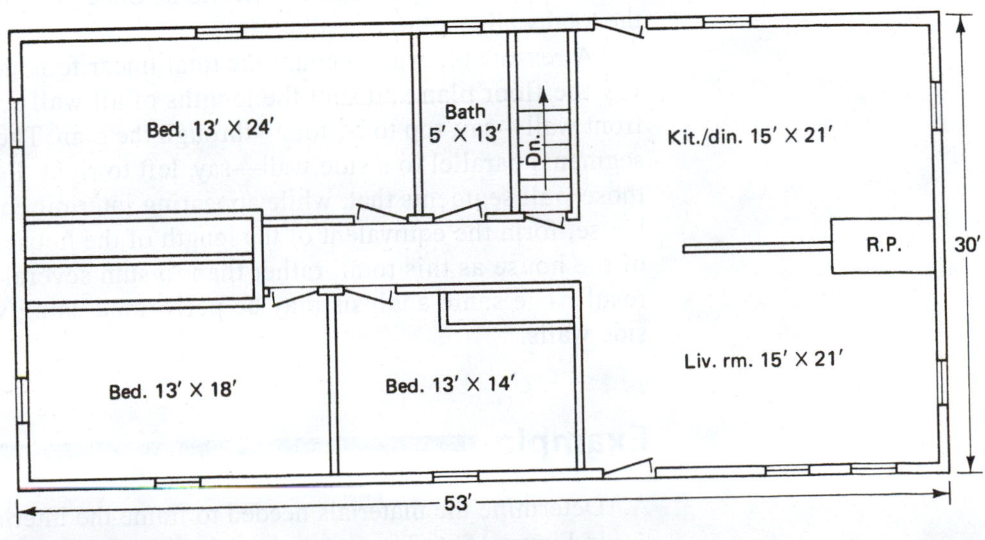

Figure 15–6

$$P = 2(53 + 30) \quad \text{Determine the perimeter.}$$
$$= 2(83)$$
$$= 166 \text{ ft}$$

shoe + plates = 3 × perimeter or 3 × 166 = 498 if of 2 × 6s.

$$NS = \frac{3}{4} \times 166 \quad \text{Determine the basic count.}$$
$$= 124.5, \text{ or } 125 \text{ studs}$$

The adjustments are as follows:

Starters and corners	8 studs
Odd lengths	4 studs
11 windows	22 studs
Interior partitions	8 studs
Total adjustments	42 studs
Total studs	167 studs

The student may now wish to review and verify the summary steps outlined at the beginning of this section on exterior walls.

15.3 Interior Wall Framing

To estimate the materials needed for interior partitions,

1. Determine the total linear feet of interior walls.
2. Multiply this figure by 3. This is the total linear feet of shoe and plate material.
3. The number of studs will be equal to the number of linear feet of walls.

Determining an accurate count of the number of studs needed to frame the interior partitions would consume enough time to call the value of such an exercise into question. Reasonable results may be obtained by totaling the linear footage of walls, closets included, and allowing one stud for each linear foot. Openings are ignored, as are corners, starters, and partition intersections. Shoe and plate materials will be three times the total wall length.

A reasonable way to count the total linear footage of interior partitions is to survey the floor plan and sum the lengths of all wall segments running parallel to the front wall—say, top to bottom—through the plan. Then sum the lengths of those wall segments parallel to a side wall—say, left to right. Particular note might be taken of those wall segments that, while appearing intermittently throughout the length of the house, form the equivalent of the length of the house. It is simpler to take the length of the house as this total, rather than to sum several lengths that produce the same result. The same analysis may be performed using wall segments that parallel the side walls.

Example

A. Determine the materials needed to frame the interior partitions for the plan shown in Figure 15–6. The material used will be 2 × 4s placed 16 in. o.c.

It is left to the student to verify the following estimates of total wall length.

Parallel to the front (53 ft) wall:	93 ft
Parallel to the side wall:	75 ft
Total length of interior partitions:	168 ft

This would amount to 3 × 168 = 504 linear feet for shoes and plates and 168 studs.

15.4 Headers

The final consideration in this chapter is estimating the material needed for window and door headers (refer to Figure 15–4). The sizing of headers depends on the span and load they are to bear. Refer to structural tables for proper sizing. As to the length of materials needed, the total width of openings serves as a guide. If headers are to be constructed by doubling the spanning material (2 × 6s, 2 × 8s, and so on), the sum of the widths would, of course, be doubled. The length of materials ordered should be chosen so as to minimize waste. For average-sized doors and windows, material ordered in 12-ft lengths is often economical.

A final comment needs to be made here. Our dependence on foreign sources of oil and gas, coupled with increased awareness of energy conservation, has led to variations in conventional framing methods. The objective is to retain a desired inside temperature (warm or cool) with a minimum fuel cost. Builders interested in energy-conserving construction methods will certainly want to make cost comparisons of various framing methods. Among the techniques that have been tried are

- conventional 2 × 4-16 in. o.c. framing with both fiberglass and rigid insulation,
- 2 × 6 framing with fiberglass (and sometimes rigid, as well) insulation,
- double 2 × 4 framing with two walls spaced to allow 12 in. of fiberglass insulation.

The relative costs of various framing methods will be considered in the exercises that follow.

REVIEW EXERCISES

1. Determine the number of studs placed 16″ o.c. needed to frame a wall segment that is 26 ft 8 in. long.
2. Answer exercise 1 if the studs are paced 24″ o.c.
3. Determine the number of studs, 16″ o.c. needed for each of the following wall segments of buildings:
   a. A wall 24 ft long.
   b. A wall 43 ft long.
   c. A rectangular building measuring 26 ft by 48 ft.
   d. A rectangular building measuring 32 ft by 44 ft.
4. Determine the number of linear feet needed for the shoe and double top plates,
   a. Of the building described in exercise 2 c.
   b. Of the building in 2 d above.

5. In example C of this chapter, the NS placed 24″ o.c., as well as the shoe and plate materials, was determined for the plan shown in Figure 15–3. Assume that the framing is to be done with 2 × 4s placed 16″ o.c. Suggest economic lengths and number of each.
   a. Determine the materials needed for the shoe and plates.
   b. Determine the number of studs.
6. Referring to the plan shown here, determine the following:

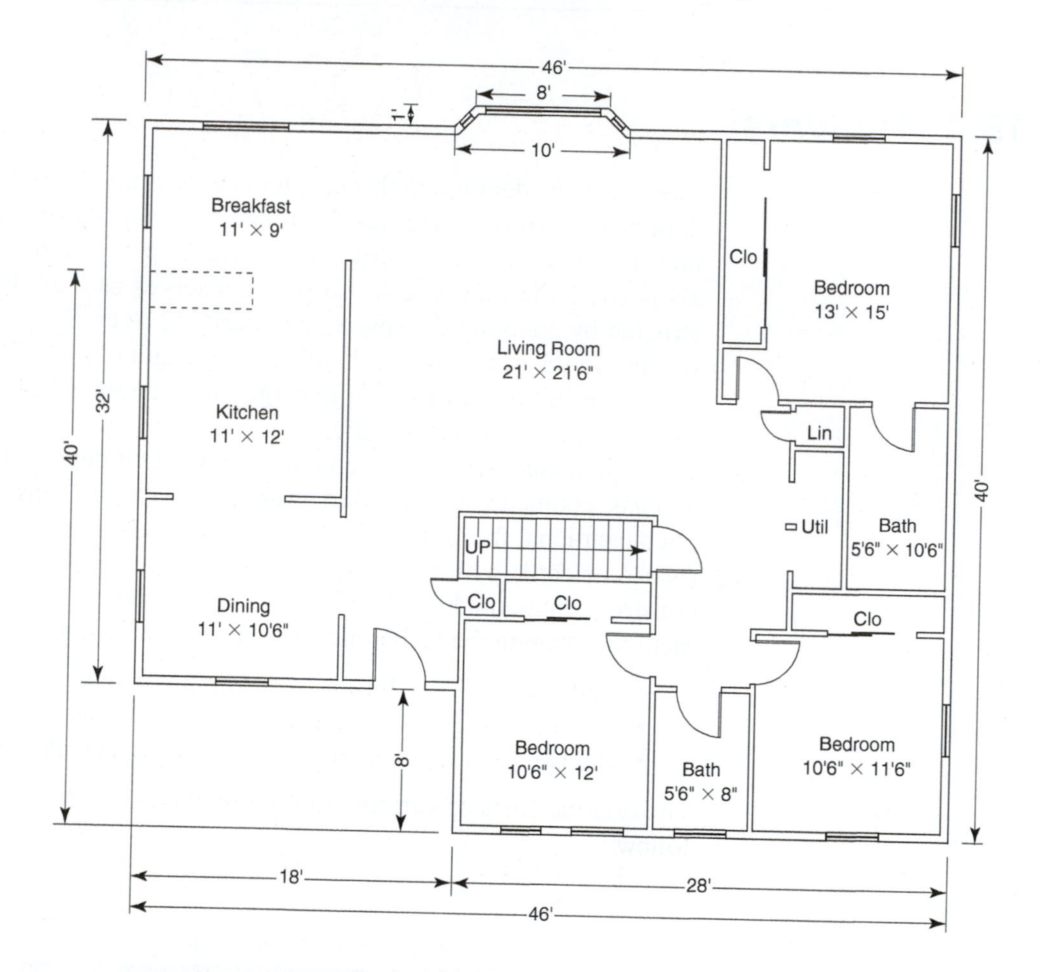

   a. Materials for the shoes and plates of the exterior walls.
   b. Studs for the exterior walls, including adjustments for openings, corners, starters, odd-length walls, and interior partition intersections. Use 2 × 6s 16″ o.c.
   c. Shoes and plates for interior walls. Framing will be 2 × 4s, 16″ o.c.
   d. Studs for the interior partitions.
   e. 2 × 6 material needed for door and window headers.
7. Assuming framing lumber costs $0.49 per board foot, compare the cost of framing with 2 × 4s 16″ o.c. to the cost of framing with 2 × 6s 24″ o.c. Use the results found in example C of this chapter and those found in exercise 5.
8. One method of framing exterior walls mentioned in this chapter is to have two separate walls of 2 × 4s placed 16″ o.c. With this technique, the outermost wall is framed conventionally, except that the 2 × 6 stud used for an interior wall intersection is omitted. The inner wall is placed 12″, outside-to-outside, and 6 in. of fiberglass insulation is placed in each wall, providing a total of

12 in. of insulation (R-38). Interior partition 2 × 6 studs are placed on the inner wall. Headers for openings in the inner wall may be of 2 × 4 material and serve primarily as fillers; the roof load is carried by the outer wall.

Determine the extra materials needed for this method of framing the exterior walls from the plan in Figure 15–6. Use the price of $0.49 per board foot.

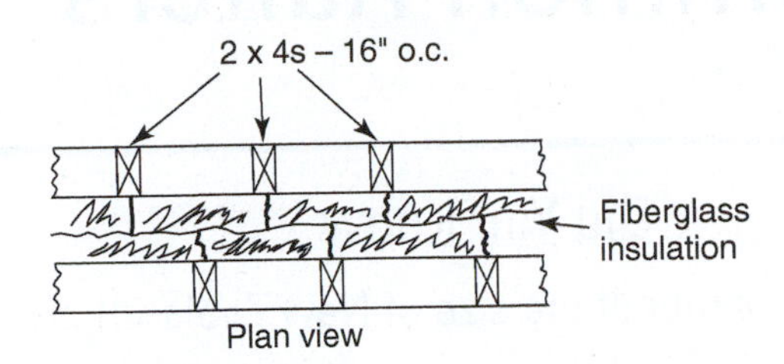

9. A building 30 ft by 52 ft, with 4 windows and one 3-ft door along one 52-ft wall, and 4 windows and one 8-ft sliding glass door along the other 52-ft wall. The 30-ft walls have 2 windows each. 2 × 6 studs are placed 16″ o.c.
   a. Find the material needed for the shoe and double top plates. Provide an economical list of lengths and number of pieces.
   b. Find the number of studs.
   c. Determine the amount of 2 × 6s needed for door and window headers.
   d. Find the cost of materials in a, b, and c if 2 × 6s are priced at 58 cents per linear foot.
10. As a group or class discussion project, compare the cost of framing the exterior walls of the house as shown in exercise 6 by two methods: (1) Using 2 × 6s 16″ o.c. and (2) using a double wall of 2 × 4s placed 12 in. apart. Discuss the cost difference between these two methods with respect to the amount of money that may be saved with the added insulation provided by the double-walled construction. What are some other reasons why one method might be better than the other? If the 2 × 6 method is used, how can the insulating qualities of that type of wall be improved? (Students may wish to revisit this problem after studying Chapter 26.)

CHAPTER 16

Roofs I: Common Rafters

Did You Know

One percent of forests (about the size of New Zealand) are consumed by fire each year.

OBJECTIVES

Upon completing this chapter, the student will be able to:

1. Calculate the length of a common rafter.
2. Calculate the distance from the ridge to the bird's-mouth cut along a rafter.
3. Calculate the distance from the rafter tail to the bird's-mouth cut.
4. Determine the total rise.
5. Determine unit rise and unit line length.
6. Determine the slope of an existing roof.

Common rafters, *hip* rafters, *valley* rafters, *jack* rafters—there are many types of rafters. *Gable* roofs, *gambrel* roofs, *hip* roofs, *saltbox* roofs, *shed* roofs—there are many types of roofs that may be formed by some of these rafters. (See Figure 16–1.) Some terms important to calculating the length and cuts on a rafter are as follows:

- *Slope triangle:* Right triangle found on a blueprint along the roof line; it describes the number of inches of rise in a roof for 12 in. of run.
- *Unit rise (URi):* Vertical leg of the slope triangle; it describes the number of inches of rise for each 12 in. of run.
- *Unit run (URu):* Horizontal leg of the slope triangle; it always equals 12.
- *Unit line length (ULL):* Hypotenuse of the slope triangle; it describes the number of inches of travel along a rafter for each 12 in. of run.

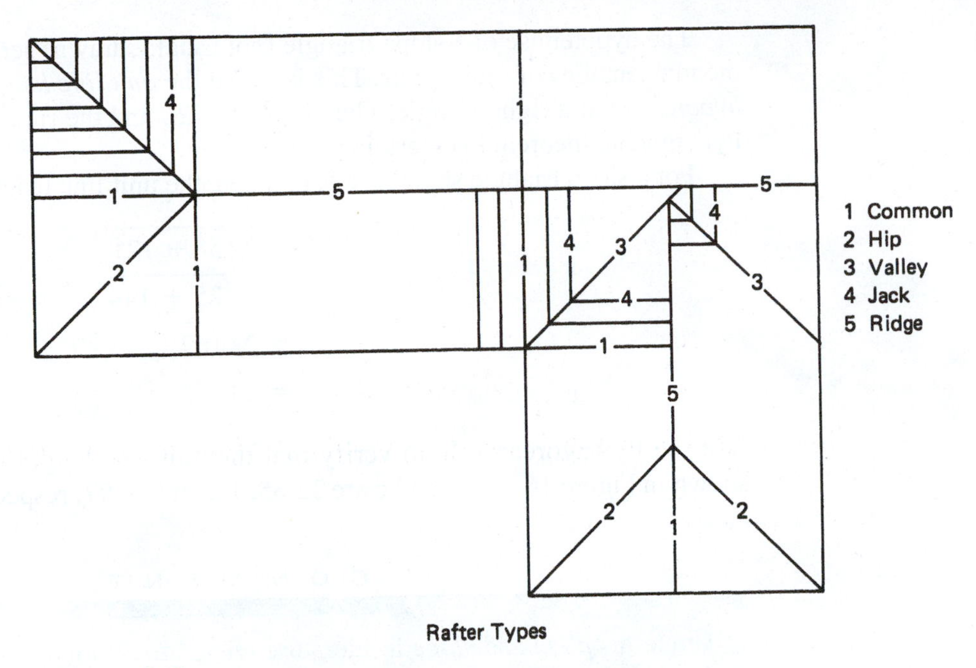

Figure 16–1

16.1 The Slope Triangle

The slope of a roof is usually described on a blueprint by means of a *slope triangle,* which is displayed along a roof line. Figure 16–2 shows a slope triangle for a roof with a "5/12 slope." This means that the roof rises 5 in. for every 12 in. or 1 ft of run. The run is always a horizontal distance. The slope triangle is a right triangle. *Unit run,* abbreviated *URu* is the horizontal leg of the slope triangle and is always 12 in. *Unit rise,* abbreviated *URi* (labeled 5 in Figure 16–2) is the vertical leg of the slope triangle and describes the number of inches of rise in 12 in. of run. So a roof having a 5/12 slope rises 5 in. for every foot (12 in.) of run. Figure 16–3a, b, and c show slope triangles having unit rises of 4, 9, and 12, respectively. Notice that each has a run of 12.

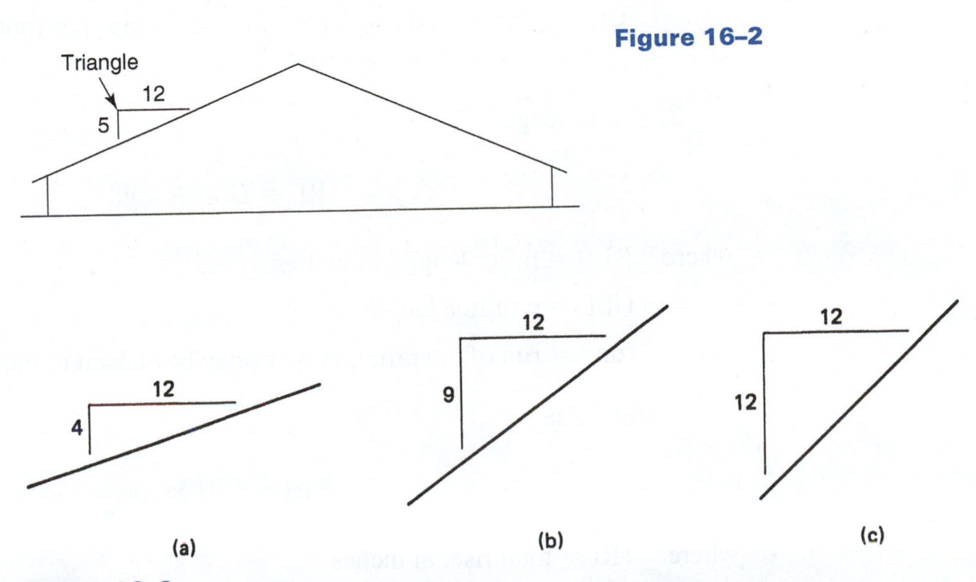

Figure 16–2

Figure 16–3

The hypotenuse of a slope triangle (not usually drawn) serves an important role in the mathematics of roof rafters. This is called the *unit line length, ULL.* The *ULL* is the *hypotenuse* of a right triangle. The unit rise, *URi,* and the run, 12, are the *legs.* Use the Pythagorean theorem to determine ULL.

For a slope triangle showing a 5/12 slope, the unit line length will be

$$\begin{aligned} \text{ULL} &= \sqrt{5^2 + 12^2} \\ &= \sqrt{25 + 144} \\ &= \sqrt{169} \\ &= 13 \end{aligned}$$

Use the Pythagorean rule to verify that the unit line lengths for the slope triangles shown in Figure 16–3a, b, and c are 12.65, 15, and 16.97, respectively.

COMMENT

The terms *pitch* and *slope* in literature related to building construction are sometimes defined differently from how they are used in the trade. "Pitch" is defined as the amount of rise divided by twice the span. "Slope" is defined as the amount of rise per foot of run. For a conventional gable roof, a 5/12 slope would mean that the pitch is 5/24.

Formulas for a common rafter

1. Unit line length

$$\text{ULL} = \sqrt{\text{URi}^2 + 12^2}$$

where ULL = unit line length

URi = unit rise, the number of inches of rise per foot of run and 12 is the unit run

2. Rafter length

$$\text{RL} = \text{ULL} \times \text{run}$$

where RL = rafter length, in inches

ULL = unit line length

run = run of the rafter, in feet (may be chosen to include the overhang)

3. Total rise

$$\text{TRi} = \text{URi} \times \text{run}$$

where TRi = total rise, in inches

URi = unit rise, in inches

run = run of the rafter, in feet

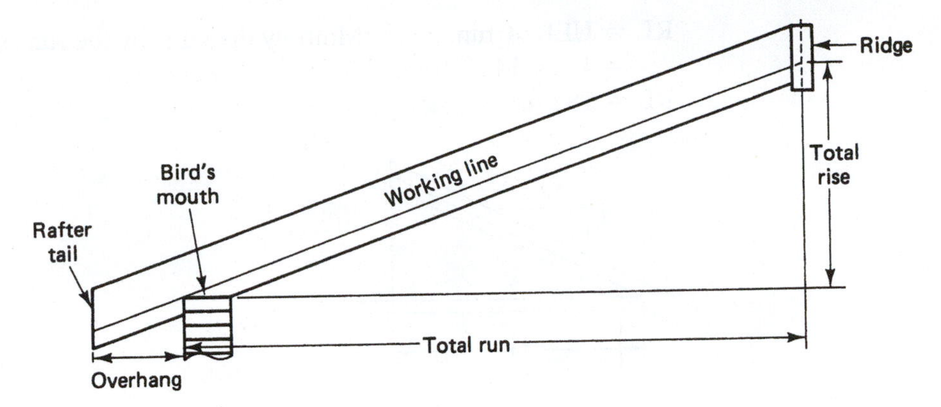

Figure 16–4

16.2 Rafter Dimensions

Study the rafter shown in Figure 16–4 and identify each of the following:

Working line: The line on the rafter along which the length is laid out. Note how it runs from the outer corner of the top wall plate to the center of the ridge.

Bird's mouth: The seat cut, a notch in the rafter to allow it to rest on the plate.

Total run: Horizontal distance from the outer edge of the wall to a vertical line through the center of the ridge.

Total rise: Vertical distance from the plane across the top of the plate to the point where the working line meets the ridge centerline.

Overhang: Horizontal distance from the outer edge of the wall to the rafter tail.

16.3 Rafter Calculations

To calculate a rafter length,

1. Calculate the ULL
2. Multiply by the run, in feet

This is the rafter length in inches as measured from the outer wall (vertical cut of the bird's mouth) to the center of the ridge.

Examples

A. Determine the common rafter length in a roof having a 5/12 slope and a run of 14 ft. Follow the steps just outlined.

$ULL = \sqrt{URi^2 + 12^2}$	Use the ULL formula.
$= \sqrt{5^2 + 12^2}$	Substitute in the proper values.
$= \sqrt{25 + 144}$	Square both terms.
$= \sqrt{169}$	Add.
$= 13$	Take the square root.

$$\begin{aligned} RL &= ULL \times run \\ &= 13 \times 14 \\ RL &= 182 \text{ in.} \end{aligned}$$

Multiply the ULL by the run.

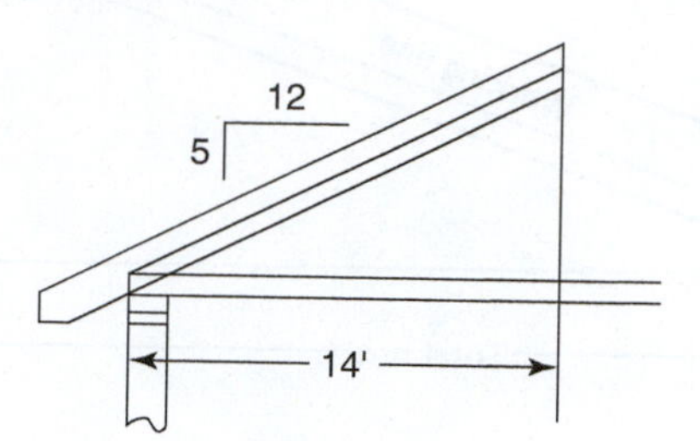

Convert this result to feet and inches:

$182 \div 12 = 15.167$ ft Divide inches by 12 to get feet.
$0.167 \times 12 = 2$ Multiply the decimal feet by 12 to get inches.

The rafter is 15 ft 2 in. long.

B. Determine the rafter length and the total rise for a roof having a 6/12 slope and a run of 15 ft 8 in.
First, determine the unit line length.

$ULL = \sqrt{URi^2 + 12^2}$	Use the ULL formula.
$= \sqrt{6^2 + 12^2}$	Substitute in the proper values.
$= \sqrt{36 + 144}$	Square each term.
$= \sqrt{180}$	Add.
$ULL = 13.416$	Take the square root.

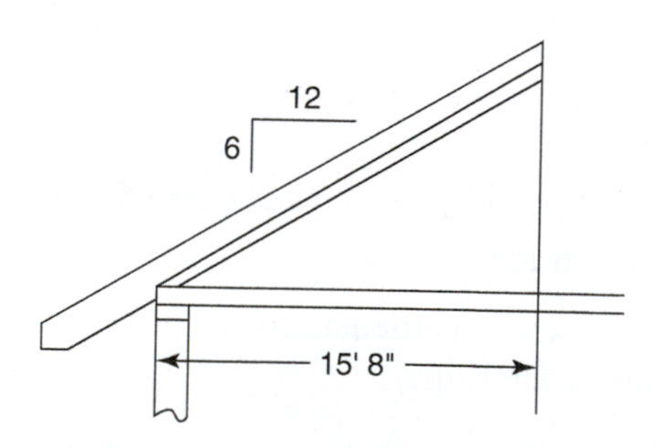

Second, determine the rafter length.

$RL = ULL \times run$	Use the rafter length formula.
$= 13.416 \times 15.667$	Substitute in the proper values.

$RL = 210.188$ in., or 17.516 ft, or 17 ft $6\frac{3}{16}$ in.

Last, determine the total rise.

$TRi = URi \times run$	Use the total rise formula.
$= 6 \times 15.667$	Substitute in the proper values.
$TRi = 94$ in.	

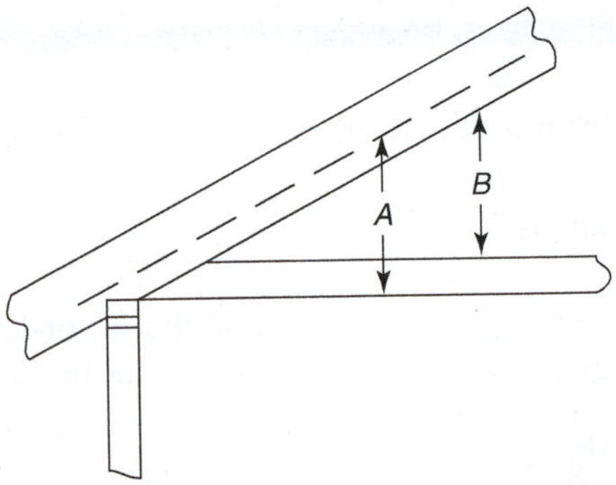

Figure 16–5

A — Theoretical headroom
B — Actual headroom

An examination of Figure 16–5 will reveal that total rise does not reflect *actual* headroom. Headroom is decreased both by the amount of material between the working line along the rafter and its lower edge and by the ceiling joists (and other materials making up possible flooring), which lie above the elevation of the top plate.

16.4 Overhangs

Overhang must be considered when calculating rafter length. We discuss overhangs more thoroughly in Chapter 18. The term *overhang* means the *horizontal projection* of the overhang. The horizontal projection is measured from the vertical plane of the plumb cut on the bird's mouth to the vertical plane of the plumb cut on the tail of the rafter. See Figure 16–6. That length of rafter extending from the corner of the bird's-mouth cut to the tail of the rafter may be determined using the rafter-length formula given earlier; that is,

$$RL = ULL \times run$$

In this case, "run" will be the overhang, expressed in feet.

Figure 16–6

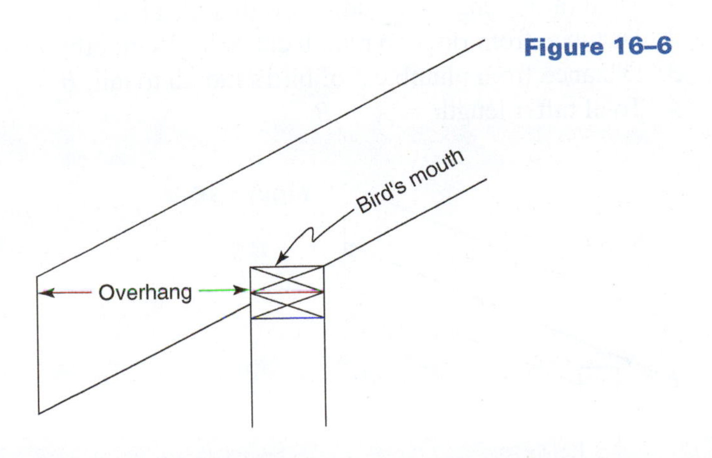

Example

A. Determine the additional length of rafter needed for an 18-in. overhang on a 6/12 slope roof.
Notice that 18 in. equals 1.5 ft.

$RL = ULL \times run$	Use the rafter-length formula.
$= 13.416 \times 1.5$	Substitute in the proper values.
$RL = 20.125$, or $20\frac{1}{8}$ in.	Multiply and convert decimal inches to a fraction.

Another approach to this problem is to include the overhang as part of the run and calculate the total rafter length to include the overhang. The run of the rafter will be added to the overhang, both being expressed in feet. That is,

$$RL = ULL \times (run + overhang)$$

To illustrate the use of this formula, refer to the last two examples, in which ULL = 13.416, run = 15.667, and overhang = 1.5. The formula above becomes

$RL = 13.416 \times (15.667 + 1.5)$	Substitute in run of common plus overhang.
$RL = 230.313$, or $230\frac{5}{16}$ in.	Add, then multiply and convert.

This is the same result when the answers to the two examples are added. That is,

$$210.188 + 20.125 = 230.313$$

An Important Check

When laying out a rafter pattern, three lengths should be calculated and checked against one another. These are

1. Total rafter length (including overhang), TRL
2. Distance from ridge to plumb cut of bird's mouth, A (see Figure 16–7)
3. Distance from plumb cut of bird's mouth to tail, B
4. Total rafter length $= A + B$

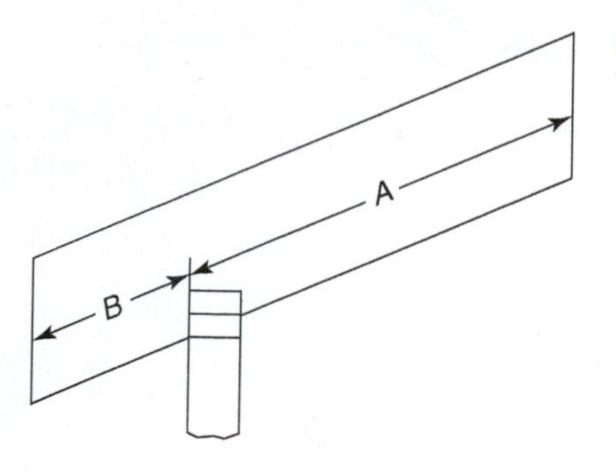

Figure 16–7

16.5 Applying the Numbers; Adjustments

After determining the lengths just described, a rafter pattern is laid out and cut and it becomes a pattern for laying out the other common rafters. When the pattern is cut, certain adjustments are considered.

Adjustment for Ridge Board

One adjustment to consider is the position of the plumb cut at the ridge. If a ridge board is used, the plumb cut is moved to a position parallel to its original position and back toward the tail of the rafter a horizontal distance one-half the thickness of the ridge board. In other words, the plumb cut is moved back one-half the ridge-board thickness as measured along a line perpendicular to the plumb cut. See Figure 16–8.

Figure 16–8

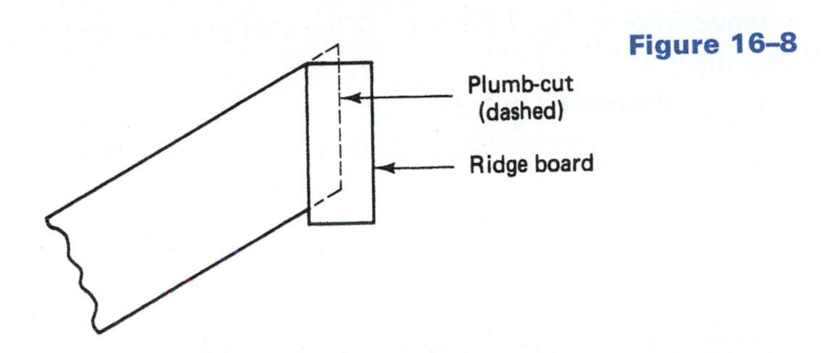

Adjustment at the Rafter Tail

Another adjustment may take place at the tail of the rafter. A plumb cut is usually made here, but the design of the overhang may dictate other types of cuts at the tail. Some carpenters prefer to defer the tail cuts until the rafters are in place. Then a chalk line may be snapped to ensure a straight overhang and compensate for any irregularities caused by cutting or placing the rafters.

The Bird's-Mouth Cut

Finally, consideration should be given to the *depth* of the bird's-mouth cut. Too much depth will weaken the overhang support. The seat cut should be of sufficient length to allow adequate nailing of the rafter to the plate. On the other hand, the corner of the bird's mouth should not project into the rafter more than one-third of its width. See Figure 16–9.

Figure 16–9

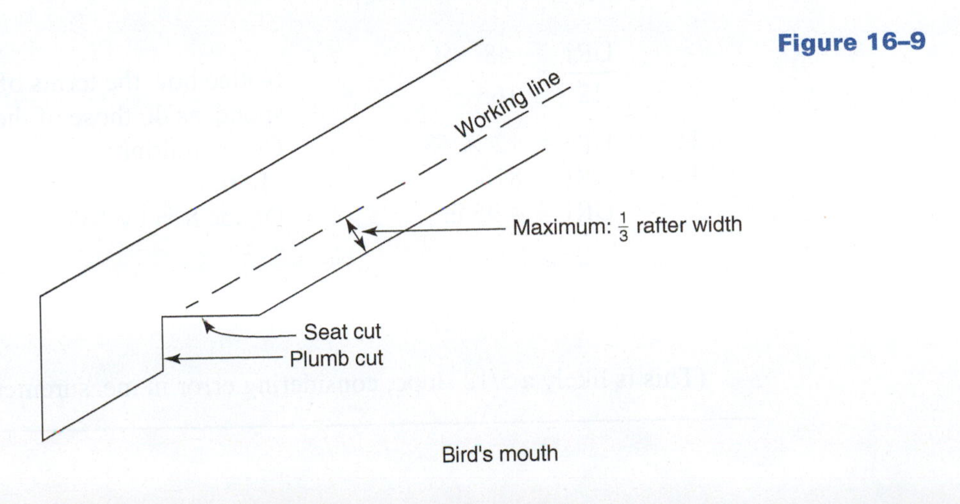

SUMMARY

1. Determine the total rafter length, the distance from the ridge to the bird's mouth, and the distance from the bird's mouth to the tail.
2. Establish the depth of the bird's mouth and draw the working line along the rafter so that it passes through the corner of the bird's mouth.
3. At the ridge end of the rafter, draw the plumb-cut line.
4. From the intersection of the working line with the ridge plumb cut, lay out the distance to the bird's mouth.
5. Draw the seat cut and plumb cut of the bird's mouth so that they intersect on the working line.
6. From the point of intersection of the bird's mouth with the working line, measure along the working line the distance to the rafter tail.
7. Draw the plumb-cut line for the tail to intersect the working line the proper distance from the bird's mouth. (This step may be postponed until the rafters are in place.)
8. Check the overall rafter length.
9. Make any compensating cuts at the ridge and/or tail.

16.6 Determining Slope of an Existing Roof

The unit rise or slope of a roof is not a known factor in every situation. In particular, suppose that an addition to an existing structure is planned and the roof lines are to be matched. The slope of the existing roof must be determined. This may be done by either direct or indirect measurement.

One technique is to measure the total run and total rise of the existing roof. With this information, the total rise and total run can be proportioned to the unit rise and unit run, with the unit rise serving as the unknown.

Example

A. The total rise and total run of a common rafter have been measured and found to be 5 ft 8 in. (or 68 in.) and 13 ft 9 in. (or 165 in.), respectively. Find the unit rise.

Set up a proportion as follows:

$$\frac{\text{URi}}{12} = \frac{68}{165}$$

Notice how the terms of both numerators correspond, as do those of the denominators.

$$165 \times \text{URi} = 12 \times 68$$

Cross-multiply.

$$165 \times \text{URi} = 816$$

Multiply.

$$\text{URi} = 4.95 \text{ in.}$$

Divide 816 by 165

4.95 in. converts to $4\frac{15}{16}$ in.

(This is likely a 5/12 slope, considering error in measurement.)

Notice that the total rise and total run were expressed in inches here. They could just as well have been expressed in decimal feet with no change in the outcome. Measuring the total run and rise may not be practical.

Another method of determining the slope of an existing roof is to use a framing square and level, with one blade of the square positioned so that the 12-in. mark coincides with the lower edge of the rafter. Holding the square in this position, place the level along this blade and move it to the level position. Read the unit rise on the corresponding edge of the other blade where it contacts the lower edge of the rafter. See Figure 16–10.

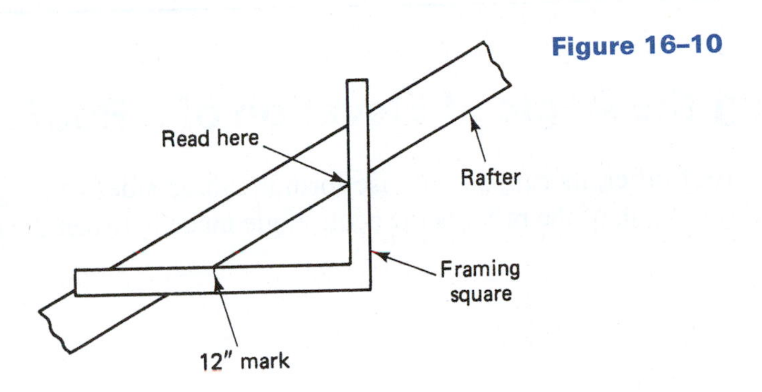

Figure 16–10

To conclude this chapter, a final example is considered involving an unknown roof slope. In this case, suppose that a roof is to be constructed with an off-center ridge—typical of a saltbox house. Assume that the slope on the front side of the roof is known, as is the horizontal distance from the front of the building to the vertical plane of the ridge. The problem is to find the slope of the back side of the roof so that rafter lengths may be calculated and proper cuts made.

Example

B. A small building 12 ft. wide is to be constructed having a roof with its ridge located 4 ft from the front of the building. The short side of the roof is to have a 10/12 slope. Determine the unit rise on the rear section. See Figure 16–11.

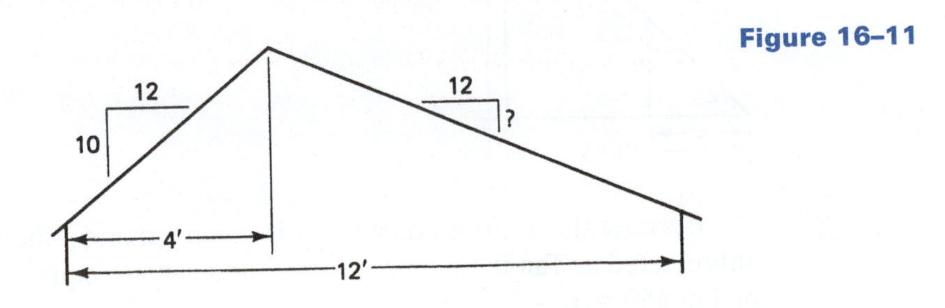

Figure 16–11

First, determine the total rise for the short roof segment.

$TRi = URi \times run$	Use the total rise formula.
$= 10 \times 4$	Substitute in the proper terms.
$TRi = 40$ in.	Multiply.

Second, set up and solve a proportion relating total rise and total run to unit rise and 12, the unit run.

$$\frac{\text{URi}}{12} = \frac{\text{TRi}}{\text{TRu}}$$ Notice that both numerators correspond, as do the denominators.

$$\frac{\text{URi}}{12} = \frac{40}{96}$$ Substitute in the proper terms. Both TRi and TRu must be in the same units.

$$96 \times \text{URi} = 12 \times 40$$ Cross-multiply.

$$\text{URi} = 5 \text{ in.}$$ Multiply 12 by 40 and divide by 96.

16.7 Determining the Angle of Elevation of a Roof

A roof rafter, its run, and its rise form the three sides of a right triangle. The angle of elevation, θ, of the rafter is the acute angle made between the rafter and its run.

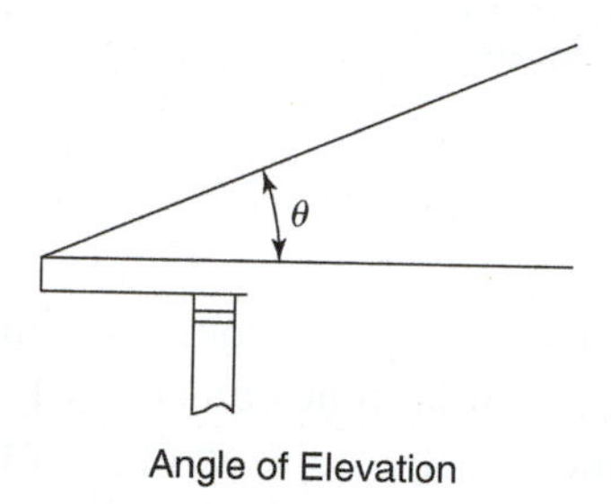

Angle of Elevation

This angle of elevation is directly related to the rise and run or slope of the rafter. For any given rise and run, this angle is always the same, regardless of the actual length of the run or amount of the rise. If the rise and run are equal, for example, as in a 12-in. slope, the angle of elevation is 45°. The actual length of the run or amount of rise does not matter, so long as they are equal, $\theta = 45°$.

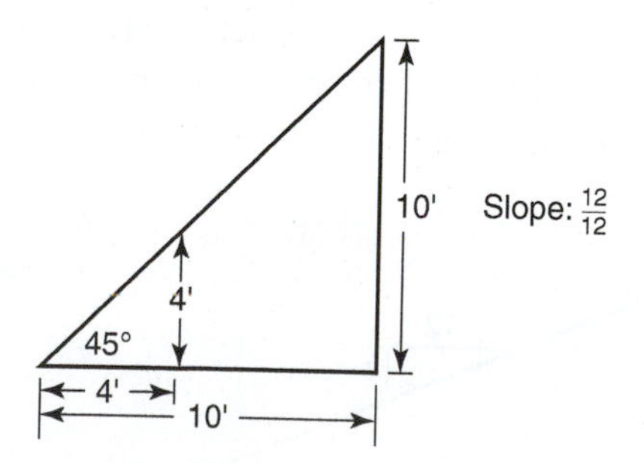

Because this ratio is constant, it is given a name—the tangent of the angle. This is abbreviated as Tan θ. So when the rise and run are equal, we can say Tan 45° = 12/12 or Tan 45° = 1.

For other slopes, the tangent function will give the size of the angle. Using a scientific calculator, enter 45, and press "TAN", and "1" will appear. For this application, this is telling us that an angle of elevation has a slope of 1. That can mean a rise-to-run ratio of 12/12.

Conversely, if the slope is known and you wish to find the angle of elevation, enter the slope as a decimal, then press "INV", "TAN". (On some calculators, instead of "INV" the button may be labeled "2nd F".)

A slope of 5/12 has an angle of elevation of 22.6°. You can verify this by entering 5 + 12 = 0.41666, and then "INV", "TAN". The result shows 22.6°.

If the roof angle of elevation is known, the slope can be found. For example, suppose the angle of elevation is 30°. To find the slope, enter "30", then "TAN", and you see 0.57735. Use this to determine what number divided by 12 gives 0.57735.

$X/12 = 0.57735$ and $X = 12 \times 0.57735 = 6.928$. Rounding this to 7, we have a 7/12 slope.

In summary: To find the angle of elevation given the slope,

1. Convert the slope to a decimal
2. Press TAN.

To find the unit rise given the angle of elevation,

1. Enter the angle
2. Press "INV", "TAN"
3. Convert this decimal value to a fraction having 12 as a denominator.

Additional discussion of trigonometry will be found in the Appendix.

REVIEW EXERCISES

1. Listed below are various values for unit rise. For each, calculate the unit line length value correct to three decimal places.

   Unit rise

   a. 3
   b. 7
   c. $4\frac{1}{2}$
   d. 10
   e. 9
   f. 15
   g. 12

2. How many decimal places should be used in the unit line length to obtain a tolerance of plus or minus $\frac{1}{16}$ in. in a rafter length?
3. Calculate the rafter length for each of the following unit rises and runs. Express answers in feet and inches, correct to the nearest $\frac{1}{16}$ in.

Unit rise	Run
a. 5	15 ft
b. $7\frac{1}{2}$	13 ft 6 in.
c. 8	14 ft 8 in.
d. 4	11 ft 5 in.
e. 12	5 ft 7 in.
f. $9\frac{1}{4}$	10 ft $6\frac{1}{2}$ in.
g. $3\frac{1}{2}$	7 ft 4 in.

4. The total rise and total run (exclusive of overhang) are given for several roofs. In each case, calculate the unit rise, correct to the nearest $\frac{1}{8}$ in.

	Rise	Run
a.	6 ft 3 in.	13 ft 5 in.
b.	4 ft 9 in.	14 ft
c.	14 ft	14 ft
d.	9 ft 8 in.	16 ft 10 in.
e.	5 ft 5 in.	11 ft 4 in.

5. The shed roof shown here spans 20 ft. The front elevation is 3 ft 8 in. higher than the rear. What is the unit rise?

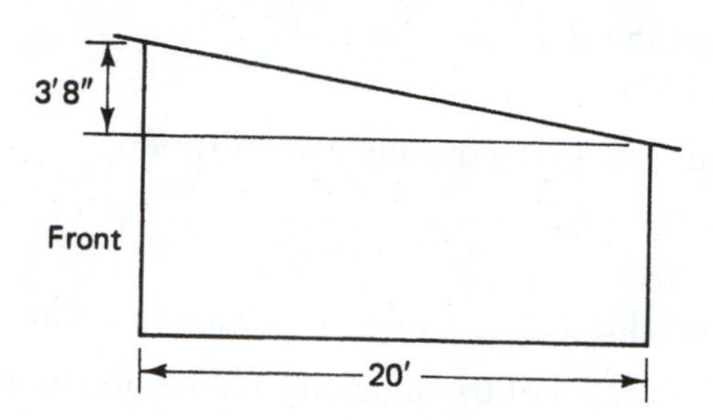

6. Refer to the building shown in the accompanying figure. The building has an off-centered ridge located 8 ft from the front. The front roof section has a 9/12 slope. The total building width is 22 ft.

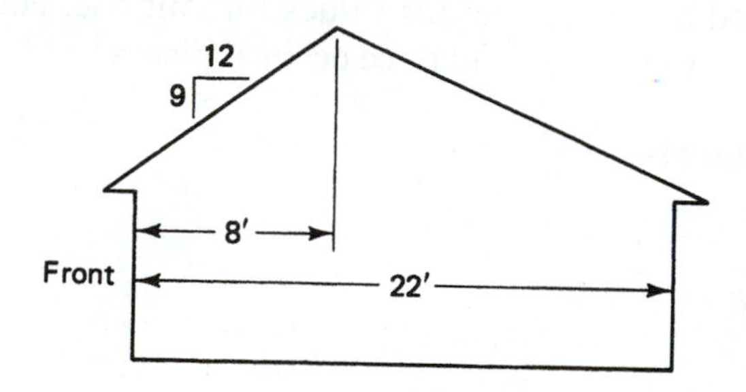

Determine the following dimensions:
a. Unit rise for the rear roof section
b. Unit line length for the rear section
c. Total rafter length for the rear section
d. Total rafter length for the front section

7. Refer to the accompanying figure. The front section of the roof has a 10-in. unit rise. Notice the different elevations of the front and rear walls.

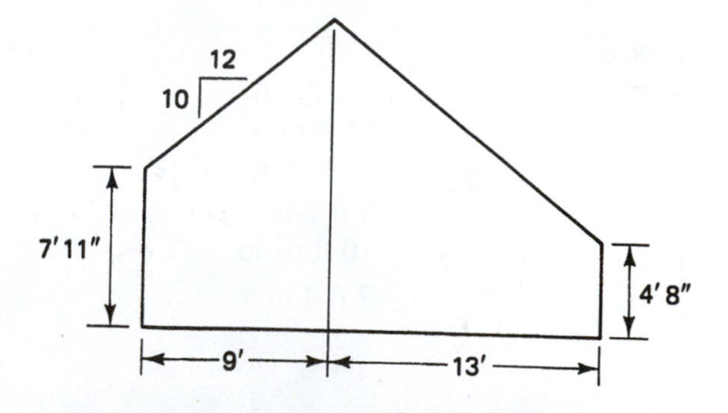

Determine the following dimensions:

a. Total rise for both the front and rear rafters
b. Unit rise for the rear rafter
c. Unit line length for both the front and rear rafters
d. Total rafter lengths for both front and rear (no overhang; answer in feet and inches, to the nearest $\frac{1}{16}$ in.)

8. The second story of a Cape Cod-style house is to be framed with knee walls 6 ft high. The width of the house is 28 ft and the roof has a 10-in. slope. Assume that 1 ft of height is occupied by the second floor joists and that part of the rafter that is below the working line. How wide will the room be?

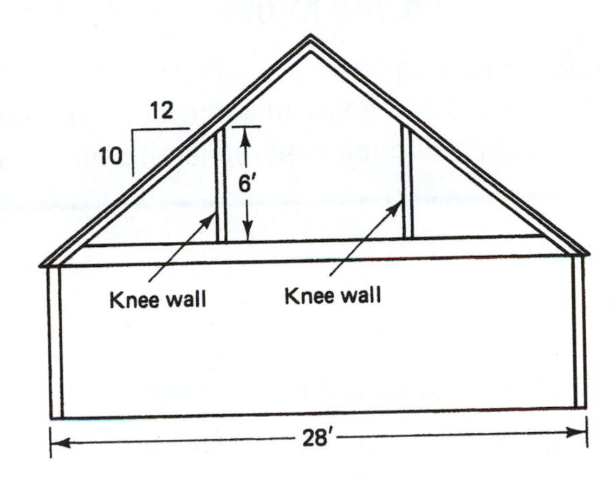

9. Refer to exercise 8. Suppose that it is desired to have a width of 20 ft for the second-story room. Assuming that the same 1 ft of height is lost from the theoretical rise, what will be the height of the knee wall?
10. For each of the following slopes, find the angle of elevation of the rafters, correct to the nearest 0.5°.
    a. 2/12 b. 3/12 c. 6/12 d. 9/12 e. 10/12
11. For each of these angles of inclination, find the unit rise, correct to the nearest $\frac{1}{4}''$.
    a. 12° b. 36° c. 20° d. 30° e. 60°
12. A shed-roofed building has walls that are 8-ft high on the front side and 12-ft high on the rear side. It is 14 ft from front to rear.
    a. Find the unit rise.
    b. What is the angle of elevation?

CHAPTER 17

Introduction to the Framing Square

Did You Know

3,288 SunPower photovoltaic panels installed in western North Carolina will generate 1.6 million kilowatt hours of electricity per year—enough for 1,000 homes. The panels occupy a retired landfill on a 7-acre site.

OBJECTIVES

Upon completing this chapter, the student will be able to:

1. Use the framing square to locate the unit line length for a given slope.
2. Apply the unit line length to find a rafter length.

The framing square is the carpenter's assistant in many ways. It is made up of two legs called the *blade* (or body) and the *tongue.* The blade is 2 in. wide and 24 in. long, and the tongue is $1\frac{1}{2}$ in. wide and 16 in. long. See Figure 17–1. These dimensions allow for the framing square's use in laying out and marking the spacing along a shoe or plate for rafters or studs. For example, the $1\frac{1}{2}$-in. width of the tongue is an aid in marking positions of $1\frac{1}{2}$-in. studs or rafters.

On the blade of the framing square can be found tables and scales. Much of the information needed in calculating rafter dimensions is found here. A carpenter proficient in mathematics can calculate the data that is given on the framing square or that same information can be read from the square. It should be noted, however, that the framing square data does not cover all situations. In some cases, an understanding of the mathematical concepts is needed to get the necessary information.

Let us take a closer look at the framing square. In Figure 17–2, notice the labels along the rows beginning at the left-hand end. Here we see "LENGTH COMMON RAFTERS PER FOOT RUN." Below that line are five more rows of figures headed "LENGTH HIP OR VALLEY RAFTERS . . . ," etc. Look at the top row of numbers headed "LENGTH OF COMMON RAFTERS PER FOOT RUN." Move along the blade of the square to the 9-in. position (see Figure 17–3). Beneath the 9 in the top row is found 15. What this means is that if a roof has a 9/12 slope, that is, if the roof rises

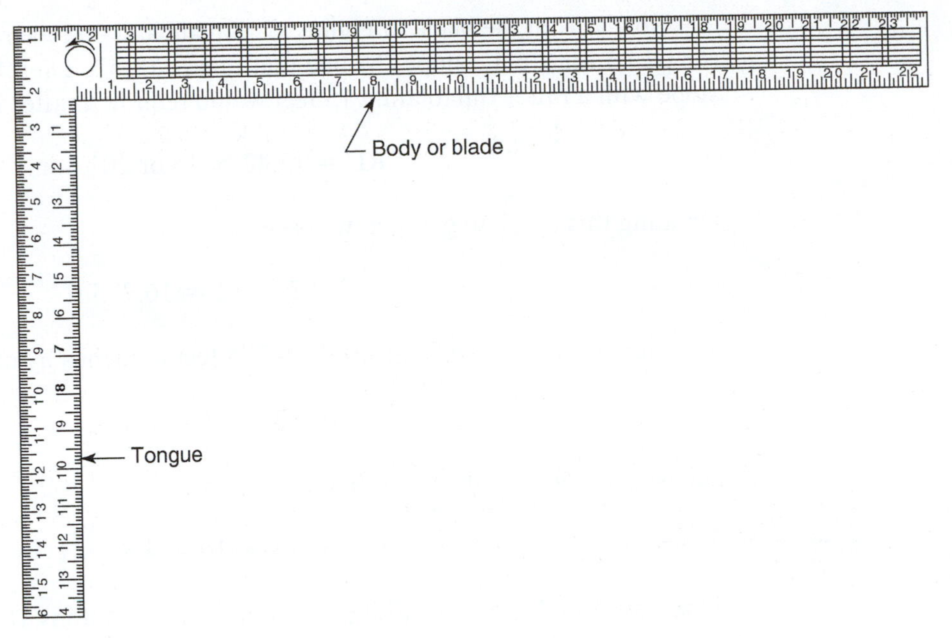

Figure 17–1

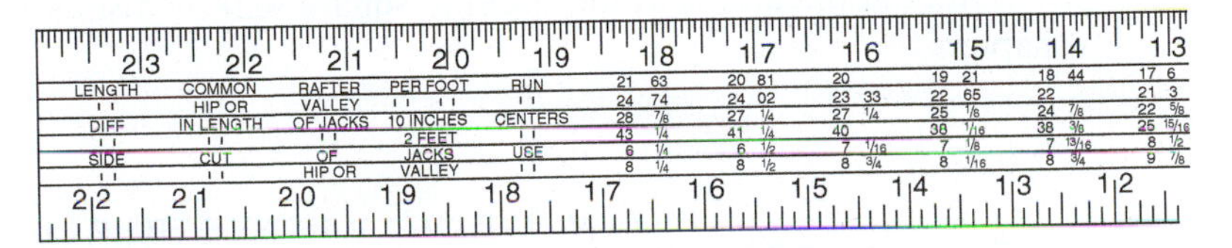

Figure 17–2

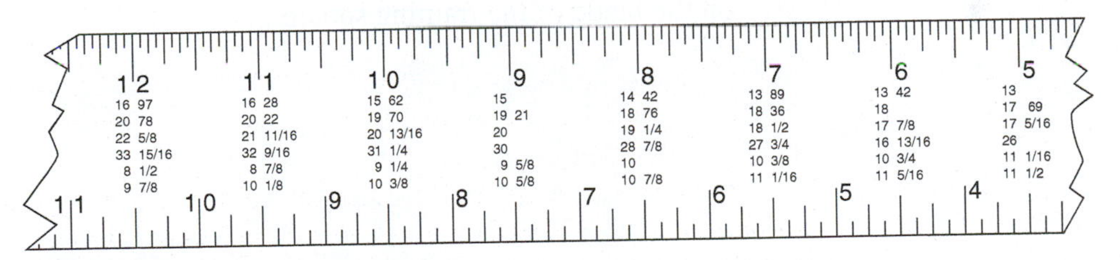

Figure 17–3

9 inches for each foot of horizontal run, the rafter length is 15 in. per foot of run. Therefore, a building with a conventional gable roof and a width of 28 feet, so that the ridge of the roof is centered at the 14-ft mark, and having a 9/12 slope will need a rafter length of 15×14, or 210 in.

It is helpful to understand how the number 15 was determined. This is the *unit line length (ULL)*. The student can verify this by using the Pythagorean rule with legs of 9 and 12. The hypotenuse will be 15.

Now look at the blade in Figure 17–3 again and find the rafter length per foot of run (or ULL) for a 6/12 slope. Notice that this is 13.42. Therefore, a roof having a 6/12 slope with a rafter run totaling 15 feet would require a rafter length of

$$RL = 13.42 \times 15 \text{ or } 201.3 \text{ in.}$$

Dividing this by 12 to get feet, we have

$$201.3 \div 12 = 16.775 \text{ ft}$$

Set aside the 16 feet and convert the 0.775 feet to inches by multiplying by 12.

$$0.775 \times 12 = 9.3 \text{ in.}$$

Set aside the 9 in. and convert 0.3 in. to 16ths.

$$0.3 \times 16 = 4.8$$

which we round to 5 and call it $\frac{5}{16}$ in. So the rafter length is 16 ft $9\frac{5}{16}$ in.

Here again the student may wish to use the Pythagorean rule to verify that for a 6/12 slope the ULL is 13.42.

Other applications of the framing square will be discussed as needed in subsequent chapters.

REVIEW EXERCISE

1. In exercise 1 in Chapter 16, you calculated the unit line length for several given values of unit rise. Verify each of your answers by finding that same information on the blade of the framing square.

CHAPTER 18

Overhangs

Did You Know

100 million trees are cut each year to make junk mail. This is the equivalent of clear-cutting Rocky Mountain National Park every four months.

OBJECTIVES

Upon completing this chapter, the student will be able to:

1. Calculate the vertical projection of an overhang for a given horizontal projection.
2. Calculate the horizontal projection of an overhang for a given vertical projection.
3. Determine the amount of overhang necessary for a given solar exposure.
4. Determine the amount of material needed to frame and enclose an overhang.

18.1 Purposes of Overhang

Overhang on a building serves various functions. Included among these are

- Eye appeal
- Solar exposure and shading
- Rainwater control

The appearance of a building can be altered considerably by the style and amount of overhang. A little extra money spent on expanding the overhang of a small ranch house can greatly enhance its eye appeal. On the other hand, the classic Cape Cod-style house is quite attractive with very little overhang. The steepness or slope of a roof affects the amount of overhang. In general, the steeper the roof, the shorter the overhang, avoiding interference with window and door tops. Today's emphasis on solar considerations in building design has resulted in considering the sun's angle at various times of the year. The amount of overhang may be designed so as to minimize solar exposure during summer months and maximize it during winter months.

Proportions and the material in Chapter 16 are the important mathematical concepts that will be used in this chapter as we answer questions related to overhangs. In Chapter 16 we established the mathematics related to problems considered here.

18.2 Calculating Overhang and Materials

Recall that the unit rise, URi, and the unit run (which is always 12) control the roof angle. So, too, do these values control the overhang.

Example

A. A roof has a 5/12 slope. If a 16-in. horizontal projection of the overhang is desired, what will be the vertical projection?

Note that the term *horizontal projection* will refer to the distance that the overhang projects outward perpendicular to the framed exterior wall. The term *vertical projection* will refer to the distance the overhang projects vertically, downward from the top plate (see Figure 18–1).

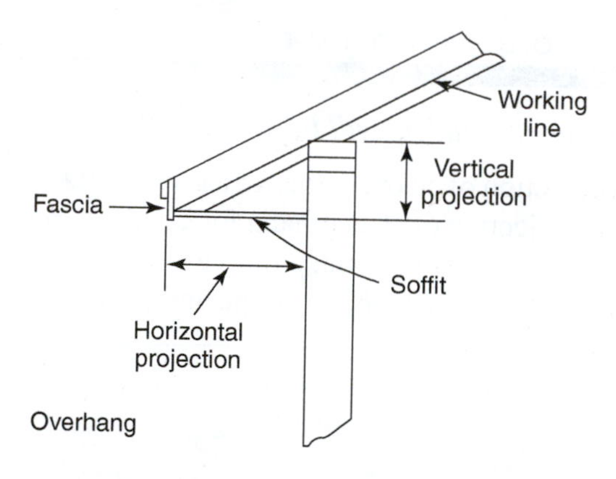

Figure 18–1

Set up a proportion relating the unit rise and unit run to the vertical and horizontal projections, respectively.

$$\frac{\text{URi}}{12} = \frac{\text{vertical projection}}{\text{horizontal projection}}$$ Use a related proportion.

$$\frac{5}{12} = \frac{\text{vertical projection}}{16}$$ Substitute in the proper values.

$$12 \times \text{vertical projection} = 5 \times 16$$ Cross-multiply.

$$\text{vertical projection} = 6.67, \text{ or } 6\frac{11}{16} \text{ in.}$$ Multiply and then divide by 12.

A careful examination of Figure 18–1 reveals that the vertical projection is actually measured with relation to the *working line* of the rafter. Different types of cuts made on the tail of the rafter may necessitate adjustments to calculated projection lengths.

The material used to cover the underside of an overhang is called the *soffit*. The material used to cover the vertical face of the overhang is called the *fascia* (see Figure 18–1). If the distance between door tops and/or window tops and the overhang is under consideration, the thickness of the soffit material must be included in the vertical projection of an overhang.

On certain types of houses, the exterior appearance is enhanced if the horizontal projection of the overhang returns to the front of the building at the door and window tops. If this is a design requirement, the elevation of the top of the door frame trim must be determined. The difference between this elevation and the elevation of the top plate will dictate the vertical projection, soffit thickness included (see Figure 18–2).

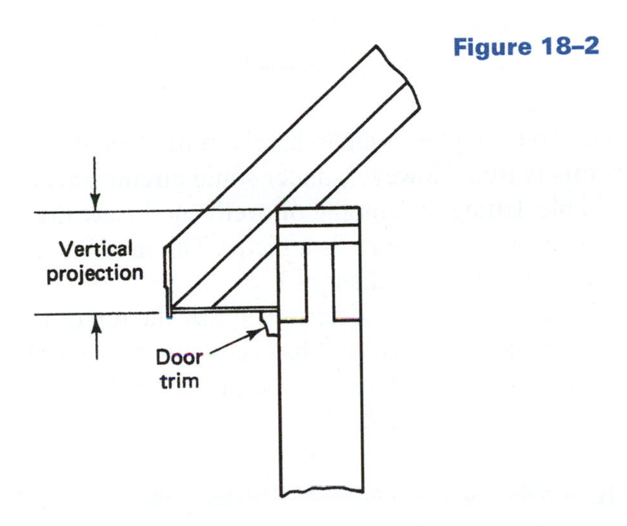

Figure 18–2

Example

B. The top of the exterior door trim will be $12\frac{3}{4}$ in. below the top plate of the framed exterior wall. The soffit material will be $\frac{3}{8}$-in A-C exterior-grade plywood. The roof carries a 6-in. slope. Find the horizontal projection of the overhang (see Figure 18–3).

The vertical projection as determined from the working line of the rafter will be $12\frac{3}{8}$ in. This is the result of subtracting the soffit thickness from the $12\frac{3}{4}$ in. available for the vertical projection. A proportion will determine the horizontal projection.

$$\frac{\text{URi}}{12} = \frac{\text{vertical projection}}{\text{horizontal projection}}$$ Set up the proportion.

$$\frac{6}{12} = \frac{12.375}{\text{horizontal projection}}$$ Substitute in the proper values.

$$\text{horizontal projection} = 24.75, \text{ or } 24\frac{3}{4} \text{ in.}$$ Cross-multiply and divide.

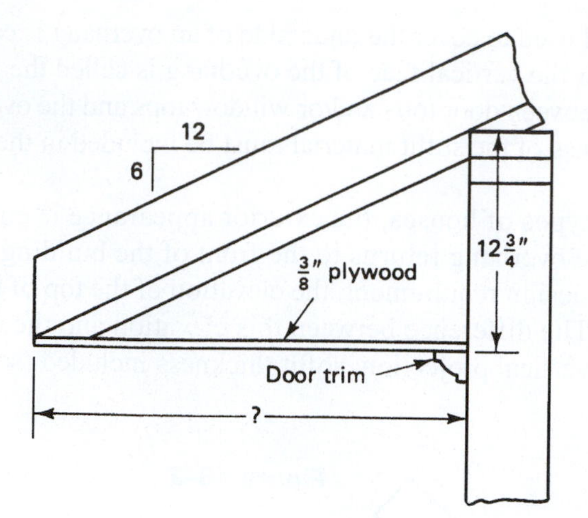

Figure 18–3

The first two examples assume that the roof slope dictates the overhang design. In most cases this is true. However, under some circumstances, design experimentation may be desirable, letting the amount of overhang dictate the slope of the roof. This may be the case when solar exposure is a factor. The mathematics does not change in this situation—only the unknown changes.

For example, if it has been determined that the top of the door trim will be met by the horizontal projection return and that a certain amount of horizontal projection is optimal for solar exposure, these are the terms that will be proportioned to the unit rise and unit run. The unit rise will be the unknown and the slope is thus determined.

Example

C. Solar analysis has determined that a horizontal projection of 18 in. will optimize exposure to the sun during the various seasons. Aesthetic quality requires the horizontal projection to return 4 in. above the door trim elevation. Analysis of the framing plan reveals that the horizontal projection meets the wall $10\frac{1}{2}$ in. below the top plate. Find the unit rise.

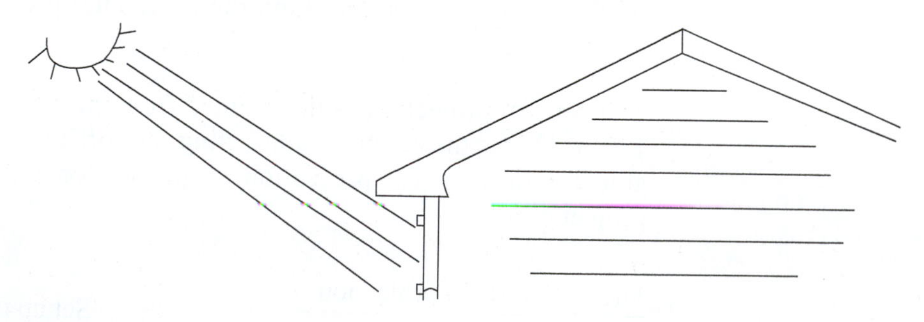

A proportion is set up as shown.

$$\frac{\text{URi}}{12} = \frac{10.5}{18}$$

Notice how 10.5, a vertical measure, corresponds to URi, while 18 corresponds to the horizontal distance or run. Be careful to set up any proportion in this way, with similar units occupying corresponding positions in each fraction.

URi = 7 — Cross-multiply and divide.

Therefore, this design calls for a 7/12 slope.

SUMMARY

Problems dealing with overhangs are readily solved using a proper proportion. Care should be taken in setting up the proportion so that the terms are consistent; that is,

- Terms describing *rise* should correspond with *vertical distances*
- Terms describing *run* should correspond with *horizontal distances*
- In general, the following relation is used:

$$\frac{\text{URi}}{12} = \frac{\text{vertical projection}}{\text{horizontal projection}}$$

where URi = unit rise

Materials needed to frame and enclose overhangs vary with type and style of overhang. This makes the subject difficult to generalize about. Some house styles, such as a steep-roofed cape, require only that the rafter ends be covered with a fascia board. A ranch-style house might require more elaborate construction of a horizontal return that is to be covered with both soffit and fascia material. There may or may not be an overhang along the sloping eave of a roof edge.

Despite these and other variables, once the style of overhang has been determined, the efficient estimator will make use of many predetermined measurements and calculations to help in determining the materials needed.

Two of the more frequently used measurements in the estimation of overhang materials are *the length of the building* and the *rafter length. Overhang width* is used in conjunction with these dimensions to determine the total square feet of soffit material needed. The same measurements are used to determine amounts of fascia material. The width of fascia material is dictated by the length of the plumb cut on the end of a roof rafter.

Another approach to determining the relationship of the sun's angle and overhang is to use the tangent function, as described in Chapter 16. When designing an overhang that will provide shading from the sun or exposure to it, the angle of elevation of the sun at certain times of the day can be a consideration.

As you can see from the figure below, if the angle of elevation of the sun is, say, 40°, the amount of overhang will determine the amount of shading provided.

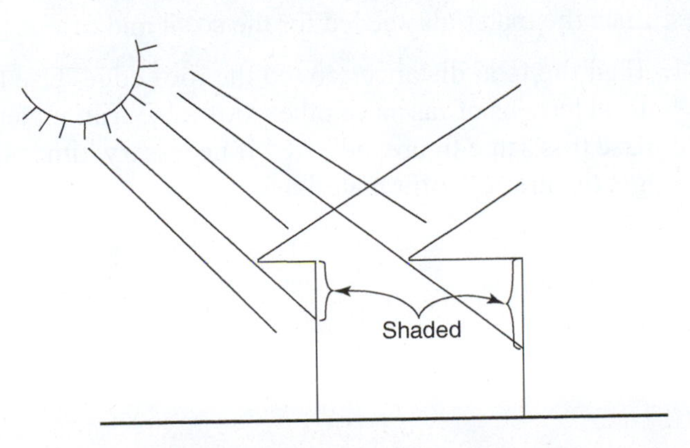

In the northern hemisphere, it could be desirable to allow ample sunshine into a room during the winter months.

Example

D. Suppose a roof has a 16-inch horizontal projection of overhang. If the sun has an angle of elevation of 30°, how many inches of shade will there be below the horizontal return?

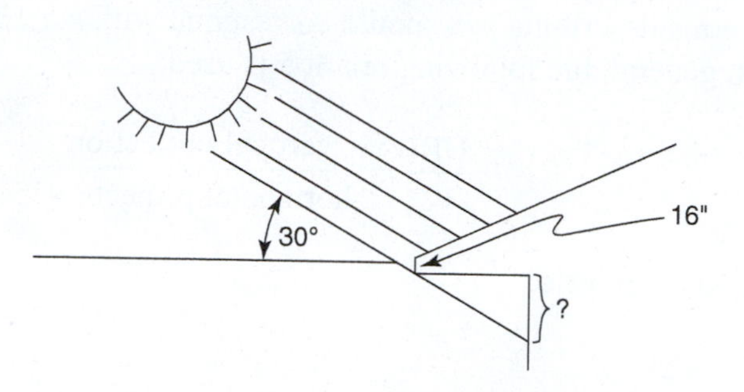

As seen from the drawing above, the angle of elevation, θ, also appears in the right triangle having as it two legs the horizontal projection and the depth of shading on the face of the house. Using the tangent function,

$$\text{Tan } 30° = X/16, \text{ where } X = \text{the depth of shading.}$$

$$\text{So } X = 16\text{Tan}30° \text{ and } X = 9.238 \text{ in. or } 9\tfrac{1}{4} \text{ in.}$$

Example

E. With the angle of elevation of the sun at 40°, if it is desired to allow the maximum amount of sunshine into a room, and the window top is 4 inches below the horizontal projection return, how much horizontal overhang should there be?

$$\text{Tan } 40° = 4/H \text{ and } H = 4/\text{Tan } 40°. \text{ So } H = 4/0.839 = 4.767 \text{ or } 4\tfrac{3}{4} \text{ inches.}$$

SUMMARY

To estimate the materials needed for the soffit and fascia,

1. Find the total distance around the roof edge. Use this figure to determine total lengths of fascia or other roof-edge trim materials.
2. Use this same figure, adjusted if necessary, times the width of the soffit to get the area of soffit material.

REVIEW EXERCISES

1. Listed below are various roof slopes and a given horizontal projection of overhang. In each case, determine the vertical projection.

Slope	Horizontal Projection
a. 5/12	16 in.
b. 7/12	9 in.
c. 12/12	24 in.
d. 12/12	5 in.
e. 10/12	8 in.
f. 6/12	2 ft 4 in.

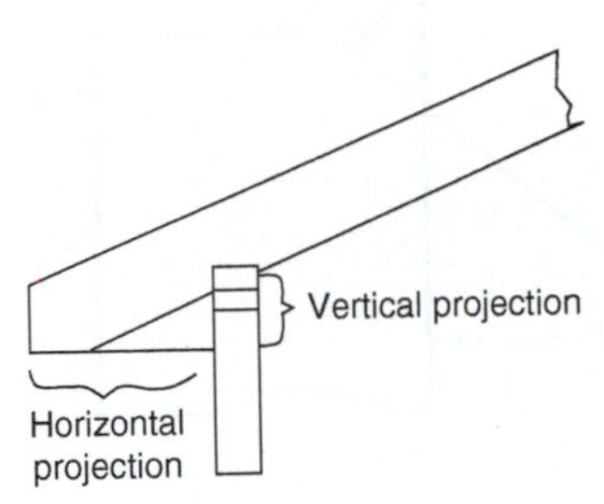

2. Listed below are various roof slopes and a given vertical projection of overhang. In each case, determine the horizontal projection.

Slope	Vertical Projection
a. 8/12	12 in.
b. 9/12	16 in.
c. 4/12	2 ft 6 in.
d. 15/12	5 in.
e. 10/12	8 in.
f. 5/12	1 ft 9 in.

3. Plans call for the return of the horizontal projection of an overhang to meet the wall at a point $11\frac{3}{8}$ in. below the top plate. The roof has an 8-in. slope. Find the length of the horizontal projection of the overhang.
4. The top of the exterior trim around a door is $7\frac{5}{8}$ in. below the top plate. The soffit is $\frac{3}{8}$-in. A-C plywood. The overhang is to return to the top of the door trim. The roof carries a 6/12 slope. What is the length of the horizontal projection of the overhang?
5. Plans call for the horizontal projection of the overhang to meet the exterior wall at a point $8\frac{1}{2}$ in. below the top plate. Determine the roof slope for the following horizontal projection overhang measures.
   a. 6 in.
   b. 10 in.
   c. 18 in.
   d. 24 in.
6. Plans specify a 10/12 slope and a 6-in. horizontal projection of the overhang. The rafter tails are to be left uncut, except for proper length. The rafter material

is 2 × 10, with an actual measurement of $9\frac{1}{2}$ in. The working line is located one-third of the distance from the rafter bottom. If the lowest point on the rafter tail is projected horizontally back to the wall, how far below the top plate will it fall? See Figure 18–4.

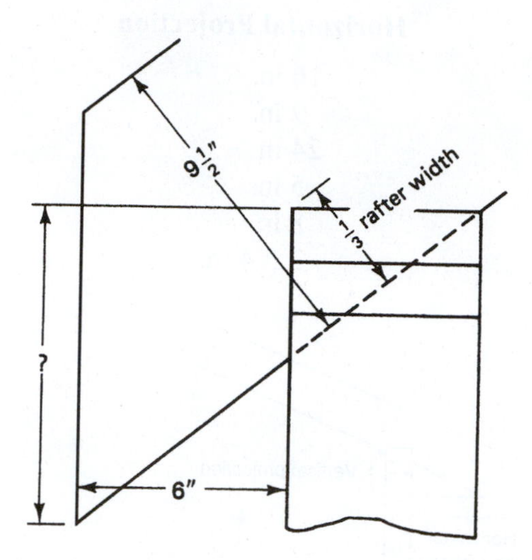

Figure 18–4

7. Repeat exercise 6 if the roof has a 7/12 slope, the horizontal projection measures 16 in., and the rafters are 2 × 8s, measuring $7\frac{1}{2}$ in.
8. Here are given several angles of elevation and a desired number of inches of shading below the horizontal return. Determine the amount of horizontal projection, correct to the nearest $\frac{1}{4}$ inch.

Angle of Elevation	Inches of Shading
a. 30°	4″
b. 45°	6″
c. 35°	18″
d. 26°	12″
e. 50°	9″

9. Here are given several amounts of horizontal projection of overhang and the sun's angle of elevation. Determine the amount of vertical shading that will result.

Horizontal Projection	Angle of Elevation
a. 12″	30°
b. 8″	40°
c. 16″	25°
d. 24″	60°

CHAPTER 19

Roofs II: Hip Rafters—The Conventional Case

Did You Know

Trees can be conserved by using products made from recycled wood from demolished buildings.

OBJECTIVES

Upon completing this chapter, the student will be able to:

1. Calculate the length of a conventional hip rafter.
2. Determine the angles for cuts at the ridge, bird's mouth, and tail of conventional hip rafters.

Hip rafters may be classified as *conventional* or *unconventional. A conventional hip rafter makes a 45° angle* with the two adjacent walls that form the corner of the building. The angle referred to lies in the plane across the top plates. Think of it as the shadow the hip rafter casts when the sun is directly overhead (see Figure 19–1). An unconventional hip rafter makes an angle with the two adjacent walls that is something other than 45°. Discussion of the unconventional case is deferred to a later chapter.

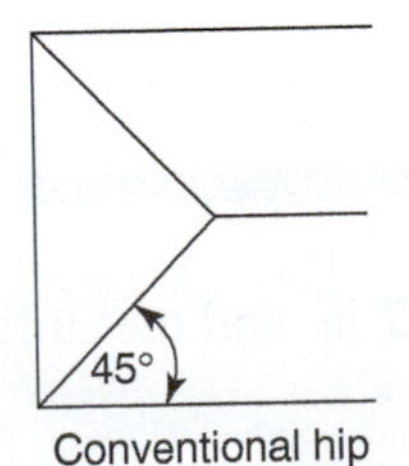

Conventional hip

Figure 19–1

A conventional hip rafter is preferred in construction—both for simplicity and for eye appeal. Some situations, however, may dictate the use of an unconventional hip rafter. For example, in attaching a porch to the face of a house, the roof design may require an unconventional hip to avoid interference with a second-story window.

We will consider two methods of determining hip rafter length. The first method will work for either a conventional or unconventional hip rafter, whereas the second method applies only to the conventional case.

19.1 Calculating Hip Rafter Length (HRL)—Method 1

The advantage of this method is that it works for any hip—conventional or unconventional.

Setback (SB) is the distance along the top plate from the corner to the position of the first full-length common rafter (see Figure 19–2). Note that in the conventional case,

$$SB = \text{run of the common rafter (CRu)}$$

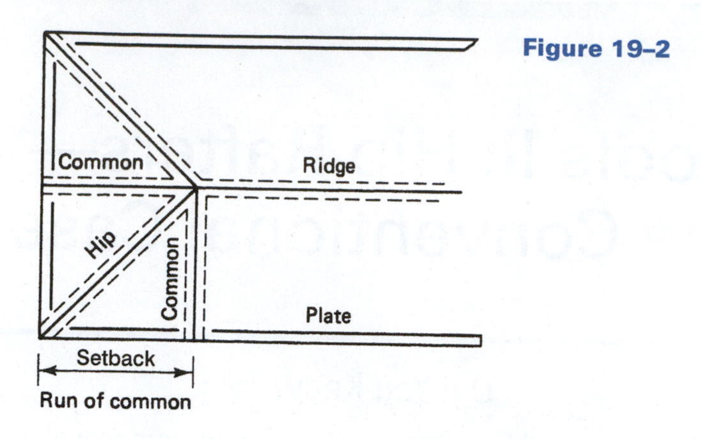

Figure 19–2

Notice that a right triangle is formed with

$$\text{hypotenuse} = \text{HRL}$$
$$\text{legs} = \text{CRu and SB}$$

Using the Pythagorean rule,

$$\text{HRL} = \sqrt{\text{CRL}^2 + \text{SB}^2}$$

where

HRL is the hip rafter length,
CRL is the common rafter length,
SB is the setback of the common rafter (or run of the common rafter in the conventional case).

Note: All units must be the same, but can be either feet or inches.

The hip rafter will extend from the corner of the building to the point on the ridge where the common rafter is attached. (Bear in mind that all references to distances are with respect to the working lines on the rafters. In a plan view of a roof, the working lines appear along the center of the rafter's top edge.)

Example

A. Find the length of a hip rafter if the common rafter measures 15 ft 2 in. and the run of the common measures 14 ft.

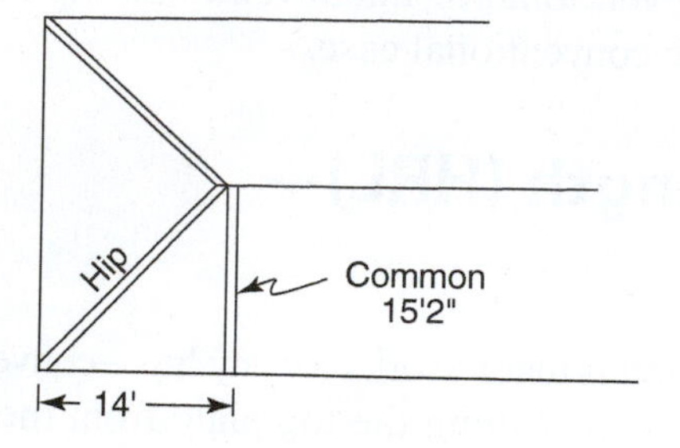

Use the right-triangle approach:

$HRL = \sqrt{CRL^2 + SB^2}$	Use the HRL formula (Pythagorean rule).
$= \sqrt{15.167^2 + 14^2}$	Substitute in the proper values.
$= \sqrt{230.038 + 196}$	Square each term.
$= \sqrt{426.038}$	Add.
$HRL = 20.641$ ft, or 20 ft $7\frac{11}{16}$ in.	Take the square root and convert.

To determine *unit rise for the hip rafter (URih),* for a conventional hip rafter, URih = URic ÷ 17, where

URih = unit rise for the hip rafter
URic = unit rise for the common rafter

For example, if a common rafter has a unit rise of 5 (that is, a 5/12 slope), then the hip rafter will have a 5/17 slope. This means that while the common rafter rises 5 in. for each foot of run, the hip will rise 5 in. over 17 in. of run.

The reason for this can be seen by observing the relationship between the run of the common rafter and the run of the hip rafter. These runs lie in the plane across the top of the plates. They form a right triangle, with the run of the hip acting as the hypotenuse and the setback and run of the common as legs. But the setback in the conventional case equals the run of the common. Therefore, these three lines form a 45° right triangle. Recall that the ratio of the sides in a 45° right triangle is $1 : 1 : \sqrt{2}$. Multiplying the unit run for the common by 1.414 (which is $\sqrt{2}$) will result in the unit run for the hip. Thus,

$$1.414 \times 12 = 16.968$$

Many carpenters round 16.968 to 16.97 or even 17. For purposes of using a framing square to draw the lines for cuts on a hip rafter, the ratio 5/17 is sufficiently accurate. However, when calculating a unit line length (ULL) to be used in determining rafter length, 16.97 should be used. A significant error is incurred if 17 is used instead.

Example

B. A conventional hip roof is to have a 6/12 slope. The run of the common rafter is 12 ft 6 in. Determine (a) the length of the common rafter, (b) the length of the hip rafter, and (c) the slope of the hip rafter.

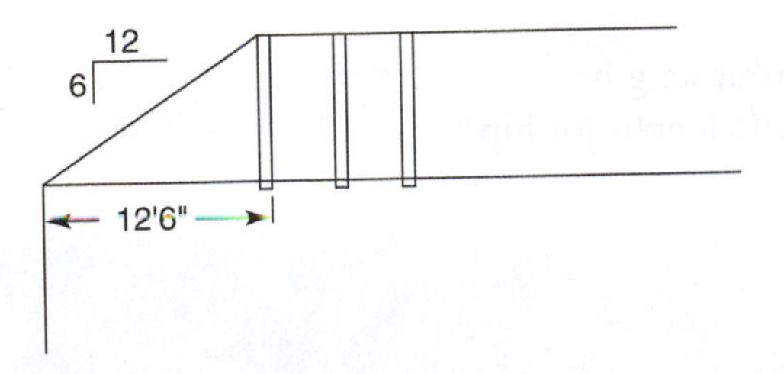

a. For a 6/12 slope, the unit line length is

$$\text{ULL} = \sqrt{6^2 + 12^2} = 13.416$$

The common rafter length, CRL, is

$$\begin{aligned} \text{CRL} &= \text{ULL} \times \text{run} \\ &= 13.416 \times 12.5 \\ \text{CRL} &= 167.7 \text{ in., or } 13 \text{ ft } 11\tfrac{11}{16} \text{ in.} \end{aligned}$$

b. The length of the hip rafter (HRL) is found by solving the right triangle. The hip rafter is the hypotenuse and the common rafter and its run (setback) are the legs.

$\text{HRL} = \sqrt{\text{CRL}^2 + \text{SB}^2}$	Use the HRL formula.
$= \sqrt{13.975^2 + 12.5^2}$	Substitute in the proper values.
$= \sqrt{351.551}$	Square and then add.
$\text{HRL} = 18.750$ ft, or 18 ft 9 in.	Take the square root.

c. The slope of the hip rafter is 6/17. Remember that with a conventional hip rafter the unit rise is the same as that for the common and the unit run becomes a constant of 17 (that is, $\sqrt{2} \times 12$).

19.2 Calculating Hip Rafter Length (HRL)—Method 2

This method works only in the conventional case. First, determine the unit line length ULL for the hip rafter.

$$\text{ULL} = \sqrt{\text{URi}^2 + \text{URu}^2}$$

where

ULL = unit line length for the hip rafter
URi = unit rise for the hip (same as the common)
URu = 16.97

Second,

$$\text{HRL} = \text{ULL} \times \text{CRu}$$

where

HRL = hip rafter length
ULL = unit line length for hip
CRu = run of the common rafter

Example

C. Referring to the preceding example, the unit line length for the hip rafter is

$\text{ULL} = \sqrt{\text{URi}^2 + \text{URu}^2}$	Use the ULL formula.
$= \sqrt{6^2 + 16.97^2}$	Substitute in the proper values.
$= \sqrt{323.98}$	Square and then add.
$\text{ULL} = 18$	Take the square root.

The unit line length is now multiplied by the run of the common, CRu.

$$\text{HRL} = \text{ULL} \times \text{CRu}$$
$$= 18 \times 12.5$$
$$\text{HRL} = 225 \text{ in., or } 18 \text{ ft } 9 \text{ in.}$$

REMINDER

Method 1 is more general and works for any type of hip rafter. Method 2 works only for the conventional hip rafter.

The Framing Square

The necessary information calculated in the example above could have been obtained from the blade of the framing square. In Figure 19–3, locate on the blade of the framing square the second row labeled "Length Hip or Valley per Foot of Run." Now read across that row to the 6, representing the 6-in. mark. This 6 also represents a 6-in. slope. Beneath the 6 in the second row, find 18. This is the *ULL* for a conventional hip rafter on a roof with a 6/12 slope.

Use this information to calculate the length of the hip rafter, as in the preceding example. That is,

$$\text{HRL} = \text{ULL} \times \text{CRu}$$

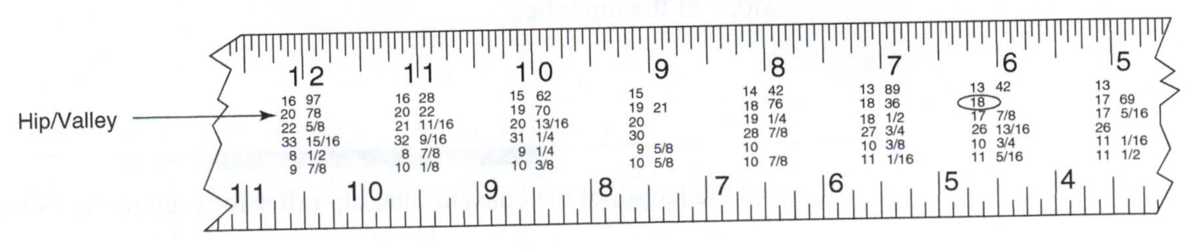

Figure 19–3

In words, to find the length of the hip rafter using the framing square:

1. Find the slope of the roof on the blade of the framing square.
2. Read below the slope in the second row (ULL for hip).
3. Multiply this figure by the CRU.
4. This is the HRL.

End Cuts and Bird's Mouth Cuts

The cuts necessary at the hip rafter ends is a subject best described in the shop with tools in hand. Nevertheless, the following points need to be considered (see Figure 19–4, a top or plan view):

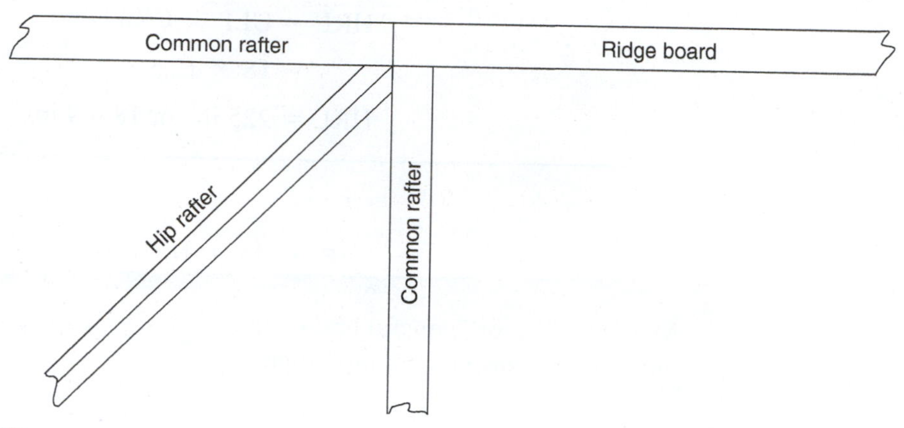

Figure 19–4

1. At the upper end of the hip rafter, the vertical cut will be scribed at the slope of the hip, that is, URi/17.
2. This vertical cut will be backed off by 1.414 (which is $\sqrt{2}$) times one-half the thickness of the ridge board. This is approximately 1 in. for a ridge board made of "2-by" material and approximately $\frac{1}{2}$ in. for "1-by" material.
3. Measurements are made along a line at the midpoint of the top edge of the rafter and along the theoretical working line on the side of the rafter.
4. In a case where the hip rafter meets two adjoining common rafters, a 45° angle cut will be needed.
5. Cuts at the tail of the rafter will depend on the style of the overhang. Nevertheless, a vertical cut will have the same angle as that at the ridge or upper end.
6. The bird's mouth cut will also be made using the same angle dictated by the slope of the hip rafter.

REVIEW EXERCISES

1. Find the length of the conventional hip rafter for each of the following cases.
   a. Common rafter length: 13 ft; run of common: 12 ft
   b. Common rafter length: 17 ft 6 in.; run of common: 14 ft

c. Slope: 6/12; run of common: 13 ft 6 in.
d. Slope: 8/12; run of common: 15 ft 4 in.
e. Slope 10/12; run of common: 9 ft 8 in.

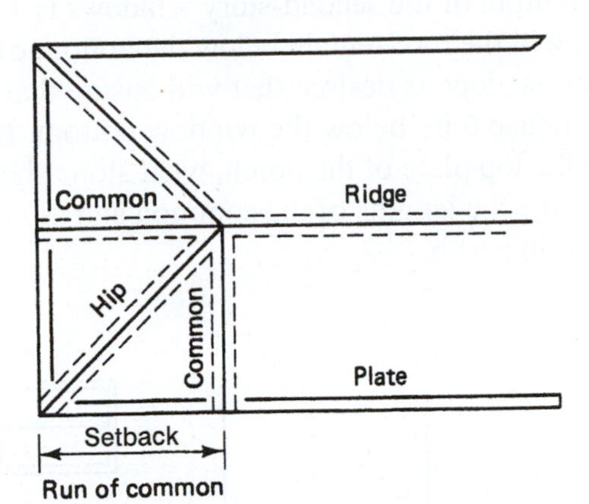

2. For each hip rafter in exercise 1, give the unit rise and unit run.
3. For each of the following hip rafters and ridge board thickness, by how much should the rafter length be reduced to accommodate the ridge board?

Uri/17	Ridge Board Thickness
a. 3/17	$\frac{1}{2}''$
b. 5/17	$1\frac{1}{2}''$
c. 7/17	$1\frac{1}{2}''$

4. Describe a situation in which calculating a hip rafter length by the formula given as

$$\text{HRL} = \sqrt{\text{CRL}^2 + \text{SB}^2}$$

will not give a correct HRL.
5. A house measuring 30 ft by 48 ft will have a hip roof. The slope of the common rafter will be 5/12, and the horizontal projection of the overhang will be 12 inches.
   a. Find the overall length of the common rafter.
   b. What is the distance from the bird's mouth to the ridge for the common rafter?
   c. Find the overall length of the hip rafter.
   d. Find the distance from the bird's mouth to the ridge for the hip rafter.
   e. If the ridge board is $1\frac{1}{2}$-in. thick, by how much will the common rafter be reduced at the ridge?
   f. How much will the hip rafter be reduced at the ridge?

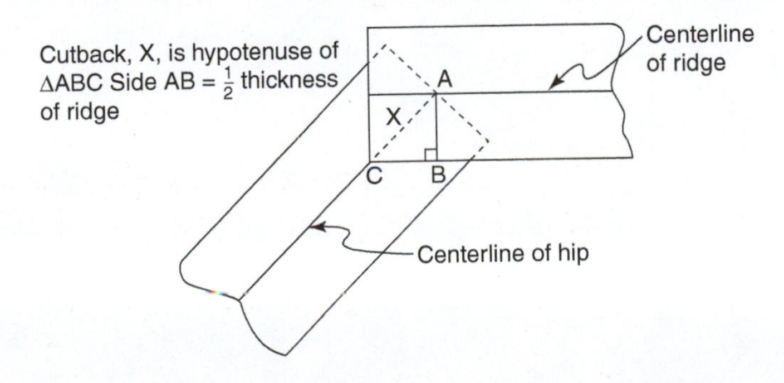

6. The accompanying figure shows a front elevation and plan view of a porch that is to be attached to the front of a house. A 5/12 slope is planned. The hip will be conventional.
   a. If the bottom of the second-story windows is 10 in. above the top plate of the porch, will the roof remain below the bottom of the windows?
   b. Suppose a slope is desired that will ensure that the porch roof meets the face of the house 6 in. below the window bottom. If the window bottom is 22 in. above the top plate of the porch, what slope should be used?
   c. Determine the lengths of the common and hip rafters using the description of the roof in part b.

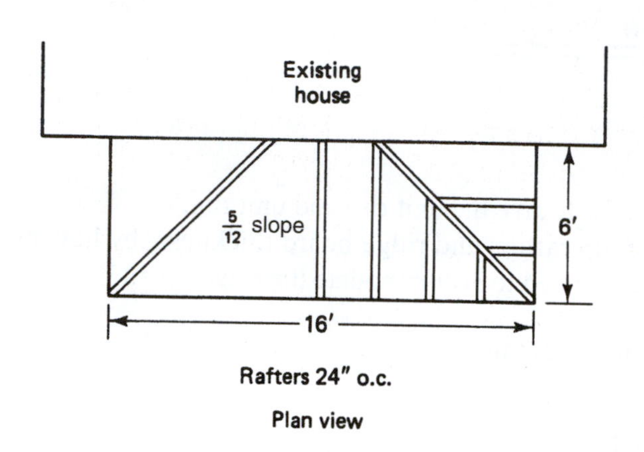

Plan view

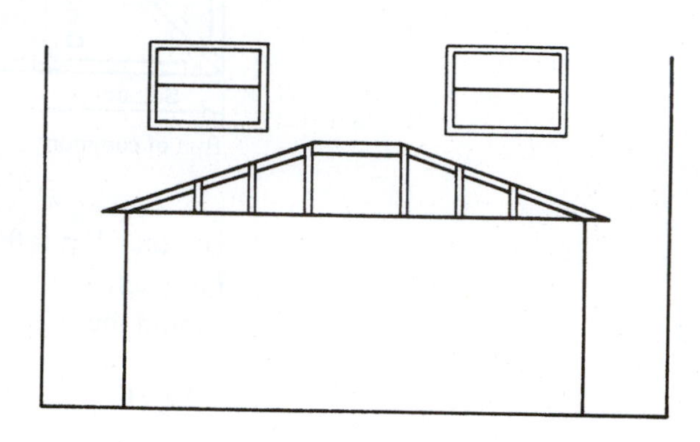
Front elevation

CHAPTER 20

Roofs III: Valley Rafters—The Conventional Case

Did You Know

Temperature difference between inside and outside is an important factor in controlling the atmosphere inside a house. At a depth of about six feet into the earth, the year round temperature is 50°C to 55°C. Consider building in the earth.

OBJECTIVES

Upon completing this chapter, the student will be able to:

1. Calculate the length of a valley rafter.
2. Determine the angles for cuts at the ridge, bird's mouth, and tail of a valley rafter.

Valley rafters are placed where two intersecting roofs meet. There are ways of constructing this intersection so that no valley rafter is needed. Essentially, this is accomplished by completing one roof plane, properly supported, to the point where it has been sheathed. The intersecting roof is then constructed and extended to meet the face of the sheathed section. However, this is not always practical.

20.1 Types of Valley Rafters

Assuming that valley rafters are to be used, the method for determining their length depends on whether the adjacent roofs have the same slope. If this is the case, the valley rafter will have a run that makes a 45° angle with the plate, as did the hip rafter. This is called a *conventional valley rafter.* Nevertheless, some differences from the hip rafter calculations occur. In cases where the adjacent roofs have different slopes, a right-triangle approach is used. This is called an *unconventional valley rafter,* a subject to be discussed in a later chapter.

The conventional valley rafter is most common and also has a more pleasing appearance than the unconventional valley rafter. If the adjacent building sections have the same width, their ridges will have the same elevation and will meet at a point. On the other hand, different widths will cause one ridge to run into the face of the other roof or rise above and terminate at a point above the other ridge. See Figure 20–1.

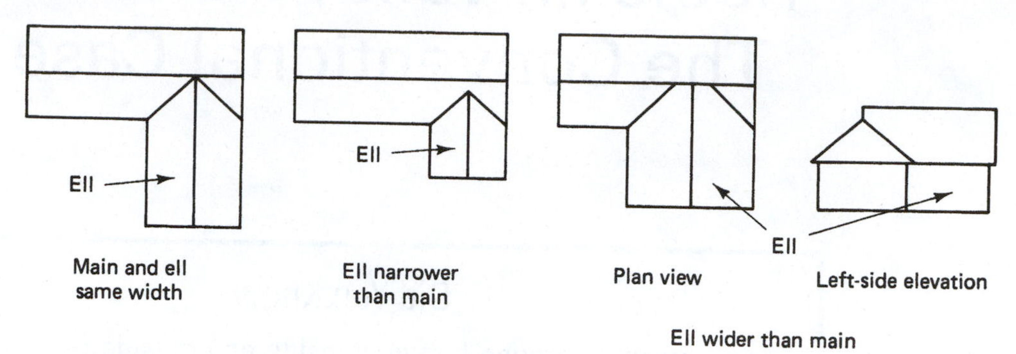

Figure 20–1

A standard way of placing the conventional valley rafters when the ell is narrower than the main building is shown in Figure 20–2. Notice that a long valley rafter extends to the main ridge, while a valley rafter of normal length meets the long valley at the ridge of the ell. The longer valley rafter must extend to meet the main ridge. This will happen if the same approach is used as for a conventional hip rafter.

20.2 Calculating the Length of a Conventional Valley Rafter

To calculate the valley rafter length, VRL, use

$$\text{VRL} = \text{ULL} \times \text{CRu}$$

where

$\text{ULL} = \sqrt{\text{URi}^2 + 16.97^2}$
URi = unit rise of the *common* rafter
CRu = run of the common on the main section

For the shorter valley rafter, use the same formula, except

CRu = run of the common on the ell.

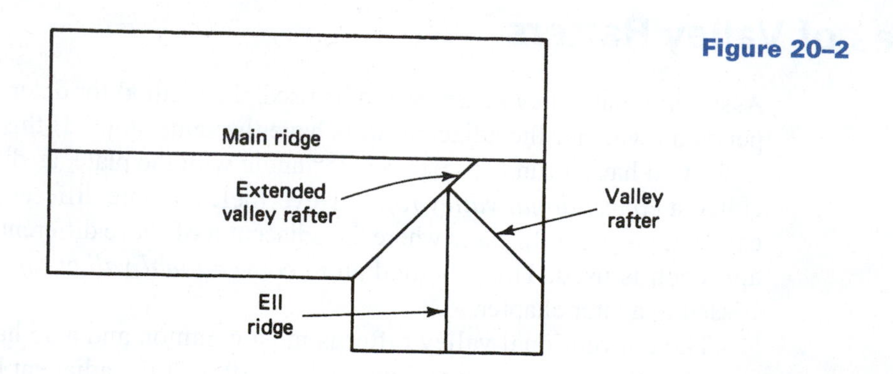

Figure 20–2

Example

A. Referring to Figure 20–2, suppose that the main building is 26 ft wide and the ell is 20 ft wide. Both roofs have a 5/12 slope. Find the lengths of both the long and the short valley rafters.

Determine the unit line length for the valley rafters:

$\text{ULL} = \sqrt{\text{URi}^2 + 16.97^2}$	Use the unit line length formula.
$= \sqrt{5^2 + 16.97^2}$	Substitute in the proper values.
$= \sqrt{312.98}$	Square and then add.
$\text{ULL} = 17.69$	Take the square root.

For the long valley rafter,

$\text{VRL} = \text{ULL} \times \text{CRu}$	Use the valley rafter length formula.
$= 17.69 \times 13$	Substitute in the proper values.
$\text{VRL} = 229.97$ in., or 19 ft 2 in.	Multiply and convert.

For the short valley rafter,

$\text{VRL} = \text{ULL} \times \text{CRu}$	Use the valley rafter length formula.
$= 17.69 \times 10$	Substitute; notice that 10 is the run of the common on the ell.
$\text{VRL} = 176.9$ in., or 14 ft $8\frac{7}{8}$ in.	Multiply and convert.

Note: Look again at the blade of the framing square. Notice that the second row provides information for both the hip and the valley rafters.

20.3 End Cuts

At the ridge, the long valley rafter will have two types of angles making up the plumb cut (see Figure 20–3):

- The plumb cut is made using the 5/17 line.
- But the rafter meets the main ridge at a 45° angle. Therefore, the plumb cut travels at a 45° angle through the lumber.

The length of this rafter must be reduced by one-half the thickness of the ridge board. This reduction is measured parallel to the 45° line. The short rafter meets the long rafter at a 90° angle. Only the 5/17 angle is scribed; however, one-half the thickness of the long rafter is cut from the length of the short valley rafter.

Valley rafter lengths are extended for overhangs by applying the formulas described here, using the horizontal projection of the overhang in place of the run of the common.

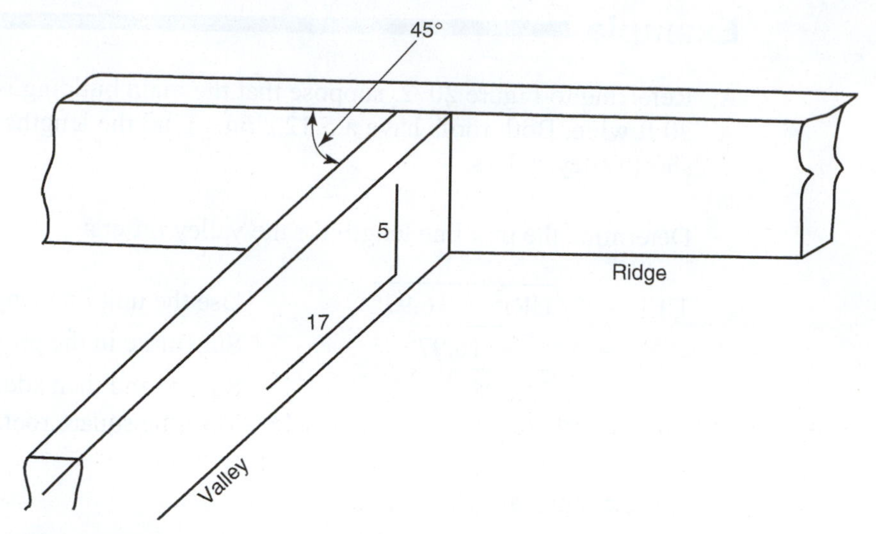

Figure 20–3

REVIEW EXERCISES

1. An ell is attached to the main house in the manner shown in Figure 20–2. Descriptions of several variations on this design follow. In each case determine (1) the common rafter length, (2) the valley rafter length(s), and (3) the unit rise and unit run for the valley rafter.
   a. Main section width: 22 ft; ell width: 22 ft; slope of both sections: 6/12
   b. Main section width: 30 ft; ell width: 24 ft; slope of both sections: 5/12
   c. Main section width: 28 ft; ell width: 16 ft; slope of main section: 8/12; ridge of ell to intersect ridge of main section
2. An ell is attached at a right angle to the main house. Its roof is framed after the main roof has been framed and sheathing applied. The slope and widths of the main and ell are given for several cases. In each case, determine (1) common rafter length, (2) valley rafter length, (3) unit rise and unit run for the valley rafter.
   a. Main: 32 ft wide; ell: 24 ft wide; slope of both: 9/12.
   b. Main: 32 ft wide; ell: 32 ft wide; slope of both: 4/12.
   c. Main: 24 ft wide; ell: 20 ft wide; slope of both: 7/12.
3. An ell is to be wider than the main section of a house. The main roof has an 8/12 slope. If the ridge of the main and ell are to have the same elevation, can the slope of the ell also be 8/12? Explain.
4. An ell-shaped house is framed 26 ft wide. Each segment of the ells is framed 24 ft long at the extreme sides. The diagram below shows a plan view of the roof. The roof carries an overhang with an 8-in. horizontal projection of the overhang. Find the length of the valley rafter (a) from the birds mouth to the ridge and (b) overall.

CHAPTER 21

Roofs IV: Jack Rafters

Did You Know

Forests cover one-third of the planet's land—just under 16 million square miles, or one football field for every person on earth.

OBJECTIVE

Upon completing this chapter, the student will be able to:

1. Calculate the decrease in length of jack rafters.

Jack rafters are segments of common rafters whose length has been reduced by a hip or valley rafter cutting diagonally through their path. Figure 21–1 shows hip and valley jack rafters.

21.1 Calculating Jack Rafter Decrease

Because the spacing of rafters is uniform, the decrease in length of successive jack rafters is constant. Should it be necessary to place a jack rafter in a position that changes the spacing, the amount of decrease will change. However, the method for calculating the amount of decrease does not change.

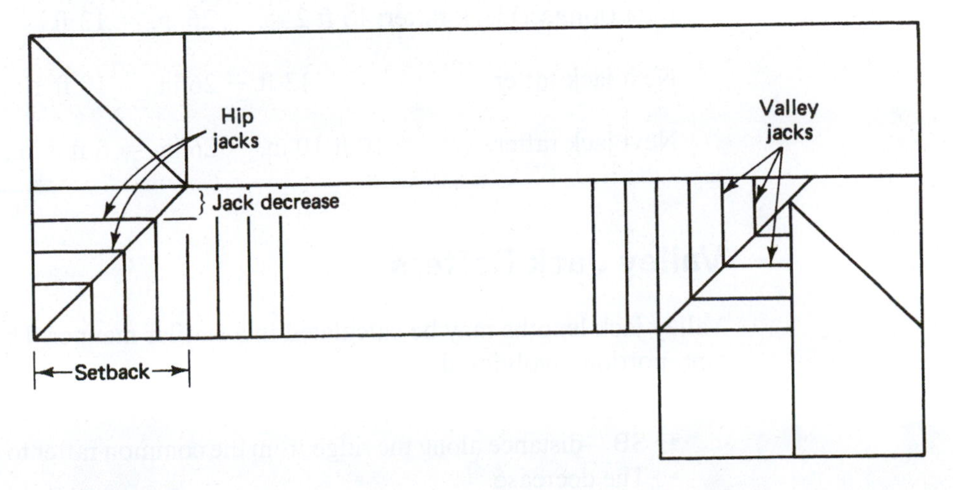

Figure 21–1

1. Determine the following:
   - Common rafter length (CRL)
   - Setback (SB)—distance along the top plate from a corner to the position of the first full-length common rafter
   - Spacing between jack rafters, in inches
2. Set up a proportion as follows:

$$\frac{\text{CRL}}{\text{SB}} = \frac{\text{decrease}}{\text{spacing}}$$

As in the case with any proportion, *attention to the units is important.* The common rafter length and the setback must be expressed in the *same units*—both in feet or both in inches. The spacing will likely be in inches (e.g., 16 in. on center (o.c.) or 24 in. o.c.). With the *spacing in inches,* the *decrease will be in inches.*

Example

A. The full-length common rafter on a hip roof measures 15 ft 2 in. (15.17 ft) and the setback is 14 ft. The rafters are spaced 24 in. o.c. Determine the decrease in length for the jack rafters. Refer to Figure 21–1.

Referring to the formula, CRL = 15.17, SB = 14, and spacing = 24. Solve for the decrease:

$$\frac{\text{CRL}}{\text{SB}} = \frac{\text{decrease}}{\text{spacing}}$$ Use the proportion.

$$\frac{15.17}{14} = \frac{\text{decrease}}{24}$$ Substitute in the proper values.

$14 \times \text{decrease} = 15.17 \times 24$ Cross-multiply.

Decrease = 26 in. Multiply and then divide by 14.

This means that with the last full-length common rafter measuring 15 ft 2 in.,

First (longest) jack rafter: 15 ft 2 in. − 26 in. = 13 ft

Next jack rafter: 13 ft − 26 in. = 10 ft 10 in.

Next jack rafter: 10 ft 10 in. − 26 in. = 8 ft 8 in., and so on

Valley Jack Rafters

Valley jack lengths may be calculated in a similar manner. The problem is solved using a proportion involving the

- CRL
- SB—distance along the ridge from the common rafter to the valley rafter
- The decrease
- The spacing

Example

B. Determine the decrease in length of the valley jacks on the main roof described in Figure 21–2. Assume that the rafters are spaced 16 in. o.c.

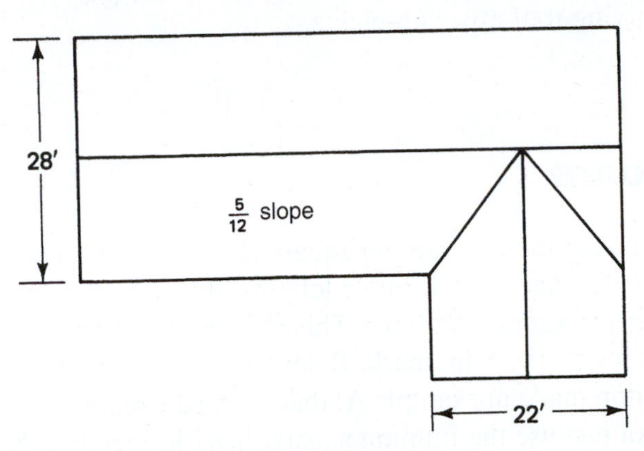

Figure 21–2

CRL is 15.17 ft; the distance along the ridge from the last common rafter to the valley is 11 ft (this replaces SB in the case of the hip roof); the spacing is 16 in. o.c.

$$\frac{\text{CRL}}{\text{SB}} = \frac{\text{decrease}}{\text{spacing}}$$ Use the proportion.

$$\frac{15.17}{11} = \frac{\text{decrease}}{16}$$ Substitute in the proper values.

$$11 \times \text{decrease} = 15.17 \times 16$$ Cross-multiply.

$$\text{decrease} = 22.065 \text{ or } 22\frac{1}{16} \text{ in.}$$ Multiply and then divide by 11, and convert.

Here again, the length of the common rafter is used for reference.

First (longest) valley jack rafter $= \text{CRL} - 22\frac{1}{16}$ in.

$$= 182 - 22\frac{1}{16} \quad (182 = 12 \times 15.167)$$

$$= 159\frac{15}{16} \text{ in.}$$

Next valley rafter $= 159\frac{15}{16} - 22\frac{1}{16}$

$$= 137\frac{7}{8} \text{ in.}$$

and each other jack rafter will be $22\frac{1}{16}$ in. shorter than the one before.

It should be noted that the valley jacks on the ell will have a different decrease than those on the main (see exercise 3).

SUMMARY

The Pythagorean rule may be used to solve most roof problems. This, along with a properly set up proportion, is the most useful mathematical tool for the carpenter in solving roof rafter problems.

21.2 The Framing Square

Once again examine the framing square (Figure 21–3). The third and fourth rows provide the "difference in jack rafter lengths" for both the 16 in. o.c. and the 24 in. o.c. cases. Refer to example A above. This roof has a 5/12 slope. On the blade of the framing square, locate the 5-in. mark. Below this, in the fourth row, read 26. This confirms the calculation made in example A; that is, the decrease in jack rafter length is 26 in.

Why not just use the framing square then? Remember that the framing square data is for conventional roofs only. Given an unconventional hip or valley rafter (where the setback is different than the run of the common rafter), the information is not available on the framing square and the carpenter must resort to the computations just illustrated.

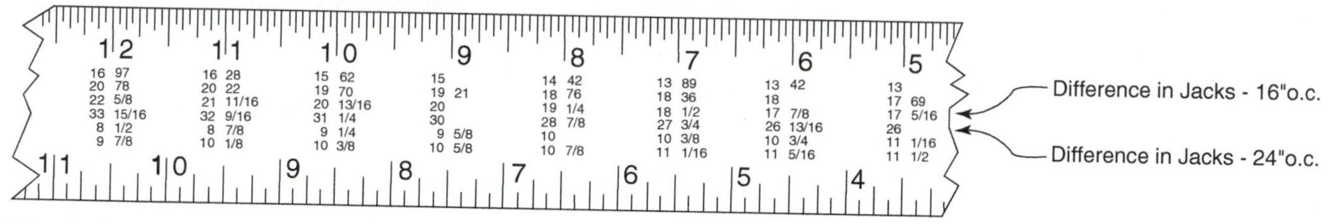

Figure 21–3

REVIEW EXERCISES

1. Previously you found the length of the conventional hip rafter for each of the following cases. Now determine the jack rafter decrease for each. Assume the rafter spacing is 16 in. o.c.

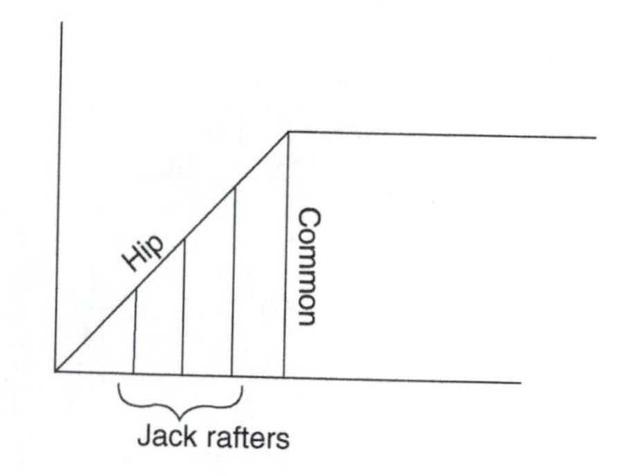

   a. Common rafter length: 13 ft; run of common: 12 ft
   b. Common rafter length: 17 ft 6 in.; run of common: 14 ft
   c. Slope: 6/12; run of common: 13 ft 6 in.
   d. Slope: 8/12; run of common: 15 ft 4 in.
   e. Slope: 10/12; run of common: 9 ft 8 in.

2. Refer back to exercise 6b in Chapter 19. Determine the jack rafter decrease for both the front and side jacks. Note that the spacing is 24 in. o.c.
3. Refer to Figure 21–2. Determine the decrease in length for the valley jack rafters on the ell. Assume 16 in. o.c. spacing.
4. Refer once again to exercise 1 in Chapter 20, in which you previously determined the common and valley rafter lengths, along with the unit rise and run for the valley rafter. Now find the jack rafter decrease for both the main and ell roof sections. The information is given again here.
   a. Main section width: 22 ft; ell width: 22 ft; slope of both sections: 6/12; 24″ o.c.
   b. Main section width: 30 ft; ell width: 24 ft; slope of both sections: 5/12; 16″ o.c.
   c. Main section width: 28 ft; ell width: 16 ft; slope of both sections: 8/12; ridge of ell to intersect ridge of main section; 24″ o.c.
5. In Chapter 20, exercise 2, the following main roofs and ells were described. In each case, determine the difference in length for each successive jack rafter on the ell. Assume rafter spacing is 16″ o.c.
   a. Main: 32 ft wide; ell: 24 ft wide; slope of both: 9/12.
   b. Main: 32 ft wide; ell: 32 ft wide; slope of both: 4/12.
   c. Main: 24 ft wide; ell: 20 ft wide; slope of both: 7/12.
6. Refer to Chapter 20, exercise 5. Find the change in length of the valley jack rafters.

CHAPTER 22

Roofs V: Hip and Valley Rafters—The Unconventional Cases

Did You Know

One-third of the earth's forests are used to produce wood and other forest products. Deforestation, mainly converting forests to agricultural land, decreased by 0.22% from 1990 to 2000, and by 0.18%.

OBJECTIVES

Upon completing this chapter, the student will be able to:

1. Calculate the length of an unconventional valley rafter.
2. Calculate the length of an unconventional hip rafter.

Unconventional hip and valley rafters are created by several situations. The common ingredient that is present in all unconventional hips and valleys is a different slope on the adjacent roof sections.

22.1 Calculating Unconventional Rafters

One situation in which an unconventional valley rafter can occur is when an ell has a different width than the main building, while the ridge on the ell meets the ridge of the main.

Example

A. A 22-ft-wide ell is attached to a 28-ft-wide building having a 5/12 slope so that the roof ridges meet at the same elevation (see Figure 22–1). The total rise of the main roof is

$$\begin{aligned} \text{TRi} &= \text{URi} \times \text{run} \\ &= 5 \times 14 \\ &= 70 \text{ in., or } 5.83 \text{ ft} \end{aligned}$$

Figure 22–1

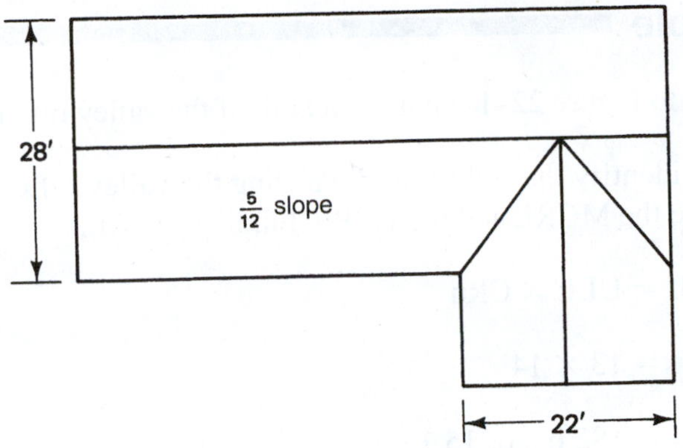

28′
$\frac{5}{12}$ slope
22′

The ell must have the same total rise but with a total run of 11 ft. Proportioning this to determine the unit rise for the ell,

$$\frac{\text{URi}}{\text{URu}} = \frac{\text{TRi}}{\text{TRu}}$$ Set up the proportion.

$$\frac{\text{URi}}{12} = \frac{5.83}{11}$$ Substitute in the proper values.

$$\text{URi} = 6.36, \text{ or } 6\frac{3}{8} \text{ in.}$$ Cross-multiply and divide by 11.

Thus, while the main section has a 5/12 slope, the ell has a 6.36/12 slope.

Unconventional Valley Rafter Length

To calculate an unconventional valley rafter length (VRL),

1. Identify a right triangle in which the valley rafter serves as the hypotenuse.
2. Use the Pythagorean rule with the legs as main common rafter length (MCRL) and run of the ell common rafter (ECRu).

See Figure 22–2.

$$\text{VRL}^2 = \text{MCRL}^2 + \text{ECRu}^2$$

Figure 22–2

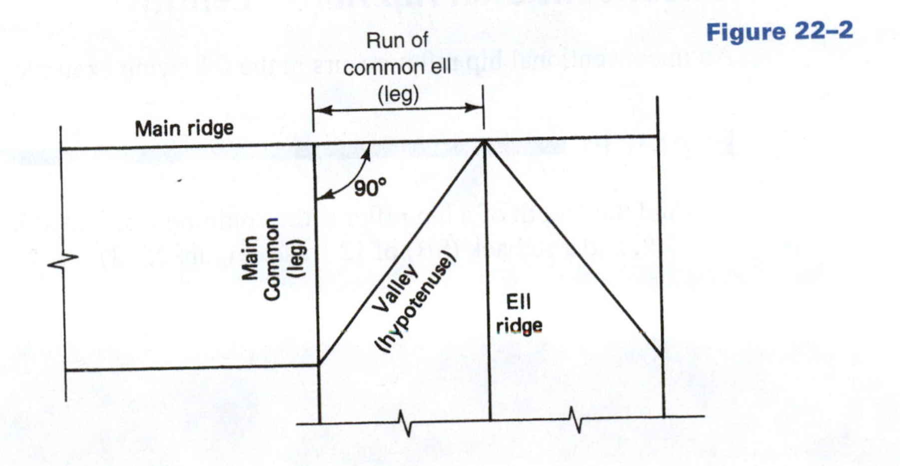

Example

B. Refer to Figure 22–1. Find the length of the valley rafter.

First, identify the right triangle having the valley rafter as its hypotenuse. The legs will be the MCRL and the ECRu. Finding MCRL,

$$\text{MCRL} = \text{ULL} \times \text{CRu}$$

$$= 13 \times 14$$

$$\text{MCRL} = 182 \text{ in., or } 15.17 \text{ ft}$$

Note that the run of the common on the ell is 11 ft.

Now use the Pythagorean rule:

$$\text{VRL}^2 = \text{MCRL}^2 + \text{ECRu}^2$$

$$\text{VRL}^2 = 15.17^2 + 11^2$$ Substitute in the proper values.

$$= 230.13 + 121$$ Square each term.

$$= 351.13$$ Add.

$$\text{VRL} = \sqrt{351.13}$$ Take the square root.

$$\text{VRL} = 18.738 \text{ ft, or } 18 \text{ ft } 8\frac{7}{8} \text{ in.}$$ Convert.

The importance of correctly identifying the parts of a right triangle in calculating any rafter length is worthy of mention. In general, more than one right triangle can be identified and used. In the preceding example, the legs could have been the common rafter for the ell and the run of the common for the main. (Still another right triangle exists. Can you find it?) In all cases, it is the rafter itself (valley, hip, or common) that serves as the hypotenuse of the right triangle.

Unconventional Hip Rafter Length

An unconventional hip rafter occurs in the following example.

Example

C. Find the length of a hip rafter if the common rafter is to have a 9/12 slope, a run of 15 ft, and a setback (SB) of 12 ft (see Figure 22–3).

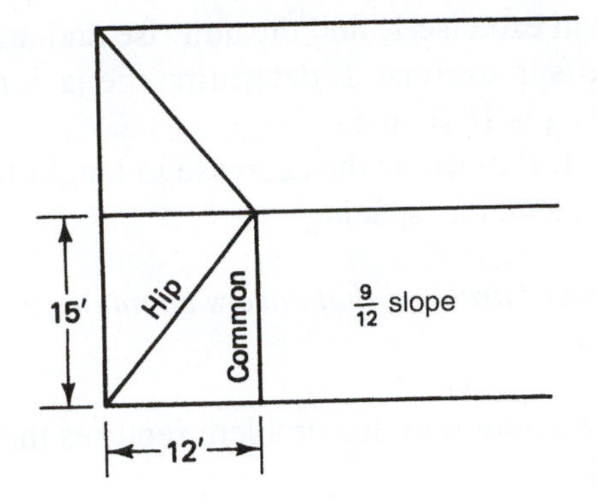

Figure 22–3

First, the length of the common rafter is found. For a 9/12 slope, the unit line length (ULL) is 15. With a run of 15 ft, the common rafter length is

$HRL = \sqrt{CRL^2 + 12^2}$	Use the Pythagorean rule.
$HRL = \sqrt{18.75^2 + 12^2}$	Substitute in the proper values.
$HRL = \sqrt{495.56}$	Take the square root.
HRL = 22.261 ft, or 22 ft $3\frac{1}{8}$ in.	Convert.

The student is reminded that the framing square does not contain tables for an unconventional case. The mathematics described here will yield the proper results.

REVIEW EXERCISES

1. Find the length of the unconventional hip rafter for each of the following cases.

Slope	Run of Common	Setback
a. 7/12	13 ft	10 ft
b. 9/12	14 ft 8 in.	12 ft
c. 12/12	10 ft 9 in.	8 ft
d. 6/12	4 ft	3 ft 6 in.

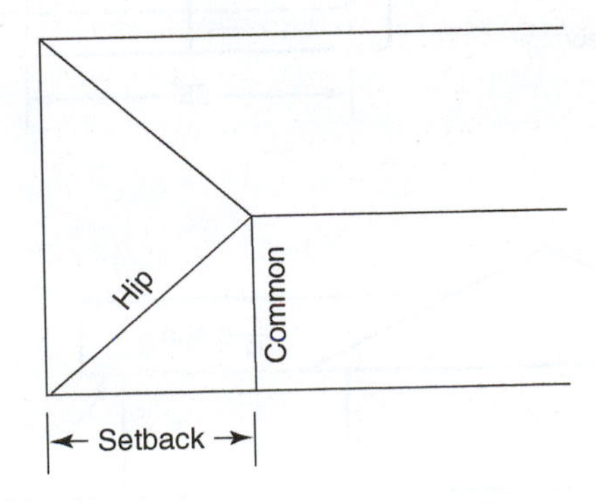

2. For each hip rafter in exercise 1, find the unit rise and unit run.
3. For each of the roofs in exercise 1, determine the jack rafter decrease. Assume that the rafter spacing is 16 in. o.c.
4. Refer to Figure 22–1. Determine the decrease in length for the valley jack rafters on the ell. Assume 16 in. o.c. spacing.

The following is a summary problem that covers common, hip, valley, and jack rafter calculations.

5. *Note:* The complete solution to this problem requires the use of some trigonometry. (See Appendix.)

   The figure below shows the plan view and right-side elevation of a roof. Dimensions, slopes, and overhang on the main are shown. Determine the following:

   a. Common rafter lengths on both the main and ell. Include distances overall, from bird's mouth to ridge, and from bird's mouth to tail.
   b. Overall short valley rafter length.
   c. Overall long valley length.
   d. Overall hip rafter length.
   e. Jack rafter decrease on (1) front of main, (2) end of main, (3) main at long valley, (4) ell at long valley.

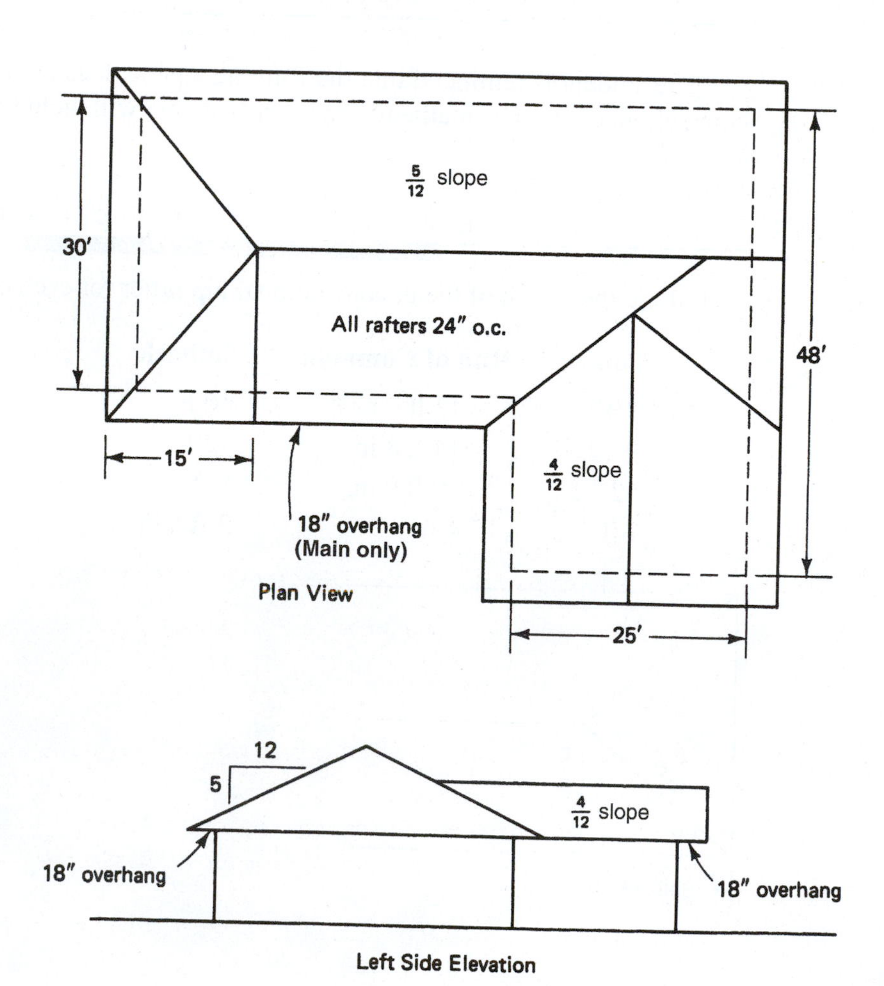

CHAPTER 23

Stairs

Did You Know

Approximately one-sixth of the wood delivered to a home construction site is wasted.

OBJECTIVES

Upon completing this chapter, the student will be able to calculate for a conventional set of stairs, as well as for a set of stairs having restricted landing areas and for stairs with intermediate landings,

1. The number of risers and runs.
2. The unit rise and the unit run.
3. The stairwell opening.
4. The total run.

Several factors must be considered when laying out a set of stairs. Included among these are the following:

- Total distance between floors or landings
- Head clearance
- Number of steps
- Height of each step
- Width of each stair tread
- Total horizontal distance covered by the set of stairs

In some situations, certain restrictions may be imposed, such as

- The length of the stairwell opening
- Space available for a landing area

Space available for the run and landing area may necessitate an intermediate landing with a 90° or 180° turn. Spiral or other configurations may be dictated by other constraints. A thorough understanding of the mathematics related to stairs will enable the carpenter to work within any reasonable restrictions and construct a set of stairs that satisfies most design standards.

Listed below are some of the design standards for stairs:

Minimum finished width: 36 in.
Minimum landing area: 3 ft × 3 ft

Minimum head clearance: 6 ft 8 in.
Maximum unit rise: $8\frac{1}{4}$ in.
Minimum tread width: 9 in.

A cardinal rule to be observed in designing a set of stairs is that all steps have the same height and tread width.

(In some older houses a "surprise step" was constructed part way up the set of stairs. The surprise was that the rise was different for that one step. This was designed to catch an intruder who might trip while climbing the steps in the dark.)

For more details regarding specifications, consult the standards in effect in your locale.

The following symbols will be used in explaining the mathematics related to a set of stairs. All dimensions are in inches (see Figure 23–1).

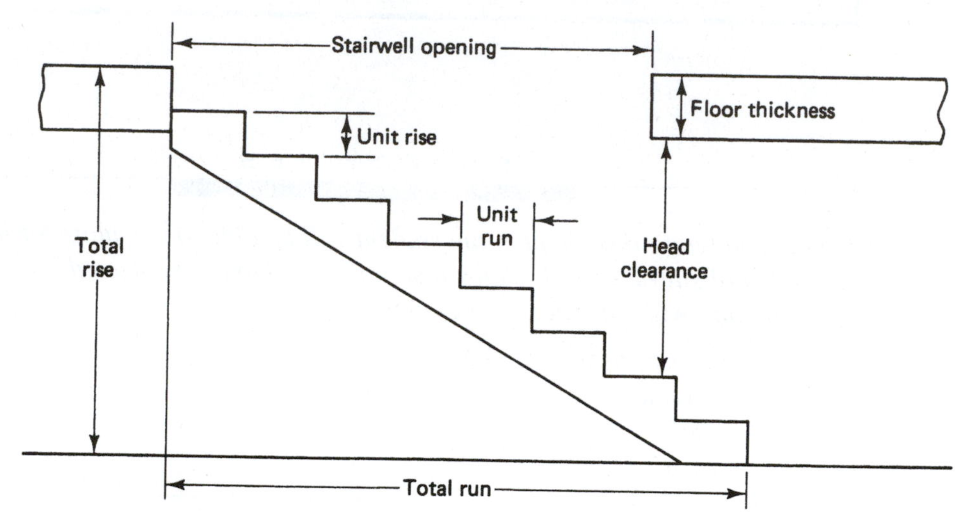

Figure 23–1

- *Total rise, TRi:* distance between finished floors
- *Total run, TRu:* horizontal distance spanned by the set of stairs
- *Unit rise, URi:* height of each step
- *Unit run, URu:* width of each step (this is the framed width and does not refer to the finished tread width)
- *Number of risers, NRi:* number of steps
- *Number of runs, NRu:* usually one less than the number of risers (i.e., NRu = NRi − 1)
- *Head clearance, HC:* vertical distance between the bottom of the floor joist (or stairwell header) and the stair tread below
- *Stairwell opening, SO:* length of the stairwell opening
- *Floor thickness, FT:* vertical distance taken up by the floor joists and all layers of flooring

The calculations involved in a set of stairs may be approached in three stages:

1. Determine the NRi and URi.
2. Determine the stairwell opening, SO.
3. Determine the total run, TRu.

23.1 Determining Number of Risers (NRi) and Unit Rise (URi)

1. Find the number of risers, NRi.

$$\text{NRi} = \text{TRi} \div 8\text{; round up to next whole number}$$

TRi is expressed in inches.

COMMENT

8 is used because it is close to the desired URi of 8 in. We could divide by 9 or 7 if conditions impose a greater or smaller URi on the set of stairs. Of course, the greater the divisor, the smaller the quotient. In this case, that means a greater NRi will result in a smaller (and possibly more desirable) unit rise.

2. Find the URi.

$$\text{URi} = \text{TRi} \div \text{NRi}$$

Example

A. Find the NRi and URi for a set of stairs having a total rise of 98 in.

$$\text{NRi} = 98 \div 8$$
$$\text{NRi} = 12.25, \text{ which we round up to } 13$$
$$\text{URi} = 98 \div 13$$
$$= 7.538 \text{ in.}$$

Therefore, there will be 13 risers of 7.538 or $7\frac{9}{16}$ in. each.

Example

B. The vertical distance from a cellar floor to the bottom of the 2 × 10 floor joist is 7 ft 8 in. The first floor is covered with $\frac{1}{2}$-in. plywood and $\frac{5}{8}$-in. particleboard. Determine the NRi and the URi for a set of stairs from the first floor to the cellar (see Figure 23–2).

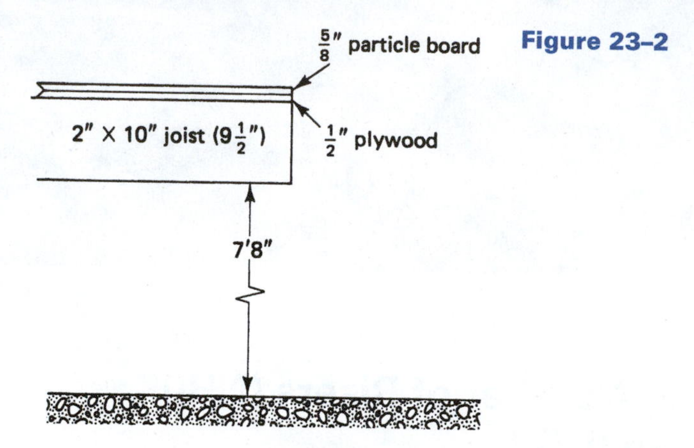

Figure 23–2

The total rise consists of the floor thickness and the distance from the cellar floor to the bottom of the floor joist.

$$\text{TRi} = 7 \text{ ft } 8 \text{ in.} + 9\frac{1}{2} \text{ in.} + \frac{1}{2} \text{ in.} + \frac{5}{8} \text{ in.}$$
$$= 92 + 9.5 + 0.5 + 0.625, \text{ converting to inches}$$
$$= 102.625, \text{ or } 102\frac{5}{8} \text{ in.}$$

The NRi becomes

$$\text{NRi} = 102.625 \div 8$$
$$\text{NRi} = 12.83, \text{ which we round up to } 13$$

Now, the URi becomes

$$\text{URi} = 102.625 \div 13$$
$$\text{URi} = 7.894, \text{ or } 7\frac{7}{8} \text{ in.}$$

Thus, there will be 13 risers of $7\frac{7}{8}$ in. each.

23.2 Determining the Stairwell Opening (SO)

When the run of the stairs is not restricted by space limitations, the SO is determined by the horizontal distance needed to descend far enough to give HC. The FT must be considered, as well as the desired HC. The general procedure is

1. Determine the NRi needed to descend a known distance.
2. Translate this into a number of runs.
3. Then multiply by the unit run.

Example

A. Determine the stairwell opening necessary to provide a HC of 6 ft 8 in. The FT is $10\frac{5}{8}$ in. and the URi has been determined to be $7\frac{5}{8}$ in.

The vertical distance that must be descended is 6 ft 8 in. plus the floor thickness, $10\frac{5}{8}$ in.—a total of 90.625 in. Divide this by the URi, $7\frac{5}{8}$, to get the NRi needed to descend 90.625 in.

$$90.625 \div 7.625 = 11.89, \text{ which we round up to } 12.$$

This means that on the twelfth riser, starting from the top, one will have descended slightly more than $90\frac{5}{8}$ in. and, considering the floor thickness, will have 6 ft 8 in. HC.

The steps taken here may be summarized as

$$\frac{\text{HC} + \text{FT}}{\text{URi}}$$

In words:

1. Add the HC and FT.
2. Divide by the URi.

In a conventional set of stairs, the NRu is one less than the NRi (see Figure 23–3). That is,

$$\text{NRu} = \text{NRi} - 1$$

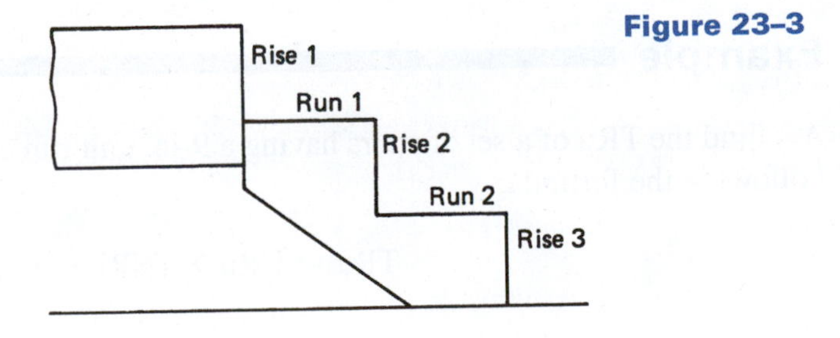

Figure 23–3

Since it takes 12 risers to descend the desired distance, the NRu will be 11.

$$\begin{aligned}\text{NRu} &= 12 - 1 \\ &= 11\end{aligned}$$

Multiplying the number of runs by the unit run will result in the stairwell opening. That is,

$$SO = NRu \times URu$$

Assuming a unit run of 9 in., the stairwell opening in this example becomes

$$SO = 11 \times 9 = 99 \text{ in.}$$

The steps needed to calculate the stairwell opening are summarized in the following formula:

$$SO = URu \times \left[\frac{HC + FT}{URi} - 1\right]$$

Or more simply,

$$SO = NRu \times URu$$

In words, to calculate the stairwell opening,

1. Add the FT and HC.
2. Divide by the URi.
3. Subtract 1.
4. Multiply by the NRu.

23.3 Determining the Total Run (TRu)

The total run (TRu) of a set of stairs is found by multiplying the URu by the NRu. That is,

$$TRu = URu \times NRu$$

The TRu can also be expressed in terms of the NRi.

$$TRu = URu \times (NRi - 1)$$

Example

A. Find the TRu of a set of stairs having a 9-in. unit run and 13 risers. Following the formula,

$$\begin{aligned} TRu &= URu \times (NRi - 1) \\ &= 9 \times (13 - 1) \\ &= 9 \times 12 \\ &= 108 \text{ in.} \end{aligned}$$

SUMMARY

The procedure and formulas related to laying out a set of stairs are as follows:

1. Determine the NRi and the URi.

$$\text{NRi} = \text{TRi} \div 8\text{, rounding up to the next whole number}$$
$$\text{URi} = \text{TRi} \div \text{NRi}$$

2. Determine the stairwell opening.

$$\text{SO} = 9 \times \left[\frac{\text{HC} + \text{FT}}{\text{URi}} - 1\right]$$

3. Determine the TRu.

$$\text{TRu} = \text{URu} \times (\text{NRi} - 1) \quad \text{or} \quad \text{TRu} = \text{URu} \times \text{NRu}$$

Example

B. The distance from a cellar floor to the bottom of the floor joists above is 6 ft 11 in. The floor joists are 2 × 8s (actual width is $7\frac{3}{8}$ in.). The subflooring is $\frac{3}{4}$-in. boards and a layer of $\frac{5}{8}$-in. hardwood makes up the finished floor. Find the NRi, the URi, the SO, and the TRu for a set of stairs going from the cellar to the first floor. Assume a HC of 6 ft 6 in. and a URu of $8\frac{1}{2}$ in.

1. Find the total rise.

$$\text{TRi} = 6\text{ ft }11\text{ in.} + 7\frac{3}{8}\text{ in.} + \frac{3}{4}\text{ in.} + \frac{5}{8}\text{ in.}$$
$$= 83\text{ in.} + 7.375\text{ in.} + 0.75\text{ in.} + 0.625\text{ in.}$$
$$\text{TRi} = 91.75\text{ in.}$$

2. Find the NRi.

$$\text{NRi} = 91.75 \div 8$$
$$\text{NRi} = 11.47\text{, which we round to 12}$$

3. Find the URi.

$$\text{URi} = 91.75 \div 12$$
$$= 7.646\text{, or } 7\frac{5}{8}\text{ in.}$$

4. Find the SO.

$$SO = 8.5 \times \left[\frac{78 + 8.75}{7.646} - 1\right]$$
$$= 8.5 \times (11.3 - 1)$$
$$= 8.5 \times (12 - 1)$$
$$= 8.5 \times 11$$
$$SO = 93.5 \text{ in.}$$

5. Find the TRu.

$$TRu = 8.5 \times (12 - 1)$$
$$= 93.5 \text{ in.}$$

23.4 Space Limitations

Consider next a situation in which the distance available for the total run is limited. This can occur due to the location of a wall that crowds the lower landing.

Example

A. The head of a set of stairs is to be located a horizontal distance 10 ft 6 in. from the vertical plane of a lower-level wall, toward which the stairs will run (see Figure 23–4). The total rise is $92\frac{1}{4}$ in., $10\frac{1}{4}$ in. of which is floor thickness. Design a set of stairs to fit in this space, determining the number of risers, the unit rise, and the unit run.

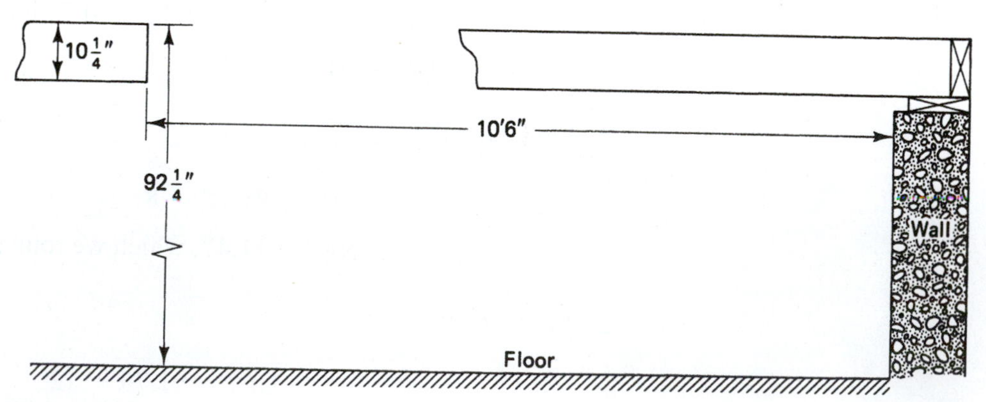

Figure 23–4

The NRi is

$$NRi = 92.25 \div 8$$
$$NRi = 11.53, \text{ which we round to } 12$$

Determine the URi.

$$\text{URi} = 92.25 \div 12$$

$$\text{URi} = 7.688, \text{ or } 7\frac{11}{16} \text{ in.}$$

There are 12 risers, so there will be 11 runs. If a distance of 36 in. is reserved for the lower landing, 90 in. remain for the total run. This results in a unit run of

$$\text{URu} = 90 \div 11$$

$$\text{URu} = 8.182, \text{ or } 8\frac{3}{16} \text{ in.}$$

Because $8\frac{3}{16}$ in. is less than the desired 9-in. minimum unit run, adjustments should be considered. A decrease in the length of the landing area should result in a corresponding increase in the total run and therefore a greater unit run. To accommodate a 9-in. unit run, 99 in. of total run would be needed ($9 \times 11 = 99$). This would leave only 27 in. for the length of the landing area. Some trial-and-error calculations might be made in this case with the objective of establishing a reasonable compromise between the landing area and the unit run.

In any situation involving limited space, the amount of HC should be explored, as well as the topics discussed previously. Limited space for the stairwell opening may result in restricted HC. The stair design is incomplete without considering the HC.

In cases where a trade-off between landing area and unit run is not feasible, an intermediate landing with a 90° or 180° turn could be considered. Of course, space must be available for the turn under consideration. In general, a landing replaces a step. Therefore, the NRi does not change. On the other hand, the total run is extended by the landing. These layouts do not conserve space; they merely reorient the stairs to take advantage of available space in another direction (see Figure 23–5).

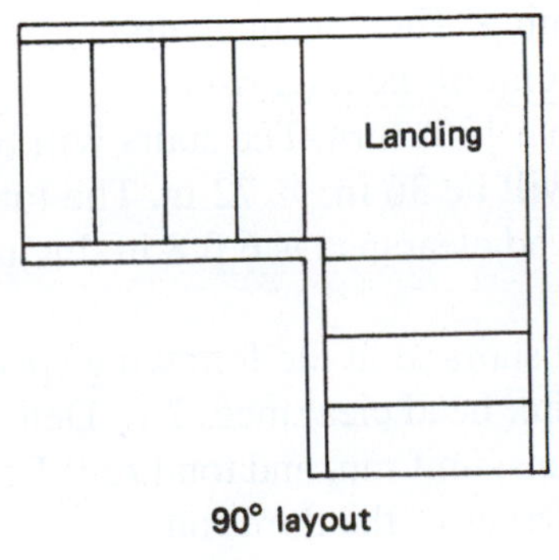

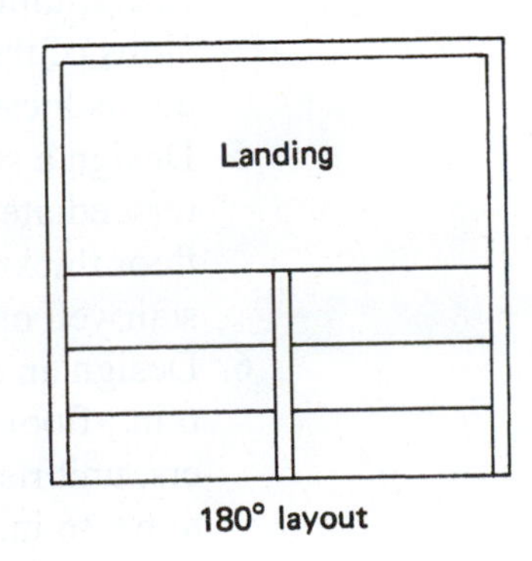

Figure 23–5

REVIEW EXERCISES

1. A set of stairs is to be constructed from the first to the second story. The vertical distance from the first floor to the bottom of the second floor joist is 7 ft $6\frac{1}{2}$ in. The floor thickness is $10\frac{3}{8}$ in. Assume a unit run of $9\frac{1}{2}$ in. Determine the following dimensions.
   a. Total rise
   b. Number of risers
   c. Unit rise
   d. Number of runs
   e. Total run
   f. Stairwell opening
   g. Head clearance
2. Design a set of stairs for each of the following situations. Dimensions given are actual size. Assume a unit run of 9 in. Determine total run, number of risers, unit rise, number of runs, stairwell opening, and head clearance.

Total Rise	Floor Joist	Subflooring	Finish Flooring
a. $94\frac{3}{4}$ in.	$7\frac{3}{8}$ in.	$\frac{3}{4}$ in.	$\frac{5}{8}$ in.
b. $93\frac{5}{8}$ in.	$9\frac{1}{2}$ in.	$\frac{1}{2}$ in.	$\frac{5}{8}$ in.
c. $95\frac{11}{16}$ in.	$5\frac{1}{2}$ in.	$\frac{7}{16}$ in.	$\frac{3}{4}$ in.
d. $92\frac{3}{4}$ in.	$11\frac{1}{2}$ in.	$\frac{1}{2}$ in.	$\frac{3}{4}$ in.
e. $107\frac{1}{8}$ in.	$9\frac{1}{4}$ in.	$\frac{1}{2}$ in.	$\frac{3}{8}$ in.

3. The horizontal distance available for a set of stairs is 11 ft 8 in. The landing area at the bottom of the stairs is to be 36 in. The total rise is 8 ft $5\frac{1}{4}$ in. and the floor thickness is $9\frac{7}{8}$ in. The head clearance required is 6 ft 8 in. Determine the number of risers, the unit rise, the number of runs, the unit run, the total run, and the stairwell opening for a set of stairs to fit these conditions.
4. Design an L-shaped set of stairs to fit the following conditions: total rise, 8 ft 2 in.; floor thickness, $10\frac{1}{4}$ in.; head clearance, 6 ft 6 in. Determine the number of risers, unit rise, number of runs, unit run, and total run. Also, give the dimensions of the stairwell opening. The intermediate landing is to be 36 in. × 36 in. and is located at the midpoint of the total rise.
5. Design a set of stairs with a 180° turn. The stairs will be 36 in. wide and the intermediate landing area will be 36 in. × 72 in. The total rise is 7 ft $11\frac{1}{2}$ in. The floor thickness is $6\frac{1}{4}$ in. Head clearance is 6 ft 8 in. Include the dimensions of the stairwell opening.
6. Design an L-shaped set of stairs to fit the following specifications: total rise, 8 ft 6 in.; floor thickness, 10 in.; head clearance, 7 ft. Determine the number of risers, unit rise, number of runs, unit run, and total run. The intermediate landing is to be 36 in. × 36 in. and serves as the third run.

CHAPTER 24

Framing and Covering Gable Ends; Exterior Trim

Did You Know

Some methods of construction that can save wood: Wall studs, floor joists, and roof framing can be spaced 24 inches on center, instead of 16 inches on center; overall sizes planned in 2-feet increments; side panels made of rigid insulation instead of plywood; solid wood rafters and floor beams replaced with trusses made of small, interconnected triangular frames.

OBJECTIVES

Upon completing this chapter, the student will be able to:

1. Determine the materials needed to frame gable ends.
2. Determine the materials needed to sheath gable ends.
3. Determine the materials needed for siding gable ends.
4. Determine the materials needed for exterior trim.

The term *gable ends* as used in this chapter will mean that portion of a house between the slope of the roof and the line extending across the top plates. It is not meant to be restricted to the triangular ends of a gable-framed roof. The discussion here will apply to all types of roofs—gable, gambrel, and so on.

24.1 Framing Gable Ends

The length of jack studs needed to frame a gable end depends on the total rise of the roof. The total rise may be determined by multiplying the unit rise by the total run. That is,

$$\text{TRi} = \text{URi} \times \text{run}$$

where TRi is the total rise, in inches, URi is the unit rise, in inches per foot, and run is measured in feet.

Example

A. A roof has a 6/12 slope. The common rafter has a run of 15 ft. Find the total rise. Using the established formula,

$$\begin{aligned} \text{TRi} &= \text{URi} \times \text{run} \\ &= 6 \times 15 \\ \text{TRi} &= 90 \text{ in., or } 7 \text{ ft } 6 \text{ in.} \end{aligned}$$

Assume that the roof described in example A is a gable roof on a ranch-type house (see Figure 24–1). The gable ends form isosceles triangles (two equal sides). An effective way to determine the number and length of studs needed to frame the gable ends is to picture both triangular ends together. These two triangles taken together form the equivalent of a rectangle having a width equal to the width of the house and a height equal to the total rise.

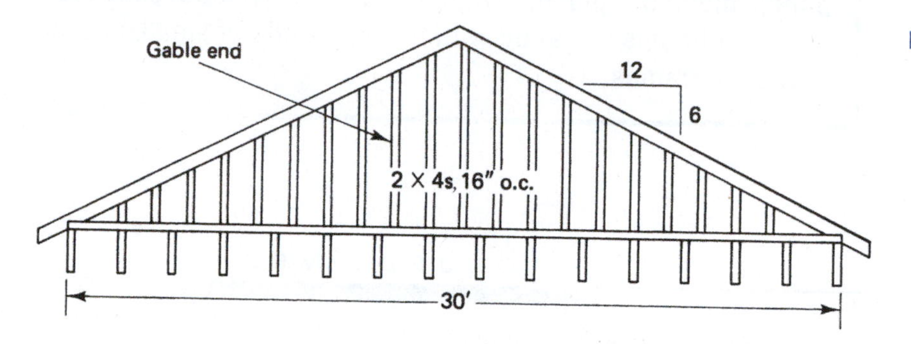

Figure 24–1

The length of materials for framing the gable ends is determined by the total rise. In this case, 8-ft studs will suffice. The maximum length is 7 ft 6 in. Consider the succession of studs starting at the center of the gable end and moving outward. The amount cut from each stud will be usable at the shorter side of the triangle.

As to the number of studs, the spacing has to be considered. Assuming 16 in. o.c. in the case described in example A, the number of studs needed, NS, to cover a distance, *L*, is

$$\begin{aligned} \text{NS} &= \frac{3}{4} \times \text{L} \\ &= \frac{3}{4} \times 30 \\ \text{NS} &= 22\frac{1}{2} \end{aligned}$$

(Recall from earlier discussions that the factor $\frac{3}{4}$ stems from the reduced fraction $\frac{12}{16}$, multiplication by 12 converts the length from feet to inches, and dividing by 16 counts the number of 16-in. spaces.) Round this to 23 studs. There is no need for a "starter" as was the case in wall framing. 23 studs 8 ft long will be sufficient to frame both gable ends.

Adjustments to the basic count should be considered where needed. Adjustments would be needed in cases where openings are planned for windows, louvered vents, and so on.

SUMMARY

To estimate materials needed to frame gable ends,

1. Find the total rise, TRi = URi × Ru.
2. Find the number of studs, NS = $\frac{3}{4}$ × L.
3. For two gable ends, use this number of studs of length given by TRi. For one gable end, use half this number of studs of length given by TRi. For other style ends, use these same principles as they apply to the shape of the end.

24.2 Sheathing and Siding Gable Ends

The amount of sheathing for a gable end is determined by the area to be covered. If plywood or a similar product is to be used, the area, in square feet, of the gable end is divided by 32, the number of square feet in a sheet of plywood. That is,

$$\text{number of sheets} = \text{area} \div 32$$

In the case of a ranch house, the gable ends should be considered together, as was suggested for the stud count. Instead of two separate triangles, the two ends form the equivalent of one rectangle.

Example

A. Determine the number of sheets of $\frac{1}{2}$-in. CDX plywood needed to cover the gable ends of a ranch house 30 ft wide with a 6/12 roof slope.

The building width is 30 ft and the total rise is 7 ft 6 in. The gable ends considered together form the equivalent of a rectangle measuring 30 ft × 7.5 ft. This is a total area of

$$\begin{aligned} \text{area} &= 30 \times 7.5 \\ &= 225 \text{ ft}^2 \end{aligned}$$

Dividing by 32, the number of square feet in a sheet of plywood,

$$\begin{aligned} \text{number of sheets} &= 225 \div 32 \\ &= 7.03125\text{, or 7 sheets} \end{aligned}$$

If only one gable end is being considered, the procedures described above can be used with slight modification. Instead of two triangles that form the equivalent of a rectangle, a single triangle exists. The number of studs is simply one-half the number previously determined. The area to be covered is also one-half that of the rectangle. Indeed, the formula for finding the area of a triangle is

$$\text{area} = \frac{1}{2}BH$$

where B is the base of the triangle and H is its height (altitude).

If boards are to be used for closing-in the gable ends, some adjustment to the quantity will be needed to account for the waste factor. Generally, adding 10% to 15% to the area will provide sufficient boarding material.

SUMMARY

To estimate materials needed for sheathing and siding a gable end,

1. Determine what type of plane figure the ends represent (that is, triangle, rectangle, trapezoid, and so on.)
2. Use the appropriate area formula to calculate its area.
3. Divide by 32 and get the number of 4′ × 8′ sheets of sheathing, if that is the material to be used.

The procedures for determining materials needed to frame and cover the ends created by other types of roofs are similar to those described here. The estimator should take care to note sections that appear elsewhere in the structure that can be combined in some geometric form in order to simplify calculations. The effort will be rewarded in faster, more efficient, and more accurate estimating.

Example

B. The building in Figure 24–2 shows a gambrel roof. The 15/12 slope has a run of 6 ft; the $\frac{4}{12}$ slope has a run of 12 ft; the building width is 36 ft. Determine the following materials for both ends of the building: (a) number of studs and their lengths to frame the ends; (b) number of sheets of $\frac{1}{2}$-in. CDX sheathing to close in the ends; (c) number of squares of cedar shingles to side the ends.

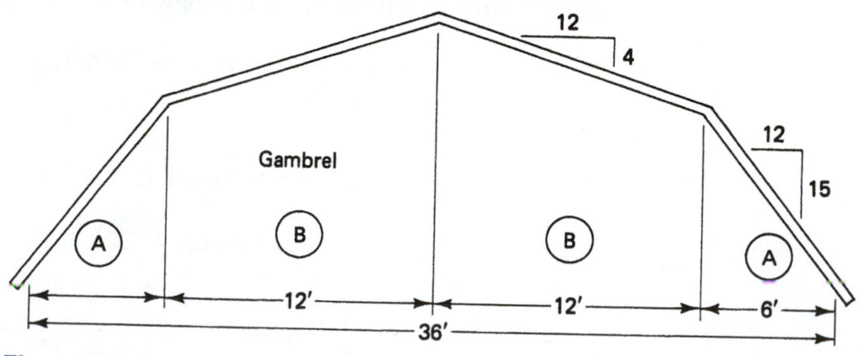

Figure 24–2

a. Assume that the stud spacing is 16 in. on center (o.c.). There are four triangular sections like that labeled A, two on each end. Two of these sections may be considered together to form a rectangle. The width of this rectangle is 6 ft. Its height is the total rise of that section of roof, namely,

$$\begin{aligned} \text{TRi} &= \text{URi} \times \text{run} \\ &= 15 \times 6 \\ \text{TRi} &= 90 \text{ in., or } 7.5 \text{ ft} \end{aligned}$$

Material length of 8 ft is suggested.

As to the number of studs,

$$\text{number of studs} = \frac{3}{4} \times 6$$
$$= 4.5, \text{ rounded to } 5$$

Five 8-ft studs will frame two of these triangular sections. A total of 10 studs will be needed for both ends.

Section B appears four times. The maximum length needed here is found by determining the additional rise for this section and adding it to the 7.5-ft rise of section A.

$$\text{TRi} = \text{URi} \times \text{run}$$
$$= 4 \times 12$$
$$\text{TRi} = 48 \text{ in., or } 4 \text{ ft}$$

Added to the 7.5-ft rise of section A, the maximum length is

$$7.5 + 4 = 11.5 \text{ ft}$$

The question as to what material lengths to order should not consume much time. It is possible to determine accurately a number of 8-, 10-, and 12-ft pieces needed for this job. It is probably just as efficient simply to order based on the maximum length needed. In this example, no more than 4 ft of material would be cut from any particular piece. This would likely find a use as blocking or spacing material.

The number of 12-ft studs needed to cover the 24-ft width of section B is

$$\text{number of studs} = \frac{3}{4} \times 24$$
$$= 18$$

Including the other end, a total of 36 studs each 12 ft long are needed.

b. The total area of the ends is needed to determine the number of sheets of sheathing for closing in. For section A, two sections together form a rectangle and there are two such rectangles. Find the area of one rectangle and double it.

$$\text{area} = 2 \times 6 \times 7.5$$
$$= 90 \text{ ft}^2$$

Section B is a trapezoid having parallel sides of 7.5 and 11.5 ft and a height of 12 ft. (Remember that the term *height* as used in the formula for area of a trapezoid refers to the perpendicular distance between the parallel sides.) For one section B, the area is

$$A = \frac{B + b}{2} \times H$$

where B and b represent the parallel sides and H the height.

$$A = \frac{7.5 + 11.5}{2} \times 12$$
$$= \frac{19}{2} \times 12$$
$$= 9.5 \times 12$$
$$A = 114 \text{ ft}^2$$

There are four of these sections, so the total area is

$$4 \times 114 = 456 \text{ ft}^2$$

Adding this to the total for section A, the combined area of the ends is

$$90 + 456 = 546 \text{ ft}^2$$

The number of sheets of sheathing required is

$$\text{number of sheets} = 546 \div 32$$
$$= 17.0625, \text{ or } 17 \text{ sheets}$$

c. The number of squares of cedar shingles is directly determined from the total area to be covered. A "square" of shingles is 100 ft^2. Dividing the total area by 100 results in

$$546 \div 100 = 5.46 \text{ squares}$$

Since most cedar shingles are bundled with three bundles to the square (that is, each bundle covers $33\frac{1}{3}$ ft^2), $5\frac{2}{3}$ squares, or 17 bundles, will suffice.

24.3 Determining Materials for Exterior Trim

Quantities of material needed for exterior trim are difficult to generalize about due to the various types of siding and methods of application. Aluminum and vinyl siding are specialty items. Jobbers specializing in the sale and application of these types of siding have knowledge of techniques for estimating materials needed for corners, door and window starter strips, soffit and fascia, and so on. Some other types of siding will require the use of corner boards and/or trim boards immediately beneath the soffit or bordering the roof line on a gable end.

Estimating materials for trim boards is a simple task involving the summation of wall heights at corners and possibly building perimeters and distances along roof edges. The efficient estimator will make use of previously calculated rafter lengths where appropriate.

REVIEW EXERCISES

1. Listed below are the building width and roof slope for several typical ranch-type houses. For both gable ends of each house, determine (1) the number of studs and their lengths and (2) the number of sheets of sheathing needed to close in the ends.

Building Width	Slope
a. 24 ft	5/12
b. 26 ft 8 in.	6/12
c. 25 ft 2 in.	8/12
d. 32 ft 6 in.	7/12
e. 38 ft	5/12

2. The ends of several buildings are shown with dimensions and roof slopes noted. In each case, for both ends, determine (1) the number of studs and their lengths, (2) the number of sheets of sheathing needed for closing in, (3) as an alternative, the number of board feet of boards needed to close in, and (4) the number of bundles of cedar shingles needed to cover the ends (3 bundles = 1 square).

a.

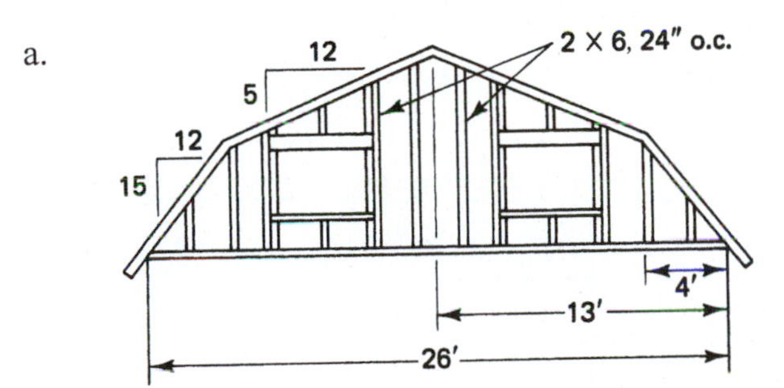

b.

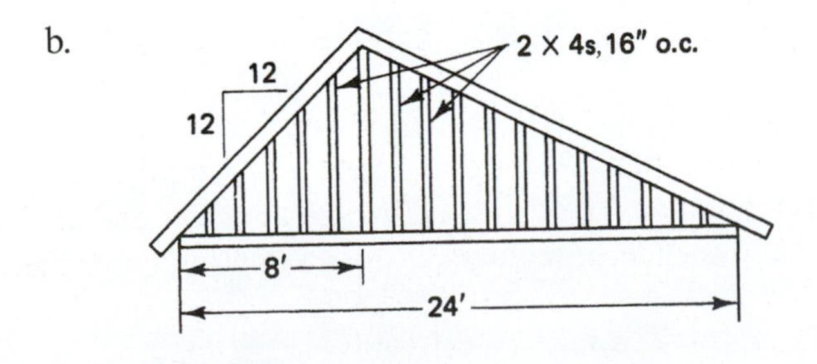

c.

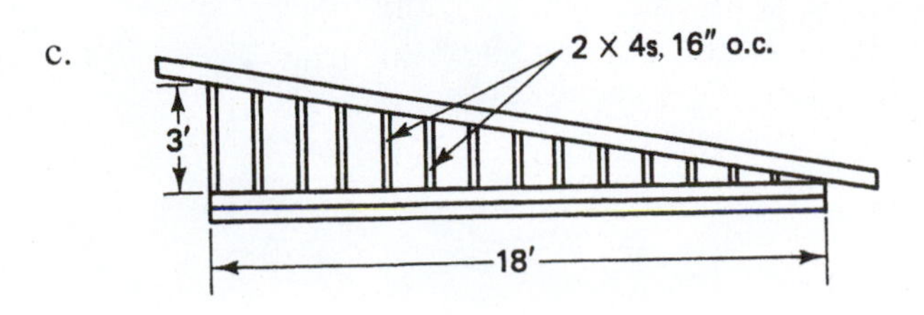

d.

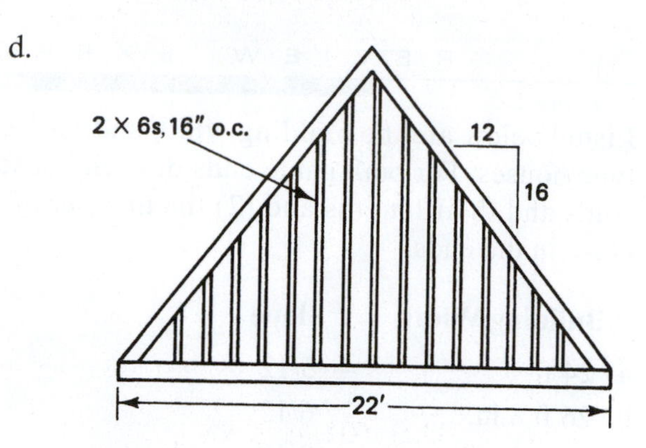

3. For each building in exercise 2 above, estimate the number of linear feet of fascia boards needed on both roof ends.

CHAPTER 25

Wall and Roof Covering

Did You Know

Laminated veneer lumber (LVL) is engineered lumber with high design values for bending, stiffness, shear strength. It has a high strength-to-weight ratio, about 50 times that of solid wood products. It also resists shrinkage, warping, splitting, checking, and moisture damage. LVL generally can be obtained in thicknesses from $1\frac{3}{4}$ in. to $3\frac{1}{2}$ in., depths from $9\frac{1}{4}$ in. to 18 in., and lengths from 24 ft to 60 ft.

OBJECTIVES

Upon completing this chapter, the student will be able to:

1. Determine the amount of materials needed to close in the exterior walls.
2. Determine the amount of various types of siding needed to cover the exterior of a house.
3. Determine the amount of drywall needed for the interior of a house.
4. Determine the amount of materials needed to sheath a roof.
5. Determine the amount of roofing material (felt paper, drip edge, shingles) needed for any roofing job.

When an estimator begins the process of determining materials needed to construct a house, one of the measurements of the structure that will serve repeatedly is its perimeter. The value of

$$\text{perimeter} \times \text{exterior wall height}$$

equals an area that can be used to determine

- exterior sheathing
- siding
- insulation
- drywall

The perimeter is also used to estimate materials needed to frame the exterior wall. It is easy to understand why these measurements are so important to the estimator and should be readily available.

25.1 Estimating Exterior Sheathing

Sheathing for the exterior walls may be determined using either

$$\text{(a) perimeter} \div 4$$
$$\text{or (b) (perimeter} \times \text{wall height)} \div 32$$

In (a) we are assuming that 4 × 8 sheathing will be laid with the long side vertical. We are determining the number of 4-ft-wide sheets required. In (b) we are determining the exterior wall area (and dividing by 32, the number of square feet in one 4 × 8 sheet of sheathing). In case (a), however, it usually requires more than 8 ft to cover from approximately 1 in. below the foundation top to the top plate. Therefore, a few sheets of sheathing must be added to cover the lower strip.

Example

A. A single-story conventional ranch house, 28 ft × 44 ft, is to have a ceiling height of 7 ft 6 in. How many sheets of $\frac{1}{2}$-in. CDX plywood are needed to cover the exterior walls to at least 1 in. below the foundation top? Assume that the floor joists are 2 × 10s.

Method (a) is used here. The student may wish to use method (b) for comparison. The perimeter of the house is

$P = 2(28 + 44)$ Find the perimeter.

$= 2(72)$

$P = 144$ ft

The number of sheets, NS, is equal to

$\text{NS} = 144 \div 4$ Divide by 4 to find NS.

$= 36$

The distance from 1 in. below the foundation top to the top plate is found by totaling the following: 7 ft 6 in. wall height, $\frac{1}{2}$-in. subflooring, $9\frac{1}{2}$-in. floor joists, $1\frac{1}{2}$-in. sill plate; and adding 1 in. below the foundation top. This gives a total of

$$7\text{ ft }6\text{ in.} + \frac{1}{2}\text{ in.} + 9\frac{1}{2}\text{ in.} + 1\frac{1}{2}\text{ in.} + 1\text{ in.} = 8\text{ ft }6\frac{1}{2}\text{ in.}$$

Assuming that the sheathing is applied to be level with the top plate, an additional $6\frac{1}{2}$ in. would remain to be covered around the lower perimeter. This is an area of

$\frac{6.5}{12} \times 144 = 78\text{ ft}^2$ Convert $6\frac{1}{2}$ in. to feet and multiply by the perimeter to find the area in square feet.

and would require extra sheathing amounting to

$78 \div 32 = 2.4$, or 3 sheets — Divide by the number of square feet in a 4×8 sheet of sheathing.

Thus, a total of 39 sheets will be needed to cover the exterior walls from 1 in. below the foundation top to the top plate.

In example A, if the interior wall height was to be 8 ft instead of 7 ft 6 in., the border perimeter would have required in excess of 4 sheets; a total of 41 sheets would therefore be necessary. In general, the estimator is on safe ground by adding from 3 to 6 sheets of sheathing to the basic count.

Materials needed to cover the gable ends (and other types of roof ends) can be determined by using the area of the ends to be covered (see Chapter 24). The area is determined and divided by 32, the number of square feet in a sheet of 4×8 sheathing.

Example

B. The 28 ft × 44 ft ranch house described in example A has a 5/12 slope with 16-in. overhangs front and rear. Determine the number of sheets of sheathing needed to cover the gable ends.

The two ends together form the equivalent of a rectangle. The height of the rectangle is the total rise of the roof, as measured from the rafter tail. Including the overhang, the run is

$$16 \text{ in.} + 14 \text{ ft} = 15.33 \text{ ft}$$

The total rise, TRi, is

$$\begin{aligned} \text{TRi} &= \text{URi} \times \text{run} \\ &= 5 \times 15.33 \\ &= 76.65 \text{ in., or } 6.39 \text{ ft} \end{aligned}$$

Applying the concept that the two triangular ends form an equivalent rectangle, the total area, *A,* for both gable ends is

$$\begin{aligned} A &= 6.39 \times 2(14 \text{ ft} + 16 \text{ in.}) \\ &= 6.39 \times 30.67 \\ &= 195.98 \text{ ft}^2 \end{aligned}$$

The number of sheets of sheathing needed to cover this area is

$$\begin{aligned} \text{NS} &= 195.98 \div 32 \\ &= 6.12 \text{ sheets} \end{aligned}$$

It is likely that 6 sheets will be sufficient.

In those situations where boards are to be used, a waste factor should be applied—usually, 10% to 15% is sufficient. The areas of gable and other roof ends are the basis for determining the amount of sheathing needed to cover those sections.

Window and door areas are not usually subtracted when determining exterior sheathing. In the construction process, as wall segments are framed and covered on the deck for erection in place, it is more efficient simply to cover the entire wall with sheathing and cut the openings later. The material covering the openings is not usually wasted. It can be used for gusset plates if trusses are constructed on site, or it can be used as shelving in the house.

25.2 Siding

In determining the amount of siding required to cover the exterior of a house, different methods are used for different types of siding. Applicators of aluminum and vinyl siding use special techniques to determine amounts of siding, corners, J-channel, and so on. Wood clapboards and wood shingles or shakes may be estimated using exterior area as a basis for calculation. The many varieties of plywood siding may be estimated using the same techniques as described here for sheathing. Vertical board siding and board-and-batten siding may be estimated using exterior area.

In general, all types of siding will be estimated based on exterior area.

The estimator should make judgments as to waste factors, but there is ordinarily little waste in most types of siding. The estimator should also make judgments as to when area for windows and doors are subtracted and when they are ignored. When estimating aluminum or vinyl siding, the openings will be considered. Special materials are used to surround these openings. In cases where plywood-type siding or other solid sheets of siding are to be applied, openings should be ignored in order to provide continuity to the visual effects of the building. On the other hand, the requirements for clapboard and wood shingle siding may be reduced by the total area of openings.

When the total area of openings is to be subtracted from the total exterior area, it is not necessary to be precise. Exterior doors usually occupy an area measuring approximately 3 ft by $6\frac{2}{3}$ ft, for a total of $3 \times 6\frac{2}{3}$ or 20 ft^2. Sliding doors measure approximately $6\frac{2}{3}$ ft high and commonly are 6 or 8 ft wide—areas of 40 or 53 ft^2. A picture window occupies 36 to 40 ft^2. Most other windows can be estimated based on their glass size or rough-opening size available from order books. In older, existing houses a typical window might be measured and its area used to represent the others.

In summary, areas to be subtracted for typical openings may be estimated based on typical openings in the house. Some average opening areas are

Door	20 ft^2
Sliding door	40 to 53 ft^2
Picture window	40 ft^2
Large window	18 ft^2
Average window	12 ft^2

Example

A. The 28 ft × 44 ft ranch house described in the first two examples is to be covered with cedar shingles. The house has two entry doors, a picture window, and nine average-size windows. Determine the quantity of cedar shingles needed for the siding.

In the previous examples it had been determined that the sheathing required 36 sheets for the walls, 3 sheets for the lower trimming, and 6 sheets for the gable ends—a total of 45 sheets. The efficient estimator will use either this or a previously calculated exterior area total as the basis for determining siding needs. 45 sheets of sheathing is equivalent to a total area, A, of

$A = 45 \times 32$ Multiply the number of sheets by the square feet in one sheet

$= 1440 \text{ ft}^2$

The total area of all openings is

Doors: $2 \times 20 = 40 \text{ ft}^2$
Picture window: 40 ft^2
Windows: $9 \times 12 = 108 \text{ ft}^2$
Total opening area: 188 ft^2

Determining the net area:

Gross area: 1440 ft^2
Less area of openings: 188 ft^2
Siding area: 1252 ft^2

Cedar shingles ordinarily come 4 bundles per square. The estimator might round this to 1300 ft^2 and order 13 squares, or 52 bundles for the job.

25.3 Drywall

Drywall applicators generally prefer to get an accurate count of the number of sheets needed for a job after the interior framing has been completed. However, reliable estimates can be made based on areas, perimeter, and/or total wall lengths.

Drywall for ceilings: For horizontal ceilings, use floor area.

This will provide the number of square feet to be covered. The applicator will decide the size of the sheets to be used. If it is to be 4×12, the area will be divided by 48, for example, to get the number of sheets.

Drywall for exterior walls: Use perimeter × wall height.

Again, the area will be divided by the number of square feet in a sheet of drywall, depending on the size of the sheets to be used.

Drywall for interior walls: Use* twice *the number of linear feet of walls, determined in a manner similar to that described in Chapter 15.

The number of linear feet of walls is doubled since both sides of the walls are covered. This number of linear feet may be multiplied by the wall height to get area, and then divided by the area of the sheet of drywall that will be used.

Thus a total area to be covered by drywall is determined by totaling the ceiling area, the exterior wall area, and the interior wall area. Drywall applicators will sometimes estimate a cost based on square footage and sometimes based on a number of sheets. Cost figures are available from any applicator.

Example

A. Determine the number of 4 × 12 sheets of drywall required to cover the ceilings and walls of the 28 ft × 44 ft house described in the previous examples. Assume that the interior wall height is 8 ft, and assume that the interior partitions total 154 linear feet. Ignore all openings.

Determine the total area to be covered, then divide by 48 square feet per sheet.

$$\text{Ceilings: } 28 \times 144 = 1232 \text{ ft}^2$$
$$\text{Exterior walls: } 144 \times 8 = 1152 \text{ ft}^2$$
$$\text{Interior walls: } 2 \times 154 \times 8 = 2464 \text{ ft}^2$$
$$\text{Total area} = 4848 \text{ ft}^2$$

The number of 4 × 12 sheets needed to cover 4848 ft^2 is

$$4848 \div 48 = 101 \text{ sheets}$$

If a total cost is desired, assuming a rate of \$0.35 per square foot this job would cost

$$4848 \times 0.35 = \$1696.80$$

25.4 Roof Covering

The last two objectives of this section, dealing with sheathing and shingling a roof, may be considered together. Both rely on roof area.

A study of the various roof planes in the chapters on roofs will show that most of the common roof sections are made up of one or more plane figures. Included among these are rectangles, parallelograms, triangles, and trapezoids. Other roofs may be divided into sections that form one or more of these figures. In certain unusual configurations, roof area may be approximated by one of the common plane figures. Study the roof shown in Figure 25–1 and identify sections that represent each of the plane figures mentioned here: rectangle, triangle, parallelogram, trapezoid.

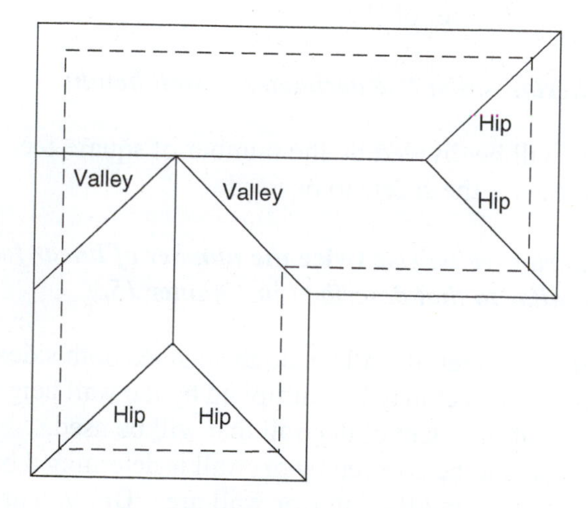

Figure 25–1

Figure 25–2 shows each of the plane figures just mentioned, as well as the area formula for that figure. An examination of Figure 25–2 will reveal that in each formula the value h is the length of a common rafter. This should suggest to the efficient estimator another value of which to make special note as calculations are being made.

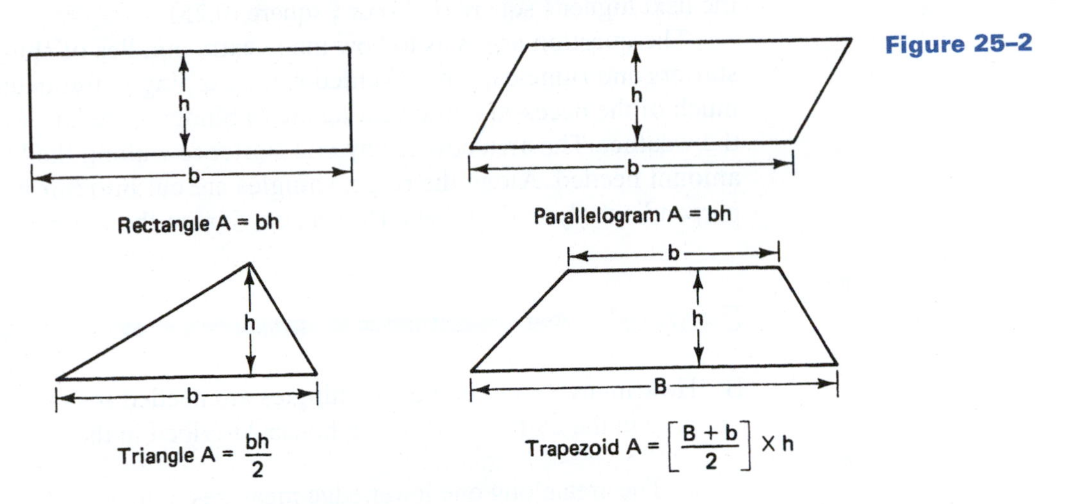

Figure 25–2

Example

A. The 28 ft × 44 ft ranch house in earlier examples has a 5/12 slope, a 16-in. overhang front and rear, and a 12-in. overhang on the ends. Determine (a) the number of sheets of $\frac{5}{8}$-in. CDX sheathing and (b) the number of squares of shingles needed to cover the roof.

First, determine the overall length of the common rafter, including overhang.

$$\text{RL} = \text{ULL} \times \text{run}$$
$$= 13 \times 15.33$$

A 5/12 slope has ULL = 13; run is 14 ft for the building plus 16 in. or 1.33 ft for the overhang—a total of 15.33 ft.

$$\text{RL} = 199.29 \text{ in., or } 16.61 \text{ ft}$$

The plane of the roof is a rectangle with a length, b, equal to the building length plus 1 ft of overhang on each end—a total of 46 ft. Using the area formula gives us

$A = bh$	Use the formula for the area of a rectangle.
$= 46 \times 16.61$	Substitute in the proper values.
$A = 764.06 \text{ ft}^2$	Multiply.

This figure is doubled to account for both sides of the roof.

$$\text{total area} = 1528.12 \text{ ft}^2$$

a. Determine the amount of $\frac{5}{8}$-in. CDX sheathing, with each sheet covering 32 ft^2.

$$1528.12 \div 32 = 47.75 \text{ sheets}$$

b. A “square” of shingles covers 100 ft^2. The number needed is found by dividing the roof area by 100 or, more simply, moving the decimal point in the area two places left. Thus,

$$1528.12 \div 100 = 15.2812 \text{ squares}$$

Most shingles are packaged 3 bundles to the square. Certain extra-thick shingles are packaged 4 bundles to the square. This means that shingle orders may be rounded to the next higher $\frac{1}{3}$ square (0.33) or $\frac{1}{4}$ square (0.25).

The question arises as to how many extra bundles of shingles should be added for starters and ridge capping. (Vented ridge capping is frequently used today, providing much of the necessary attic venting area.) Shingles are laid with approximately 5 in. to the weather. The area covered by one starter row along the lower edge determines the amount needed. Along the ridge, shingles are cut into thirds (12 in. each) and placed perpendicularly to the others. Here again, 5 in. to the weather is the norm.

Example

B. How many extra bundles of shingles are needed for (a) starters and (b) ridge capping in the 28 ft × 44 ft ranch house described in the previous example?

a. The area along one lower edge measures 5 in. × 46 ft. There are two edges, for a total of

$$A = \frac{5}{12} \times 46 \times 2 \qquad \text{5 in. is converted to feet}$$

$$= 38.33 \text{ ft}^2$$

This represents slightly more than 1 bundle.

b. The length of the ridge will be covered at the rate of 5 in. per piece, with each piece 12 in. or 1 ft wide. Essentially, this amounts to covering an area equal to the length of the ridge and 1 ft wide. Therefore, the length of the ridge also equals the number of square feet to be covered. In this case, 46 ft of ridge length means 46 ft^2 or 0.46 square of shingles. All three quantities of shingles are added to the order. This roof called for 15.28 squares for the main roof, 0.38 square for the starters, and 0.46 square for capping—a total of

$$15.28 + 0.38 + 0.46 = 16.12 \text{ squares}$$

Assuming that the shingles are packaged 3 bundles per square, 49 bundles would be ordered.

In summary, the concept of area plays a most important role in estimating quantities of materials needed for covering walls, ceilings, and roofs of buildings. Perimeter multiplied by a wall height or divided by a width of material to be applied to walls is an easy way to determine material quantities. The efficient estimator will maintain ready access to those building dimensions that determine the necessary perimeters and areas.

We have made no special study of insulation as was done in an earlier version of this text. Determination of heat loss calculations, *R*-values, and so on, usually are the responsibilities of a heating/air-conditioning contractor. Even the installation of insulation has become a specialty job. However, the estimator can easily determine quantities of insulating materials needed by using wall and ceiling areas.

REVIEW EXERCISE

1. Sufficiently detailed plans for several houses are shown. For each house, determine the following materials requirements.
   1. Number of sheets of sheathing needed to close in the exterior walls.
   2. Number of square feet of siding (a) disregarding openings and (b) deducting for openings. (Assume windows to be average size. Sliding doors are 8 ft wide.)
   3. Number of sheets of 4 × 8 sheetrock for walls and ceilings.
   4. Number of sheets of roof sheathing.
   5. Number of squares of shingles. (Assume that aluminum ridge cap is to be used on all except the ranch house; include a starter strip for all.)
   6. Metal drip edge comes in lengths of 10 ft. Assume that drip edge is to be applied to all roof edges along the rake as well as the eaves. Determine the number of pieces of drip edge needed for each roof.

a.

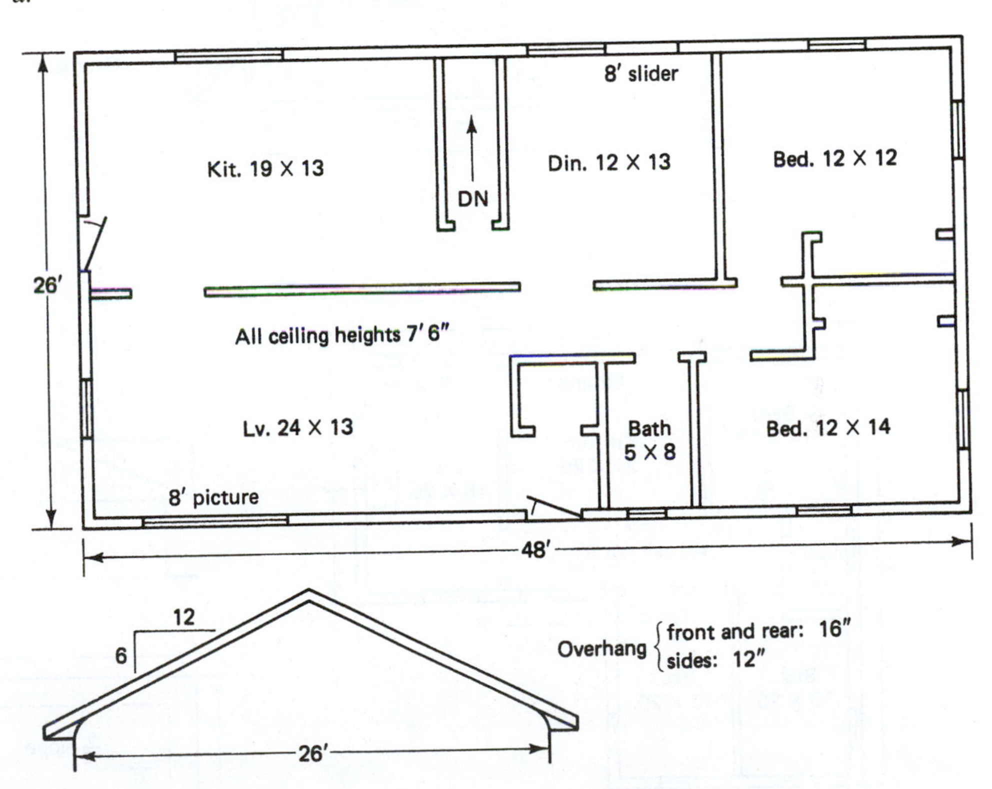

b.

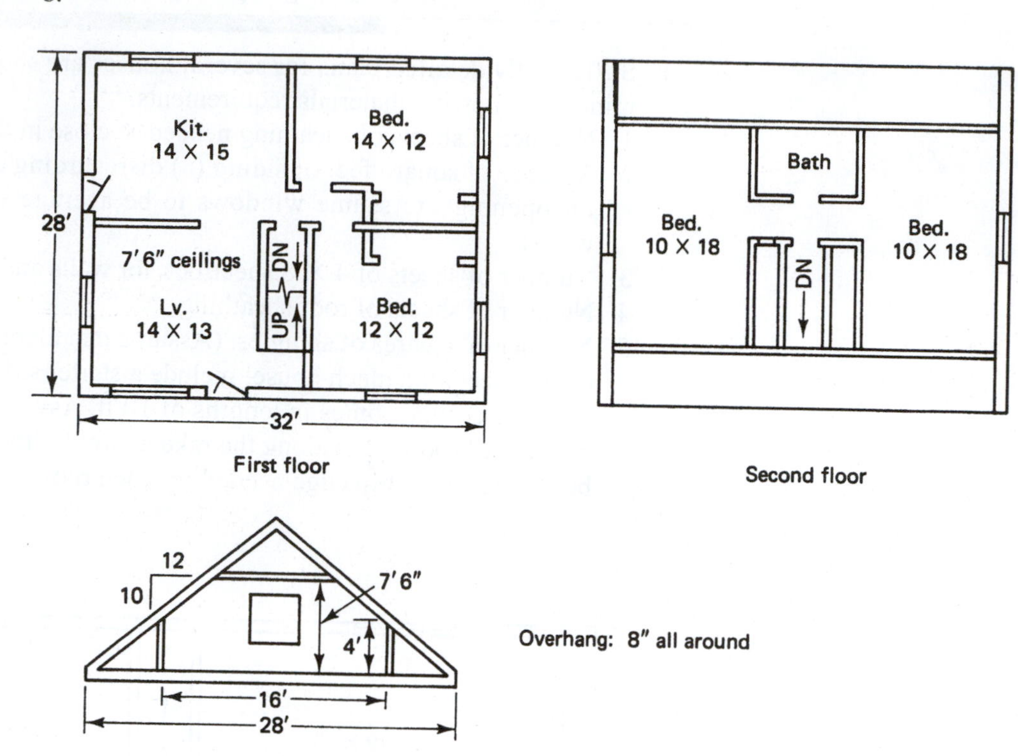
Kit.
14 X 15
Bed.
14 X 12
28′
7′6″ ceilings
DN
UP
Lv.
14 X 13
Bed.
12 X 12
32′
First floor
Bath
Bed.
10 X 18
Bed.
10 X 18
DN
Second floor
12
10
7′6″
4′
16′
28′
Overhang: 8″ all around

c.

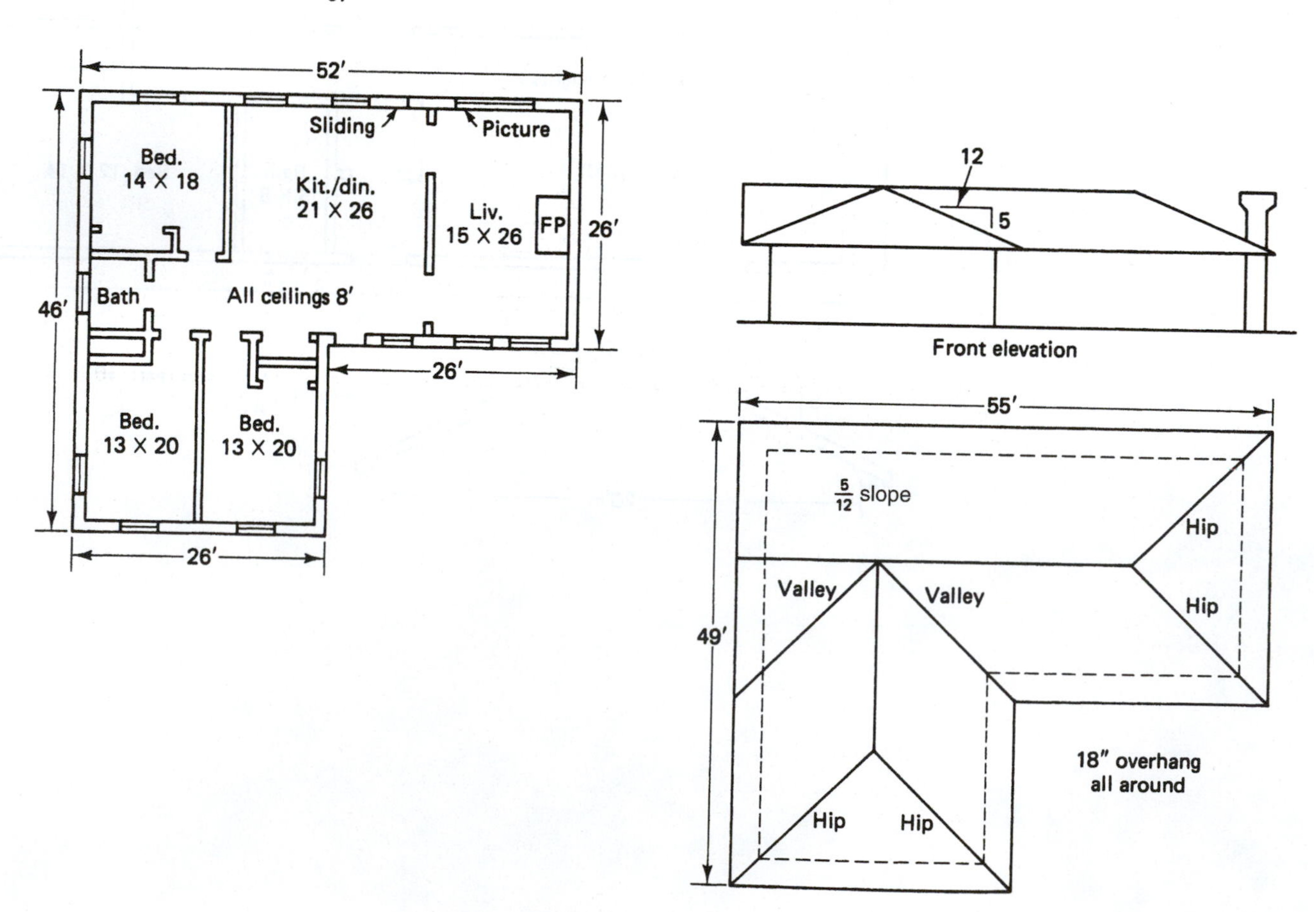
52′
Sliding
Picture
Bed.
14 X 18
Kit./din.
21 X 26
Liv.
15 X 26
FP
26′
Bath
All ceilings 8′
46′
26′
Bed.
13 X 20
Bed.
13 X 20
26′
12
5
Front elevation
55′
5/12 slope
Hip
Hip
Valley
Valley
49′
Hip
Hip
18″ overhang
all around

d.

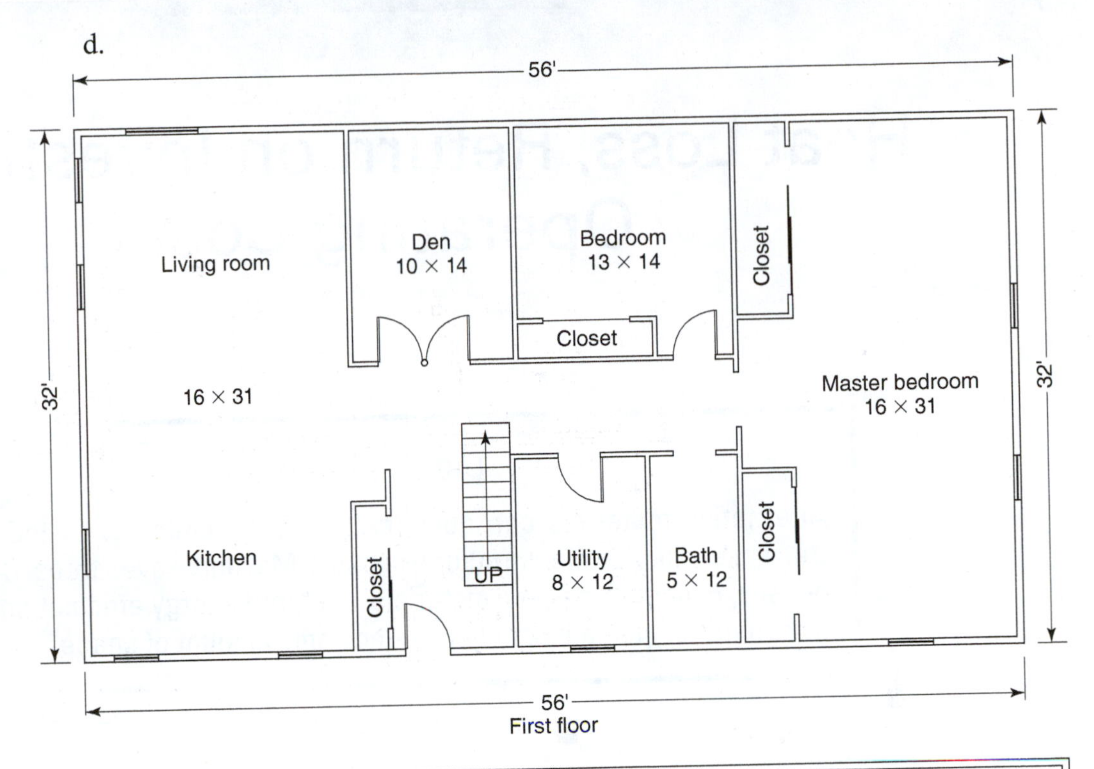

First floor

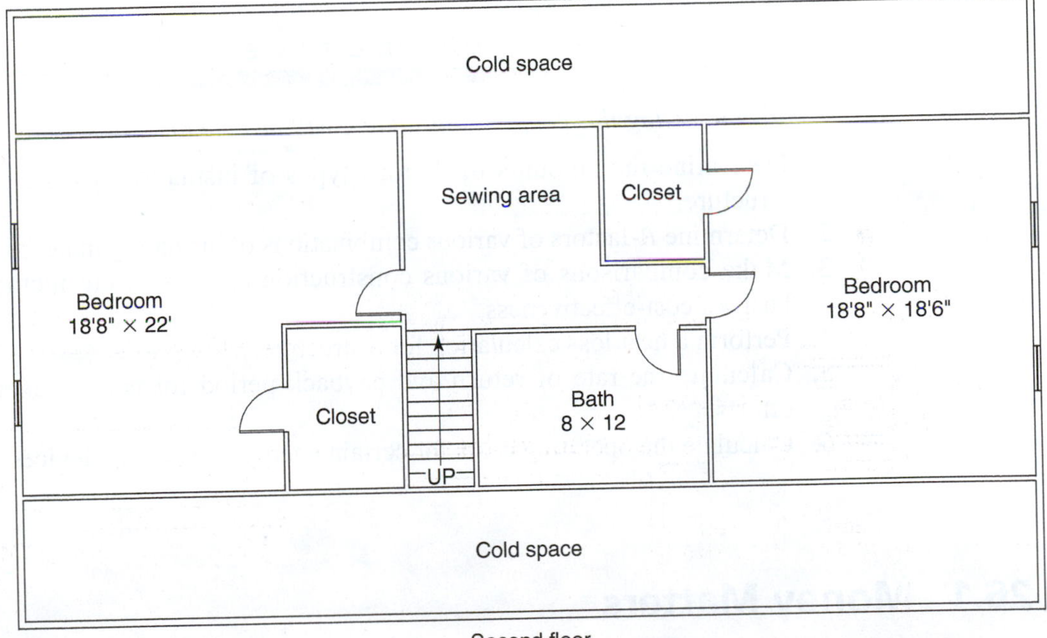

Second floor

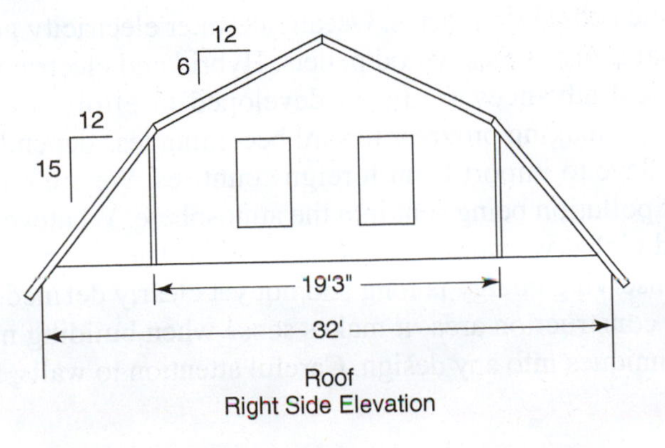

Roof
Right Side Elevation

CHAPTER 26

Heat Loss, Return on Investment, Operating Costs

Did You Know

Alternative materials can conserve in house building. Culled dead trees might be used for structural support. Masonry saves trees. Domed houses, built from stacked materials, are more energy efficient and use less material than a box providing the same amount of space.

OBJECTIVES

Upon completing this chapter, the student will be able to:

1. Determine the amounts of various types of insulation needed for a given structure.
2. Determine *R*-factors of various combinations of insulating materials.
3. Make comparisons of various construction and insulating methods on the basis of cost-effectiveness.
4. Perform a heat loss calculation for a structure.
5. Calculate the rate of return and payback period for certain energy-saving efforts.
6. Calculate the operating costs for certain energy-consuming devices.

26.1 Money Matters

Photovoltaic cells. Wind farms. Ocean tidewater electricity generation. Solar collectors. Geothermal heat sources. Wood pellets. Hybrid and electric vehicles. All this and more technological advances are being developed in efforts to "green" America and the world. We are making progress toward becoming less dependent on fossil fuels, most of which we have to import from foreign countries. We are making efforts to reduce the amount of pollution being sent into the atmosphere. Whatever contribution man is making toward global warming is being addressed. There is much work to be done and the road to conserving energy is long and not yet clearly defined.

In the construction area, it makes sense when building new to incorporate energy-saving techniques into any design. Careful attention to walls, roofs, doors and windows, indeed, the entire structure, should be given to minimize the cost of heating and cooling. The same consideration should be given to existing structures, while recognizing the increased difficulty of working with sections already closed with wall and ceiling materials.

The cost of energy-saving devices and methods must be given serious consideration. The idea is that dollars invested today in energy-saving methods will be returned over time. This brings us to the concepts of **payback periods** and **return on investment (ROI)**.

If you are able to invest some money, say $1,000, someplace and find that you will earn $50 interest in one year, that is a rate of return of 5%. You have earned 5% interest on your money. 50 ÷ 1000 = 0.05 or 5%. ROI is said to be 5%. In addition, the **payback period** is 20 years. That is, it will take 20 years for the $1,000 to be paid back at the rate of $50 per year. 1000 ÷ 50 = 20.

In simple terms, the ROI (also called the rate of return) and the payback period are two important factors when considering any investment. They are of particular importance when considering whether to spend extra dollars for more insulation, better-insulated windows, a more efficient heating system, solar panels, water heating methods, and so on. We look at the *cost* of the item, determine our best estimate of the amount of *money saved* in reduced energy, and then calculate the ROI.

Suppose your electric bills have been averaging $100 per month. You determine that you can replace your electric storage water heater with an on-demand or instantaneous water heater (one that heats water only when the faucet is opened, as opposed to maintaining water at a fairly constant temperature, regardless of use). The on-demand system will cost $1120, but will reduce your electricity bill by $28 per month. What is the payback period and what is the ROI? Over the course of one year, saving $28 per month means you save 12 × 28 per year, or $336. By investing $1120, a savings of $336 is realized in one year. 336 ÷ 1120 = 0.3. This means a ROI of 30%. The payback period would be 1120 ÷ 336, or 3.33 years. So you would have all of your money back in $3\frac{1}{3}$ years. That would be quite an attractive investment. Of course, these are assumed numbers and in a real case, the actual costs should be more carefully determined.

To find a **rate of return or ROI**, divide the money invested into the annual return in dollars and convert to a percent.

ROI = (Dollars earned or saved in one year ÷ Dollars invested) × 100

To find a payback period, divide the amount invested by the annual return in dollars. This is the number of years required to return your investment.

Payback period = Dollars invested ÷ Dollars earned or saved per year

Note: Payback period can also be determined by dividing the ROI into 100. So a 20% ROI would give a payback period of 100 ÷ 20 = 5, or 5 years.

Example

Suppose you are going to replace an aging boiler in a hot water heating system. The new boiler operates at a much greater rate of efficiency and it has been estimated that your No. 2 heating oil consumption will drop by 30% as a result. Average oil consumption for each of the past three years has been 2400 gallons. This means that the expected consumption going forward will be 30% less, or 1680 gallons.

(2400 − .30 × 2400 = 2400 − 720 = 1680.) If No. 2 fuel costs $2.35 per gallon, and the new furnace costs $7,200, find the ROI and the payback period.

Solution:

2400 gallons at $2.35 per gallon would cost $5,640. 1680 gallons at $2.35 would cost $3,948. This is a savings of $1,692, and the ROI is 1692 ÷ 7200 = 0.235 or 23.5%. The payback period is 7200 ÷ 1692 = 4.25 years. In other words, making this investment of $7200 in a new boiler will give back your investment in about 4 years and 3 months. (Note that 100 ÷ 23.5 = 4.25, illustrating the other method of determining payback period.)

Of course, there are assumptions here that may not hold true. The price of oil will likely change. The amount of oil consumed depends on the severity of the winter, how the thermostat is set, and so on. But it is fair to say that making the change is probably a good investment.

What determines if the ROI or payback period is favorable? More than one factor can come into play. One is the average rate of interest earned on money, or the average interest rate on borrowed money. These are two different factors, but if one can earn, say, 3 or 4 percent, any ROI greater than that makes a good investment. Some people are guided as much by such things as contributing to the overall concept of energy conservation, without regard to dollar amounts. People weigh these factors individually.

26.2 Heat Loss

Proper insulation contributes greatly to controlling indoor climate. In the north, we wish to retain as much heat as possible within our homes. In the south, we wish to prevent as much heat as possible from entering our homes. The basic concept in both cases is to minimize the ability of heat to transfer through our protective outer walls, roofs, and floors. Insulation is the key.

The insulating quality of a substance is given by its **R-value**. The R-value is a measure of the ability of a substance to resist the transfer of heat through it. The higher the R-value, the greater the ability to resist the transfer of heat. Material with an R-value of 11 is a better insulator than material with an R-value of 3. Heat tends to transfer from a warm place to a cooler place. Substances with a high R-value tend to resist this transfer, keeping warm air in when that is desired, or cool air inside when in a hot climate.

For example, each inch thickness of concrete has an R-value of 0.08, showing that it is an excellent conductor of heat; and therefore a very poor insulator. Most wood has an R-value of approximately 1.25 per inch, and rigid foam has an R-value of about 5 per inch. For a more complete listing of the R-values of various materials, see Appendix C in the back of the book.

In most sections of the country the standard of wall framing has become 2 × 6s placed 16 in. o.c. This allows fiberglass insulation to be placed between studs that provides an R-value of 19. In addition to that, sometimes a layer of 1-in. rigid foam sheeting is placed on the wall, adding R-5, for a total of R-24. Other methods of framing have been used to increase the R-value of walls. A double wall made of two walls of 2 × 4s, 16″ o.c., separated by 12 in. outside-to-outside can provide R-38, using 12 in. of fiberglass insulation. The outer wall is the support wall, and the studs of the two walls are staggered so that there is not a continuous gap from inside to outside. (See Figure 26–1.)

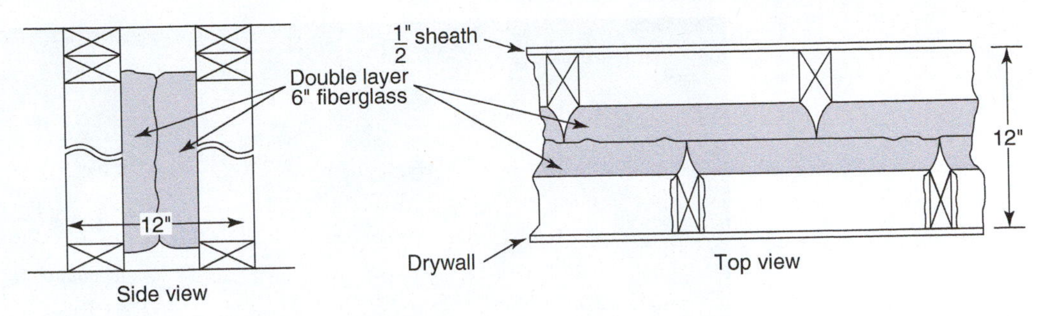

Figure 26–1

With the more conventional single-studded wall, wrapping the shell with a layer of rigid foam might be considered. With every stud, the R-value is reduced to approximately 6 or 7, because wood has a much lower R-value than fiberglass. So especially for the studded area, the R-value can be increased by R-5 by adding rigid foam.

In any structure, analysis can show where the greatest heat loss occurs. Infiltration is often the biggest source of heat loss. Cracks around openings—doors, windows, through vents, chimneys, electrical outlets—all contribute to infiltration. Since heat rises, ceilings are the next largest source of heat loss. Then come walls, followed by floors. All this is stated in general terms and specific cases may show different results as to which source is greatest for heat loss. Infrared devices can determine where the greatest heat loss is occurring in a structure. Suffice it to say that all areas should be given careful consideration when it comes to insulating. This is easy to do with new construction, more difficult, but worth considering when retrofitting an existing building.

What about other areas of the structure? Let's start with the foundation. A poured concrete foundation wall can be insulated on the exterior by applying, for example, 2-in. of rigid foam to the walls before backfilling. Mastic is applied to adhere the foam to the wall. After backfilling, any exposed foam has to be protected from the deteriorating effects of the sun's rays. Protective coatings have been developed for this. Another method is to stud the interior walls of the cellar and insulate between the studs with fiberglass. Insulating foam spray is also very effective. For a structure built on a concrete slab, rigid foam can be laid beneath the slab before the concrete is placed. Slab edges should also be considered. Rigid foam can be applied there to the thickness of the slab, and a bit lower. Foundations built of foam blocks, called insulated concrete forms (ICFs), provide good insulation by the wall itself. Here the foundation is built by laying up hollow foam blocks that have been molded to fit tongue-and-groove to each other. A great many steel reinforcing bars are used both vertically and horizontally. Once in place and properly braced, the foam blocks are filled with concrete.

Whether a floor is insulated is a function of how the basement is to be used. If the cellar is to be heated on a zone separate from the upper level, insulation can help to preserve the temperature difference desired. With a house built on a poured concrete foundation, the *area covered* by the floor joist headers should be insulated with sprayed foam, rigid foam, or fiberglass.

Ceilings can be insulated with layers of fiberglass insulation, easily building up to a very high R-value. However, with a cathedral ceiling, the space available for insulation is limited and calls for use of insulating materials other than fiberglass.

Proper installation of vapor barriers must also be considered. The number of air exchanges to maintain a healthy atmosphere is important. In this text, we are primarily

concerned with calculating any relevant values, so we will leave the reader to explore in detail various types of insulation and methods of installing them.

Let us now look at a few examples.

Example 26–1

Specifications for a 28 ft $\times$ 44 ft single-story ranch house call for $6\frac{1}{4}$ in. of fiberglass insulation in the exterior walls, 1 in. of rigid foam wrapped around the exterior walls, extending from the top plate to 5 ft below the foundation top, and a double layer of $6\frac{1}{4}$ in. fiberglass insulation over the ceiling. The two ceiling layers will be at right angles to each other to minimize air gaps. The exterior walls are framed with $2 \times 6 - 16$ in. o.c. The door and window openings total 210 ft^2. Determine the amount of each type of insulation needed. (*Note:* A roll of foil-faced $6\frac{1}{4} \times 23\frac{1}{2}$ insulation will cover approximately 75 ft^2. A roll of foil-faced $6\frac{1}{4}$ in. $\times$ $15\frac{1}{2}$ in. insulation will cover approximately 50 ft^2. Rigid foam 1 in. thick is available in sheets measuring 2 ft $\times$ 8 ft and 4 ft $\times$ 8 ft.)

Solution:

The exterior walls will require $6\frac{1}{4} \times 15\frac{1}{2}$ fiberglass. The wall area equals the perimeter multiplied by the wall height, minus the area of the openings. The perimeter is

$$\begin{aligned} P &= 2(28 + 44) \\ &= 2(72) \\ &= 144 \text{ ft} \end{aligned}$$

$$\begin{aligned}\text{Wall area} &= \text{perimeter} \times \text{wall height}\\ &= 144 \times 8\\ &= 1152\ \text{ft}^2\end{aligned}$$

$$\begin{aligned}\text{Insulated area} &= \text{wall area} - \text{area of openings}\\ &= 1152 - 210\\ &= 942\ \text{ft}^2\end{aligned}$$

Thus 942 ft^2 of foil faced fiberglass insulation is needed for the walls. At 50 ft^2 per roll, the number of rolls required would be

$$942 \div 50 = 18.84 \text{ or } 19 \text{ rolls}$$

The amount of rigid foam needed for the exterior can be found by adding the net wall area of 942 ft^2 to the area to be insulated below that. This depends on the size of the floor joists and sill plate. Let us assume that they total approximately 1 ft. Added to the 5 ft of the foundation that is to be covered, we have 6 more feet around the perimeter.

$$6 \times 144 = 864\ \text{ft}^2$$

Added to the net wall area yields

$$864 + 942 = 1806\ \text{ft}^2$$

to be covered. That would require

$$1806 \div 16 = 112.875$$

or 13 sheets of 2 × 8 rigid foam.

The ceiling area measures 28 ft × 44 ft, and

$$28 \times 44 = 1232\ \text{ft}^2.$$

A double layer of $6\frac{1}{4}$ in. insulation is needed, which brings the total to

$$2 \times 1232 = 2464\ \text{ft}^2$$

With 75 ft^2 per roll, the order would be placed for

$$2464 \div 75 = 32.85 \text{ or } 33 \text{ rolls of } 6\tfrac{1}{4} \text{ in.} \times 23\tfrac{1}{2} \text{ in. unfaced fiberglass.}$$

Example 26–2

Compare the cost of materials used in framing the exterior walls of the 28 ft × 44 ft house described in example 26-1 by each of the following methods: (a) 2 × 4s – 16 in. o.c. (b) 2 × 6 – 16 in. o.c. (c) a double wall of 2 × 4s – 16 in. o.c. Use a price of $525 mbf for all sizes used. Ignore studs needed for openings.

Solution:

(a) Materials needed for shoe and double plate total three times the perimeter of

$$3 \times 144 = 432 \text{ linear feet.}$$

Placing studs 16 in. o.c. will require

$$\frac{3}{4} \times 144 \text{ or } 108 \text{ studs.}$$

Adding one for each corner and one for each starter brings the total to 116 studs. The number of board feet of 2 × 4s used

$$\text{For the studs: } \frac{2}{3} \times 108 \times 8 = 576$$

$$\text{For the shoe and plates: } \frac{2}{3} \times 432 = 288$$

$$\text{Total: } 576 + 288 = 864 \text{ board feet.}$$

$$\text{The cost is } 864 \times \$.525 = \$453.60.$$

(b) Framing with 2 × 6 − 16 in. o.c. will require the same number of linear feet for the shoe and plates, as well as for the studs. So, determining the number of board feet, we have

432 lf of shoe and plates plus 116 × 8 = 928 lf of studs from a total of 1360 lf.
Since 1 lf of 2 × 6 equals 1 bf, we have 1360 bf.

$$\text{The cost is } 1360 \times \$.525 = \$714.00$$

(c) The double wall construction would simply be double the cost found in part (a), or $907.20.

Example 26–3

As a follow-up to the previous example, the three methods described there allow for various amounts of insulation. Using only fiberglass bats between studs, the R-values that can be obtained are R-11 in (a), R-19 in (b), and R-38 in (c). Find the cost of insulating the walls in each of these three cases. Assume that fiberglass insulation costs $.34 per ft^2 for R-19 and $.28 per ft^2 for R-11.

Solution:

The wall area is the same in the first two cases, and doubled in the third case. There were 942 ft^2 of net wall area. (See example 26–1.)

For the 2 × 4 − framed walls, 942 × $.28 = $263.76.
For the 2 × 6 − framed walls, 942 × $.34 = $320.28.
For the double 2 × 4 − framed walls, 2(942) × $.34 = $640.56.

Note: Of course, the double walls cost the most to build and the most to insulate. The important long-term consideration is whether the extra dollars will produce an attractive ROI and a reasonable payback period. A more thorough heat loss calculation and the resulting savings in heating fuel consumption are needed to reach a decision. However, it is safe to say that the reduced energy consumption of a few hundred dollars will likely be repaid within a reasonable period of time.

Insulation needs vary in each part of the country. Some R-values recommended by the U.S. Department of Energy are shown in the table below. A more complete listing of R values may be found at www.ornl.gov/sci/.

	Attic	Cathedral Ceiling	Wall Cavity	Floor
Northeast	49–60	30–60	13–21	25–30
Central East	38–60	30–38	13–15	25–30
Florida	30–60	22–38	13–15	13

Heat-Loss Calculations

The amount of money saved by certain insulating techniques is determined by calculating the amount of heat lost from a structure. Heating requirements for a building are measured in **Btu**. Comparisons of Btu requirements by various insulating methods can be made and the expected heating costs calculated. A source of heat can be priced in terms of the Btu output it provides. These figures can then be projected to approximate the annual heating costs.

Several different methods for determining heat loss are in use, with varying results from each. Some of the differences can be attributed to certain assumptions made about the structure. Nevertheless, a certain amount of validity lies in each result.

Regardless of the method used to calculate heat loss, certain basic data must be collected. One of these assumptions is the temperature difference inside and outside the building. A desired inside temperature is assumed and the average annual outside temperature is determined. (This is available from collectors of weather data in any region.) This temperature difference (called ΔT, or delta T) is used in each calculation. Another assumption is the number of air exchanges of inside with outside air within a certain time period, usually a day. Air exchanges occur due to infiltration losses as well as door and window openings and closings. Other factors that affect heat loss are the size and number of windows and doors, materials used in construction, and type of construction. In older buildings it is not always easy to accurately determine some of these factors, but resulting errors should not seriously affect the conclusions.

All materials have some insulating capability. Appendix B includes R-factors of many materials. When choosing materials, comparisons of their R-factors should be scrutinized. Glass, for example, comes with various R-values. Listed here are the R-factors for a single pane, double-paned insulating glass, and triple-paned insulating glass. Figures are winter figures and assume a $\frac{1}{4}$-in. air space between panes.

Single glass	0.89
Two panes	1.55
Three panes	2.13
Low emissivity (Low E)	3.23

When analyzing alternatives, the cost of each additional pane must be included. *Remember:* The primary concern lies in expected return on dollars invested. When considering a third pane, an alternative might be to install it as a separate insert. This would have the advantage of providing added resistance to infiltration losses. Also, the amount of space between panes affects the R-factor. In the list above, a $\frac{1}{2}$ in. space with three panes increases the R-factor from 2.13 to 2.78.

To calculate the heat requirements for a structure, certain facts must be assembled, such as data related to the type of structure and assumptions related to the environment. The following facts are needed about the type of structure.

1. Type of interior wall covering material.
2. Type of exterior wall covering material.
3. Type of exterior siding material. Wall stud size and spacing.
4. Type of foundation material.
5. Type of ceiling material.
6. Amount and type of insulation in walls, ceilings, and floors.
7. Type of vapor barrier material.
8. Building dimensions: length, width, ceiling height.
9. Area of openings: doors and windows.
10. Adequacy of attic ventilation.

Environment-related data needed are as follows:

1. Interior temperature.
2. Exterior temperature.
3. Rate of infiltration loss.

Once the necessary data have been collected and design assumptions are made, calculations of the heat loss rate are made for each segment of the structure. The heat loss rate is measured in British Thermal Units per Hour (Btus/hr). The sum of R-factors, R_t, of materials affecting heat loss is found. This sum is then divided into the product of the design temperature difference, ΔT, and the affected area, A. That is,

$$\text{Btu/hr} = (\Delta T \times A) \div R_t$$

After the calculation of heat loss has been completed for each section, these values are totaled. The result is the total heat loss for the structure measured in Btu/hr.

Example 26–4

Calculate the heat loss for the structure described below, given the stated assumptions.

Single story 28 ft $\times$ 44 ft ranch
Foundation: 4-in. slab on grade
Wall height: 8 ft
Exterior wall framing: 2 $\times$ 6 $-$ 16″ o.c.
Interior wall covering: $\frac{1}{2}$-in. drywall
Exterior wall covering: $\frac{1}{2}$-in. plywood; $\frac{3}{4}$-in. $\times$ 10 in. lapped siding
Ceiling covering: $\frac{1}{2}$-in. drywall
Attic ventilation: adequate
Windows: all double-pane, insulated glass with $\frac{1}{4}$-in. air space; total window area of 120 ft^2
Doors: two insulated steel doors; total area 40 ft^2
Insulation: Walls $-$ $6\frac{1}{4}$-in. foil faced fiberglass
Ceilings: $12\frac{1}{2}$-in. unfaced fiberglass
Slab: 1-in. Styrofoam within 4 ft of perimeter
Vapor barrier: 4-milpolyethylene on ceiling

Design assumptions:

Interior design temperature: 70°F
Exterior design temperature: –10°F
Design wind speed: 15 mph
Slab design temperature: 0°F for outer 4 ft; 35°F center
Infiltration losses: one air exchange per hour

Comments: Temperatures inside will vary from ceiling to floor. The assumption is based on the average between ceiling and floor temperatures. Infiltration losses vary with the tightness of the structure and the number of people entering and leaving each hour.

Solution:

1. Heat loss due to infiltration

$$\text{Btu/hr} = V \times \Delta T \times 0.02$$

$$\text{Volume} = 28 \times 44 \times 8 = 9856 \text{ ft}^3$$

$$9856 \times 80 \times 0.02 = 15{,}770 \text{ Btu/hr}$$

Comment: This step represents a deviation from the formula stated above. The factor 0.02 represents the number of Btu per hour per square foot to heat dry air. Calculations for other segments will be done by the Btu/hr formula that uses area.

2. Heat loss through the walls
   a. Studded portion

Inside air film	0.68
$\frac{1}{2}$-in. drywall	0.45
Stud	6.88
$\frac{1}{2}$-in. plywood	0.63
Siding	1.05
Outside air film	0.17
Total (R_t)	9.86

Total area of studs:

$$144 \times \tfrac{3}{4} \text{ studs} \times \left(\tfrac{1.5}{12}\right) \text{ ft wide} \times 8 \text{ ft high} = 108 \text{ ft}^2$$

$$(\Delta T \times A) \div R_t = (80 \times 108) \div 9.86 = 876 \text{ Btu/hr}$$

   b. Unstudded portion

Inside air film	0.68
$\frac{1}{2}$-in. drywall	0.45
$6\frac{1}{4}$ in.glass	19.00
$\frac{1}{2}$-in. plywood	0.63
Siding	1.05
Outside air film	0.17
R_t	21.98

$$\text{Area} = \text{wall area} - (\text{stud area} + \text{window area} + \text{door area})$$

$$= (144 \times 8) - (108 + 20 + 40) = 884 \text{ ft}^2$$

$$(\Delta T \times A) \div R_t = (80 \times 884) \div 21.98 = 3217 \text{ Btu/hr}$$

c. Heat loss through doors

Inside air film	0.68
Door	7.30
Outside air film	0.17
R_t	8.15

(80 × 40) ÷ 8.15 = 393 Btu/hr

d. Heat loss through windows.

Inside air film	0.68
Glass	1.55
Outside air film	0.17
R_t	2.40

(80 × 120) ÷ 2.4 = 4000 Btu/hr

e. Heat loss through ceiling

Inside air film	0.68
$\frac{1}{2}$-in. drywall	0.45
$12\frac{1}{2}$-in. fiberglass	38.00
Outside air film	0.17
R_t	39.30

(80 × 1232) ÷ 39.30 = 2508 Btu/hr

f. Heat loss through the slab

Total area = 28 × 44 = 1232 ft²
Uninsulated: 20 × 36 = 720
Insulated 512 ft² (by subtraction)

(1) Uninsulated area

Inside air film	0.68
4-in. concrete	0.32
R_t	1.00

(40 × 720) ÷ 1 = 28,800 Btu/hr

(2) Insulated area

Inside air film	0.68
4-in. concrete	0.32
1-in. Styrofoam	5.00
R_t	6.00

(65 × 512) ÷ 6 = 5547 Btu/hr

Total heat loss:

Infiltration	15,770
Walls	
Studded	876
Unstudded	884
Doors	393
Windows	4000
Ceiling	2508
Floor	
Uninsulated	28,800
Insulated	5547
Total	58,778 Btu/hr

Note: It is suggested in the exercises that variations in heat loss for this structure be explored. Vapor barriers and permeability have not been discussed here—not because the topic is unimportant, but because it is not an objective of this chapter. Vapor barriers do play a significant role in proper insulation. References have been included in the bibliography.

26.3 Operating Costs

When considering any particular method of supplying energy, the operating costs need to be considered. The factors that determine annual operating costs are the cost of fuel per Btu or per Therm (one Therm = 10,000 Btus), and the energy factor* (EF) of the fuel. The fuel may be natural gas, propane, electric, coal, oil, or wood. For example, consider the annual operating cost of heating water in an on-demand natural gas system, given that natural gas has an EF = 0.57 and the cost of natural gas is \$0.00000619/Btu. This is found by using a formula for gas and oil water heaters. Annual cost = 365 × 41405/EF × Fuel cost per Btu, or it could be used as Annual cost = 365 × 0.41405/EF × Fuel cost per Therm. Here, Annual Cost = $365 \times \frac{41405}{0.57} \times \$0.00000619 = \$164$. Using the Therm factor helps with the decimal places. Annual cost = $365 \times \frac{0.41405}{0.57} \times \$0.619 = \$164$. There is a different formula for determining the annual operating cost by electric water heaters and heat pumps: Annual cost = 365 × 12.03/EF × Electricity cost (Kwh). For more information, see www.energysavers.gov.

REVIEW EXERCISES

1. Below are examples of amounts saved, amounts invested, and time factors. For each, calculate the ROI and the payback period.

	Amount Invested	Monthly Savings	Annual ROI	Payback Period In Years
A	\$250	\$8		
B	\$1125	\$18		
C	\$4250	\$15		
D	\$16,000	\$80		
E	\$3200	\$10		

2. A new energy-efficient water heater will cost \$895 but should reduce the monthly electric costs by \$7.50. Calculate the ROI and payback period.
3. A ground water heat pump installer claims that his system will result in an annual reduction in heating costs by 20%. Heating costs for the past three years were \$2350, \$2475, and \$2940. The heat pump system will cost \$11,375 installed. Using the average of the three years heating costs and assuming his savings prediction is accurate, calculate the:
   (a) expected annual heating cost.
   (b) annual savings.
   (c) ROI.
   (d) payback period.

*Energy Factor, EF, is a measure of the overall efficiency based on heat produced per unit of fuel consumption.

4. A single-story house measures 32 ft by 40 ft. The walls are 8 ft high. Ignoring door and window openings, calculate the cost of framing the exterior walls, including the amount of insulation as specified in each of the following cases. Assume all framing materials cost $0.65 per board foot, 6.25 in. fiberglass costs $0.38 per sq ft, and 1 in. rigid foam insulation costs $0.59 per sq ft.
   (a) $2 \times 6 - 16$ in. o.c. with $6\frac{1}{4}$ in. of fiberglass.
   (b) Same as (a) but with a wrap of 1 in. rigid foam insulation added.
   (c) A double 2×4 wall, spaced 12 in. apart, using a double layer of $6\frac{1}{4}$ fiberglass insulation.
5. Compute the total heat loss in Btu/hr for the $32' \times 40'$ house described in exercise 4, assuming no insulation has been installed. Assume a total door area of 40 ft^2 and a window area of 120 ft^2.
6. Recalculate the preceding problem where necessary to determine the total heat loss if $6\frac{1}{4}''$ of fiberglass insulation is added to the walls and $12\frac{1}{2}''$ is added to the ceilings. How does this compare to the heat loss without insulation?
7. By how much more would the heat loss be reduced if 1 in. of Styrofoam insulation (R-5) were added to the exterior walls?
8. Calculate the percentage heat loss distribution for the house in exercise 26–4. That is, what percent is lost through: infiltration, walls, doors, windows, ceilings, and floors? Use the results after fiberglass was added to the walls and ceilings.
9. Study the results from Problem 26–8. Through what part of the unheated house does the greatest heat loss occur? The least? What priorities should be followed in insulating a house?

   (The following problems are lengthy and may be assigned as projects to be done outside of class or in class as a group project.)
10. Research the heating costs per Btu for various sources of heat (No. 2 oil, electric, wood, electric, natural or propane gas; choose one or more.) Then determine the added costs, ROI, and payback period for various insulating methods. For example, you may compare insulating with fiberglass to fiberglass with an additional Styrofoam wrap.
11. Select a house of your choice and perform a heat loss calculation. (You may choose one of the plans in Appendix D, or another of your choice.)

CHAPTER 27

The Estimating Process

Did You Know

In general, priority for insulating a house should be the attic, including a hatch cover, followed by under the floors above unheated spaces, then around walls in a heated cellar, and finally around the edges of slab on grade.

OBJECTIVE

Upon completing this chapter, the student will be able to:

1. Prepare a materials list for construction of a residence.

The purpose of this chapter is to illustrate the process of accurately and efficiently estimating the materials cost for a residence. Concepts and procedures described in preceding chapters are applied.

The estimating process results in figures that do just what the term implies—*estimate* the cost and quantity of materials. A certain amount of error will ordinarily exist in the estimate. It is unfair to both the contractor and the customer to present estimates that grossly vary from the actual amount. It is incumbent on the estimator to minimize error insofar as he has control over the variables.

Despite an accurate forecast by an estimator as to the amount and cost of materials, several factors could alter the figures. Included among these are unforeseen delays in delivery of materials, wasted materials due to poor judgment by workers, materials stolen from the site, defective materials, and unforeseen price changes. The effective estimator will consider these factors and prepare the estimate accordingly. To fend off the potential losses incurred by the unforeseen, it is wise to use a "fudge factor," a percent multiplier that is added to a total.

Whether a job is being performed under a contract that specifies a price or under a cost-plus agreement, an accurate estimate is to be expected. Actual results should generally fall within plus or minus 5% of an estimate; an outer limit of variance should be 10%.

The accuracy of the estimate will be somewhat assured if the estimator uses good mathematics. A potential problem stemming from omission of materials can be avoided by using a checklist. The checklist should be written so that materials are listed in the order of use. This provides a logical flow for the estimator to follow. In listing materials, the estimator can mentally picture the construction process and allow materials to flow onto paper in the same order in which they will be used on the job site. Figure 27–1 shows a sample list for a single-story ranch house. Certain jobs will require modifications to this list. Other possible contingencies should be considered as well. For example, ledge might be encountered in the excavation process. Extra costs would be incurred for its removal. Clauses are usually inserted in a contract to cover events beyond the control of the contractor.

An estimator usually makes an estimate starting with a worksheet on which calculations are made. The careful estimator will keep the worksheets for future reference in checking for errors. The worksheet in Figure 27–2 shows the calculations related to the house shown in Figure 27–3.

Each item has been numbered to correspond to the estimate summary sheet. A partially complete estimate summary sheet is shown in Figure 27–4.

A take-off list summarizing the materials for a project is usually needed for convenience in pricing and for submission to suppliers wishing to bid on a materials package. This list also brings together similar materials that appear in various places throughout the estimate summary sheet. The take-off list is easily generated from the estimate summary sheet. A sample take-off list for the house in Figure 27–3 is shown in Figure 27–5.

Labor hours and related costs are one of the more difficult areas to estimate. Different workers work at different rates. Different types of jobs (framing versus finish work, for example) require different degrees of labor intensity. But it is primarily individual differences that make projections of labor time difficult.

For a contractor working steadily with the same crew, the best source of data on which to base labor requirements is experience. A few minutes at the end of each workday spent recording jobs done and number of labor-hours consumed will provide the most reliable resource for future estimates of labor requirements. For the carpenter working for a contractor, but with the ambition to become a contractor, time spent performing the same record keeping will be time well invested.

ESTIMATE SUMMARY

Project

Location ____________

Sheet no. _____ of _____

Item	Size/ number/ length	Bd. ft	Unit cost	Labor	Others	Total
1. Site preparation						
2. Excavation						
3. Footing						
4. Foundation						
5. Drain tile						
6. Foundation coating						
7. Water or well						
8. Septic or sewer						
9. Exterior cellar insulation						
10. Backfill						
11. Gravel						
12. Cellar slab						
13. Girder supports						
14. Girder						
15. Sill seal						
16. Sill plates						
17. Floor joists						
18. F.j. headers						
19. Bridging						
20. Subfloor						
21. Shoe and plates						
22. Studs						
23. Headers and sills						
24. Wall sheathing						
25. Stairs						
26. Cross-ties						
27. Rafters						
28. Roof sheathing						
29. Jack studs						
30. Gable end sheathing						
31. Soffit and fascia						
Framing						
Covering						
32. Drip edge						
33. Asphalt shingles						
34. Ridge vent						
35. Window schedule						

Figure 27–1

Item	Size/ number/ length	Bd. ft	Unit cost	Labor	Others	Total
36. Door schedule (ext.)						
37. Exterior trim						
38. Siding						
39. Interior framing						
Shoe and plates						
Studs						
40. Wiring						
41. Plumbing						
42. Insulation						
43. Drywall						
44. Masonry						
45. Kitchen						
46. Bath(s)						
47. Heating						
48. Second-floor covering						
49. Door schedule (int.)						
50. Molding						
51. Paint, caulk						
52. Flooring						
53. Driveway						
54. Landscaping						
55. Nails						
56. Sales tax						
57. Markup						
Totals						

Figure 27–1 (*Continued*)

WORKSHEET

Project location ____________________ Sheet no. ______ of ______

1. Site preparation
2. Excavation

$$\frac{34 \times 50 \times 6}{27} = 378 \text{ yd}^3$$

3. Footing

$$\frac{\frac{8}{12} \times \frac{20}{12} \times 144}{27} = 5.9 \text{ or } 6 \text{ yd}^3$$

Figure 27–2

4. Foundation

$$\frac{7 \times \frac{10}{12} \times 144}{27} = 31 \text{ yd}^3$$

5. Drain tile and filter paper
   $144 \times 1.4 = 200 \text{ ft}^2$ paper
   144 lf tile
6. Foundation coating
   $5 \times 144 = 720 \text{ ft}^2$ at 100 ft^2 per gallon approx. two 5-gal cans
7. Well—as per subcontract
8. Septic system—as per subcontract
9. Cellar insulation
   1 in. Styrofoam to 5 ft below grade, or approx. 6 ft below foundation
   $6 \times 144 = 864 \text{ ft}^2/16 = 54$ sheets 1 in. by 2 ft by 8 ft
10. Backfill
    Suitable excavated site material
11. Gravel
    Crushed stone for drain tile: $\frac{1}{2} \times 1 \text{ ft} \times 1 \text{ ft} \times 144 = 72 \text{ ft}^3$

$$\frac{72}{27} = 2.7 \text{ or } 3 \text{ yd}^3$$

    Cellar slab base: average 6 in. depth

$$\frac{\frac{6}{12} \times 1232}{27} = 22.8 \text{ or } 23 \text{ yd}^3$$

12. Cellar slab

$$\frac{1232 \times \frac{3}{12}}{27} = 11.4 \text{ yd}^3$$

13. Girder supports
    Adjustable steel columns every 8 ft—5 posts
14. Girder
    6×10 built-up; $\frac{6}{16s}, \frac{3}{12s}$
15. Sill seal
    144 lf
16. Sill plates
    2×6: $\frac{4}{14s}, \frac{2}{12s}, \frac{4}{16s}$
17. Floor joists
    2×10: $\frac{70}{14s}$
18. Floor joist headers
    2×10: $\frac{4}{16s}, \frac{2}{12s}$
19. Bridging
    $3 \text{ ft/space} \times 44 \text{ spaces} \times 2 \text{ rows} = 264$ lf
20. Subflooring
    1 in. CDX plywood: $\frac{1232}{32} = 38.5$ or 39 sheets
21. Shoes/Plates
    2×6: $3 \times 144 = 432$ lf or $\frac{36}{12s}$
22. Studs
    2×6–16 in. o.c. $\frac{3}{4} \times 144 = 108 = 4$ starters + 4 corners + 18 windows + 1 partitions = 145 studs
23. Headers/sills
    2×6: $\frac{7}{12s}, \frac{4}{10s}$

Figure 27–2 ***(Continued)***

24. Exterior wall sheathing
    $\frac{144}{4} = 36$ sheets of $\frac{1}{2}$ in. CDX
25. Stairs
    2 × 12 risers: $\frac{2}{14s}$
    2 × 12 treads: 11 risers × 3 ft = 33 ft or $\frac{3}{12s}$
26. Cross-ties
    2 × 8s—$\frac{66}{14s}$
27. Rafters
    2 × 8s—$\frac{68}{16s}$
28. Roof sheathing
    $\frac{5}{8}$ in. CDX: $\frac{13 \times 15.5}{12} = 16.79$ ft rafter length
    $16.9 \times 2 \times 46 = \frac{1545 \text{ ft}^2}{32} = 49$ sheets
29. Jack studs (gable ends)
    TRi = URi × run = 5 × 14 = 70 in. or 6 ft
    $\frac{3}{4} \times 28 = 21$ studs − 6 ft or $\frac{11}{12s}$
30. Gable end sheathing
    $\frac{6 \times 28}{32} = 5.25$ or 6 sheets
31. Overhang
    End rafters: 2 × 6: $\frac{4}{18s}$
    Overhang framing:
        3 ft × 25 = 75 lf or $\frac{6}{12s}$
        Face plate: $\frac{6}{12s}$, $\frac{2}{10s}$
    $\frac{3}{8}$-in. A-C exterior plywood
    $\frac{18 \text{ in.} \times 46 \text{ ft} \times 2}{32} = 4.3$
    $\frac{12 \text{ in.} \times 17 \text{ ft} \times 4}{32} = 2.1$
    Total 6.4 or 7 sheets
32. Drip edge
    Eaves: 46 × 2 = 92 lf
    Rakes: 16.8 × 4 = 67.2 lf
    Total 159.2 lf or 16 lengths
33. Asphalt shingle/felt
    Main: 16.8 × 2 × 46 = 1545.6 ft^2
    Starter: $\frac{5}{12}$ × 46 = 38 ft^2
    Total 1584 ft^2 or 16 squares
    15-1b felt: $\frac{1584}{432} = 3.7$ or 4 rolls
34. Ridge vent
    46 ft or $\frac{4}{10s}$
35. Window schedule
    7–2 ft 9 in. × 4 ft 1 in. double hung
    1–2 ft 9 in. × 3 ft 5 in. double hung
    1–9 ft 5 in. × 4 ft 5 in. picture with 2 ft flanking double hung
    1–3 ft 4 in. × 3 ft 4 in. casement
    1–7 ft l0 in. × 5ft 1 in. bay

Figure 27–2 ***(Continued)***

36. Exterior door schedule
    1–2 ft 8 in. × 6 ft 8 in. insulated steel entry
    1–8 ft 0 in. × 6 ft 8 in. insulated glass wood-framed slider
37. Exterior trim
    No. 2 pine corner boards: 1 × 5: $\frac{4}{10s}$
    1 × 6: $\frac{4}{10s}$
38. Siding
    Sides: 8.5 × 144 = 1224 ft^2
    Gable ends: 6 × 28 = 168 ft^2
    Subtotal 1392 ft^2
    Less openings
    Windows: 180 ft^2
    Doors: 74 ft^2
    Total 254 ft^2
    Net area 1138 ft^2 or 12 squares
39. Interior framing
    Shoe/plates: 178 ft × 3 = 534 lf or $\frac{45}{12s}$
    Studs: 2 × 4: $\frac{534}{8s}$
40. Electrical—as per subcontract
41. Plumbing—as per subcontract
42. Insulation
    Ceiling: double-layer $6\frac{1}{4}$ in. × $23\frac{1}{2}$ in. unfaced f.g.
    $\frac{2464\ ft^2}{75\ ft^2}$ per roll = 33 rolls
    Walls: $6\frac{1}{4}$ in. × $15\frac{1}{2}$ in. foil-faced f.g.
    $$(144 \times 7.5) - 254 = \frac{826\ ft^2}{50} = 17 \text{ rolls}$$
    1 in. Styrofoam, 2 ft × 8 ft sheets:
    $$(8 \text{ ft } 8 \text{ in.} \times 144) - 254 = \frac{994\ ft^2}{16} = 62 \text{ sheets}$$
43. Drywall
    Exterior walls: 144 × 8 = 1152 ft^2
    Ceilings: 1232 ft^2
    Interior walls: 178 × 8 × 2 = 2848 ft^2
    Total area: $\frac{5232\ ft^2}{48}$ = 109 sheets of $\frac{1}{2}$ in. × 4 ft × 12 ft
44. Masonry–as per subcontract
45. Kitchen–as per bid
46. Baths–as per subcontract
47. Heating–as per subcontract
48. Second-floor covering
    $\frac{5}{8}$-in. particleboard: $\frac{1232}{32}$ = 38.5 or 39 sheets
49. Interior door schedule
    Prehung six-panel pine, split jambs: 7 sets 2 ft 6 in. × 6 ft 6 in.
    Slider six-panel pine, no jamb: 2 sets 2 ft 0 in. × 6 ft 6 in.
    Slider six-panel pine, no jamb: 1 set 4 ft 0 in. × 6 ft 6 in.
50. Moldings
    Mopboards: $3\frac{1}{2}$ in. clamshell—638 lf
    Door/window: $2\frac{1}{2}$ in. clam—544 lf
    Window sill: 48 lf
    Opening build-out: select pine—1 × 6—68 lf

Figure 27–2 ***(Continued)***

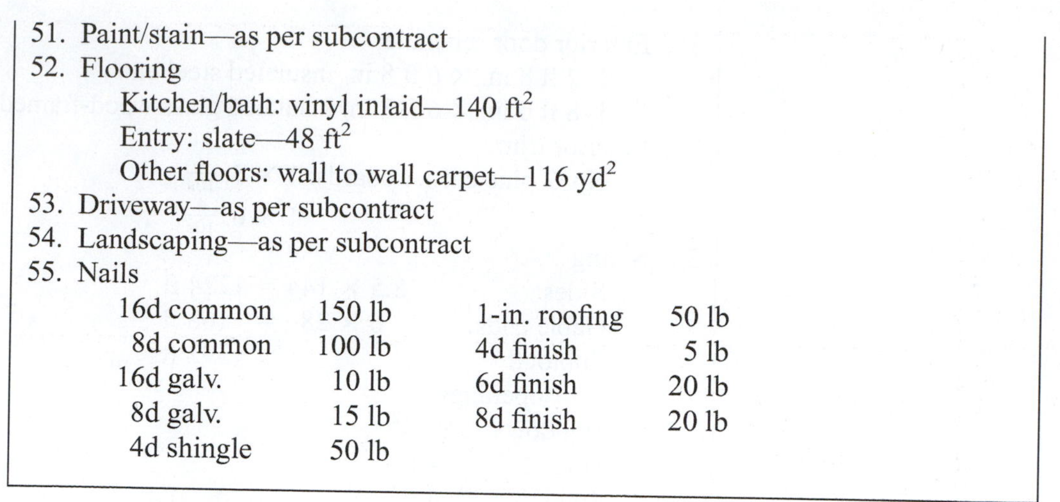

51. Paint/stain—as per subcontract
52. Flooring
    Kitchen/bath: vinyl inlaid—140 ft^2
    Entry: slate—48 ft^2
    Other floors: wall to wall carpet—116 yd^2
53. Driveway—as per subcontract
54. Landscaping—as per subcontract
55. Nails

16d common	150 lb	1-in. roofing	50 lb
8d common	100 lb	4d finish	5 lb
16d galv.	10 lb	6d finish	20 lb
8d galv.	15 lb	8d finish	20 lb
4d shingle	50 lb		

Figure 27-2 (*Continued*)

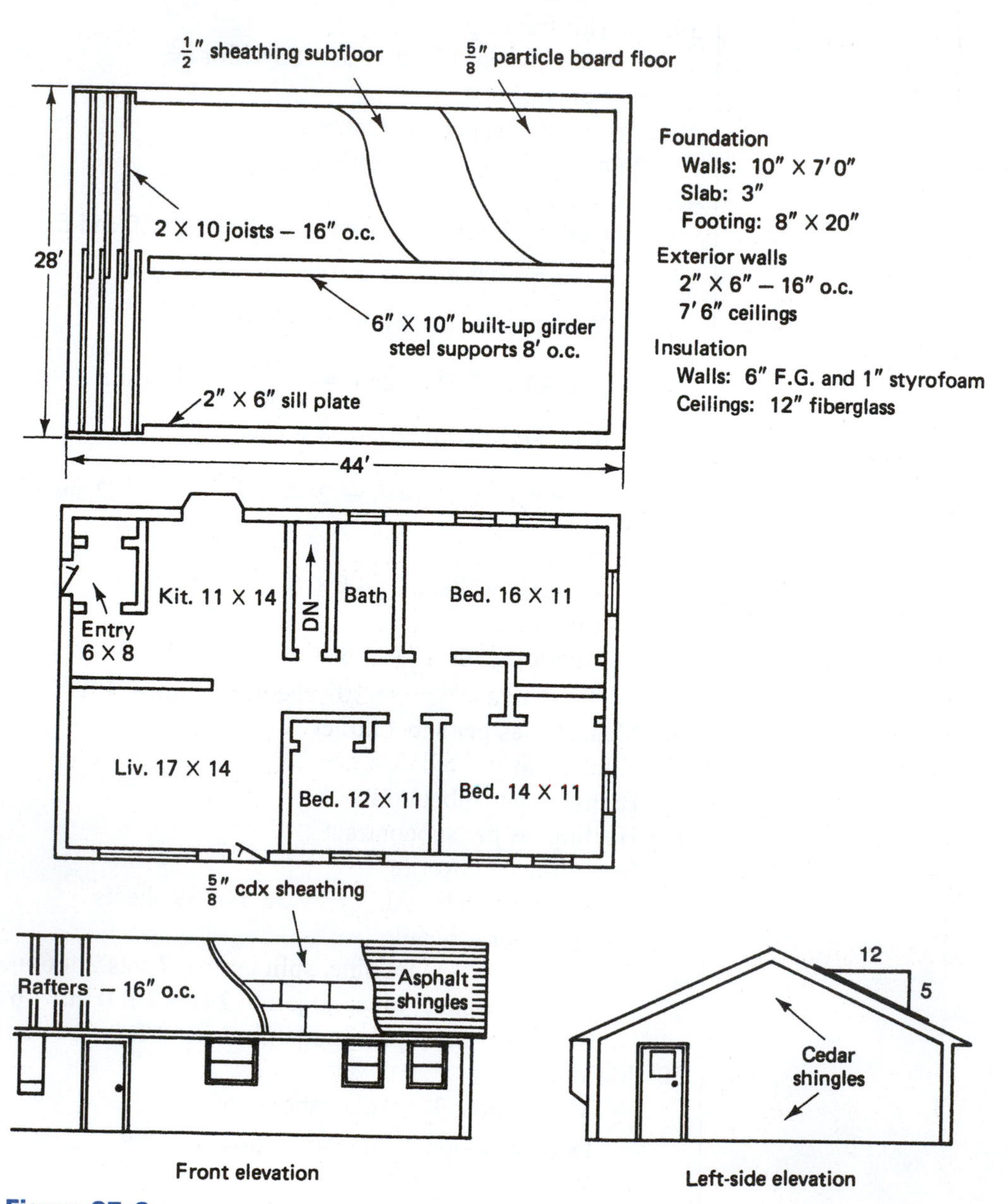

Figure 27-3

ESTIMATE SUMMARY

Project location ______________ Sheet no. ____ of ____

Item	Size/ number/ length	Bd. ft.	Unit cost	Labor	Others	Total
1. Site preparation					×	
2. Excavation					×	
3. Footing	6 yd^3					
4. Foundation	31 yd^3					
5. Drain tile	144 lf					
Filter paper	200 ft^2					
6. Foundation coating	10 gal					
7. Water or well					×	
8. Septic or sewer					×	
9. Exterior cellar insulation	54 sheets 1 in. × 2 ft × 8 ft					
10. Backfill					×	
11. Gravel	3 yd^3 crushed stone 23 yd^3 gravel					
12. Cellar slab	11.5 yd^3					
13. Girder supports	5 adj. posts					
14. Girder	2 × 10: 3/12s 6/16s					
15. Sill seal	144 lf					
16. Sill plates	2 × 6: 2/12s 4/14s 4/16s					
17. Floor joists	2 × 10: 70/14s					
18. F.j. headers	2 × 10: 2/12s 4/16s					
19. Bridging	1 × 3: 264 lf					
20. Subfloor	1/2 in. CDX: 39 sheets					
21. Shoe and plates	2 × 6: 36/12s					
22. Studs	2 × 6: 145/8s					
23. Headers and sills	2 × 6: 4/10s 7/12s					
24. Wall sheathing	1/2 in. CDX plywood: 36 sheets					
25. Stairs	2 × 12: 3/12s 2/14s					
26. Cross-ties (trusses)						
27. Rafters (trusses)	2 × 6: 46/18s 2 × 4: 46/14s 46/12s					
28. Roof sheathing	5/8 in. CDX: 49 sheets					
29. Jack studs	2 × 4: 11/12s					

Figure 27–4

Item	Size/ number/ length	Bd. ft.	Unit cost	Labor	Others	Total
30. Gable end sheathing	1/2 in. CDX: 6 sheets					
31. Soffit and fascia						
Framing	2 × 6: 4/18s					
	2 × 4: 12/12s					
	2/10s					
Covering	3/8 in. A-C ext: 7 sheets					
32. Drip edge	16 pcs.					
33. Asphalt shingles	16 squares					
Felt	4 rolls					
34. Ridge vent	4 pcs.					
35. Window schedule						
7–2 ft 9 in. × 4 ft 1 in. d.h.						
1–2 ft 9 in. × 3 ft 5 in. d.h.						
1–9 ft 5 in. × 4 ft 5 in. pict., 2 ft 0 in. flank. d.h.						
1–3 ft 4 in. × 3 ft 4 in. csmt.						
1–7 ft 10 in. × 5 ft 1 in. bay						
36. Door schedule (ext).						
1–2 ft 8 in. × 6 ft 8 in. insul. steel						
1–8 ft 0 in. × 6 ft 8 in. insul. glass slider						
37. Exterior trim	No. 2 pine:					
	1 × 5: 4/10s					
	1 × 6: 4/10s					
38. Siding	Cedar shingles 12 squares					
39. Interior framing						
Shoe and plates	2 × 4: 45/12s					
Studs	2 × 4: 534/8s					
40. Wiring					×	
41. Plumbing					×	
42. Insulation	6 × 23 unfaced f.g. 33 rolls					
	6 × 15 foil-faced f.g. 17 rolls					
	1 in. Styrofoam 62 sheets					
43. Drywall	1/2 in. × 4 ft × 12 ft: 109 sheets					
44. Masonry					×	
45. Kitchen					×	
46. Bath(s)					×	
47. Heating					×	
48. Second floor covering	5/8 in. particleboard: 39 sheets					
49. Door schedule (int.)						
7 2 ft 6 in. × 6 ft 6 in. prehung six-panel pine						
2 sets 2 ft 0 in. × 6 ft 6 in. sliders						
1 set 4 ft 0 in. × 6 ft 6 in. slider						
50. Molding	3 1/2 in. clam: 638 lf					
	2 1/2 in. clam: 544 lf					
	Window sill: 48 lf					
	1 × 6 sel. pine: 68 lf					

Figure 27–4 (*Continued*)

Item	Size/ number/ length	Bd. ft.	Unit cost	Labor	Others	Total
51. Paint, caulk					×	
52. Flooring	Vinyl: 140 ft^2 Slate: 48 ft^2 W/w carpet: 116 yd^2					
53. Driveway					×	
54. Landscaping					×	
55. Nails	16d common: 150 lb 8d common: 100 lb 16d galv.: 10 lb 8d galv.: 15 lb 4d shingle: 50 lb 1-in. roofing: 50 lb 4d finish: 5 lb 6d finish: 20 lb 8d finish: 20 lb					
56. Sales tax						
57. Markup						
Totals						

Figure 27–4 ***(Continued)***

TAKE-OFF LIST
Nails and Numbers Construction Company
Termite, Maine

To: Knothole Lumber Co.

Please submit your bid for the following materials at your earliest convenience.

2×12: $\frac{3}{12s}$, $\frac{2}{14s}$ 2×8: $\frac{66}{14s}$, $\frac{66}{18s}$

2×10: $\frac{5}{12s}$, $\frac{70}{14s}$, $\frac{10}{16s}$

2×6: $\frac{145}{8s}$, $\frac{4}{10s}$, $\frac{45}{12s}$, $\frac{4}{14s}$, $\frac{4}{16s}$, $\frac{4}{18s}$

2×4: $\frac{534}{8s}$, $\frac{2}{10s}$, $\frac{62}{12s}$

Plywood: $\frac{1}{2}$-in. CDX: 81 sheets
$\frac{5}{8}$-in. CDX: 49 sheets
$\frac{3}{8}$-in. A-C ext.: 7 sheets

No. 2 pine: 1×5: $\frac{4}{10s}$
1×6: $\frac{4}{10s}$

Strapping: 1×3: 264 lf
Particleboard: $\frac{5}{8}$ in.: 39 sheets

Figure 27–5

Roofing materials:
15-lb felt paper: 4 rolls
8-in. galv. drip edge: 16 lengths
Asphalt shingles: 16 squares
Ridge vent: 4 lengths
Drywall: $\frac{1}{2}$ in. × 4 ft × 12 ft: 109 sheets
Cedar shingles: 12 squares
Moldings:
$3\frac{1}{2}$ in. clam—638 lf
$2\frac{1}{2}$ in. clam—544 lf
Sill cap: 48 lf
Select pine: 1 × 6—68 lf
Insulation:
1 in. × 2 ft × 4 ft: 116 sheets
Sill seal: 144 lf
$6\frac{1}{4}$ in. × $23\frac{1}{2}$ in. unfaced f.g.—33 rolls
$6\frac{1}{4}$ in. × $15\frac{1}{2}$ in. foil-faced f.g.—17 rolls
Nails:

16d common:	150 lb
8d common:	100 lb
16d galv.:	10 lb
8d galv.:	15 lb
4d shingle:	50 lb
1 in. roofing:	50 lb
4d finish:	5 lb
6d finish:	20 lb
8d finish:	20 lb

200 ft^2 filter paper
144 lf 4-in. drain tile
10 gal foundation coating
5 adjustable steel columns
Window schedule: All SeeThru brand thermopane
7–2 ft 9 in. × 4 ft 1 in. D.H.
1–2 ft 9 in. × 3 ft 5 in. D.H.
1–9 ft 5 in. × 4 ft 5 in. picture with 2 ft D.H. flanking
1–3 ft 4 in. × 3 ft 4 in. casement
1–7 ft 10 in. × 5 ft 0 in. bay
Door schedule:
1–2 ft 8 in. × 6 ft 8 in. insul. steel entry
1–8 ft 0 in. × 6 ft 8 in. insul. glass wood-framed slider
7–2 ft 6 in. × 6 ft 6 in. six-panel pine prehung, split jambs
2 sets 2 ft 6 in. × 6 ft 6 in. sliding, six-panel pine, no jambs
1 set 4 ft 0 in. × 6 ft 6 in. sliding, six-panel pine, no jambs

Figure 27–5 (*Continued*)

Notwithstanding these comments, references do exist for estimating labor. (We would add that many estimating references dismiss the topic as we did in the two previous paragraphs.) One resource is the Steinberg and Stempel book cited in the Bibliography.

REVIEW EXERCISE

1. As a summary project, submit a completed work sheet, estimate summary sheet, and take-off list for one of the plans shown in Appendix D.

APPENDIX A

Right-Triangle Trigonometry

Trigonometry has been studied and developed for thousands of years. The Greeks used it in astronomy, navigation, and geography 2000 years ago, and it has become an essential part of many disciplines: physics, engineering, chemistry, other fields of mathematics, and certain areas of medicine, to name a few. (Trigonometry comes from the Greek words for triangle and measure.)

Additionally, trigonometry (trig, for short) has practical applications in surveying and building construction. Of the several types of trig, this appendix will address only right triangle trigonometry. Right triangle trig is the simplest, most straightforward, and arguably the most useful for building construction applications.

Reviewing Chapter 7, we know that if two right triangles have an equal acute angle, they are similar. In fact, all right triangles with that same acute angle are similar. (Why?) In Figure A–1, $\triangle ABC \sim \triangle ADE \sim \triangle AFG$. (~ is the symbol for "is similar to.")

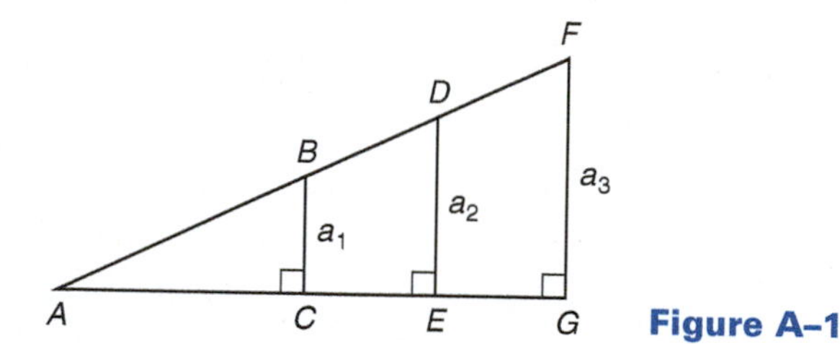

Figure A–1

Recall that similar triangles have a very important characteristic: the ratios of their corresponding sides are equal. Therefore, in Figure A–2,

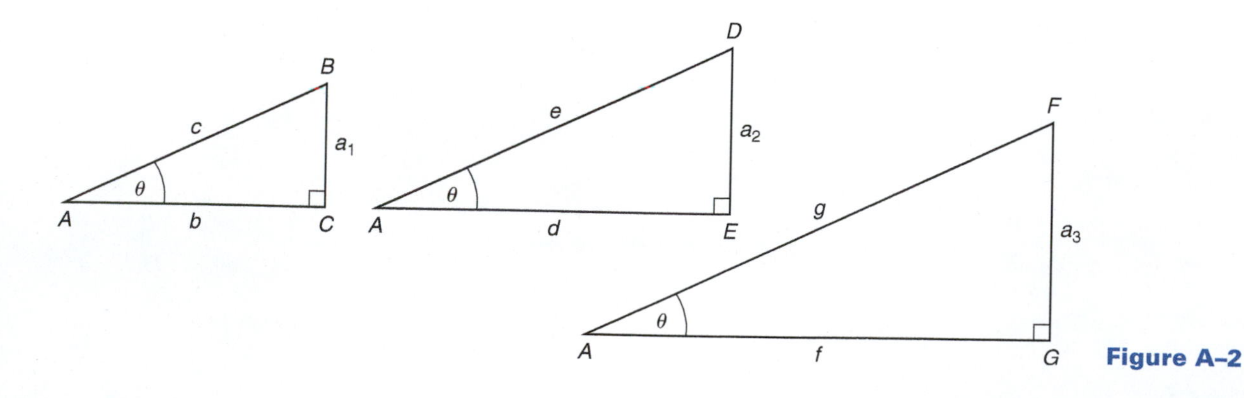

Figure A–2

$\frac{a_1}{c} = \frac{a_2}{e} = \frac{a_3}{g}$ These ratios are called the sine function when referring to $\angle A$.

$\frac{b}{c} = \frac{d}{e} = \frac{f}{g}$ These ratios are called the cosine function when referring to $\angle A$.

$\frac{a_1}{b} = \frac{a_2}{d} = \frac{a_3}{f}$ These ratios are called the tangent function when referring to $\angle A$.

In similar triangles, even though the lengths of the corresponding sides are different, the ratios of corresponding sides are not. These ratios will be constant. Because the ratios are constant, we give them names. $\triangle ABC$ in Figure A–3, when referring to $\angle A$: side a is the *opposite side*; side b is the *adjacent side*; snd side c is the *hypotenuse*.

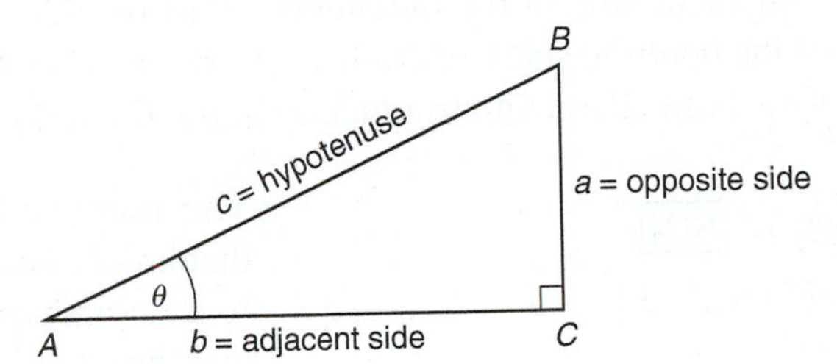

Figure A–3

(The hypotenuse is always opposite the right angle.) "θ" is used to indicate any acute angle in the triangle. Therefore,

$$\text{sine } \theta = \frac{\text{opposite side}}{\text{hypotenuse}} \qquad \text{abbreviated as } \sin\theta = \frac{\text{opp}}{\text{hyp}}$$

$$\text{cosine } \theta = \frac{\text{adjacent side}}{\text{hypotenuse}} \qquad \text{abbreviated as } \cos\theta = \frac{\text{adj}}{\text{hyp}}$$

$$\text{tangent } \theta = \frac{\text{opposite side}}{\text{adjacent side}} \qquad \text{abbreviated as } \tan\theta = \frac{\text{opp}}{\text{adj}}$$

For $\triangle XYZ$ in Figure A–4, when referring to $\angle X$:

$$\sin X = \frac{x}{z} \quad \left(\frac{\text{opp}}{\text{hyp}}\right)$$

$$\cos X = \frac{y}{z} \quad \left(\frac{\text{adj}}{\text{hyp}}\right)$$

$$\tan X = \frac{x}{y} \quad \left(\frac{\text{opp}}{\text{adj}}\right)$$

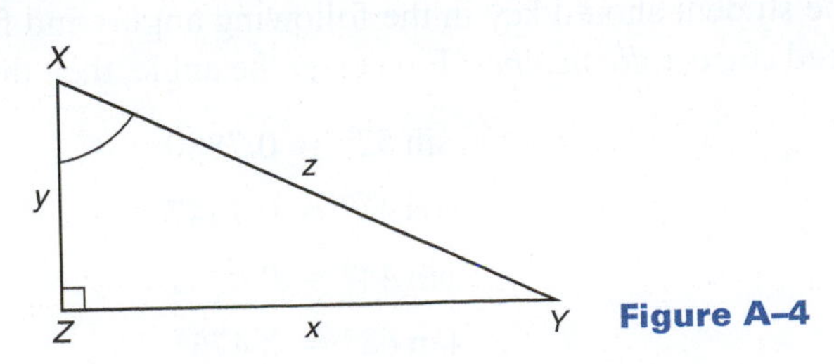

Figure A–4

In reference to $\angle Y$ in Figure A–5, note that x now becomes the adjacent side and y becomes the opposite side. The hypotenuse is always the same side regardless of the angle in reference.

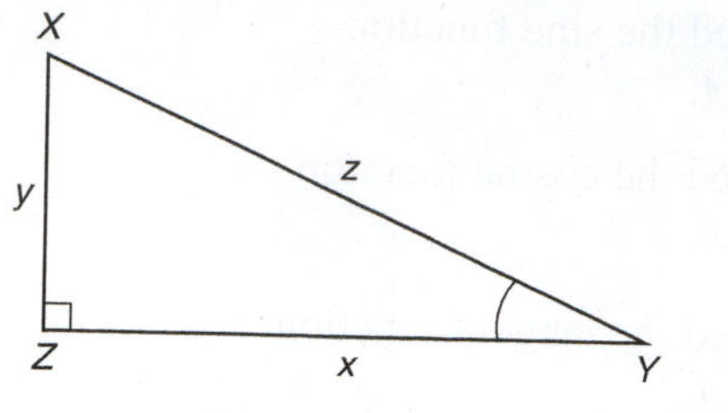

Figure A–5

$$\sin Y = \frac{y}{z} \quad \left(\frac{\text{opp}}{\text{hyp}}\right)$$

$$\cos Y = \frac{x}{z} \quad \left(\frac{\text{adj}}{\text{hyp}}\right)$$

$$\tan Y = \frac{y}{x} \quad \left(\frac{\text{opp}}{\text{adj}}\right)$$

Since every right triangle with an acute angle of 38°, for example, is similar, the ratios of their opposite side to hypotenuse (sin) will be the same: sin 38° = 0.6157. This is also true for the other ratios. No matter what size right triangle, if it has an acute angle of 38°, the ratio of the adjacent side to the hypotenuse will be a constant: cos 38° = 0.7880. Also, the ratio of the opposite side to the adjacent side is a constant: tan 38° = 0.7813. To find these ratios on a calculator with trig functions, use the following procedure:

enter: 38 [SIN]

sin 38° = .6157

The number 0.6156615 will then be displayed. This is generally rounded to four decimal places, or .6157. [Be sure that the calculator is in the degree (deg) mode.]

A right triangle with an acute angle of 38° was used as an illustration. A right triangle with an acute angle of any value follows the same principle. For example, if two right triangles of different sizes have an acute angle of 53.1°, they are similar triangles and the ratios of their corresponding sides are equal (Figure A–6).

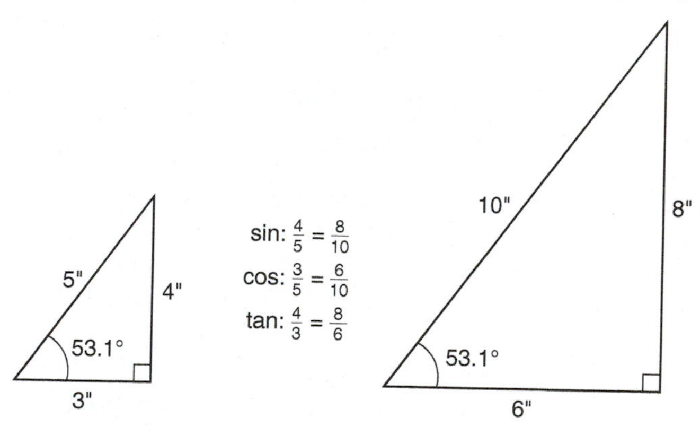

Figure A–6

The following list gives the values of the ratios (called functions) for the sine, cosine, and tangent of any right triangle with the specified angles. Using a calculator with trig functions, the student should key in the following angles and functions to ascertain that they are indeed correct. *Remember:* First enter the angle, then the sin, cos, or tan function.

sin 52° = 0.7880

cos 41° = 0.7547

sin 44° = 0.6947

tan 68° = 2.475

cos 32° = 0.8480

tan 59° = 1.664

The values for the sin and cos of an angle can never exceed 1.000. The tan of an angle can have any value.

Exercise A–1

Complete the table:

	angle	sin	cos	tan
a.	37.80°	0.6129	0.7902	0.7757
1.	43.80°			
2.	19.26°			
3.	45.00°			
4.	18.62°			

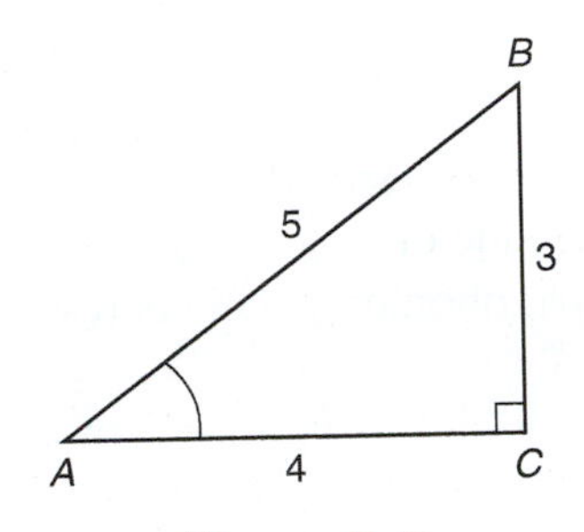

Figure A–7

In examples given above, the ratio of the sides of a triangle could be found if the angle was known. In reverse form, if the ratio of the sides of a triangle is known, the angle can be found. For $\triangle ABC$ in Figure A–7.

$$\sin A = \frac{3}{5} = 0.6000$$

$$\cos A = \frac{4}{5} = 0.8000$$

$$\tan A = \frac{3}{4} = 0.75000$$

Knowing that $\sin A = 0.6000$, $\angle A$ can be found by following this sequence on the calculator:

.6 [INV] [SIN] or .6 [2nd] [SIN^{-1}] (depending on calculator)

Knowing that $\cos A = 0.8000$, $\angle A$ can be found by following this sequence:

.8 [INV] [COS] or .8 [2nd] [COS^{-1}]

Knowing that $\tan A = 0.7500$, use this sequence to find $\angle A$:

.75 [INV] [TAN] or .75 [2nd] [TAN^{-1}]

$\angle A = 36.87°$ regardless of which of the three functions was used to find it. Ordinarily, one method is more convenient than the others, and that is the method chosen to find the unknown angle.

EXAMPLE

A. Find $\angle A$ for the triangle shown. All three methods of finding $\angle A$ will be shown.

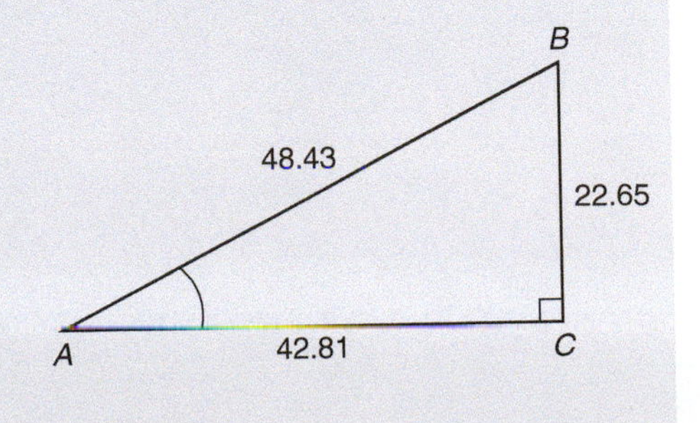

$$\sin A = \frac{22.65}{48.43}$$

$$= 0.4677$$

$$\angle A = 27.88°$$

$$\cos A = \frac{42.81}{48.43}$$
$$= 0.8840$$
$$\angle A = 27.88°$$
$$\tan A = \frac{22.65}{42.81}$$
$$= 0.5291$$
$$\angle A = 27.88°$$

The three sides and three angles of a triangle are referred to as *parts* of the triangle. In a right triangle, if any side plus one other part (another side or an acute angle) are given, the remaining three parts can be found by using trigonometry. This is referred to as *solving* the triangle.

EXAMPLES

B. Solve $\triangle ABC$.

$A = 28°$	$a =$
$B =$	$b =$
$C = 90°$	$c = 55$ in.

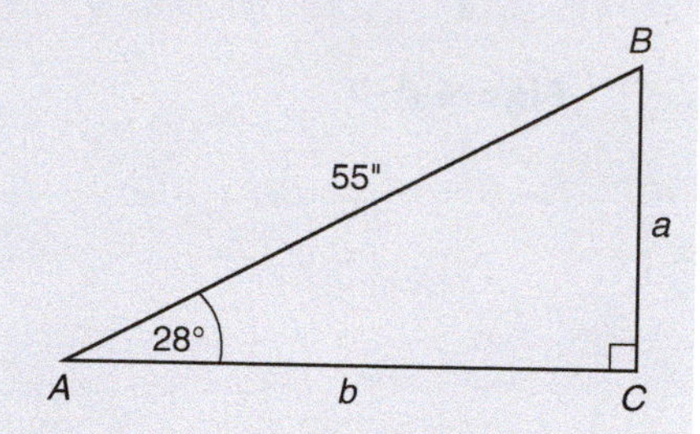

The right angle is always given, and in this example $\angle A$ and side c are also given.

$$\angle B = 90° - 28° = 62°$$

1. The two acute angles of a right triangle always add to 90°.

$$\sin 28° = \frac{a}{55 \text{ in.}}$$

2. Sin $A = a/c$. Set up as a proportion and substitute all known values.

$$\frac{\sin 28°}{1} = \frac{a}{55 \text{ in.}}$$

$$\frac{0.4694716}{1} = \frac{a}{55 \text{ in.}}$$

3. Enter 28 [SIN] to determine value of sin 28°.

$$a = 25.8 \text{ in.}$$

4. Cross-multiply and solve for a.

$$\cos 28° = \frac{b}{55 \text{ in.}}$$

5. Cos $A = b/c$. Set up as a proportion and substitute all known values.

$$\frac{\cos 28}{1} = \frac{b}{55 \text{ in.}}$$

$$b = 48.6 \text{ in.}$$

6. Cross-multiply and solve for b.

$A = 28°$	$a = 25.8$ in.
$B = 62°$	$b = 48.6$ in.
$C = 90°$	$c = 55$ in.

C. Solve $\triangle xyz$:

$X = 53°$ $x = 22$ in.

$Y =$ $y =$

$Z = 90°$ $z =$

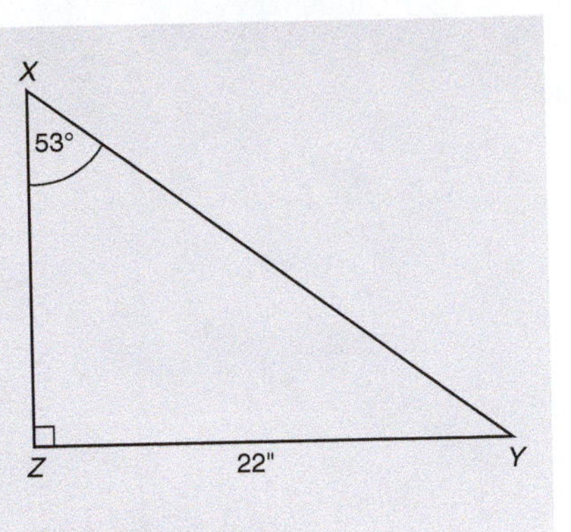

$\angle Y = 90° - 53° = 37°$ 1. Find $\angle Y$.

$\sin 53° = \dfrac{22}{z}$ 2. Set up sin function as a proportion.

$\dfrac{\sin 53°}{1} = \dfrac{22 \text{ in.}}{z}$ 3. In a proportion, the diagonal values can be swapped.

$\dfrac{z}{1} = \dfrac{22 \text{ in.}}{\sin 53°}$ 4. z is easier to solve for in this form.

$z = 27.5$ in.

$\dfrac{\tan 53°}{1} = \dfrac{22 \text{ in.}}{y}$ 5. Set up the tan value to find y.

$\dfrac{y}{1} = \dfrac{22 \text{ in.}}{\tan 53°}$ 6. Swap diagonal values so y can be found more easily.

$y = 16.6$ in.

$X = 53°$ $x = 22$ in.

$Y = 37°$ $y = 16.6$ in.

$Z = 90°$ $z = 27.5$ in.

On any triangle, the shortest side is always opposite the smallest angle, and the longest side is opposite the largest angle.

Exercise A–2

Solve the triangles.

1. $A = 42°$ $a =$
   $B =$ $b =$
   $C = 90°$ $c = 38$ ft

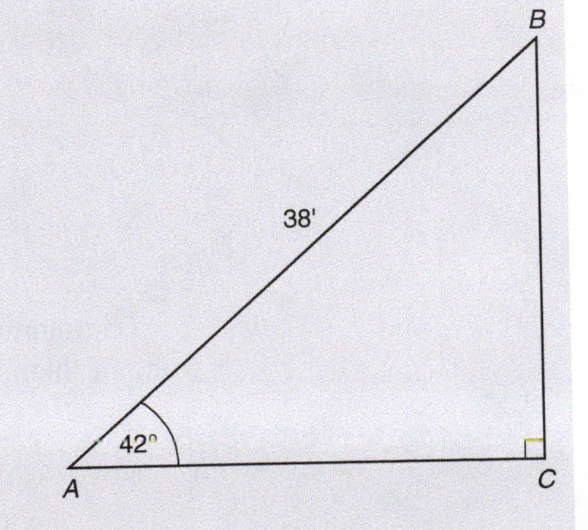

2. $X =$ $\quad$ $x = 15$ ft
   $Y = 38°$ $\quad$ $y =$
   $Z = 90°$ $\quad$ $z =$

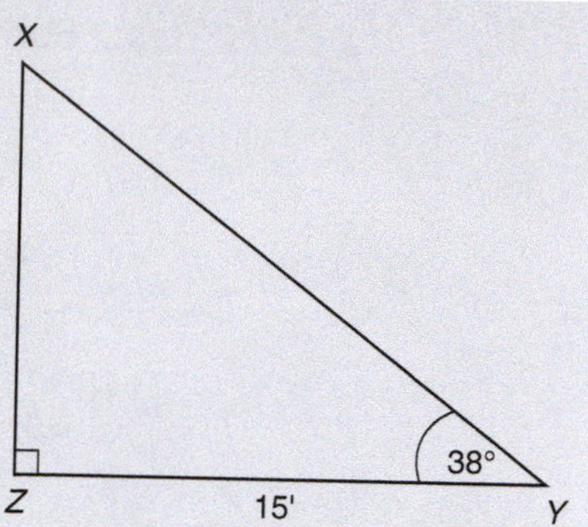

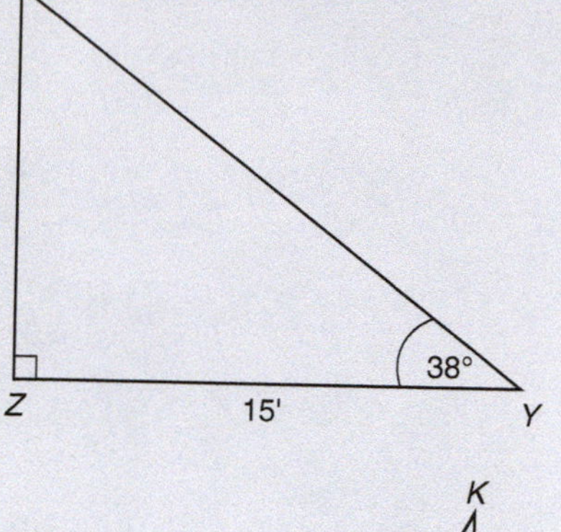

3. $J =$ $\quad$ $j = 46$ in.
   $K = 29°$ $\quad$ $k =$
   $L = 90°$ $\quad$ $l =$

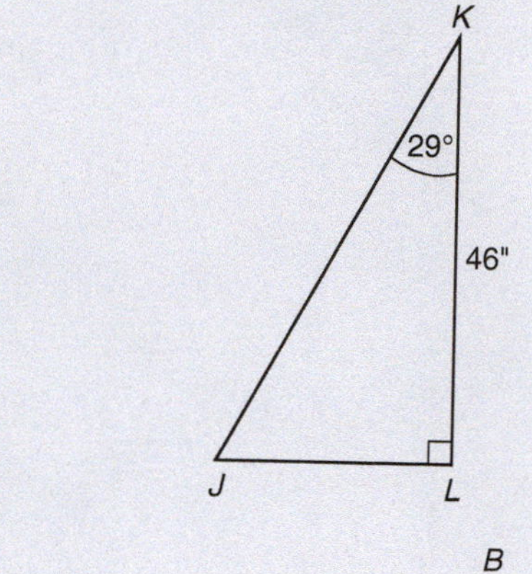

4. $A = 36°$ $\quad$ $a =$
   $B =$ $\quad$ $b =$
   $C = 90°$ $\quad$ $c = 29$ ft

5. $A =$ $\quad$ $a =$
   $B = 45°$ $\quad$ $b =$
   $C = 90°$ $\quad$ $c = 38.2$

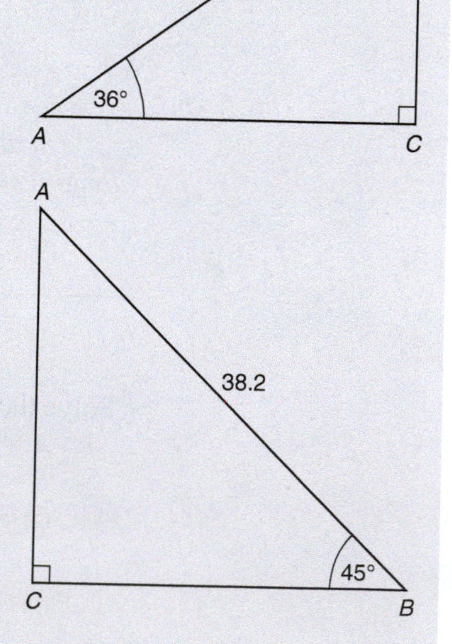

Trigonometry has many useful applications. Study the examples below to learn some of them.

EXAMPLES

D. A bridge is to be built across a river and the distance across the river is to be measured at that point. A surveyor sets up stakes 75 ft apart as shown and, with a transit, swings the angle to a stake on the opposite shore. If this angle is 58°, what is the distance across the river?

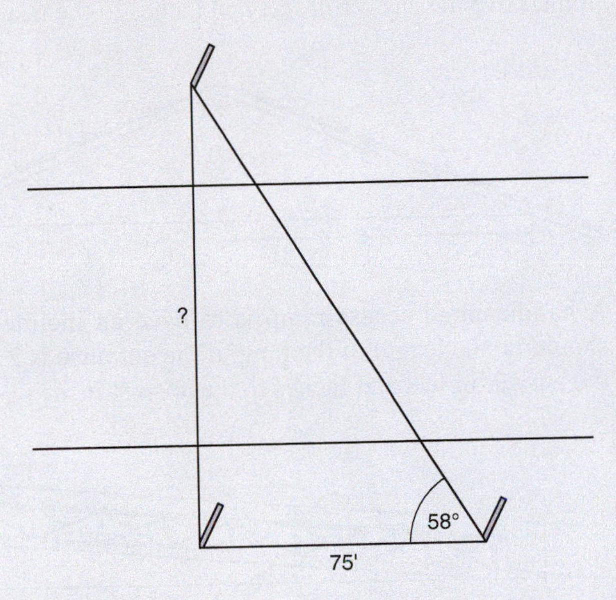

Since the side to be found is opposite the 58° angle, the tangent function is used.

$\tan 58° = \dfrac{x}{75}$	1. Set up the tangent function.
$\dfrac{\tan 58°}{1} = \dfrac{x}{75 \text{ ft}}$	2. Cross-multiply to solve for x.
$x = 120 \text{ ft}$	3. The distance across the river between the stakes is 120 ft.

E. A loading ramp 20 ft long must reach a platform 3 ft 8 in. high. What will be the angle of inclination?

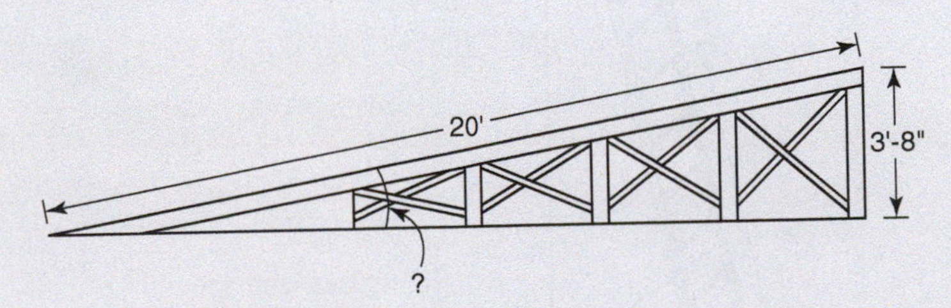

$3 \text{ ft } 8 \text{ in.} = 3.\overline{6} \text{ ft}$	1. Change the height of the platform to decimal feet.
$\dfrac{\sin \theta}{1} = \dfrac{3.\overline{6} \text{ ft}}{20 \text{ ft}}$	2. Set up a proportion using the sine function.
$\sin \theta = 0.1833$	
$\theta = 10.6°$	3. The angle of inclination is approximately 11°

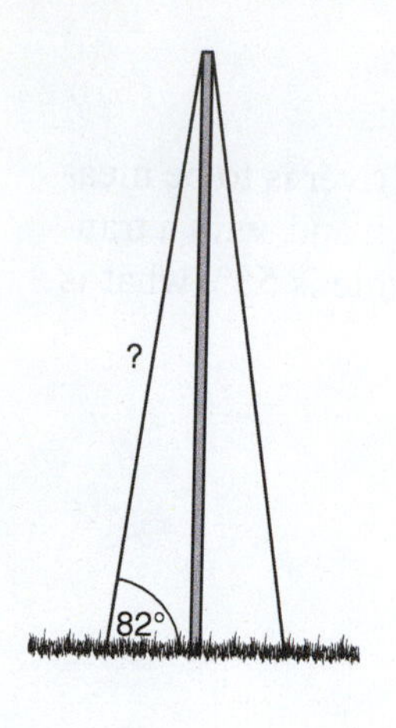

Exercise A–3

1. Guy wires supporting a 20-ft pole make an angle of 82° with the ground. How long are the guy wires?
2. A gable roof has a slope of 22°. If the run is 22 ft 6 in., find the rise at its highest point. Give the answer in feet and inches to the nearest inch.

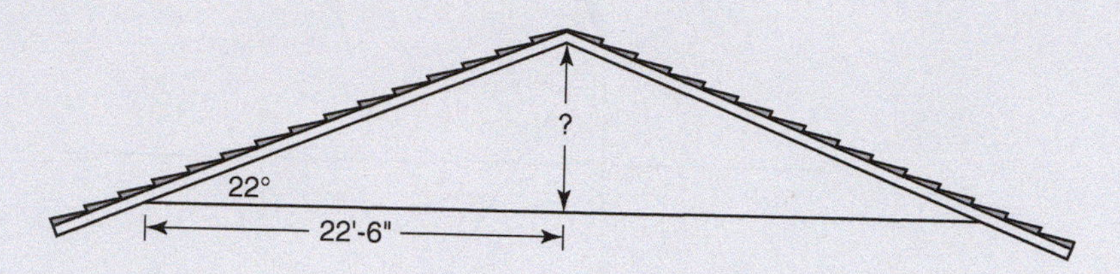

3. A handicapped access ramp is to have an incline of no more than 5°. What should be the length of the ramp if the entrance is 2 ft 9 in. off the ground? Give the answer in feet and inches to the nearest inch.

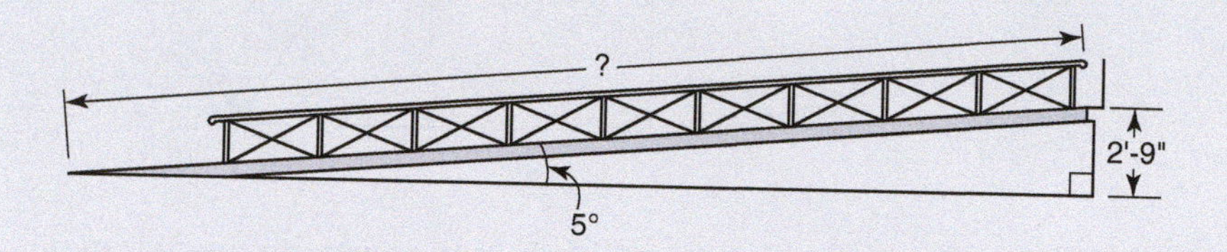

4. A 20-ft ladder is propped against a wall in such a way that the foot of the ladder is 4 ft 3 in. from the wall. What angle does the ladder make with the ground?

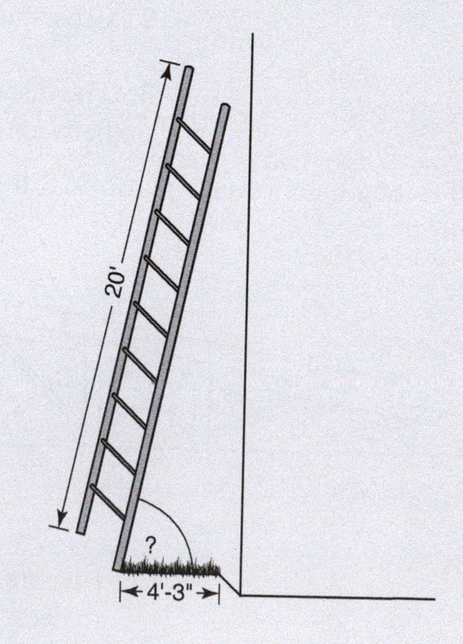

5. A surveyor who is 35 ft from the base of a fire tower finds the angle of elevation to the top of the tower to be 71°. How tall is the tower, to the nearest foot, if the man is 6 ft tall?

6. A roof has a unit rise of 5 in. (5 in. of rise for every 1 ft of run). What is the slope of the roof?
7. A board that is $5\frac{3}{4}$ in. wide is cut on a miter box at an angle of 37°. What is the length, to the nearest $\frac{1}{32}$ in., of the cut edge?

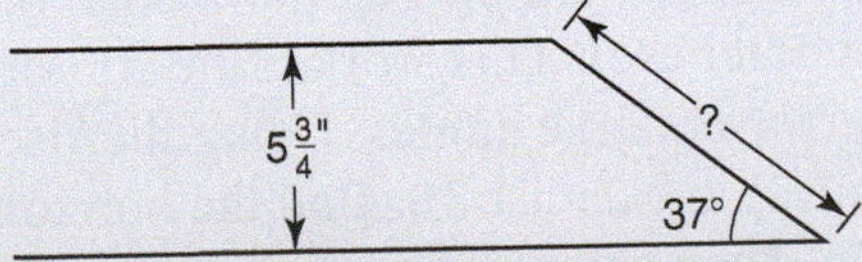

8. On the foundation shown, the diagonals are measured to ensure a perfect rectangle. What should be the angle between the length and the diagonal? What should be the angle between the width and the diagonal? What should the sum of the two angles equal?

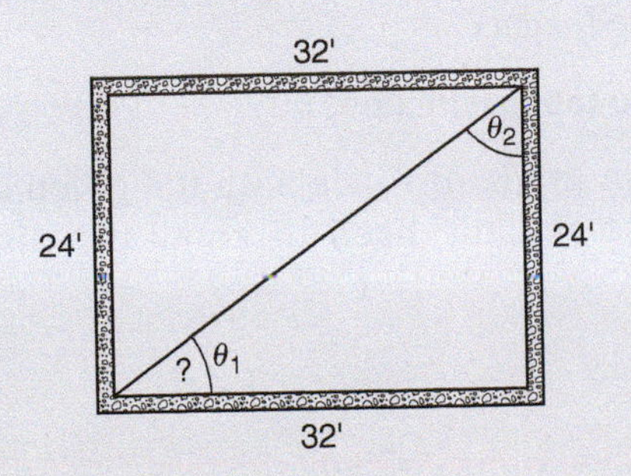

9. A solar panel 3 ft 5 in. wide has multiple adjustment options so that the top edge of the panel can be 6 in. off the horizontal, 1 ft 6 in. off the horizontal, 2 feet off the horizontal, or 2 ft 6 in. off the horizontal. Find the minimum and maximum angles the panel makes with the horizontal.

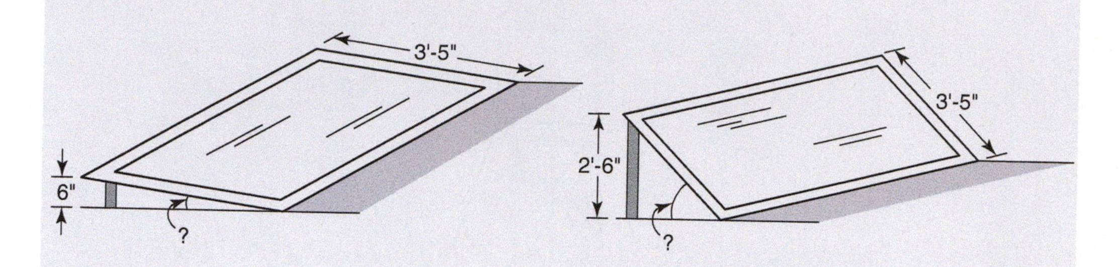

10. The angle of incline for a road can be determined if we know the distance along the road, and the rise in elevation of the road over that distance. For example, a sign indicates that the elevation at point A is 3975 ft. Two miles later, a sign indicates the elevation is 4768 ft. Find the average angle of incline for the road. *Note:* the road does not have to be straight for this calculation to work.) 1 mile = 5280 ft.

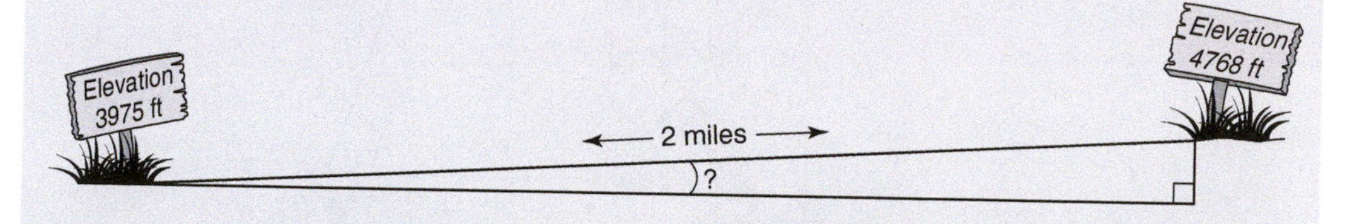

11. Percent grade is defined as the rise over the run, given as a percent. In exercise 10, the 793 foot rise in elevation over the horizontal run equals the percent grade. Most inclines are given as percent grades because the calculations can be done without trig tables or a calculator with trig functions. However, a common method for determining percent grades is actually an approximation of the percent grade. It does not use the run (horizontal distance), but rather the road distance (the hypotenuse). This works fine for small angles, but would produce significant error on large angles. Using the rise and hypotenuse (the road distance of 2 miles) in exercise 10, find the horizontal run of the distance between the two signs. Then calculate percent grade using the exact method (rise over run, then convert the decimal value to a percent). Then recalculate percent grade using the approximate method (rise over road distance).

   a. horizontal run

   b. percent grade (exact)

   c. percent grade (approximately.)

12. Find the percent grade of roads with the given angles of inclination. Notice that you do not actually need to know the distance. The first problem is

worked. You will need to use a calculator with trig functions or a trig table to work these problems.

a. angle of inclination: 5° percent grade: ?

$$\tan 5^\circ = \frac{\text{Rise}}{\text{Run}}$$

$$\tan 5^\circ = 0.0875$$

$$0.0875\% = \frac{\text{Rise}}{\text{Run}}$$

8.7% or 9% = percent grade

b. angle of inclination: 4.5° percent grade:

c. angle of inclination: 6.8° percent grade:

d. angle of inclination: 3° percent grade:

13. A $9\frac{3}{4}''$ wide strip of plywood is cut at an angle of 55°. How long is the cut edge?

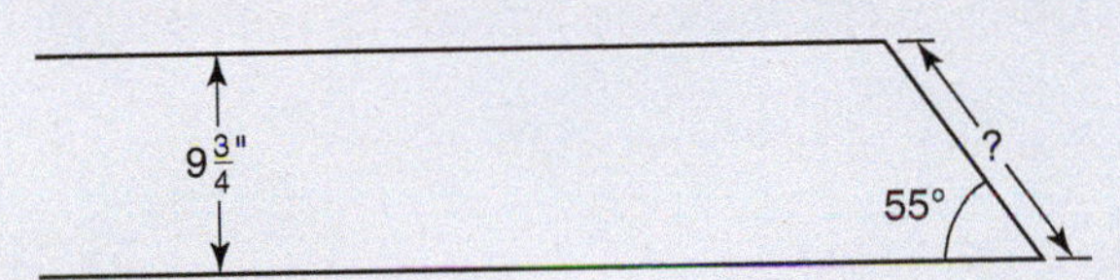

14. An A-Frame cabin has the dimensions shown. Find the angle of the roof with the ground.

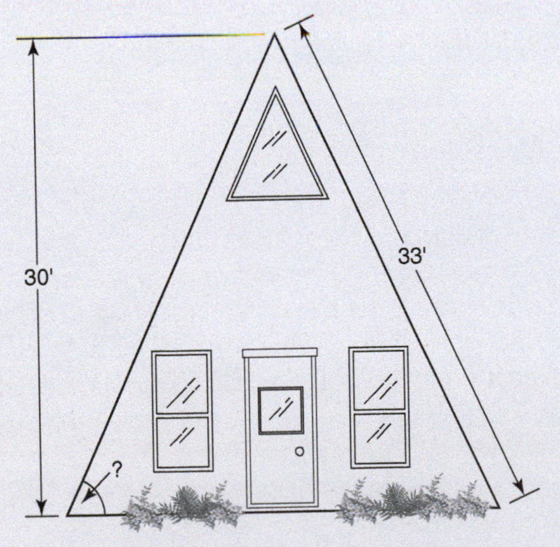

15. Find the slope (in degrees) of the roof shown:

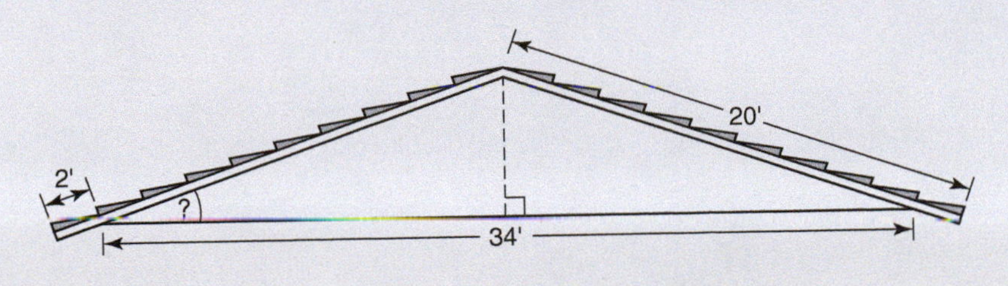

The right triangle ABC is to be used for exercises 16–20.

16. Side $b = 14.8$ and $\angle B = 31.5°$, find the hypotenuse c.
17. In exercise 16, find side a.
18. For $c = 18.5$ and $b = 13.7$, $\angle A = ?$
19. Given $\angle A = 35°$ and $\angle B = 55°$, do we have enough information to find a, b, and c? Why or why not?
20. Given side $a = 18$ and side $b = 24$, do we have enough information to find $\angle A$ and $\angle B$? If so, find angles A and B side c.

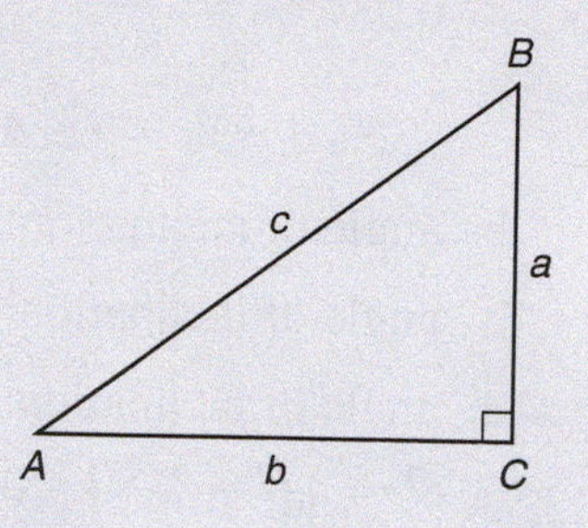

APPENDIX B

Resistances (*R*-Factors) of Various Materials

Note: Values listed are representative of the material, but are not precise. *R*-factors vary with temperature and density. Manufacturers' specifications for specific materials should be consulted.

Material	*R*-factor[a]	Material	*R*-factor[a]
Building materials		Air spaces	
Asphalt shingles	0.44[b]	Horizontal	0.87
Brick	0.20	45° slope	0.94
Carpet		Vertical	1.01
Fibrous pad	2.08[b]	Glass[b]	
Rubber pad	1.23[b]	Single	0.89
Clapboards		Insul. (2)	
$\frac{1}{2}$ in. × 8 in., lapped	0.81[b]	$\frac{1}{4}$ in. space	1.55
$\frac{3}{4}$ in. × 10 in., lapped	1.05[b]	$\frac{1}{2}$ in. space	1.72
Concrete	0.08	Insul. (3)	
Door, $1\frac{3}{4}$ in. solid wood	1.96[b]	$\frac{1}{4}$ in. space	2.13
Drywall	0.90	$\frac{1}{2}$ in. space	2.78
Felt paper	0.06[b]		
Tile, vinyl	0.05[b]	Storm	
Wood	1.25	4 in. space	1.79
Plywood	1.25		
Fiber board	2.38		
Insulating materials			
Acoustical tile	2.38		
Expanded polystyrene			
Extruded, 60°F	4.00		
Molded beads, 75°F	3.57		
Expanded polyurethane (blown, 50°F)	6.25		
Glass fiber (60°F)	3.00		
Mineral wool (60°)			
Resin binder	3.57		
Loose fill	3.70		
Wood fiber	4.00		

[a]Per inch of thickness.

[b]For material as listed.

APPENDIX C

House Plans

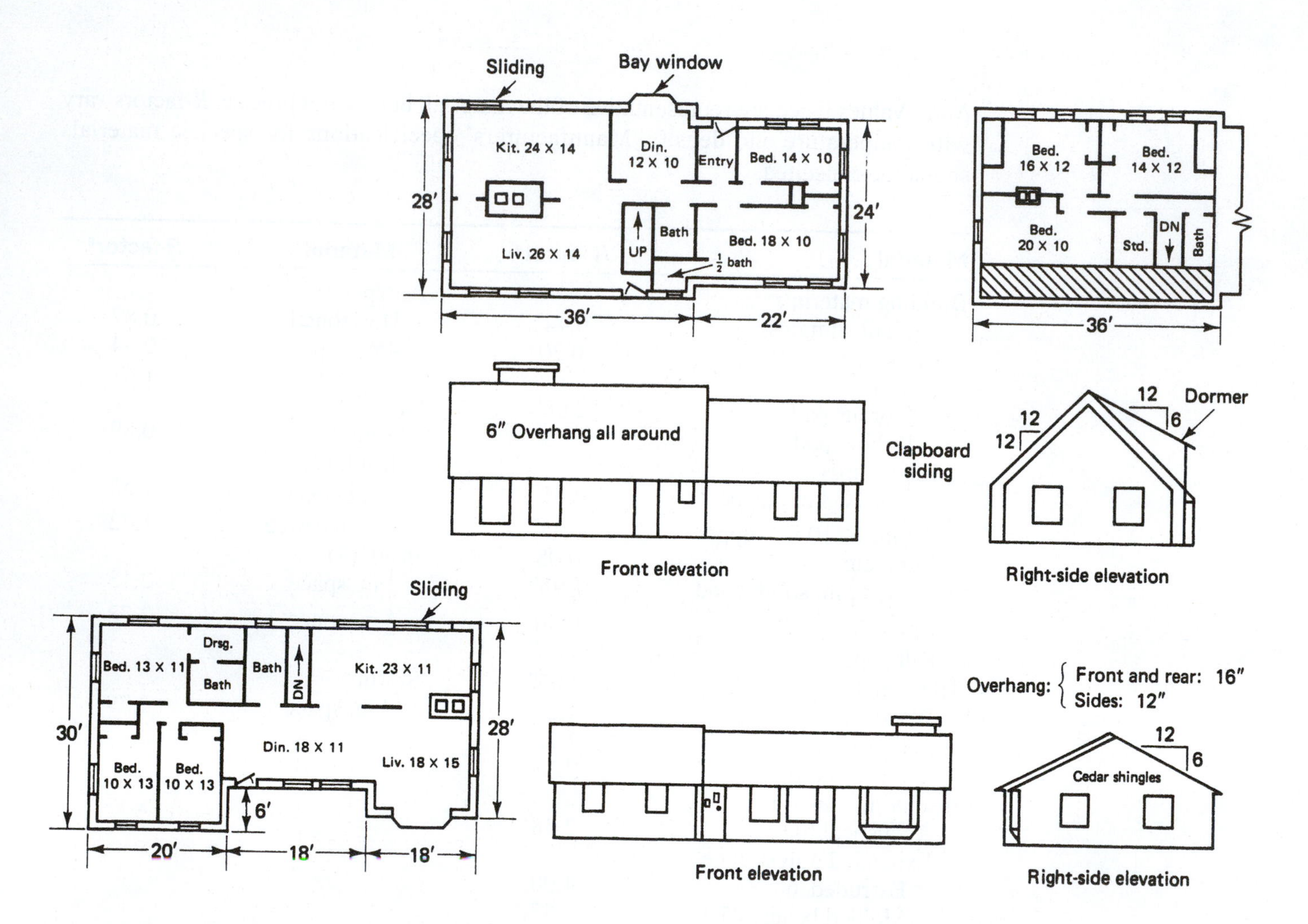

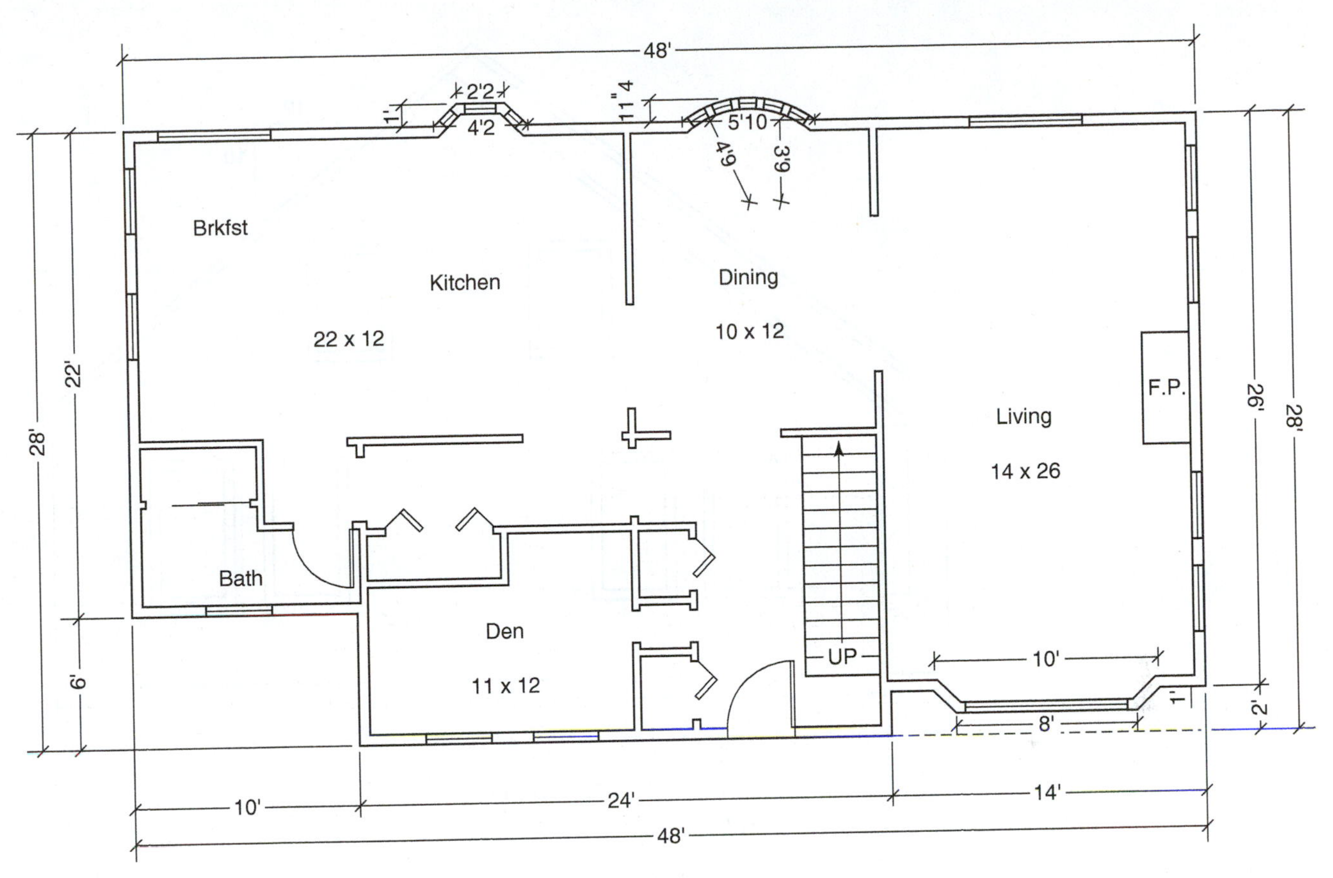

First Floor

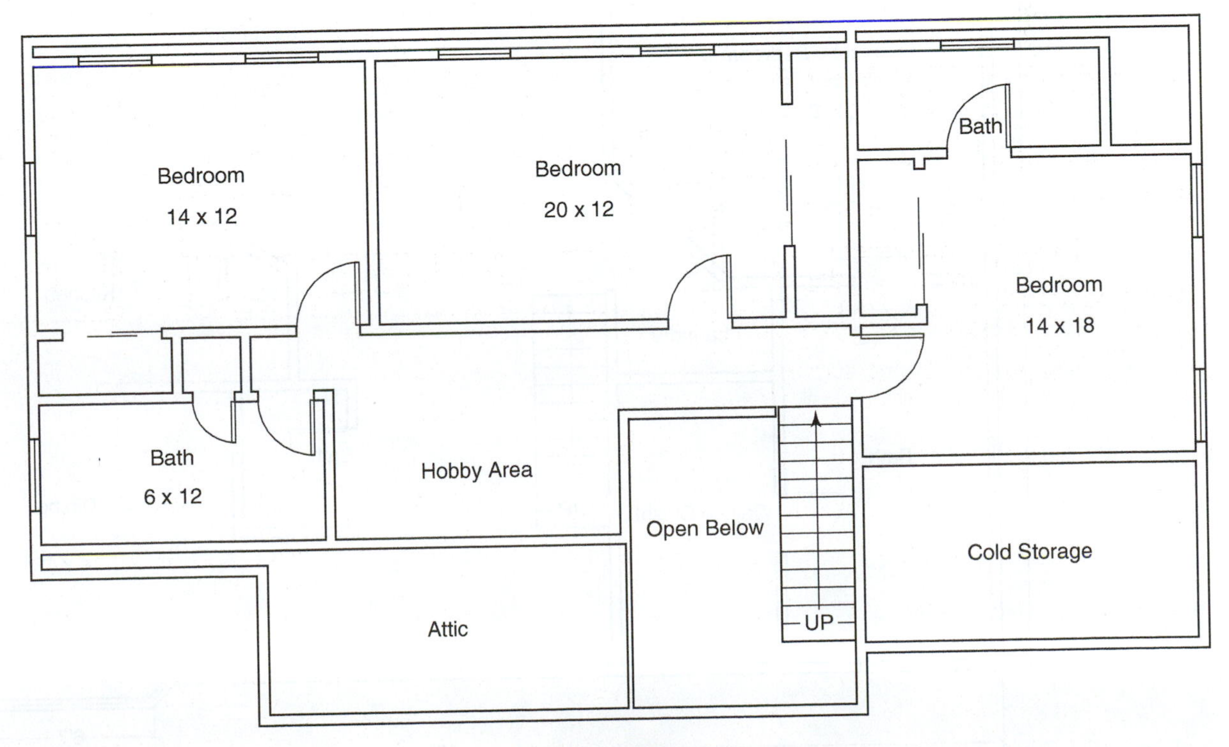

Second Floor

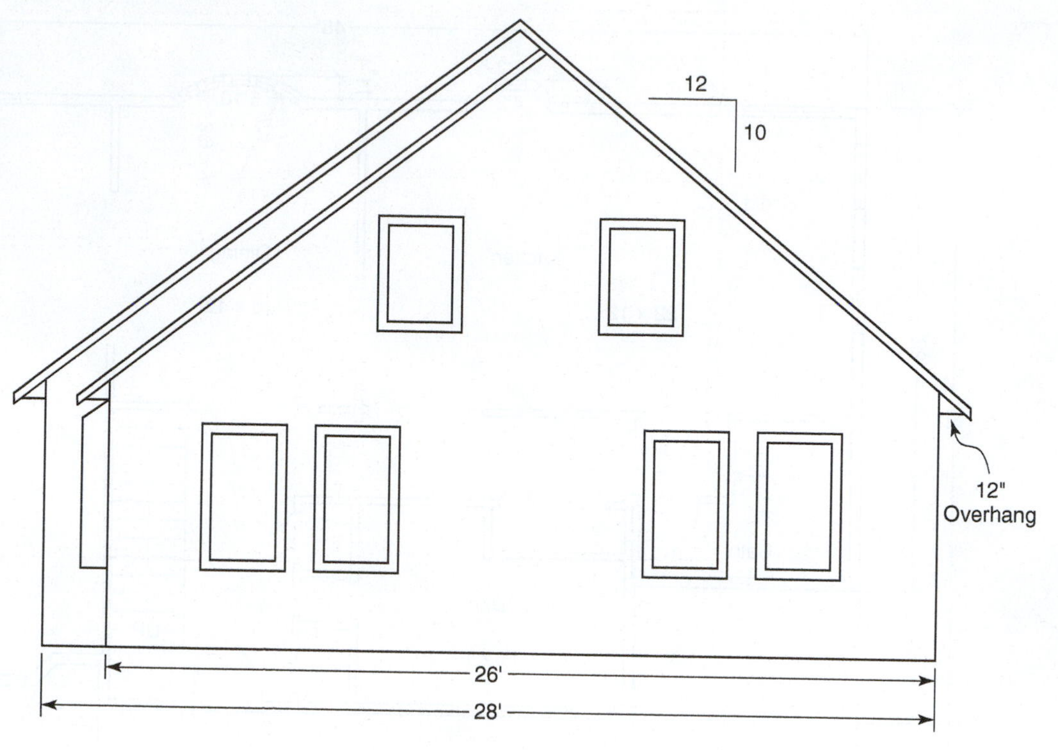

Right Side Elevation

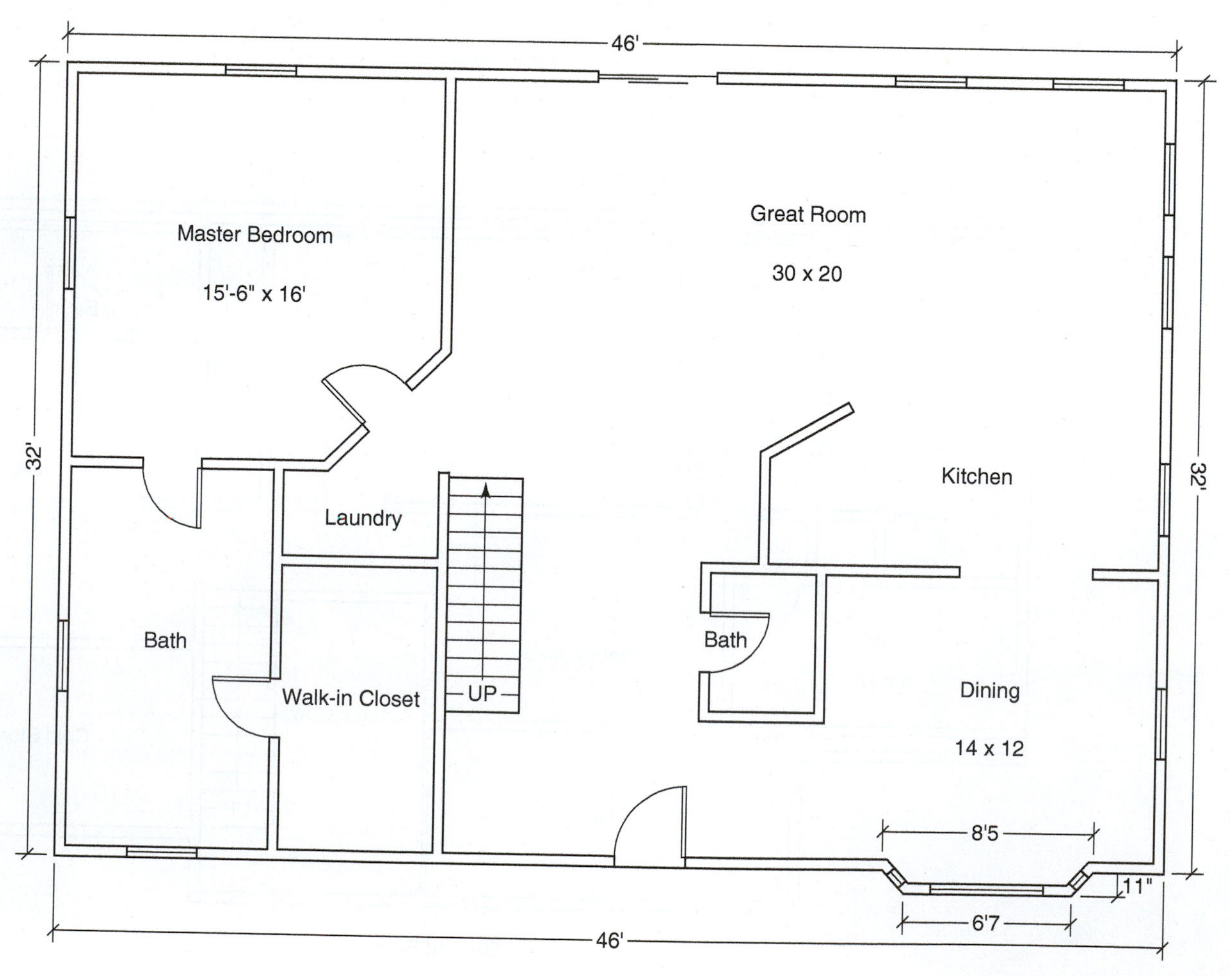

First Floor

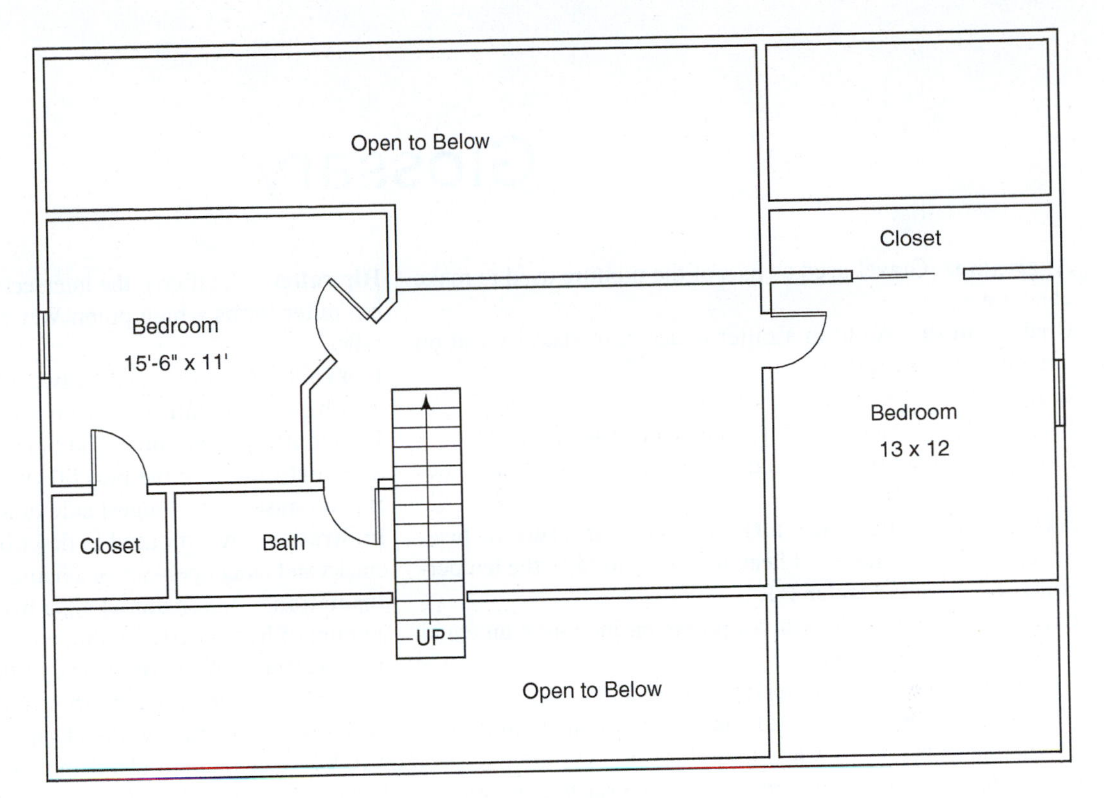

Second Floor

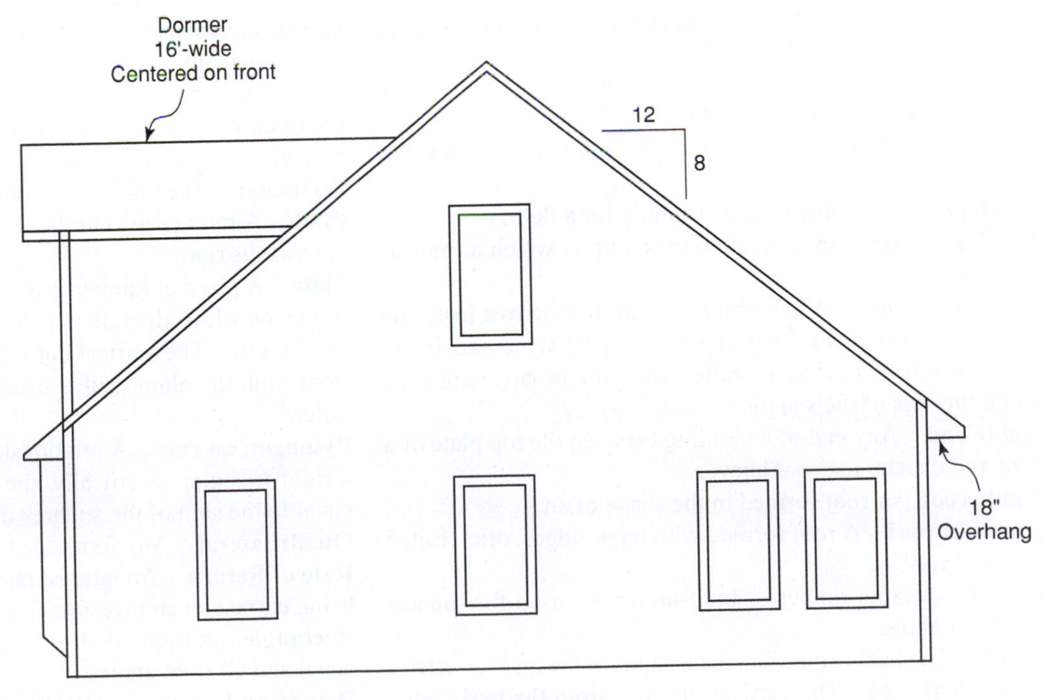

Right Side Elevation

Glossary

Aggregate Gravel used as part of the mixture used to make concrete.

Bird's mouth A cut in a rafter to accommodate its seat on the top plate of a wall.

Blade The wider and longer leg of a framing square.

Board feet A measure of wood equal to 144 cubic inches.

Bridging Strengthening and stiffening braces inserted between adjacent floor joists.

BTU (British Thermal Unit) A unit of measure of heat. Specifically, the amount of heat necessary to raise the temperature of 1 pound of water 1°F.

Circle A plane curve with all points on the curve an equal distance from the center.

Circumference The perimeter of a circle.

Common rafter A piece of the roof framing running perpendicular to the roof's edge.

Concrete A mixture of cement, sand, aggregate, and water.

Congruent triangles and polygons Figures that are identical in size and shape.

Conventional (hip or valley) rafter A rafter that forms a 45-degree angle where two adjacent roof planes meet at 90 degrees to each other.

Drywall Also called plasterboard, the material normally used to enclose the interior walls and ceilings.

Fascia The front (vertical) material used to enclose a rafter end.

Floor joists Lumber used as framing for a floor.

Footing A base, usually of concrete, upon which a foundation rests.

Framing square A carpenter's square having two legs, one 2 in. wide and 24 in. long, the other $1\frac{1}{2}$ in. wide and 16 in. long. Numbers related to rafter cuts are usually engraved along the legs of the square.

Gable end Any end of a building between the top plate of a wall and the sloping roof above.

Gable roof A roof formed in the shape of an A.

Gambrel roof A roof formed with three ridges, often called a "barn" roof.

Girder A large timber or built-up timber used for support of a floor above.

Gram The basic unit of mass (weight) in the metric system.

Head clearance The vertical distance from the lower edge of a floor frame to a step on a set of descending stairs.

Header A piece of lumber to which the ends of floor joists are fastened, or a built-up support over a window or door opening.

Heat Loss A term used to refer to the amount of heat that is transferred through sections of a house.

Hip rafter A rafter at the intersection of two roof planes. The hip rafter forms a high point. Water will run away from a hip rafter.

Hip roof A roof shaped by two adjacent planes that meet at a high point and fall away from each other.

Horizontal projection The horizontal distance a rafter extends outward from the face of a wall as part of the overhang.

Hypotenuse The longest side of a right triangle.

Infiltration A term used to describe how heat is lost through cracks and other openings in a house.

Insulation Any material that may be used to reduce the amount of heat transfer within a house.

Jack rafters A set of rafters of tapering lengths that form the roof between common rafters and a hip or valley rafter.

Jack studs Vertical studs of tapering lengths that form the wall between a floor and a sloping roof.

Liter The basic unit of volume in the metric system.

Meter The basic unit of length in the metric system.

Octagon An eight-sided, closed, plane figure.

Overhang The horizontal amount by which a roof extends beyond the outer edge of the wall on which it rests.

Parallelogram A four-sided, closed, plane figure having opposite sides parallel.

Payback Period The amount of time required to regain an amount of money invested.

Perimeter The distance around an object.

Pitch A ratio of the number of inches of rise of a roof rafter to twice its run.

Plate A piece of lumber that serves as the top piece in a wall frame, on which the rafters rest.

Plumb cut The vertical cut made on a rafter, parallel to the front wall; the plumb cut is usually made on both ends of the rafter.

Pythagorean rule A relationship between the three sides of a right triangle; specifically, the square of the hypotenuse is equal to the sum of the squares of the two legs.

Quadrilateral Any four-sided closed plane figure.

Rate of Return An interest rate showing what percentage is being earned on an investment with respect to time.

Rectangle A four-sided, closed, plane having opposite sides equal and all right angles.

Return on Investment (ROI) The amount of interest, often expressed as a percent, earned on an investment. It is found by dividing the annual amount earned by the amount invested. It can also be calculated by dividing the rate of return into 100%.

R-Factor A measure of the ability of any material to resist the transfer of heat. The higher the R-factor, the greater the ability to resist heat transfer.

Ridge The centerline of a roof, located at its highest point.
Ridge board A board used at the high point of a roof to join opposing rafters.
Risers The collection of vertical distances that form a set of stairs.
Run A horizontal distance covered by a rafter or a stair.
Salt box A roof style formed when one side wall of a house is at a higher elevation than its opposite side wall.
Setback On a hip or valley-formed roof, the distance along a top plate from a corner to the point where the first full-length common rafter is set.
Sheathing Any material used to enclose the sides or roof of a building.
Shed roof A roof of a single plane sloping from one end of a building to the opposite end.
Shoe Also called a sole plate; a piece of lumber, usually a 2 × 4 or 2 × 6, to which studs are attached and which serves as the base of a wall.
Sill plate A layer of lumber, usually pressure treated, set on top of a foundation, upon which floor joists will be set.
Similar triangles and polygons Figures that are identical in shape (that is, they have the same angles), but are not necessarily the same size.
Slope The ratio of the number of inches of rise of a roof to 12 inches.
Slope triangle A right triangle, usually shown along the drawing of a roof, showing the number of inches of rise for each foot of run.
Soffit The underside (usually horizontal) material used to enclose the overhang of a roof.
Sphere A solid curved figure with all points on the surface of the figure an equal distance from the center.
Square A four-sided, closed, plane figure having all equal sides and all right angles.
Stud A vertical piece of lumber, usually a 2 × 4 or 2 × 6, that together with other studs, forms and frames a wall.
Therm 100,000 BTUs.
Tongue The narrower and shorter leg of a framing square.
Total rise The vertical distance from the plane of the top plate to the ridge; the vertical distance a rafter occupies. Also the vertical distance covered by a set of stairs.
Trapezoid A four-sided, closed, plane figure having one pair of parallel sides.
Triangle A three-sided, closed, plane figure.
Unconventional (hip or valley) rafter A rafter that forms an angle other than 45 degrees where two adjacent roof planes meet.
Unit line length The distance along a rafter covered when measured against 12 in. of horizontal distance. Or the length of the hypotenuse of a right triangle having legs equal to the unit rise and 12 in.
Unit rise The vertical distance covered in a horizontal distance of 12 in.
Unit run 12 in.
Valley rafter A rafter forming the angle where two adjacent roof planes meet at 90 degrees to each other, so as to slope upward from each other. The valley rafter forms a low point. Water will run along a valley rafter.
Vertical projection The vertical distance from the top of a wall to the lowest point a roof rafter hangs.

Answers to Odd-Numbered Exercises

Chapter 1

Exercise 1–1A

1. ones **3.** hundred thousands **5.** millions

Exercise 1–1B

1. 1 **3.** 4 **5.** 2

Exercise 1–1C

1. 862,450 862,500 **3.** 542,000 540,000
5. 25,478,500 25,000,000 **7.** \$800 **9.** 290,000 ft^2

Exercise 1–2

1. a. yes **b.** cannot make that assumption **c.** yes

Exercise 1–3

1. 21 **3.** 6 **5.** 22 **7.** 140 miles **9.** \$710 **11.** 35 pounds

Review Exercises Chapter 1

1. \$1323
3. 1728 inches
5. 189 yd^3
7. \$34,242
9. 23,075 plugs
11. \$366/mbf
13. 9 in.
15. 20 connectors
17. 6 pieces
19. 8 in.
21. 5110 pounds
23. 154 feet—rectangle has less wall and foundation length
25. a. no
b. yes
c. 5
d. 285,042 ft^2
27. a. 864 kwh
b. 2040 kwh
c. yes
d. yes, reasonable assumption

Chapter 2

Exercise 2–1A

1. $\frac{7}{9}$ **3.** $\frac{1}{2}$ **5.** $\frac{1}{3}$ **7.** $\frac{4}{17}$ **9.** $\frac{1}{8}$ **11.** $\frac{5}{32}$ **13.** $\frac{1}{4}$ **15.** $\frac{3}{4}$ **17.** $\frac{1}{20}$

Exercise 2–1B

1. $\frac{15}{27}$ **3.** $\frac{12}{64}$ **5.** $\frac{28}{56}$ **7.** $\frac{36}{64}$ **9.** $\frac{35}{35}$ **11.** $\frac{10}{16}$ **13.** $\frac{12}{32}$ **15.** $\frac{14}{32}$

Exercise 2–1C

1. $\frac{41}{8}$ **3.** $\frac{15}{8}$ **5.** $\frac{51}{8}$ **7.** $\frac{67}{8}$ **9.** $\frac{27}{5}$ **11.** $\frac{5}{4}''$ **13.** $\frac{24}{16}''$ **15.** $\frac{3}{2}'' \times \frac{11}{2}''$

Exercise 2–1D

1. $1\frac{3}{5}$ **3.** $1\frac{1}{16}$ **5.** $1\frac{9}{16}$ **7.** $4\frac{2}{15}$ **9.** 7

Exercise 2–1E

A. $\frac{1}{16}''$ **B.** $\frac{1}{4}''$ **C.** $\frac{9}{16}''$ **D.** $\frac{7}{8}''$ **E.** $1\frac{1}{16}''$ **F.** $1\frac{9}{16}''$ **G.** $1\frac{7}{8}''$ **H.** $2\frac{3}{8}''$
I. $2\frac{15}{16}''$ **J.** $3\frac{5}{16}''$ **K.** $3\frac{3}{4}''$ **L.** $4\frac{1}{8}''$ **M.** $4\frac{7}{16}''$ **N.** $5''$ **O.** $5\frac{5}{16}''$ **P.** $5\frac{3}{4}''$

Exercise 2–1F

A. $\frac{5}{16}''$ **B.** $\frac{3}{4}''$ **C.** $1''$ **D.** $1\frac{5}{16}''$ **E.** $1\frac{9}{16}''$ **F.** $1\frac{7}{8}''$ **G.** $2\frac{1}{8}''$
H. $2\frac{9}{16}''$ **I.** $3\frac{1}{16}''$ **J.** $3\frac{1}{4}''$ **K.** $3\frac{7}{16}''$ **L.** $3\frac{15}{16}''$ **M.** $4\frac{1}{16}''$ **N.** $4\frac{11}{16}''$
O. $5\frac{3}{16}''$ **P.** $5\frac{7}{16}''$ **Q.** $5\frac{9}{16}''$ **R.** $5\frac{7}{8}''$ **S.** $5\frac{15}{16}''$ **T.** $4\frac{7}{8}''$ **U.** $4\frac{7}{16}''$

Exercise 2–2A

1. $\frac{9}{24}, \frac{20}{24}$ **3.** $\frac{8}{12}, \frac{9}{12}$ **5.** $\frac{15}{24}, \frac{16}{24}$ **7.** $\frac{9}{64}, \frac{6}{64}$ **9.** $\frac{5}{18}, \frac{12}{18}$ **11.** $\frac{3}{4}, \frac{2}{4}$
13. $\frac{1}{9}, \frac{6}{9}$ **15.** $\frac{33}{36}, \frac{20}{36}$ **17.** $\frac{5}{6}, \frac{2}{6}$ **19.** $\frac{6}{16}, \frac{5}{16}$ **21.** $\frac{12}{16}, \frac{9}{16}$ **23.** $\frac{3}{8}, \frac{4}{8}$

Exercise 2–2B

1. $\frac{15}{64}''$ **3.** $\frac{11}{16}''$ **5.** $\frac{5}{8}''$ **7.** $\frac{23}{32}''$ **9.** $\frac{51}{64}''$
11. $\frac{11}{64}'', \frac{9}{64}''$ **13.** $\frac{63}{64}'', \frac{61}{64}''$ **15.** $\frac{41}{64}'', \frac{39}{64}''$ **17.** $\frac{49}{64}'', \frac{47}{64}''$ **19.** $\frac{25}{64}'', \frac{23}{64}''$

Exercise 2–3

1. $\frac{11}{12}$ **3.** $1\frac{1}{6}$ **5.** $7\frac{1}{2}$ **7.** $14\frac{1}{2}$ **9.** $9\frac{17}{18}$
11. $382\frac{11}{12}$ ft. **13.** $4\frac{1}{4}''$ **15.** $7\frac{9}{16}''$ **17.** $4\frac{1}{8}''$

19.

$A = 21\frac{7}{8}''$	$I = 23\frac{23}{32}''$	$Q = 25''$	$Y = 26\frac{29}{32}''$
$B = 22\frac{3}{32}''$	$J = 23\frac{25}{32}''$	$R = 25\frac{11}{32}''$	$Z = 27\frac{1}{16}''$
$C = 22\frac{1}{4}''$	$K = 23\frac{31}{32}''$	$S = 25\frac{17}{32}''$	$aa = 27\frac{1}{4}''$
$D = 22\frac{15}{32}''$	$L = 24\frac{3}{32}''$	$T = 25\frac{23}{32}''$	$bb = 27\frac{1}{2}''$
$E = 22\frac{21}{32}''$	$M = 24\frac{9}{32}''$	$U = 25\frac{27}{32}''$	$cc = 27\frac{27}{32}''$
$F = 22\frac{7}{8}''$	$N = 24\frac{15}{32}''$	$V = 26''$	$dd = 28\frac{1}{4}''$
$G = 23\frac{1}{16}''$	$O = 24\frac{9}{16}''$	$W = 26\frac{3}{16}''$	$ee = 28\frac{1}{2}''$
$H = 23\frac{5}{16}''$	$P = 24\frac{23}{32}''$	$X = 26\frac{7}{16}''$	

Exercise 2–4

1. $\frac{1}{8}$ **3.** $\frac{3}{16}$ **5.** $3\frac{8}{9}$ **7.** $2\frac{5}{8}$ **9.** $\frac{27}{32}$ **11.** $39\frac{3}{16}''$
13. $47\frac{7}{16}''$ **15.** $36\frac{13}{16}''$ **17.** $1\frac{11}{16}''$ **19.** $16\frac{3}{8}''$ **21.** $\frac{39}{64}''$

Exercise 2–5

1. $\frac{2}{3}$ **3.** $\frac{15}{28}$ **5.** $59\frac{1}{2}$ **7.** $1\frac{3}{4}$ **9.** $4\frac{4}{5}$
11. $3\frac{3}{4}''$ **13.** $106\frac{7}{16}''$ **15.** $64\frac{11}{16}''$ **17.** $9\frac{1}{6}$ ft (9 ft 2 in.) **19.** 16-foot board
21. $162\frac{1}{4}''$

Exercise 2–6

1. $\frac{5}{22}$ **3.** $\frac{18}{25}$ **5.** $2\frac{4}{7}$ **7.** $\frac{1}{48}$ **9.** $\frac{4}{9}$
11. 38 boards **13.** 3 posts **15.** $\frac{11}{16}$ **17.** 4 shelves **19.** 375 bdf (board feet)
21. 1280 lineal feet

Review Exericses Chapter 2

1. $10\frac{3}{8}''$ **3.** $972\frac{3}{16}$ lbs **5.** 6-foot lengths **7.** $4\frac{1}{4}''$ **9.** $8\frac{3}{4}''$
11. $\frac{43}{64}'', \frac{11}{16}'', \frac{25}{32}'', \frac{7}{8}''$ **13.** $12\frac{1}{2}$ gallon tank **15.** $\frac{1}{8}''$ **17.** $4\frac{9}{16}''$
19. a. 1146 ft **b.** 382 ft **c.** $47\frac{3}{4}$ ft or 47 ft 9 in. **21.** 2 feet wide

Chapter 3

Exercise 3–1

1. 83.41 83.407 **3.** 43 42.6 **5.** 0.5 0.50
7. 4.6 4.60 **9.** 0 0.4 **11.** 228 ft by 116 ft

Exercise 3–2

1. 0.1 tenths 15.9 ± 0.05
3. 0.01 hundredths 14.82 ± 0.005
5. 0.1 tenths 105.2 ± 0.05

Exercise 3–3

1. 31.8 **3.** 16.47 **5.** 72.07 **7.** 74.6 **9.** 2.03
11. 0.6875″ **13.** 0.375″ **15.** $435.72 **17.** 0.7567″ **19.** 0.875″

Exercise 3–4

1. 0.615 **3.** 0.8076 **5.** 0.2045 **7.** 6.365 **9.** 0.011
11. 0.5502 **13.** 0.9785″ **15.** 5.372″ **17.** 0.4375″ **19.** 5.4″

Exercise 3–5

1. 22.684 or 22.7 **3.** 0.903 or 0.90 **5.** $29.88 **7.** 0.305 or 0.31
9. 142.6 or 143 **11.** $47.81 **13.** 0.625″ **15.** 82.95 ft
17. 272.24 lbs or 272 lbs

Exercise 3–6

1. 0.099 **3.** 2.9 **5.** 9.12 **7.** 0.02
9. 1000 **11.** 240 screws per pound **13.** 12 treads **15.** $0.42 per lineal foot
17. 19 boards **19.** 83 sheets

Exercise 3–7

1. 830,000 **3.** 50 **5.** 48 **7.** 0.038
9. 62 **11.** $0.73 **13.** $0.719 or $0.72 per bf **15.** $3190

Exercise 3–8A

1. 0.625 **3.** 0.666. . . . or $0.\overline{6}$ **5.** 1.875 **7.** $8.\overline{6}$ **9.** 19.5625

11. $\frac{33}{40}$ **13.** $\frac{41}{100}$ **15.** $15\frac{17}{40}$ **17.** $4\frac{47}{200}$ **19.** $26\frac{6}{25}$

21. $\frac{2}{3}$ **23.** $2\frac{5}{11}$ **25.** $15\frac{4}{9}$ **27.** $3\frac{2}{3}$ **29.** $35\frac{2}{9}$

Number	Decimal	Half	Fourth	8th	16th	32nd	64th
a.	2.39″	$2\frac{1}{2}$″	$2\frac{1}{2}$″	$2\frac{3}{8}$″	$2\frac{3}{8}$″	$2\frac{3}{8}$″	$2\frac{25}{64}$″
31.	15.891″	16″	16″	$15\frac{7}{8}$″	$15\frac{7}{8}$″	$15\frac{29}{32}$″	$15\frac{57}{64}$″
33.	7.299″	$7\frac{1}{2}$″	$7\frac{1}{4}$″	$7\frac{1}{4}$″	$7\frac{5}{16}$″	$7\frac{5}{16}$″	$7\frac{19}{64}$″
35.	7.629″	$7\frac{1}{2}$″	$7\frac{3}{4}$″	$7\frac{5}{8}$″	$7\frac{5}{8}$″	$7\frac{5}{8}$″	$7\frac{5}{8}$″
37.	48.584″	$48\frac{1}{2}$″	$48\frac{1}{2}$″	$48\frac{5}{8}$″	$48\frac{9}{16}$″	$48\frac{19}{32}$″	$48\frac{37}{64}$″
39.	13.622″	$13\frac{1}{2}$″	$13\frac{1}{2}$″	$13\frac{5}{8}$″	$13\frac{5}{8}$″	$13\frac{5}{8}$″	$13\frac{5}{8}$″

41. $73\frac{5}{8}$″ **43.** $\frac{5}{32}$″ **45.** $8\frac{1}{4}$″ **47.** $\frac{5}{16}$″ **49.** $2\frac{15}{16}$″

Exercise 3–8B

1. 0.3125″ **3.** 0.6875″ **5.** 19.75″ **7.** 0.109375″ **9.** 28.5″ **11.** $\frac{23}{64}$″
13. $14\frac{7}{16}$″ **15.** $30\frac{5}{8}$″ **17.** $\frac{43}{64}$″ **19.** $4\frac{1}{2}$″ **21.** $4\frac{7}{8}$″ **23.** $1\frac{7}{8}$″
25. $\frac{25}{32}$″

Exercise 3–8C

1. 5.375″
3. 7.25″
5. 0.078125″
7. 0.5625
9. 0.5375
11. $\frac{33}{40} \doteq \frac{13}{16}''$
13. $\frac{1}{32} = \frac{1}{32}$
15. $2\frac{2}{25} \doteq 2\frac{1}{16}''$
17. $3\frac{9}{16}''$
19. $\frac{7}{25} \doteq \frac{1}{4}''$
21. $8\frac{9}{10}'' \doteq 8\frac{29}{32}''$
23. $\frac{11}{16}''$
25. $\frac{11}{16}''$

Chapter 4

Exercise 4–1A

1. 2 ft 9 in.
3. 9 ft 6 in.
5. 7 ft 1 in.
7. 7 ft 1 in.
9. 6 ft 2 in.

Exercise 4–1B

1. 4 ft $2\frac{1}{8}$ in.
3. 7 ft $2\frac{3}{8}$ in.
5. 1 ft $7\frac{3}{8}$ in.
7. 7 ft $1\frac{5}{16}$ in.
9. 4 ft 9 in.
11. 3 ft $8\frac{27}{32}$ in.
13. 7 ft $4\frac{3}{8}$ in.
15. 6 ft $6\frac{7}{8}$ in.
17. 52 ft $9\frac{3}{4}$ in.
19. 4 ft 1 in.

Exercise 4–2A

1. 17 ft 11 in.
3. 2 ft 1 in.
5. 19 ft 7 in.
7. 6.992 in. or 7 in.
9. 1 ft 8 in.
11. 24 ft 8 in.
13. 3 ft $9\frac{1}{4}$ in.
15. 1 ft $5\frac{1}{4}$ in.

Exercise 4–2B

1. 43 ft 9 in.
3. 14 in. or 1 ft 2 in.
5. 0 ft 4 in. or 4 in.
7. 0 ft 6 in. or 6 in.
9. 10 ft
11. 24 ft 6 in.
13. 2 ft $1\frac{1}{2}$ in.
15. 5 ft $11\frac{7}{8}$ in.

Exercise 4–3

1. 1.54 miles
3. 70 fathoms
5. 13.75 yards
7. 297 ft
9. 285.1 ft by 355.6 ft
11. $0.0\overline{45}$ chunks
13. $23.\overline{93}$ scoops
15. same!

Exercise 4–4

1. 46,656 in^3
3. 2.228 mi^2
5. 1.26 yd^3
7. 2.1 rod^2
9. 17.4 acres
11. 55.6 or 56 yd^3
13. 0.81 acres
15. 80.7 ft^2
17. $15.\overline{1}$ ft^2
19. 4.58 ft^2
21. 17 yd^3
23. 8.13 yd^3
25. 615 acres

Exercise 4–5

1. 132,000 ft/hr
3. 214 lbs per minute
5. 0.112 oz per bf
7. 1.53 lbs/in.^2 (psi)

9. 232 lbs/min.
11. 5.6 oz/in.
13. R48 per foot
15. R = 29.4
17. 168 ft^3/hr
19. 42 mpg

Review Exercises Chapter 4

1. 4 ft $10\frac{1}{2}$ in.
3. 4 ft $1\frac{3}{16}$ in.
5. 2 ft $3\frac{5}{16}$ in.
7. 11 ft $5\frac{1}{2}$ in.
9. 13 steps
11. 160 sq rods
13. 0.825 bags
15. 5.7 containers
17. 220 yds.
19. 19 ft $7\frac{1}{2}$ in.

Chapter 5

Exercise 5–1

1. 6:1
3. 1:1
5. 1:3
7. 1:6
9. 16:1
11. 24 ft by 40 ft
13. 8:5
15. $432 and $720
17. $\frac{1}{4}$
19. 37 ft 6 in.

Exercise 5–2

1. 12
3. 15.5
5. 2.1
7. $\frac{9}{10}$
9. 0.74
11. 144 miles
13. $53\frac{1}{8}$ miles
15. $557.67
17. 200 lbs
19. $13\frac{3}{4}$ ft or 13 ft 9 in.
21. 33,680 or 34,000 lbs
23. 479 rpm
25. 104 hrs
27. $82.08
29. $\frac{6}{12}$
31. 485 mi
33. 11 ft 4 in. by 33 ft
35. $185\frac{1}{2}$ rpm
37. 6250 watts or 6.25 kw
39. $7\frac{1}{2}$ hrs

Chapter 6

Exercise 6–1

	Fraction	Decimal	Percent
1.	$1\frac{4}{5}$	1.8	180%
3.	$1\frac{41}{50}$	1.82	182%
5.	$\frac{1}{4}$	0.25	25%
7.	$\frac{1}{1000}$	0.001	0.1%
9.	$\frac{1}{200}$	0.005	$\frac{1}{2}$%
11.	$\frac{1}{40}$	0.025	2.5%
13.	$\frac{2}{3}$	$0.\overline{6}$	$66\frac{2}{3}$%
15.	$\frac{1}{40}$	0.025	2.5%
17.	$\frac{31}{50}$	0.62	62%
19.	$\frac{3}{5000}$	0.0006	0.06%

Exercise 6–2A

1. amount
3. base
5. base
7. base
9. base
11. amount
13. percent
15. base
17. amount

Exercise 6–2B

1. 12.76
3. 321
5. 625
7. $233.\overline{3}$
9. 600
11. 12.325
13. 62.5% or $62\frac{1}{2}$%
15. 138.5
17. 56.25 or $56\frac{1}{4}$
19. c. new car

Exercise 6–2C

1. 171.6
3. 478
5. 31.4%
7. 82.3
9. $530.14
11. $997.50
13. 20.3%
15. 123.5
17. 130.7
19. $277.15, no, dif. base

Exercise 6–3

1. $4372.50
3. $184.95
5. $250,000
7. $331,442
9. $161.98
11. 22.7%
13. 3 years
15. $438.45
17. 5.6%
19. 14.8%
21. $213.14
23. 2035 bdf
25. 6 ft 8 in.
27. $488.16
29. $220,640

Chapter 7

Exercise 7–1

1. acute
3. vertical angles and obtuse
5. complementary and acute
7. straight angle
9. 65°
11. 45°
13. 90° each
15. angle B must be acute

Exercise 7–2

1. $x = 3,\ y = 4,\ z = 5,\ \angle X = 37°\ \angle Y = 53°\ \angle Z = 90°$
3. $x \doteq 2.7$
5. 40 ft

Exercise 7–3A

1. 67.90
3. 2285.80
5. 86.49
7. 1764
9. 31.36
11. 5
13. 1
15. 9
17. 4.18
19. 24.49

Exercise 7–3B

1. 8.52
3. 10.41
5. 10.42
7. 47 ft 2 in.
9. 21 ft 8 in.
11. 26 ft
13. 13
15. $59\frac{1}{2}''$ or 4 ft $11\frac{1}{2}''$
17. 7 ft $9\frac{5}{8}''$ or 7.8 ft
19. Rectangle

Exercise 7–4

1. $a = 15.44,\ b = 8.92$
3. $j = 26.69,\ l = 53.38$
5. $x = 12.62,\ y = 12.62$
7. $c = 15$ ft
9. $p = 26$ ft
11. 13.86 ft or 13 ft $10\frac{1}{4}''$
13. 18 ft $5\frac{11}{16}''$
15. 24 ft

Exercise 7–5

Number	Side *a*	Side *b*	Side *c*	Perimeter
1.	8.26	4.11	9.23	21.60
3.	4.12	6.85	7.99	18.96
5.	12	5.00	13.00	30

Number	Side *a*	Side *b*	Side *c*	Area
7.	3	4	5	6
9.	3.2	6.4	7.16	10.24

Number	Side *a*	Side *b*	Side *c*	Area
11.	8.26	4.17	9.23	17.22
13.	8.3	8.3	8.3	29.8
15.	5	12	15	26.5

17. 9.45 sq units
19. 4.2 acres
21. 12 ft 8 in., or 13 ft
23. 6.9 ft^2
25. 17,829 ft^2
27. 72 ft
29. 30°: 1 ft $1\frac{1}{2}''$, 60°: 1 ft $11\frac{3}{8}''$

Chapter 8

Exercise 8–1

1. 26 in. **3.** 30 in. **5.** 9 ft^2 **7.** 25 ft^2 **9.** 66 ft **11.** 180 sq units

Number	Length	Width	Perimeter	Area
13.	10 in.	4 in.	28 in.	40 in^2
15.	15 ft 3 in.	8 ft 6 in.	47 ft 6 in.	129.6 ft^2

17. 4 ft^2
19. $26.55
21. $900.90
23. 112 ft
25. 144 ft^2
27. 15.1 ft^2
29. 11.25 ft^2

Exercise 8–2

Number	Radius	Diameter	Circumference	Area
1.	5 in.	10″	31.4″	78.5 in^2
3.	13.1 ft	26.3 ft	82.6 ft	543 ft^2
5.	8 in.	16″	50.3″	201 in^2
7.	5″	10″	31.4 in.	78.5 in^2
9.	4″	8″	25.1″	50.265 in^2

11. 1.4 ft^2 (201 $in.^2$)
13. 254 $in.^2$
15. 11 ft
17. 8 $in.^2$
19. 14 in.
21. 8.3 ft^2
23. 12 ft $6\frac{13}{16}$ in.
25. 8 ft 2 in.

Exercise 8–3

1. 159 ft^2 **3.** 14 ft **5.** 80 ft

Exercise 8–4

1. 1580 ft^2
3. 1518 ft^2
5. 195 ft
7. 63 in. (62.8 in.)
9. 29.7 ft, 63 ft^2
11. 54 in., 110 $in.^2$
13. 22.9 ft
15. 76.3 ft^2
17. 18 in^2
19. 359 ft^2
21. 18.28 ft or 18 ft $3\frac{3}{8}$ in.

The polygons in the table below are all *Regular* polygons.

	Sides	Total degrees in angles	Degrees in each angle
23.	7	900°	128.57°
25.	28	4680°	167.14°
27.	55	9540°	173.45°
29.	101	17,820°	176.44°

31. As the number of sides increase, the total degrees in the polygon and the degrees in each angle of the polygon increase. In fact, as the number of sides increases, the angles get closer to 180°, but never actually reach 180° (a straight angle); larger angles
33. square table; round table = 12.57 ft^2, square table = 16 ft^2
35. 317.5 ft^2

Chapter 9

Exercise 9–1

Number	LSA	TSA	Volume
1.	384 ft^2	432 ft^2	384 ft^3
3.	4800 ft^2	9550 ft^2	47,500 ft^3

5. 25 yd^3 concrete
7. 4620 ft^3
9. 672 ft^3 air
11. 36″
13. 20 in.
15. 828 $in.^2$

Exercise 9–2

Number	Radius	Diameter	Circumference	Area (base)	Height	Volume	LSA	TSA
1.	8 in.	16″	50.3″	201 in^2	5 in.	1005 in^3	251 in^2	653 in^2
3.	2.5″	5″	15.7 in.	19.6 in^2	5 in.	98 in^3	78.5 in^2	118 in^2
5.	11″	1 ft 10″	69″	380 in^2	1 ft 3″	5700 in^3	1035 in^2	1795 in^2

7. 375 $in.^3$
9. 294 gal
11. 67 ft^2
13. 1508 $in.^2$
15. 245 ft^2
17. 20 ft^2
19. 176 ft^2
21. 5.2 gallons
23. 4750 $in.^2$ or 33 ft^2
25. 5184 $in.^2$ or 36 ft^2

Exercise 9–3

Number	Radius	Diameter	Circumference	Volume	Surface Area
1.	8″	16″	50.3″	2145 in.^{3}	804 in.^{2}
3.	8 ft	16 ft	50.3 ft	2145 ft^3	804 ft^2
5.	3.5 ft	7 ft	22 ft	180 ft^3	154 ft^2
7.	8.6″	17.2″	54″	2664 in.^{3}	929 in.^{2}

Exercise 9–4

1. 13,090 ft^3
3. 1164 ft^2
5. 100 gallons
7. 21.8 in. or 22″
9. 4500 ft^3 = 33,750 gallons
11. 1458 ft^2 – 10% = 1312 ft^2
13. 12,210 ft^3

Chapter 10

Exercise 10–1A

1. 0.047 km
3. 830 cm
5. 25 mm
7. 4800 cm
9. 4600 mm
11. 1800 mg
13. 58 dg
15. 5 g

Exercise 10–1B

1. 0.038 m^3
3. 4.2 km^3
5. 0.082 m^2
7. 48 dm^3
9. 410 000 cm^2
11. 3200 mm^3
13. 4.2 liters
15. 378 m^3
17. $500.50
19. 9000 kg/19,800 lbs.

Exercise 10–2

1. 125.33 ft
3. 23 miles
5. 83.6 or 84 lbs.
7. 336 g
9. 7.7 km/l
11. 813 mm
13. yes, with approx. $5\frac{1}{2}$″ clearance
15. approx. 54.5 kg
17. 9 ft × 11 ft
19. 158.39 ft

Exercise 10–3

Number	a.	b.	c.	d.	e.	f.
1.	9°C	0°C	20°C	11°C	37°C	27°C

3. 113°F
5. 50°F and 95°F
7. 7.2°C–35°C
9. –128.6°F. Brrrr!

Review Exercises Chapter 10

1. 1.5 lb/ft
3. 1016 lbs
5. 406 mm
7. 7.6 kg/board
9. 29°C
11. 2787 cm^2
13. 372 km/hr
15. $\frac{3}{32}$″

Chapter 11

Review Exercises

1. $\frac{2}{3}$; 1; $\frac{4}{3}$; $\frac{5}{3}$; 2
3. **a.** $2853\frac{1}{3}$ bf; **b.** 2240 bf; **c.** 168 bf; **d.** $1013\frac{1}{3}$ bf; **e.** 504 bf
5. 224 lf
7. 312 lf
9. 35 sheets
11. 178 bf

Chapter 12

Review Exercises

1. \$0.42/bf
3. \$0.69/lf
5. **a.** \$728; \$179.52; **b.** \$282.29; \$159.12; \$85.68 for 8s, 12s, 19s respectively **c.** \$66.08; **d.** \$74.67; \$940.80 for 10s, 14s respectively
7. \$37.12
9. \$1.825 is less by \$0.10/bf

Chapter 13

Review Exercises

	Cement	Sand	Aggregate
1. a.	$4\frac{1}{2}$ ft^3	9 ft^3	$13\frac{1}{2}$ ft^3
b.	27 ft^3	54 ft^3	81 ft^3
c.	90 ft^2	180 ft^3	270 ft^3

3. 15 yd^3
5. 4.75 yd^3
7. **a.** footing: 5.0; walls: 26.3; slab: 10.2; total: 41.5 yd^3; **b.** footing: 7.6; walls: 42.6; slab: 14.6; total: 64.8 yd^3; **c.** footing: 7.1; walls: 38.9; slab: 14.2; total: 60.2 yd^3

Chapter 14

Review Exercises

1. Several designs are possible. One is $\frac{8}{16s}, \frac{2}{8s}, \frac{1}{12}$.
3.

	(a)	(b)	(c)
Girder	$\frac{3}{18s}$	$\frac{4}{8s}, \frac{2}{12s}, \frac{7}{16s}$	$\frac{2}{8s}, \frac{1}{10s}, \frac{7}{12s}, \frac{1}{18}$
Sill Plates	$\frac{2}{8s}, \frac{6}{10s}, \frac{4}{12s}, \frac{2}{14s}$	$\frac{3}{8s}, \frac{1}{10}, \frac{14}{12s}$	$\frac{2}{10s}, \frac{8}{12s}, \frac{4}{14s}$
Floor Joists	2 × 12:$\frac{47}{20s}$ (12″ o.c.); 2 × 6:$\frac{19}{10s}$ (12″ o.c.)	2 × 10:$\frac{74}{16s}$ (16″ o.c.); 2 × 8:$\frac{15}{10s}$	2 × 10:$\frac{56}{14s}$; $\frac{36}{12s}$ (16″ o.c.)
FJ Headers	2 × 12:$\frac{1}{8}, \frac{2}{10s}, \frac{3}{12s}, \frac{2}{14s}$	2 × 10:$\frac{8}{12s}$; $\frac{2}{8s}$	2 × 10:$\frac{8}{12s}, \frac{2}{10s}$
Bridging	330 lf	336 lf	348 lf
Subflooring	35 sheets	52 sheets	48 sheets

Chapter 15

Review Exercises

1. 22 studs
3. **a.** 19 studs; **b.** 34 studs; **c.** 121 studs; **d.** 160 studs
5. **a.** 480 lf 2 × 4 use $\frac{3}{10s}, \frac{3}{12s}, \frac{21}{14s}, \frac{6}{16s}$; **b.** 136 studs;
5. 498 lf for shoes/plates; 125 studs; total bf: 1498
7. 2 × 4s: \$504.37; 2 × 6s: \$599.76
9. **(a)** 492 lf; $\frac{6}{10s}, \frac{24}{14s}, \frac{6}{16s}$; **(b)** 157 studs; **(c)** 84 lf; **(d)** \$1,062.56

Chapter 16

Review Exercises

1. **a.** 12.369; **b.** 13.892; **c.** 12.816; **d.** 15.620; **e.** 15; **f.** 19.209; **g.** 16.971
3. **a.** 16 ft 3 in.; **b.** 15 ft $11\frac{1}{16}$ in.; **c.** 17 ft $7\frac{1}{2}$ in.; **d.** 12 ft $\frac{7}{16}$ in.; **e.** 7 ft $10\frac{3}{4}$ in.; **f.** 13 ft $3\frac{3}{4}$ in.; **g.** 7 ft $7\frac{11}{16}$ in.
5. $2\frac{1}{4}$
7. **a.** front: 90 in.; rear: 129 in.; **b.** $9\frac{15}{16}$ in.; **c.** front: 15.620; rear: 15.569; **d.** front: 11 ft $8\frac{9}{16}$ in.; rear: 16 ft $10\frac{3}{8}$ in.
9. 2 ft $3\frac{5}{8}$ in.

Chapter 17

Review Exercises

1. As per readings

Chapter 18

Review Exercises

1. **a.** $6\frac{11}{16}$ in.; **b.** $5\frac{1}{4}$ in.; **c.** 24 in.; **d.** 5 in.; **e.** $6\frac{11}{16}$ in.; **f.** 14 in.
3. $17\frac{1}{16}$ in.
5. **a.** $\frac{17}{12}$

   b. $10\frac{3}{16}/12$; **c.** $5\frac{11}{16}/12$; **d.** $4\frac{1}{4}/12$
7. $8\frac{3}{4}$ in.
9. **a.** $6\frac{15}{16}''$ **b.** $6\frac{11}{16}''$ **c.** $7\frac{1}{2}''$ **d.** $41\frac{9}{16}''$

Chapter 19

Review Exercises

1. **a.** 17 ft $8\frac{5}{16}$ in.; **b.** 22 ft $4\frac{15}{16}$ in.; **c.** 20 ft 3 in.; **d.** 23 ft $11\frac{11}{16}$ in.; **e.** 15 ft $10\frac{7}{16}$ in.
3. **a.** $\frac{1}{2}''$ **b.** 1″; **c.** 1″
5. **a.** 17′4″ **b.** 16′3″ **c.** 23′ $7\frac{1}{16}''$ **d.** 22′ $1\frac{3}{8}''$ **e.** $1\frac{1}{6}''$

Chapter 20

Review Exercises

1.

	a	b	c
CRL main	$12'3\frac{9}{16}''$	16′3″	$16'9\frac{15}{16}''$
CRL ell	$12'3\frac{9}{16}''$	13′	$12'3\frac{1}{2}''$
VRL long	16′6″	$22'1\frac{3}{8}''$	$18'7\frac{9}{16}''$
VRL short	16′6″	$17'8\frac{5}{16}''$	$18'7\frac{9}{16}''$
Slope	6/17	5/17	14/12

3. No. It has to be less than $\frac{8}{12}$ in order to meet the elevation of the narrower main.

Chapter 21

Review Exercises

1. **a.** $17\frac{5}{16}$ in.; **b.** 20 in.; **c.** $17\frac{7}{8}$ in.; **d.** $19\frac{1}{4}$ in.; **e.** $20\frac{13}{16}$ in.
3. $17\frac{5}{16}$ in.
5. **a.** 20″ **b.** $15\frac{5}{16}''$ **c.** $13\frac{13}{16}''$

Chapter 22

Review Exercises

1. **a.** 18 ft $\frac{13}{16}$ in.; **b.** 21 ft $10\frac{15}{16}$ in.; **c.** 17 ft $2\frac{1}{8}$ in.; **d.** 5 ft $8\frac{1}{8}$ in.
3. **a.** $24\frac{1}{16}$ in.; **b.** $24\frac{7}{16}$ in.; **c.** $22\frac{13}{16}$ in.; **d.** $15\frac{5}{16}$ in.
5.

		Overall	Ridge to BM	BM to Tail
a.	Main	$17'10\frac{1}{2}''$	16′3″	$1'7\frac{1}{2}''$
	Ell	$15'1\frac{13}{16}''$	$13'2\frac{1}{8}''$	$1'11\frac{11}{16}''$

b. $18'1\frac{7}{8}''$

c. $26'0\frac{9}{16}''$

d. 23′4″

e. (1) Hipfront: $28\frac{5}{8}''$

(2) Hip End: 24″

(3) Long Valley

Main: $22\frac{5}{8}''$

f. Long Vally 11: $36\frac{3}{8}''$

Chapter 23

Review Exercises

1. **a.** $100\frac{7}{8}$ in.; **b.** 13; **c.** $7\frac{3}{4}$ in.; **d.** 12; **e.** 114 in.; **f.** $104\frac{1}{2}$ in.; **g.** 80 in.
3. 13; $7\frac{13}{16}$ in.; 12; $8\frac{11}{16}$ in.; 104 in. ($95\frac{9}{16}''$ to get 6′8″ clearance)
5. NRi: 12; URi: $7\frac{15}{16}$ in.; NRu: 11; URu: 9 in. assumed; from top, 6 risers to landing. Opening: $83\frac{1}{2}''$ by 36″ and $64\frac{1}{2}''$ by 36″

Chapter 24

Review Exercises

1. **a.** 18/10s; 4 sheets; **b.** 20/14s; 6 sheets; **c.** 2/10s, 17/8s; 7 sheets; **d.** 24/10s; 10 sheets; **e.** 32/8s; 10 sheets
3. **a.** 65 ft; **b.** 58 ft; **c.** 37 ft; **d.** 73 ft

Chapter 25

Review Exercise

1.

		a.	b.	c.	d.
Ext. wall Sheathing		43 sheets	41 sheets	55 sheets	61 sheets
Siding	a.	1353 ft^2	1290 ft^2	1956 ft^2	1932 ft^2
	b.	1133 ft^2	1114 ft^2	1714 ft^2	1636 ft^2
Drywall	Walls	93 sheets	78 sheets	111 sheets	222 sheets
	Ceilings	39 sheets	47 sheets	59 sheets	90 sheets
Roof sheathing		50 sheets	40 sheets	74 sheets	76 sheets
Shingles		17 squares	13 squares	$24\frac{2}{3}$ squares	25 sheets
Drip edge		17 lengths	15 lengths	21 lengths	20 lengths

Chapter 26

Review Exercise

1.

	ROI in %	Payback—Years
a.	38.4	2.6
b.	14.2	5.2
c.	4.2	23.6
d.	6	16.67
e.	3.75	26.67

3. **a.** \$2070.66; **b.** \$517.67; **c.** 4.55%; **d.** 21.97 Yes
5. 173,732 Btu/HR
7. 77,387 Btu/HR
9. Floor, then infiltration (Results vary in other situations.)

Chapter 27

Review Exercise

1. Students should compare their results.

Appendix A

Exercise A-1

1.	0.6921	0.7218	0.9590
3.	0.7071	0.7071	1.0000

Exercise A-2

1. $A = 42°$ $a = 25.4$ ft
   $B = 48°$ $b = 28.2$ ft
   $C = 90°$ $c = 38$ ft
3. $J = 61°$ $j = 46$ in.
   $K = 29°$ $k = 25.5$ in.
   $L = 90°$ $l = 52.6$ in.
5. $A = 45°$ $a = 27$
   $B = 45°$ $b = 27$
   $C = 90°$ $c = 38.2$

Exercise A-3

1. 20′ 2″
3. 31′ 7″
5. 108 ft.
7. $9\frac{9}{16}''$
9. 8.4°; 47°
11. **a.** 10,530 ft
    b. 7.53% or 7.5% or 8%
    c. 7.51% or 7.5% or 8%
13. $11\frac{7}{8}''$
15. 19°
17. 24.2
19. No, many similar triangles of different sizes have those angles; a side must be among the given information

Bibliography

Dagostino, Frank R. *Estimating in Building Construction.* Reston, VA: Reston, 1973.

Jansen, Ted J. *Solar Engineering Technology.* Englewood Cliffs, NJ: Prentice-Hall, 1985.

Lewis, Jack R. *Basic Construction Estimating.* Englewood Cliffs, NJ: Prentice-Hall, 1983.

Ostwald, Phillip F. *Cost Estimating,* 2nd ed. Englewood Cliffs, NJ: Prentice-Hall, 1984.

Petri, Robert W. *Construction Estimating.* Reston, VA: Reston, 1979.

Reed, Mortimer P. *Residential Carpentry.* New York: John Wiley & Sons, 1980.

Shurcliff, William A. *Superinsulated Houses and Double Envelope Houses.* Andover, MA: Brick House, 1981.

Smith, Ronald C. *Principles and Practices of Light Construction,* 3rd ed. Englewood Cliffs, NJ: Prentice-Hall, 1980.

Wass, Alonzo, *Estimating Residential Construction.* Englewood Cliffs, NJ: Prentice-Hall, 1980.

www.buildingfoundation.umn.edu/

www.greenhousebuilding.com

Index

Continued from inside front cover

Board Feet

$\text{Bf} = \dfrac{t \times w \times L}{12}$, where t = thickness (in.), w = width (in.), L = length (ft).

Unit Line Length for Rafters

$\text{ULL} = \sqrt{\text{URi}^2 + 12^2}$, where URi = unit rise.

Rafter Length

$\text{RL} = \text{URi} \times \text{Run}$, where Run is the horizontal run of the rafter (ft).

Total Rise of a Rafter

$\text{TRi} = \text{URi} \times \text{Run}$, where URi = unit rise and Run = horizontal run of the rafter.

Hip Rafter Length (Conventional Case)

$\text{HRL} = \sqrt{\text{CRL}^2 + \text{SB}^2}$, where CRL = common rafter length, SB = setback, the distance along the plate from the last common rafter to the hip.

Unit Line Length for Hip

$\text{ULL} = \sqrt{\text{URi}^2 + \text{URu}^2}$, where URi = unit run, URu = unit rise.

Hip Rafter Length (Conventional)

$\text{HRL} = \text{ULL} \times \text{CRu}$, where ULL = unit line length,
CRu = run of common rafter.

Valley Rafter Length

$\text{VRL} = \text{ULL} \times \text{CRu}$, where ULL = unit line length,
CRu = run of common rafter.

Decrease in Length of Jack Rafter

$\dfrac{\text{CRL}}{\text{SB}} = \dfrac{\text{Dec}}{\text{Spacing}}$, where CRL = common rafter length, SB = setback,
Dec = decrease in length,
Spacing = distance between rafters.